GONDWANA FIVE

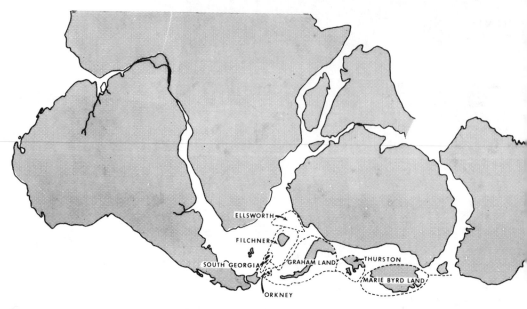

Gondwana reassembly by the participants of the Reunite Gondwanaland workshop at the University of Witwatersrand, Johannesburg 1979, and presented to the 5th Gondwana Symposium by Maarten de Wit. Lambert equal-area projection centred at 20°S, 40°E; New Zealand not yet shown. This assembly is the basis for 1 : 10 000 000 geologic, tectonic and mineral resources wall maps presently being compiled and drawn by M. de Wit and M. Jeffery. Maps are expected to be published during 1981.

PROCEEDINGS OF THE FIFTH INTERNATIONAL GONDWANA SYMPOSIUM /
WELLINGTON, NEW ZEALAND / 11–16 FEBRUARY 1980

GONDWANA FIVE

*Selected papers and abstracts of papers presented
at the Fifth International Gondwana Symposium*

Edited by

M.M.CRESSWELL & P.VELLA

Victoria University of Wellington, New Zealand

A.A.BALKEMA / ROTTERDAM / 1981

The texts of the various papers in this volume were set individually
by typists under the supervision of each of the authors concerned.

ISBN 90 6191 088 9

Distributed in USA & Canada by MBS, 99 Main Street, Salem, NH 03079

Printed in the Netherlands

Table of contents

2 Structure and paleogeology

3 Break-up of Gondwana

Foreword

Gondwana Five is not a large or pretentious volume because our resources are not large and we are too aware of our limitations to have any pretensions. The editors are very grateful to those many authors who cheerfully and gracefully accepted cuts in their scripts, sometimes outright mutilation, in order to meet space restrictions. We hope that the volume fairly represents the proceedings of the Fifth Gondwana Symposium and that the papers still contain enough information to be useful.

Every three years since 1967 symposia have been held to discuss problems and advances in Gondwana geology. The first two (Mar del Plata, 1967; Capetown and Johannesburg, 1970) were devoted to reviews of the continental stratigraphy and paleontology. The third (Canberra, 1973) spent some of its time on marine Gondwana sequences, as was to be expected because of the prevalence of marine Permian strata in Australia, and this drew attention to the margin of Gondwana. At the Canberra meeting there was considerable discussion as to the proper limits of Gondwana geology and hence as to what should be the scope of subsequent symposia, and there was evidence of a developing symbiosis between the Gondwana paleontologist-stratigrapher and the continent-assembling geophysicist. The latter is interested in the events of not only the Gondwana period but also the earlier and later periods. The fourth symposium (Calcutta, 1977), as was appropriate to its venue, reverted to the style of the first two, with emphasis on the classical Gondwana sequences.

The fifth symposium, held early this year in Wellington on the Pacific edge of Gondwana, seemed to the organisers to be a suitable occasion to develop the new approaches that had been started at Canberra. Papers were invited on a wide range of topics, including the classical biostratigraphy, the economic geology, the evolution of the margin, the early break-up history, and the reconstruction of Gondwana, and, for the benefit of overseas delegates wishing to take the opportunity of seeing some local Gondwana, the history of New Zealand. Parts of the margin have been studied in considerable detail but have seldom been viewed in the context of Gondwana. New

Zealand itself is an outstanding example. Its geology has been examined fairly thoroughly and is quite well known amongst overseas geologists, but it does not appear on many Gondwana reconstructions (cf. frontispiece of this volume), and N Zealand geologists themselves have been slow to perceive their own country as part of Gondwana. It is hoped that the fifth symposium will have had an impact on New Zealand geologists and on the reconstructors. The early break-up history, though outside the traditional domain of the Gondwana symposia, is an important aspect of the geology, with implications for basin development, igneous activity, biotic distributions, and the emplacement of economic deposits. Understanding of break-up processes has grown with plate tectonics theory, which itself began to develop near the time of the first Gondwana symposium at Mar del Plata. Elucidation of break-up history is a truly multidisciplinary exercise, requiring careful geological mapping and stratigraphy supported by data of geophysics, petrography, petrochemistry and geochronology, with conclusions tested by the paleontology which was the original evidence for the existence and eventual fragmentation of Gondwana. Each discipline was well represented at the fifth Gondwana symposium.

Another new dimension was added to Gondwana symposia by the inclusion of two papers from China. Gondwana purists may disagree, but the paleontological links between China and Gondwana have been recognised for a long time, and we welcomed the papers offered. Gondwana did not suddenly materialise at the time of the Permo-Carboniferous glaciation. It existed as a supercontinent probably from late Precambrian time. The late Mesozoic - early Tertiary "break-up" was not the only rifting phase. Turkey, Iran, Afganistan, Tibet, and Shan-Thai (East Burma - West Thailand) were probably all parts of Gondwana that rifted away in late Paleozoic times. China and Indo-China possibly separated in middle Paleozoic times. Are they to be regarded as non-Gondwana because they separated before the classical "Gondwana period"?

This volume contains texts of about half the papers presented and abstracts of some others. The order of papers is different

from that in the symposium because we see
the needs as different. In a symposium papers are arranged to form sessions of reasonably uniform interest. In a book they
need to be arranged for most convenient reference. They have been divided into three
major sections, which are distinguished only
in the list of contents--Paleontology and
stratigraphy, Structure and paleogeology,
and Break-up and reconstruction. Even within this broad framework, it is difficult to
decide the appropriate section for some papers. The first section is the largest and
has been subdivided into pure paleontology,
stratigraphy-paleontology, and sedimentology, arranged in that order. Within those
divisions the papers are arranged geographically. The geographical distribution of
papers amongst the traditional Gondwana continents is nicely even for Australia (5 full
texts and 2 abstracts), India (6 full texts
and 9 abstracts), and South America (5 full
texts). Antarctica (11 full texts) has the
lion's share; this seems appropriate because
New Zealand is one of the main jumping-off
points for Antarctica and New Zealand geologists have played a large part in the exploration of Antarctica, and because authors
from a variety of nations have contributed
papers on Antarctic geology. Africa has the
smallest representation, with only three regional papers, but the Gondwana reconstruction in the front and one of the major biostratigraphic papers have been contributed
by African geologists. New Zealand's contribution (7 full texts) shows our interest
in establishing our proper place on the map
of Gondwana.

   From our experience at Wellington, we are
confident of the future of the Gondwana symposia. We were impressed with the variety
of approaches--paleontological, petrological, geochronological--and see this as
evidence of the continuing value of meetings
in which specialists from different fields
have the chance to address the same topic.
We hope that the Wellington symposium has
set the sights of Gondwana geologists on new
horizons that will be explored more
thoroughly in future symposia.

P. J. BARRETT
P. VELLA
C. A. FLEMING
R. H. GRAPES
M. CRESSWELL

Editorial Committee
July 1980

The Organising Committee are particularly
grateful to the following organisations
and people whose financial support con-
tributed to the success of the symposium:
  The Royal Society of New Zealand
  International Union of Geological
    Sciences
  Todd Petroleum Mining Company Ltd., New
    Zealand
  Shell (Petroleum Mining) Company Ltd.,
    New Zealand
  Education Support Committee of The
    British Petroleum Co. Ltd., London
  BP (Oil Exploration) Company of New
    Zealand Ltd.
  Victoria University of Wellington Student
    Geological Society
  Professor R. Crawford

P. J. BARRETT
C. J. BURGESS
J. D. COLLEN
J. A. GRANT-MACKIE
G. W. GRINDLEY
M. G. LAIRD
J. M. DICKINS

Organising Committee

1. Paleontology and stratigraphy

World Permo-Triassic correlations: Their biostratigraphic basis

JOHN MALCOLM ANDERSON
Botanical Research Institute, Pretoria, South Africa

Four Permo-Triassic correlation charts are presented: one on Laurasia and three on Gondwana (West, East, and Circum-Gondwana). Current biostratigraphic data on four groups of
organisms--marine invertebrates, palynomorphs, megaplants, tetrapods--form the basis of
the correlations. Each chart includes from 15 to 20 columns depicting the most productive comprehensive sections in the region. An account of marine and terrestrial realms
and provinces accompanies the charts. Provinciality is introduced in so far as it
bears on the correlation potential between regions.

INTRODUCTION

The Permian and Triassic periods span the
main phase of the unique supercontinent Pangaea, and the Paleozoic/Mesozoic boundary,
acknowledged as witnessing one of the major crises in biological evolution. Even
so, there exists no coherent set of Permo-
Triassic correlation charts of global scope
relating the significant biostratigraphic
sections, marine and non-marine. Charts of
this nature are indispensable in paleogeographic reconstructions and related biogeographic studies.

For these reasons the present charts have
been compiled, and since the theme of the
volume is Gondwana, this region of Pangaea
is emphasised.

Many local or regional biostratigraphic
schemes have been published during the last
decade. Likewise, a number of correlation
charts of widely varying scope and resolution have appeared. It is the aim here to
present a general synthesis, incorporating
with a minimum of conflict, the data and
conclusions reflected in the recent literature and not to make detailed comment on
particular points of correlation.

Biostratigraphic data on the four most
ubiquitous fossil groups are considered:
These are: Invertebrates (marine--macro-
and microscopic); palynomorphs (spores,
pollen grains, and acritarchs); megaplants
(Bryophyta and higher plants); tetrapods
(amphibians and reptiles).

The charts were originally prepared and
presented at the congress as part of a more
comprehensive study on the Permo-Triassic
boundary of Gondwana. Included in addition were (a) a set of eight Gondwana maps
showing the known distribution per degree
square of each of the four fossil groups
mentioned for the Permian and Triassic;
(b) a series of Gondwana reconstructions at
stage intervals portraying latitude, topography, fossil occurrence and provinciality; and (c) a selection of reconstructions
showing climatic and vegetation zones.

In view of space considerations the correlations are presented here substantially
alone. A summary account of provinciality
gleaned from the biogeographic reconstructions is also presented in that it gives an
indication of the efficacy of correlations
between particular regions.

NOTES ON THE CORRELATION CHARTS (FIGS. 1-4)

Permian and Triassic Standards. Inter-
nationally-adopted standard reference sections (series, stages, and zones) for the
Permian and Triassic have as yet not been
attained. The Permian has proved particularly difficult with several different
schemes having been proposed over the past
decade (e.g., Furnish 1973; Anderson 1973;
Waterhouse 1976). An informal composite
scheme following Nakazawa & Kapoor (1979)
is used here. This is based on the ammonite and fusulinid biostratigraphy of the S
Urals (L Perm) and Trans-Caucasus (M-U
Perm).

For the Triassic the standard proposed by
Tozer (1967) and Silberling & Tozer (1968)
and refined in later papers (Tozer 1974,
1978, 1979) has been most widely adopted
and is followed here. It is based on the
ammonoid faunas of western and arctic Canada (particularly British Columbia and Sverdrup Basins, respectively).

Choice of Sections. Somewhat different
considerations have led to the choice of
columns representing the four regions. Concern with biogeographic reconstructions has
influenced the selection.

Laurasia: includes standard reference
sections adopted here, standards proposed in
the past, other possible competitive standards, further sections most relevant to
Gondwana/Laurasia correlations.

Circum-Gondwana: includes a selection of

FIGURE 1

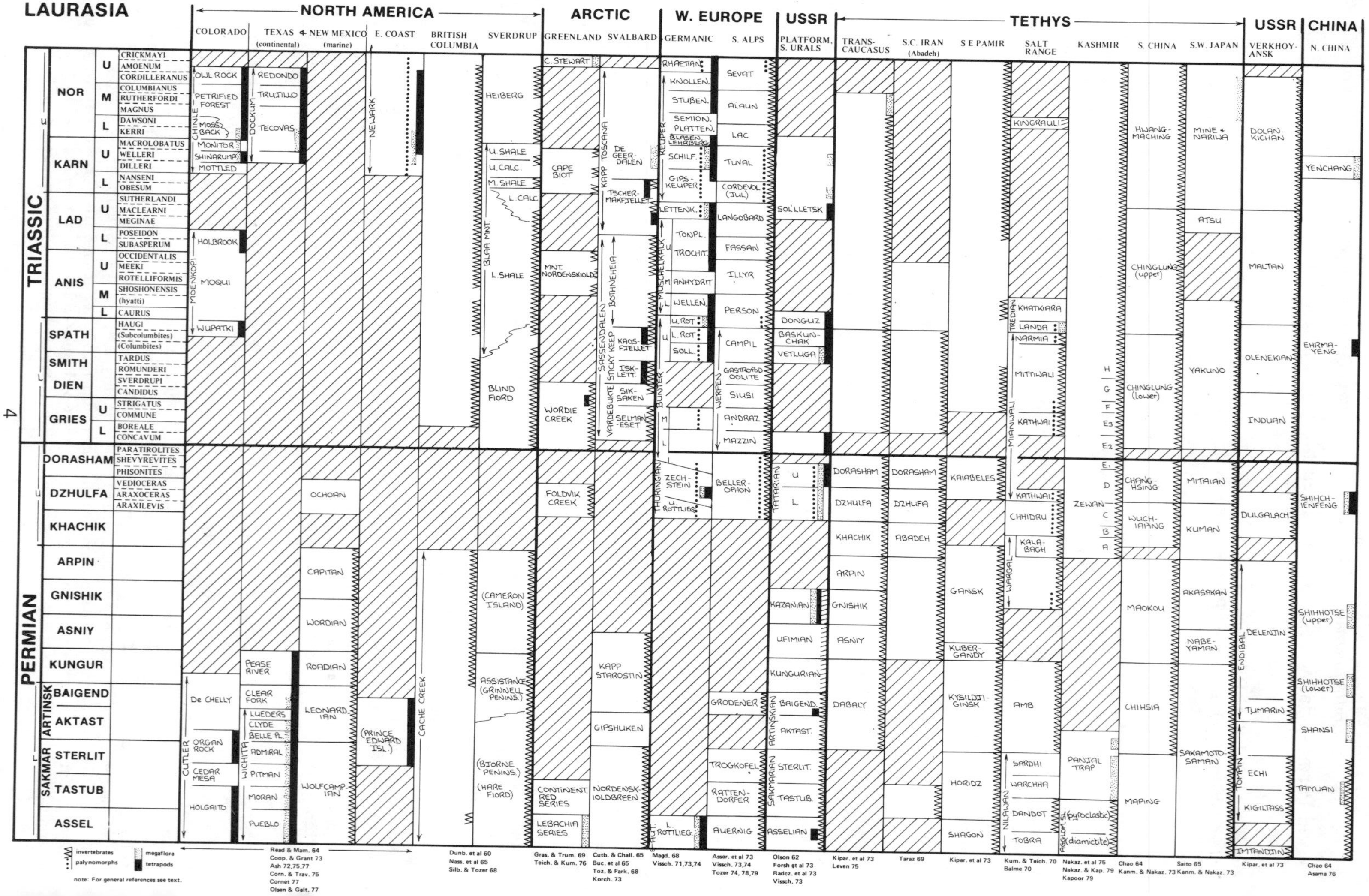

LAURASIA
NORTH AMERICA
COLORADO
TEXAS (continental)
NEW MEXICO (marine)
E. COAST
BRITISH COLUMBIA
SVERDRUP
ARCTIC
GREENLAND
SVALBARD
W. EUROPE
GERMANIC
S. ALPS
USSR
PLATFORM, S. URALS
TETHYS
TRANS-CAUCASUS
S.C. IRAN (Abadeh)
S E PAMIR
SALT RANGE
KASHMIR
S. CHINA
S.W. JAPAN
USSR
VERKHOY-ANSK
CHINA
N. CHINA
TRIASSIC
PERMIAN
NOR
KARN
LAD
ANIS
SPATH
SMITH
DIEN
GRIES
DORASHAM
DZHULFA
KHACHIK
ARPIN
GNISHIK
ASNIY
KUNGUR
ARTINSK
BAIGEND
AKTAST
SAKMAR
STERLIT
TASTUB
ASSEL
CRICKMAYI
AMOENUM
CORDILLERANUS
COLUMBIANUS
RUTHERFORDI
MAGNUS
DAWSONI
KERRI
MACROLOBATUS
WELLERI
DILLERI
NANSENI
OBESUM
SUTHERLANDI
MACLEARNI
MEGINAE
POSEIDON
SUBASPERUM
OCCIDENTALIS
MEEKI
ROTELLIFORMIS
SHOSHONENSIS
(hyatti)
CAURUS
HAUGI
(Subcolumbites)
(Columbites)
TARDUS
ROMUNDERI
SVERDRUPI
CANDIDUS
STRIGATUS
COMMUNE
BOREALE
CONCAVUM
PARATIROLITES
SHEVYREVITES
PHISONITES
VEDIOCERAS
ARAXOCERAS
ARAXILEVIS
OWL ROCK
PETRIFIED FOREST
MOSS BACK
CHINLE
MONITOR
SHINARUMP
MOTTLED
HOLBROOK
MOQUI
MOENKOPI
WUPATKI
REDONDO
TRUJILLO
TECOVAS
DOCKUM
NEWARK
HEIBERG
U. SHALE
U. CALC.
M. SHALE
L. CALC
BLAA MNT.
L. SHALE
BLIND FIORD
C. STEWART
CAPE BIOT
MNT. NORDENSKIOLD
WORDIE CREEK
KAPP TOSCANA
DE GEER-DALEN
TSCHER-MAKFJELLET
SASSENDALEN
BOTHNEHEIA
STICKY KEEP
KAOS-FJELLET
ISK-LETT
VARDEBUKTE
SIK-SAKEN
SELMAN-ESET
RHAETIAN
KNOLLEN.
STUBEN.
SEMION. PLATTEN.
BLASEN. LEHRBERG
SCHILF.
GIPS-KEUPER
KEUPER
LETTENK.
TONPL
TROCHIT.
ANHYDRIT
WELLEN.
MUSCHELKALK
U. ROT
L. ROT
SOLL.
BUNTER
SEVAT
ALAUN
LAC
TUVAL
CORDEVOL (JUL)
LANGOBARD
FASSAN
ILLYR
PERSON
CAMPIL
GASTROPOD OOLITE
SIUSI
ANDRAZ
MAZZIN
WERFEN
SOL'LLETSK
DONGUZ
BASKUN-CHAK
VETLUGA
KINGRAULI
TREDIAN
KHATKIARA
LANDA
NARMIA
MITTIWALI
KATHWAI
MIANWALI
H
G
F
E3
E2
HWANG-MACHING
CHINGLUNG (upper)
CHINGLUNG (lower)
MINE + NARIWA
ATSU
YAKUNO
DOLAN-KICHAN
MALTAN
OLENEKIAN
INDUAN
YENCHANG
EHRMA-YENG
OCHOAN
CAPITAN
WORDIAN
ROADIAN
LEONARD-IAN
WOLFCAMP-IAN
PEASE RIVER
CLEAR FORK
LUEDERS
CLYDE
BELLE P.
ADMIRAL
PITMAN
MORAN
PUEBLO
WICHITA
De CHELLY
ORGAN ROCK
CEDAR MESA
HOLGAITO
CUTLER
(PRINCE EDWARD ISL.)
CACHE CREEK
(CAMERON ISLAND)
ASSISTANCE (GRINNELL PENINS.)
(BJORNE PENINS.)
(HARE FIORD)
FOLDVIK CREEK
CONTINENT. RED SERIES
LEBACHIA SERIES
KAPP STAROSTIN
GIPSHUKEN
NORDENSK-IOLDBREEN
ZECH-STEIN
U. ROTTLIEG.
THURINGIAN
L. ROTTLIEG.
AUT.
BELLER-OPHON
GRODENER
TROGKOFEL
RATTEN-DORFER
AUERNIG
U
L
TATARIAN
KAZANIAN
UFIMIAN
KUNGURIAN
BAIGEND.
AKTAST.
ARTINSKIAN
STERLIT.
TASTUB.
SAKMARIAN
ASSELIAN
DORASHAM
DZHULFA
KHACHIK
ARPIN
GNISHIK
ASNIY
DABALY
DORASHAM
DZHUFA
ABADEH
KAIABELES
GANSK
KUBER-GANDY
KYSILDJI-GINSK
HORIDZ
SHAGON
KATHWAI
CHHIDRU
KALA-BAGH
WARGAL
AMB
SARDHI
WARCHHA
DANDOT
TOBRA
NILAWAN
E1
D
ZEWAN
C
B
A
PANJAL TRAP
AGGLOM. SL. (pyroclastic)
(diamictite)
CHANG-HSING
WUCH-IAPING
MAOKOU
CHIHSIA
MAPING
MITAIAN
KUMAN
AKASAKAN
NABE-YAMAN
SAKAMOTO-SAMAN
DULGALACH
DELENJIN
ENDIBAL
TUMARIN
ECHI
TOMPIN
KIGILTASS
IMTANDJIN
SHIHCH-IENFENG
SHIHHOTSE (upper)
SHIHHOTSE (lower)
SHANSI
TAIYUAN
invertebrates
palynomorphs
megaflora
tetrapods
note: For general references see text.
Read & Mam. 64
Coop. & Grant 73
Ash 72,75,77
Corn. & Trav. 75
Cornet 77
Olsen & Galt. 77
Dunb. et al 60
Nass. et al 65
Silb. & Tozer 68
Gras. & Trum. 69
Teich. & Kum. 76
Cutb. & Chall. 65
Buc. et al 65
Toz. & Park. 68
Korch. 73
Magd. 68
Vissch. 71,73,74
Asser. et al 73
Vissch. 73,74
Tozer 74, 78,79
Olson 62
Forsh et al 73
Radcz. et al 73
Vissch. 73
Kipar. et al 73
Leven 75
Taraz 69
Kipar. et al 73
Kum. & Teich. 70
Balme 70
Nakaz. et al 75
Nakaz. & Kap. 79
Kapoor 79
Chao 64
Kanm. & Nakaz. 73
Saito 65
Kanm. & Nakaz. 73
Kipar. et al 73
Chao 64
Asama 76

FIGURE 2

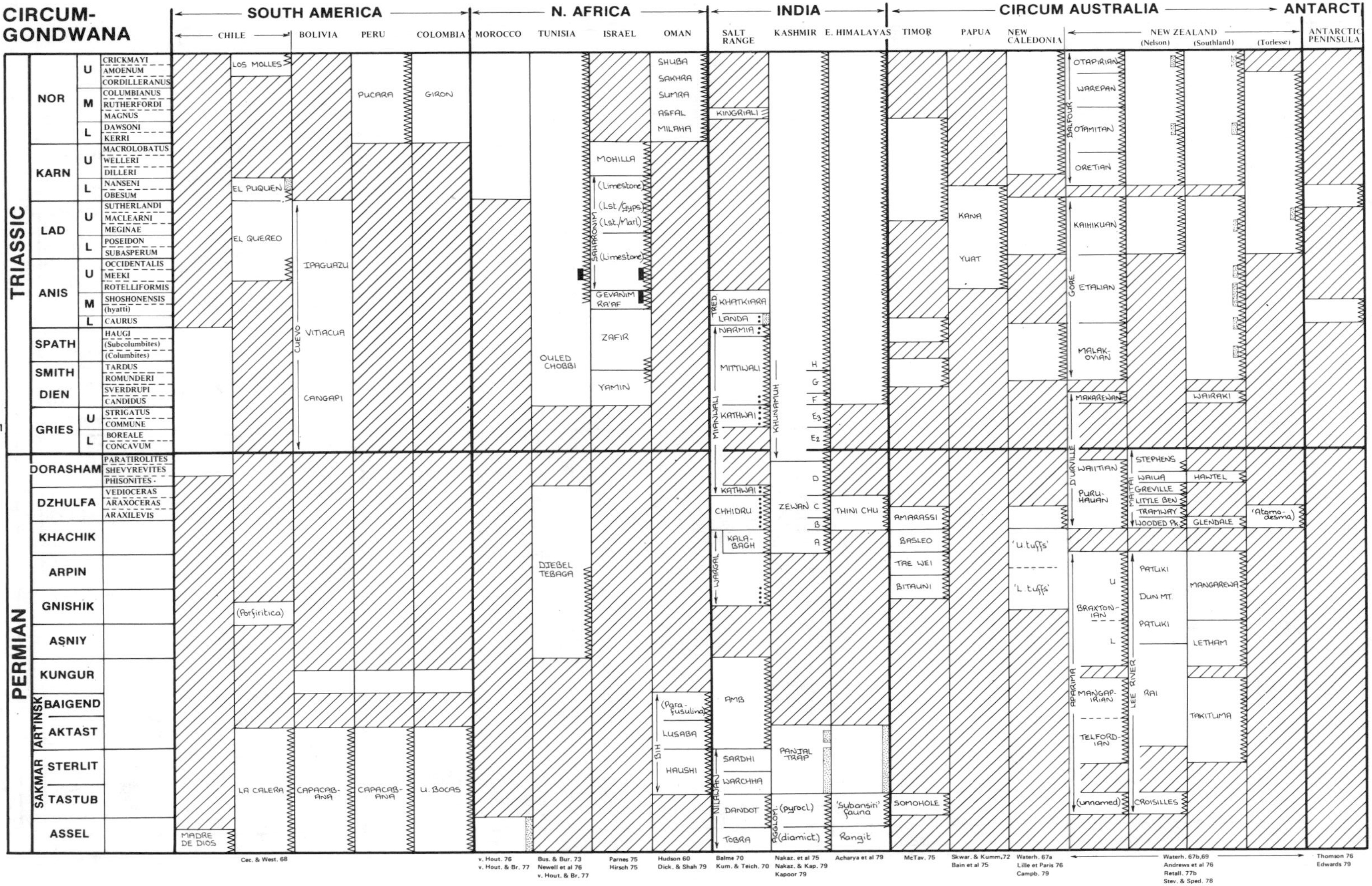
CIRCUM-GONDWANA
SOUTH AMERICA
CHILE
BOLIVIA
PERU
COLOMBIA
N. AFRICA
MOROCCO
TUNISIA
ISRAEL
OMAN
INDIA
SALT RANGE
KASHMIR
E. HIMALAYAS
CIRCUM AUSTRALIA
TIMOR
PAPUA
NEW CALEDONIA
NEW ZEALAND
(Nelson)
(Southland)
(Torlesse)
ANTARCT
ANTARCTIC PENINSULA
TRIASSIC
NOR
U
CRICKMAYI
AMOENUM
CORDILLERANUS
M
COLUMBIANUS
RUTHERFORDI
MAGNUS
L
DAWSONI
KERRI
KARN
U
MACROLOBATUS
WELLERI
DILLERI
L
NANSENI
OBESUM
LAD
U
SUTHERLANDI
MACLEARNI
MEGINAE
L
POSEIDON
SUBASPERUM
ANIS
U
OCCIDENTALIS
MEEKI
ROTELLIFORMIS
M
SHOSHONENSIS
(hyatti)
L
CAURUS
SPATH
HAUGI
(Subcolumbites)
(Columbites)
SMITH
TARDUS
ROMUNDERI
DIEN
SVERDRUPI
CANDIDUS
GRIES
U
STRIGATUS
COMMUNE
L
BOREALE
CONCAVUM
PERMIAN
DORASHAM
PARATIROLITES
SHEVYREVITES
PHISONITES
DZHULFA
VEDIOCERAS
ARAXOCERAS
ARAXILEVIS
KHACHIK
ARPIN
GNISHIK
ASNIY
KUNGUR
ARTINSK
BAIGEND
AKTAST
SAKMAR
STERLIT
TASTUB
ASSEL
LOS MOLLES
EL PUQUEN
EL QUEREO
(Porfiritica)
LA CALERA
MADRE DE DIOS
CUEVO
IPAGUAZU
VITIACUA
CANGAPI
CAPACAB-ANA
PUCARA
GIRON
U. BOCAS
OULED CHOBBI
DJEBEL TEBAGA
MOHILLA
SAHARONIM
(Limestone)
(Lst./Gyps)
(Lst./Marl)
(Limestone)
GEVANIM
RA'AF
ZAFIR
YAMIN
SHUBA
SAKHRA
SUMRA
ASFAL
MILAHA
(Para-fusulina)
LUSABA
HAUSHI
HIB
KINGRIALI
TRED.
KHATKIARA
LANDA
NARMIA
MITTIWALI
KATHWAI
MIANWALI
KATHWAI
CHHIDRU
KALA-BAGH
WARGAL
AMB
SARDHI
WARCHHA
DANDOT
TOBRA
NILAWAN
KHUNAMUH
H
G
F
E3
E2
D
ZEWAN
C
B
A
PANJAL TRAP
AGGLOM.
(pyrocl.)
(diamict.)
THINI CHU
'Subansiri' fauna
Rangit
AMARASSI
BASLEO
TAE WEI
BITAUNI
SOMOHOLE
KANA
YUAT
'U. tuffs'
'L. tuffs'
OTAPIRIAN
WAREPAN
BALFOUR
OTAMITAN
ORETIAN
KAIHIKUAN
GORE
ETALIAN
MALAK-OVIAN
MAKAREWAN
D'URVILLE
WAITIAN
PURU-HAUAN
APARIMA
BRAXTON-IAN
U
L
MANGAP-IRIAN
TELFORD-IAN
(unnamed)
STEPHENS
MAITAI
WAIUA
GREVILLE
LITTLE BEN
TRAMWAY
WOODED Pk
PATUKI
DUN MT
PATUKI
RAI
LEE RIVER
CROISILLES
WAIRAKI
HAWTEL
GLENDALE
MANGAREWA
LETHAM
TAKITIMU
'Atomodesma)
Cec. & West. 68
v. Hout. 76
v. Hout. & Br. 77
Bus. & Bur. 73
Newell et al 76
v. Hout. & Br. 77
Parnes 75
Hirsch 75
Hudson 60
Dick. & Shah 79
Balme 70
Kum. & Teich. 70
Nakaz. et al 75
Nakaz. & Kap. 79
Kapoor 79
Acharya et al 79
McTav. 75
Skwar. & Kumm.72
Bain et al 75
Waterh. 67a
Lille et Paris 76
Campb. 79
Waterh. 67b,69
Andrews et al 76
Retall. 77b
Stev. & Sped. 78
Thomson 76
Edwards 79

FIGURE 3

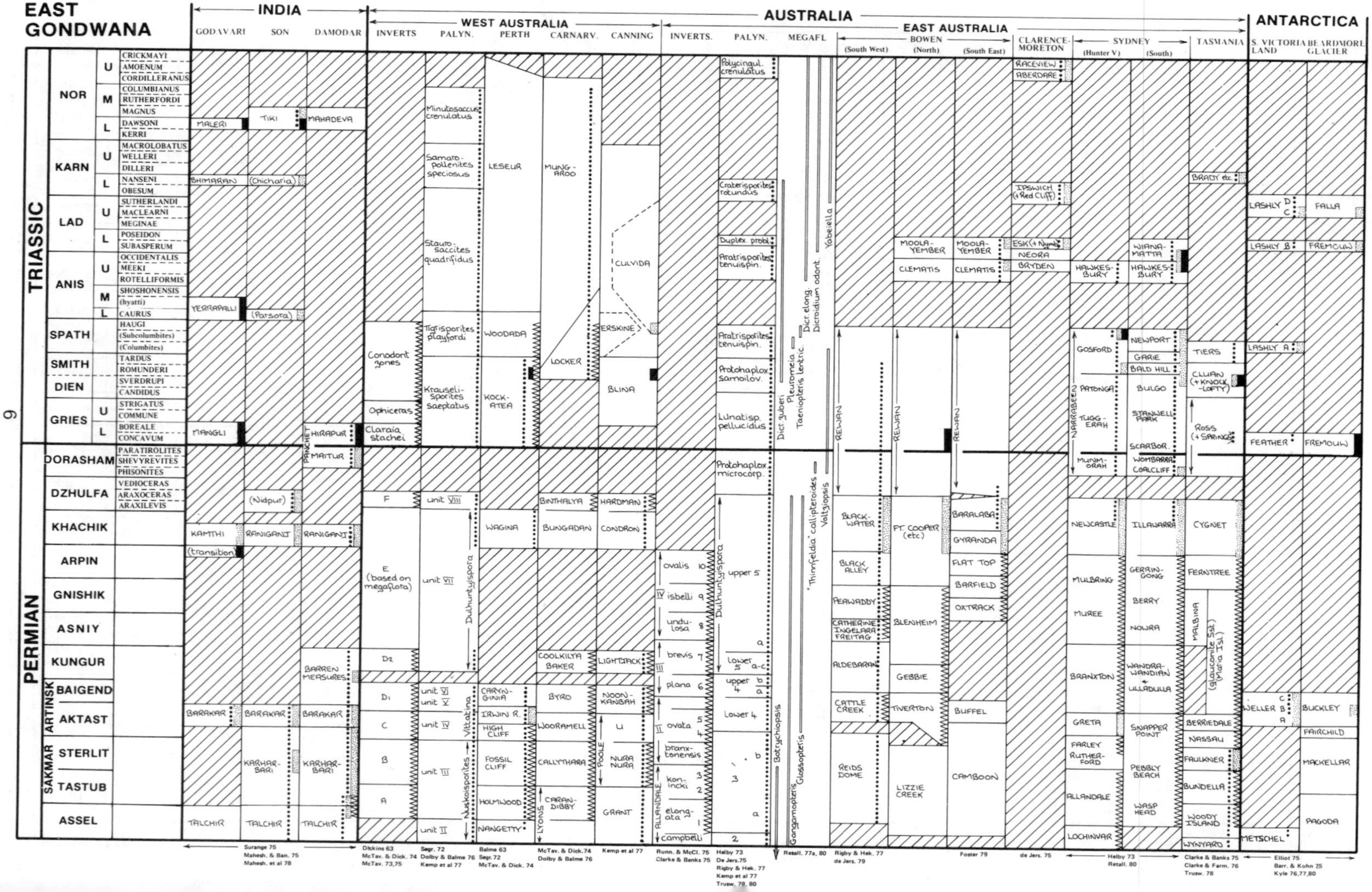
EAST GONDWANA
INDIA
GODAVARI
SON
DAMODAR
AUSTRALIA
WEST AUSTRALIA
INVERTS
PALYN.
PERTH
CARNARV.
CANNING
EAST AUSTRALIA
INVERTS.
PALYN.
MEGAFL
BOWEN
(South West)
(North)
(South East)
CLARENCE-MORETON
SYDNEY
(Hunter V)
(South)
TASMANIA
ANTARCTICA
S. VICTORIA LAND
BEARDMORE GLACIER
TRIASSIC
PERMIAN
NOR
KARN
LAD
ANIS
SPATH
SMITH
DIEN
GRIES
DORASHAM
DZHULFA
KHACHIK
ARPIN
GNISHIK
ASNIY
KUNGUR
BAIGEND
AKTAST
STERLIT
TASTUB
ASSEL
ARTINSK
SAKMAR

FIGURE 4

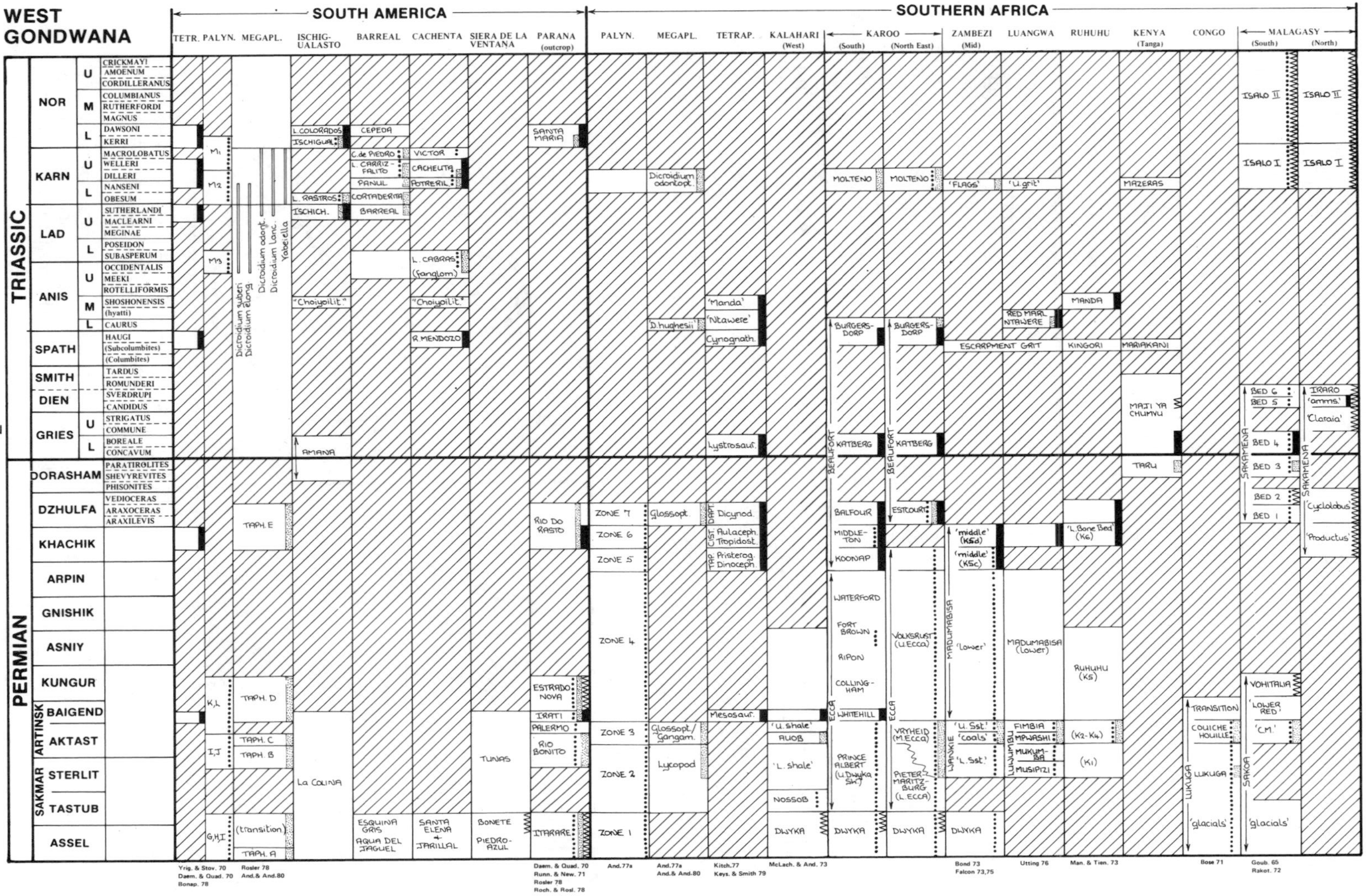
WEST GONDWANA
SOUTH AMERICA
SOUTHERN AFRICA
TETR. PALYN. MEGAPL. ISCHIG-UALASTO BARREAL CACHENTA SIERA DE LA VENTANA PARANA (outcrop)
PALYN. MEGAPL. TETRAP. KALAHARI (West) KAROO (South) (North East) ZAMBEZI (Mid) LUANGWA RUHUHU KENYA (Tanga) CONGO MALAGASY (South) (North)
TRIASSIC
NOR
KARN
LAD
ANIS
SPATH
SMITH
DIEN
GRIES
PERMIAN
DORASHAM
DZHULFA
KHACHIK
ARPIN
GNISHIK
ASNIY
KUNGUR
ARTINSK BAIGEND
AKTAST
SAKMAR STERLIT
TASTUB
ASSEL
CRICKMAYI
AMOENUM
CORDILLERANUS
COLUMBIANUS
RUTHERFORDI
MAGNUS
DAWSONI
KERRI
MACROLOBATUS
WELLERI
DILLERI
NANSENI
OBESUM
SUTHERLANDI
MACLEARNI
MEGINAE
POSEIDON
SUBASPERUM
OCCIDENTALIS
MEEKI
ROTELLIFORMIS
SHOSHONENSIS
(hyatti)
CAURUS
HAUGI
(Subcolumbites)
(Columbites)
TARDUS
ROMUNDERI
SVERDRUPI
CANDIDUS
STRIGATUS
COMMUNE
BOREALE
CONCAVUM
PARATIROLITES
SHEVYREVITES
PHISONITES
VEDIOCERAS
ARAXOCERAS
ARAXILEVIS
Dicroidium suberi
Dicroidium elong.
Dicroidium odont.
Dicroidium lanc.
Yabeiella
L. COLORADOS
ISCHIGUAL
L. RASTROS
ISCHICH.
"Choiyoilit."
AMANA
La COLINA
CEPEDA
C. de PIEDRO
L. CARRIZ-FALITO
PANUL
CORTADERITA
BARREAL
ESQUINA GRIS
AGUA DEL JAGUEL
VICTOR
CACHEUTA
POTRERIL.
L. CABRAS (fanglom)
"Choiyoilit."
R. MENDOZO
SANTA ELENA + JARILLAL
TUNAS
BONETE
PIEDRO-AZUL
SANTA MARIA
RIO DO RASTO
ESTRADO NOVA
IRATI
PALERMO
RIO BONITO
ITARARE
TAPH. E
TAPH. D
TAPH. C
TAPH. B
(transition)
TAPH. A
ZONE 7
ZONE 6
ZONE 5
ZONE 4
ZONE 3
ZONE 2
ZONE 1
Dicroidium odontopt.
D. hughesii
Glossopt.
Glossopt./Gangam.
Lycopod
'Manda'
'Ntawere'
Cynognath.
Lystrosaur.
Dicynod.
Aulaceph. Tropidost.
Pristerog. Dinoceph.
Mesosaur.
'U. shale'
AUOB
'L. shale'
NOSSOB
DWYKA
MOLTENO
BURGERSDORP
KATBERG
BEAUFORT
BALFOUR
MIDDLETON
KOONAP
WATERFORD
FORT BROWN
RIPON
COLLINGHAM
WHITEHILL
PRINCE ALBERT (L. Dwyka Sh.)
ESTCOURT
VOLKSRUST (U. Ecca)
VRYHEID (M. ECCA)
PIETERMARITZBURG (L. ECCA)
ECCA
'FLAGS'
ESCARPMENT GRIT
'middle' (K5d)
'middle' (K5c)
'Lower'
MADUMABISA
'U. Sst'
'coals'
'L. Sst'
WANKIE
'U. grit'
RED MARL
NTAWERE
MADUMABISA (Lower)
FIMBIA
MPWASHI
MUKUMBA
MUSIPIZI
LUWUMBU
MANDA
KINGORI
'L. Bone Bed' (K6)
RUHUHU (K5)
(K2-K4)
(K1)
MAZERAS
MARIAKANI
MAJI YA CHUMVU
TARU
TRANSITION
COUCHE HOUILLE
LUKUGA
'glacials'
ISALO II
ISALO I
BED 6
BED 5
BED 4
BED 3
BED 2
BED 1
SAKAMENA
IRARO
'amms.'
'Claraia'
'Cyclolobus'
'Productus'
VOHITALIA
'LOWER RED'
'CM.'
SAKOA
Yrig. & Stov. 70
Daem. & Quad. 70
Bonap. 78
Rosler 78
And.& And.80
Daem. & Quad. 70
Runn. & New. 71
Rosler 78
Roch. & Rosl. 78
And.77a
And.77a
And.& And.80
Kitch.77
Keys. & Smith 79
McLach. & And. 73
Bond 73
Falcon 73,75
Utting 76
Man. & Tien. 73
Bose 71
Goub. 65
Rakot. 72

sections around the periphery of the Gondwana landmass such as to define its changing coastline through time.

East and West Gondwana: includes sections from each continent for which the best biostratigraphic works are available, particularly those facilitating correlations and reconstructions.

Indication of Productive Strata. The productive strata in each section have been indicated. The four fossil groups considered are differentiated and are entered with varying emphasis:

Invertebrates: all productive marine sections.

Palynomorphs: Gondwana (+NE Shelf) Trias-all productive strata on which systematic or biostratigraphic works have appeared. Other - sequences on which correlations have been based.

Megaplants: Gondwana (+NE Shelf) Trias-all productive strata. Other - most productive horizons only.

Tetrapods: All productive horizons.

References. As far as possible, the latest and most authoritative biostratigraphic literature has been consulted. As space limitations make it impossible to include these citations, a 3000-word list of references may be obtained from the author. Where reasonably current accounts of a particular column or aspect thereof do not exist, earlier published correlations are followed. These are Anderson & Anderson 1970 (Gondwana Trias); Anderson 1973 (World Permian); Rocha-Campos 1973 (S America Permo-Trias); Dickins 1976 (Australia Perm.); Waterhouse 1976 (World Perm., marine); Anderson & Cruickshank 1978 (World Permo-Trias, tetrapods); Banks 1978 (Australia Trias); Nakazawa & Kapoor 1979 (Tethys Perm, marine); Anderson *et al.* 1980 (Gondwana Permo-Trias, megaplants).

PROVINCIALITY (FAUNAL AND FLORAL REALMS AND PROVINCES (FIG. 5)

Of major concern in biostratigraphic correlations is provinciality. Assuming provinces and realms were quantitatively defined and consistently applied, a measure of the correlation potential between any two regions could be judged directly from their respective provincial affinities. Sections within a single province should be readily correlated; those in the same realm but distinct provinces less easily; and those in separate realms only with difficulty. Unfortunately no such consistency has been reached.

Gordon (1976) in his work on Mesozoic ammonoid zoogeography applied the following criteria: 0-50% generic overlap - different realms; 51-75% generic overlap - same realm, different provinces: 76-100% generic overlap - same province.

He pointed out that a broad consensus exists that realms may number from one to four at any time and that boundaries based

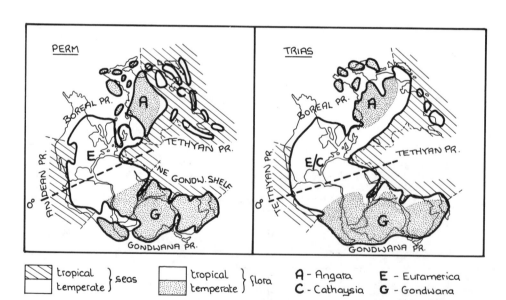

Fig. 5--Marine and floral realms and provinces. (For maps of tetrapod provinciality see Anderson & Cruickshank 1978.) Maximum extent of shelf seas shown for each period. Equator and provinces (marine and floral) indicated for middle of period.

8

on different groups of organisms cannot be
expected to correspond.

It is clear that for criteria such as
these to be effectively applied, the generic concept as well as the taxa used
need to be stabilised, and sampling needs
to be equally comprehensive. Again, no
such conditions have as yet been remotely
attained in any fossil group.

The following brief account of Permo-Triassic provinciality in the four groups of
organisms being considered must therefore
be seen as a crude approximation. The delineation and nomenclature of realms and
provinces outlined here departs, where necessary in favour of consistency, from
usage in the literature cited.

Independent of the above classification
are endemic faunas (invertebrate or tetrapod) deriving from isolated areas. Such
faunas, by virtue of their exclusive composition, deserve the status of a separate
realm but areally are of lesser significance than a province.

Invertebrates. (Sweet *et al.* 1971; Druce
1973; McTavish 1975; Runnegar & McClung
1975; Waterhouse & Bonham-Carter 1975; Gordon 1976; Ross 1979; Runnegar 1979)
Permian: (1) Laurasian Realm: (a) Boreal
Province (temp.); (b) Andean Province
(trop.); (c) Tethyan Province (trop.); (d)
NE Gonwana Shelf (temp./trop.).

(2) Gondwana Realm: (a) Gondwana Province (cool. temp.).

(3) Endemic faunas: (a) Parana Sea (Estrado Nova Fm).
Triassic: (1) Pangaean Realm: (a) Boreal
Province (temp.); (b) Tethyan Province
(trop.); (c) Gondwana Province (temp.).

The marine invertebrate realms and provinces strongly reflect ocean surface temperatures. During both the Permian and
Triassic provinciality was more pronounced
earlier on and diminished through the period.

During the Asselian (L Permian)--the final phase and demise of glaciation--the
break between the tropical provinces and
the Gondwana Realm was sharp. The Andean
faunas (Peru, Bolivia, Colombia) were vastly
different from those of the cool water Kar-
roo/Parana Sea and Chile. The cool temperate faunas extended along the NE Gondwana
Shelf as far north as the Salt Range. The
Gondwana assemblages all belong to the
characteristic *Eurydesma* fauna.

Through the remainder of the Permian--
with the northward drift of Gondwana--the
faunas of the NE Shelf from Malagasy and
the Salt Range to West Australia and Timor
came increasingly under the influence of
the southward-deflected tropical Tethyan
waters. A province transitional between

the Tethyan and Gondwana provinces can be
recognised.

Palynomorphs and Megaplants (Terrestrial
Floras). (Perm.: Chaloner & Meyen 1973;
Plumstead 1973; Kremp 1974; Anderson 1977;
Trias: Balme 1970; de Jersey 1971; Visscher
1973; Barnard 1973; Dolby & Balme 1976;
Retallack 1977; Dobruskina 1978)
Permian: (1) Angara Realm (with up to 4
provinces).

(2) Cathaysian Realm.

(3) Euramerican Realm (with up to 3 provinces).

(4) Gondwana Realm.
Triassic: (1) Laurasian Realm: (a) Angara
Province; (b) Euramerica/Cathaysia Pro-
vince.

(2) Gondwana Realm: (a) Gondwana Province.

Terrestrial floras made their earliest
appearance in the Silurian. Thereafter
followed a pattern of global homogeneity
through the Devonian, gradual development
of provincialism through the Carboniferous,
rapidly reaching a maximum towards the
middle of the Permian, with a reversal to
more uniformity in the Triassic and Jurassic. The spread of the extensive southern
ice cap about the time of the Permo-Carboniferous boundary, with the resulting increased differentiation of climatic belts,
may explain the mid-Permian maximum in
diversification.

The Gondwana Realm was markedly distinctive with the Glossopteridophyta (seed
ferns) dominant through most of the Permian and *Dicroidium* (seed ferns) dominant
through most of the Triassic. Both remain
unknown outside Gondwana. Following the
global crisis near the Paleozoic/Mesozoic
boundary and occupying the interval (± L
Triassic) between the demise of the *Glossopteris* flora and the rise of the *Dicroidium*
flora, there spread a global flora (particularly noticeable in coastal associations)
suggesting a single realm.

Tetrapods. (Anderson & Cruickshank 1978)
Permian (Lowland Terrestrial Faunas):
(1) Pangaean Realm (no provinces recognised).
Permian (Aquatic Faunas): (1) Faunas of
shelf seas and shoreline unknown.

(2) Endemic faunas (isolated seas): (a)
Karroo / Parana Sea (Whitehill Fm); (b) C
European Sea (Zechstein Gp); (c) Malagasy/
Kenya Sea (M Sakamena Gp).
Triassic (Lowland Terrestrial Faunas):
(1) Pangaean Realm (no provinces recognised).
Triassic (Aquatic Faunas): (1) Pangaean
Realm (shelf seas and shoreline).

The tetrapods, first the amphibians and
later the reptiles, made their evolutionary
debut during the Carboniferous along the
equatorial belt of Euramerica. In the
earlier stages of the Permian the faunas

were still restricted entirely to the equatorial belt of Euramerica. Through the remainder of the Permian and Triassic their colonisation of Pangaea is witnessed in the scattered and interrupted fossil record. During the colonisation five biotic crises and renewed evolutionary pulses can be seen in the terrestrial lowland faunas. Present information suggests that during each of the successive evolutionary pulses the lowland communities were essentially similar throughout their known Pangaeic range. No distinction into realms or provinces can be made. However, coeval with the lowland faunas occurred at times distinctive aquatic faunas. These developed either in isolated epicontinental seas or in open shelf seas and fringing coastlines (lagoonal, estuarine and deltaic).

BIOSTRATIGRAPHIC REFERENCE SECTIONS

Listed are the most comprehensive fully-studied sequences critical in effecting intra- and extra-Gondwana correlations. For the NE Gondwana Shelf and Gondwana Realm the sequences noted would be suitable for erection as regional standards. The Laurasian sections are those with which the southern standards are most readily compared.

Invertebrates. *Laurasia:* L Perm.: USSR (S Urals); M-U Perm.: Tethys (Trans-Caucasus); Trias: W Europe (S Alps).
NE Gondwana Shelf: L Perm. - L Trias: W Australia (Perth, Carnarvon, Canning basins); L Perm. - L Trias : Himalayas (Salt Range, Kashmir).
Gondwana Realm (Province): Perm.: E Australia (Bowen, Sydney, Tasmania basins); Trias: N.Z. (Nelson, Southland synclines).

Palynomorphs. *Laurasia:* U Perm - U Trias: W Europe (Germanic basin, S Alps).
NE Gondwana Shelf: L Perm - U Trias: W Australia (Perth, Carnarvon, Canning basins); U Perm. - L Trias: Himalayas (Salt Range).
Gondwana Realm (Province): Perm.: E Australia (Bowen, Sydney basins); Trias: E Australia (Bowen, Sydney, Clarence-Moreton basins).

Megaplants. *Laurasia:* nil.
NE Gondwana Shelf: nil.
Gondwana Realm (Province): Perm.: No particular sequence is clearly most suitable--E Australia (Sydney B.), India (Damodar), S Africa (NE Karroo), S America (Parana); Trias: E Australia (Clarence-Moreton, Sydney basins).

Tetrapods (Lowland Terrestrial Faunas).
Laurasia: L Perm.: N America (Texas); U Perm. - M Trias: USSR (Platform); Trias: W Europe (Germanic Basin); U Trias: N America (Colorado, Texas, E coast).
NE Gondwana Shelf: nil.

Gondwana Realm (Province): U Perm. - L Trias: S Africa (Karroo Basin); U Trias: S America (Ischigualasto, Barreal, Cacheuta basins).

CONCLUDING COMMENTS

(a) Reliable correlations depend on reliable biostratigraphy based on comprehensive sampling and sound taxonomy. The latter are subject to an order of improvement. The present set of charts reflects very variable correlation accuracy ranging from zone (±1-2 m.y.) to series (±15 m.y.) level, probably averaging somewhat short of the stage (±5 m.y.) level.

(b) The NE Gondwana Shelf, considering its transitional biogeographic position, is critical in correlations between the Gondwana Province and the Tethyan (marine) or Euramerican (continental) provinces.

(c) In the Gondwana Realm (exclusive of the NE Shelf), the E Australian basins generally yield the most comprehensive invertebrate, palynomorph and megaplant biostratigraphic sequences. In view of this unique triple role and its proximity to the NE Shelf, E Australia provides the ideal opportunity for a composite biostratigraphic standard facilitating intra- and extra-Gondwana correlations.

(d) Southern Africa and Argentina on the other hand yield the sequences suitable for a terrestrial tetrapod biostratigraphic standard.

(e) This work has been concerned with the role of biostratigraphy in correlations. Lithostratigraphy (rates of deposition and formation contacts) and geophysics (paleomagnetism and radiometric dating) should be incorporated in future compilations.

(f) The study on the Permo-Triassic biogeography of Gondwana, prepared and presented originally together with the correlation charts, will be published elsewhere.

ACKNOWLEDGMENTS

I wish to express my thanks to the organising committee of the symposium for inviting me to present this paper and for providing part of the funds needed, to the Dept. of Agricultural Technical Services, South Africa, for providing the remaining funds; to my wife, Heidi, and father-in-law, Rolf Schwyzer, for considerable help during preparation of the paper; to numerous colleagues attending the symposium and others visited in N.Z. and Australia for advice and comment; to Mrs Loots for typing the manuscript and Nienke Van der Meulen for drafting the charts; and finally to a number of friends who gave of their time to help in various ways.

Non-marine evidence for Paleozoic/Mesozoic Gondwana correlations: Update

PAUL TASCH
Wichita State University, Kansas, USA

Old correlations within and between Gondwana continents have been firmed up and new dispersal tracks linking Gondwana and USSR and China inferred. The evidence consists of accumulating Paleozoic/Mesozoic fossil conchostracan and palynomorph data from Africa, India, China, South America, and Antarctica.

Triassic *Paleolimnadia* links West and East Australia basins; Knocklofty Fm, Tasmania; Panchet of India; Santa Maria Fm of Brazil. *Cornia* relates the Panchets of India, the lower Cassanje Series (Northern Angola), the Cave Sandstone sequence (Lesotho), the USSR Triassic-Jurassic and the Antarctic Jurassic. Triassic *Paleolimnadiopsis* links Australia and China; China and USSR; South America and Africa. *Gabonestheria* (U. Permian, Rio do Rasto Fm, Brazil) points to Gabon, equatorial Africa, as a probable source. The requirement for non-marine dispersal tracks during the Paleozoic/Mesozoic between the indicated Gondwana, USSR, and China areas, is a control that can improve global reconstructions.

Estheriella

Four African conchostracan genera have become important in Gondwana Mesozoic correlations: *Estheriella* (Marlière 1950), *Paleolimnadia*, *Estheriina* and *Cornia*, the last three (previously unreported) from the Angola or Belgian Congo Karroo System (Tasch & Osterlen 1977). In India Ghosh and Shah (1978) described *Estheriella taschi* and a new estheriellid-related genus (Tasch *et al.* in prep.). It is also in the late Mesozoic Botucatu Fm of Brazil (Almeida 1960); a relative, *Graptestheriella*, is in the lower Cretaceous Japoata Fm, Bahia (Cardosa 1966). This indicates an India-Africa-South America dispersal track (Fig. 1A).

Estheriina

The Lower Triassic non-marine deposits of Australia contain species of *Estheriina* (Tasch & Jones 1979a). In the Upper Triassic there are several: Kapel'ka's Yenissei River (near Krasnozarsk) species-- several Carnic or Noric ones (Kobayashi 1975); one from Brazil from the Santa Maria Fm (Pinto 1965).

Estheriniids continued into the Jurassic and Cretaceous in India, Kota Fm, Jurassic (Tasch *et al.* 1975), and the lower Cretaceous Inhlas Fm, Bahia (Cardosa 1966). They are also known from the Jurassic and Cretaceous of Mongolia (Kobayashi 1975) (Fig. 1C).

This suggests a later Mesozoic Asian dispersal.

Paleolimmadia

Paleolimmadia (Fig. 1B) are found in NW and western Australia (Tasch & Jones 1979

a, b). Paleolimnadiids from the Knocklofty Fm of Tasmania (Tasch 1975) have verified correlations with the Blina Shale established by vertebrate fossils. Others are known from the Lower Triassic Panchet Fm of India (Tasch *et al.* 1975), the Upper Triassic of Angola, the Santa Maria Fm of Brazil, and the Russian Triassic and Permo-Carboniferous (Novojilov 1970: 105).

Others occur in the Lower Jurassic of Yunnan and Szechuan (Chen pei-chu 1974), the Kota Fm (Tasch *et al.* 1975), and the Antarctic Jurassic (Storm Peak and Mauger Nunatak)

There can be little doubt of the existence of a paleolimnadiid dispersal track.

Paleolimnadiopsis

In South America *Paleolimnadiopsis* occurs in the Upper Permian of Brazil (Mendes 1954) and in the Lower Cretaceous (Cardosa 1966). An Upper Cretaceous species reported from the Bauru Fm (Sao Paulo) (Messalera 1975) is questionable. In Zaire, the genus occurs in beds ranging from Upper Triassic to Upper Jurassic (Kimmeridgian) (Defretin-Le Franc 1967). Chang and Chen (1964) described two species of *Paleolimnadiopsis* from the Nunjiang Fm of western Jilin (Upper Cretaceous).

Others are found in Australia (Tasch 1975; Webb 1978). In the USSR, paleolimnadiopsids occur from the Carboniferous through the Lower Triassic (Novojilov 1958) (Fig. 1E). This is of special interest, since Martynov (1928) thought 21 species of insects from the European part of the USSR originated from precursors in the Permian (Leonardian) Wellington shale of Kansas from which I describe a new *Paleo-*

limnadiopsis species.

Dispersal potential appears to have existed between the USSR, North America, and South America during the Permian and subsequently was extended to China, Australia, and Africa.

Leaia pruvosti, known for decades from the Rio do Rasto Fm, had been correlated with the *Leaia* zone in the Mt Glossopteris Fm, Antarctica, as well as with leaiids in South Africa, southern Rhodesia, and New South Wales (Tasch 1969).

A Permian leaiid bed in the Tasa Cuna Fm of Argentina contains ribbed forms close to the Antarctic forms and those of Brazil (Lequizamon 1974) and adds to an Antarctica (Carapace Nunatak) - Argentina Jurassic tie suggested by both plant and cyziciid conchostracan fossils (Tasch & Volkheimer 1970).

Gabonestheriids are also known from the Kansas Permian and the Permian, Triassic, and Cretaceous of the USSR. Gabonestheriid species from the Middle Triassic of China were assigned to a different genus (Chang *et al*. 1976, pl.13, pl.14 fig.12).

Again, these records point to dispersal between USSR and China during the Triassic and between equatorial Africa and Brazil during the Permian.

Cornia

Pennsylvanian (Conemaugh) corniids are known from Ohio, the Permian of Kansas, USA, the Permo-Carboniferous of the USSR (Novojilov 1970, table 136), and Australia (Tasch 1970), Lower Triassic Panchet Fm of the Raniganj and East Bokharo Coal Fields of India, the Upper Triassic Cave Sandstone at Thabaneng, Lesotho (pers. coll.), and the Angolan Upper Triassic.

Corniids are also known from the Triassic and Jurassic of the USSR (Novojilov 1970, Table 136), in my collections from the Jurassic at Blizzard Heights, Transantarctic Mountains, and in the late Mesozoic of China where they have been assigned to a different genus (Chang *et al*. 1976, pl. 14) (Fig. 1D).

Mesozoic *Cornia* dispersal between the USSR, India, Africa and Antarctica Triassic-Jurassic, and between the USSR and China, in the late Mesozoic, may be postulated.

AFRICA, SOUTH AMERICA, ANTARCTICA

Because the Cretaceous distribution of paleolimnadiopsids indicates continued ligature of Africa and South America (as with tetrapods; see Colbert, this volume), the occurrences of estheriinids and estherellids in the two continents during the Mesozoic can be explained readily.

The conchostracan link between the Gondwana continents has been importantly strengthened by the discovery of a new species of *Gabonestheria* (Marlière 1950) in the Upper Permian Rio do Rasto (Tasch 1978).

CHINA, AUSTRALIA, USSR

An important discovery of ribbed conchostracans (leaiids) was made in the upper part of the Middle Devonian Guitou Gp of South China (Shen 1978; Chang *et al*. 1976) From my own observations and collections at two sites in Lechang I was satisfied that the leaiids and related forms are *in situ*. Associated fossils include the Devonian fish *Bothriolepis*; immediately overlying sediments contain a Devonian stromatoporoid, and still higher sediments contain the Devonian-Carboniferous brachiopod, *Cyrtospirifer*, originally described by Grabau from China. The leaiids from the Lechang sites are definitely Devonian.

The fossils include species of *Leaia* (not previously known below the Carboniferous), *Hemicycloleaia* (reassigned as a subgenus of *Leaia* (Tasch & Jones 1979a), *Trileaia* and *Rostroleaia*. Shen concluded that leaiids among other ribbed types first appeared in eastern Asia at the middle of the Devonian and radiated south and west.

Shen's conclusion can be expanded with regard to *Rostroleaia* and *Hemicycloleaia*; the former is known from the Nikolaevsk Gorge in the Ural Mountains, and from the lower Carboniferous (late Visean, possibly early Namurian) of the Canning Basin (Tasch & Jones 1979a). There is good reason therefore to postulate southward dispersal to Canning Basin during the Carboniferous and northward dispersal during the Permian to the Urals.

Hemicycloleaia is abundantly represented in the Carboniferous of the USSR (Lower Stephanian of the Donetz Basin and Westphalian of the Karagand Coal region) and in the Lower Carboniferous Anderson Fm, Canning Basin (Tasch & Jones 1979a), again signifying a southward and a northward dispersal from eastern China. *Trileaia* and *Leaia* both occur in the upper Paleozoic of Australia and support the inference of a southern dispersal from eastern China.

There were equivalent non-marine dispersal tracks between Australia, China and the USSR during Mesozoic time for the genus *Paleolimnadiopsis* (Tasch 1970).

SUMMARY

Global reconstructions that fail to incorporate the requirement for non-marine

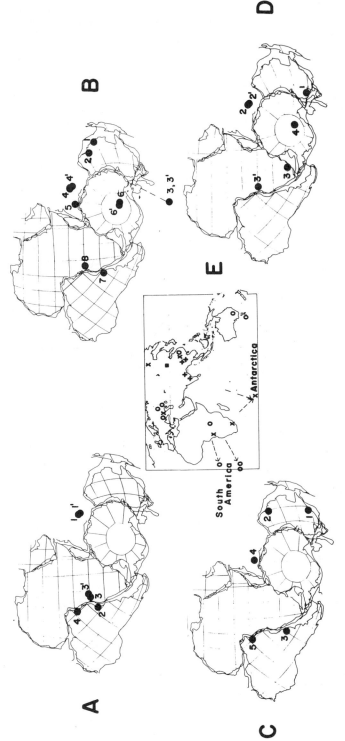

Fig. 1.--Distribution of selected Paleozoic/Mesozoic Conchostraca (Gondwana, China, USSR). A. *Estheriellids* 1,1'. Panchet Fm, East Bokaro Coal Field and Raniganj Basin. New genus. 2. Botucatu Fm, Sao Paulo and S. Mato Grosso. *Estheriella*. Late Mesozoic. 3, 3'. "Phyllopod Beds" Karroo System: Angola: Malange and Quela. *Estheriella*. Upper Triassic. 4. Japoata Fm, Bahia. *Graptestheriella*. Lower Cretaceous. B. *Paleolimnadia*: 1. Well cuttings, Bonaparte Gulf Basin. Lower Triassic. 2. Blina Shale, Canning Basin. Lower Triassic. 3, 3'. Knocklofty Shale. Tasmania: Paotina and Knocklofty. Lower Triassic. 4, 4'. Panchet Fm. Lower Triassic. 5. Kota Fm, Pranhita-Godavari Valley. Lower Jurassic. 6, 6'. Blizzard Heights and Mauger Nunatak. Jurassic. 7. Santa Maria Fm. Upper Triassic. 8. "Phyllopod Beds" C. *Estherina*: 1. Rewan Gp, Bowen Basin. Lower Triassic. 2. Blina Shale. 3. Santa Maria Fm, Rio Grande do Sul. Upper Triassic. 4. Kota Fm. 5. Ilhas Fm, Bahia. Lower Cretaceous. D. *Cornia*: 1. Newcastle Coal Measures. 2, 2'. Panchet Fm. 3, 3'. Cave Sandstone, Lesotho. Upper Triassic. 4. Blizzard Heights. Jurassic. E. *Paleolimnadiopsis*. 6, 6. Rio do Rasto Fm, Poco Preto. (6') Kazahkstan. Upper Permian. 7. Kuznetz Basin. Lower Triassic. 8,8'. Knocklofty Fm. Tasmania; Hawkesbury Sandstone. NSW: Sydney Basin. Middle Triassic. 9. Stanleyville Series and others(Africa): Zaire. Upper Triassic/Upper Jurassic. 10, 10'. Ilhas Fm. Lower Cretaceous. Area near Jilan, China. Upper Cretaceous.
+ = *Paleolimnadia*   x = *Cornia*   0 = *Paleolimnadiopsis*   ◼ = *Estherina*

13

dispersal tracks are unacceptable. An example, is the separation of India and Australia seen on many global reconstructions (but not in Curray & Moore 1974).

ACKNOWLEDGMENTS

Field research in southern Africa and the People's Republic of China; Laboratory research at the Nanking Institute of Geology and Paleontology, and the SEM laboratory of the Geological Survey of India (Calcutta); and museum study of conchostracan types on deposit at the Australian Museum, Sydney, were financed by the Office of Polar Programs, National Science Foundation, Grant OPP-77204490.

REFERENCES

CARDOSA, R.N. 1966. Conchostrácos do Grupo Bahia. Escola Federal de Minas de Oura Preto, Minas Gerais, Brasil. 111 pp.

CHANG, W.T.; P.C. CHEN; Y.B. SHEN 1976. Fossil Conchostraca of China. Peking: Science Press. 268 pp.

---; ---; 1964. New Cretaceous conchostracans from Jilin and Heilongjiang. Acta Palaeontol. Sinica 12 (1): 1-10.

CURRAY, J.R.; D.G. MOORE 1974. Sedimentary and tectonic processes in the Bengal Deep-Sea Fan and geosyncline. In C.A. Burke, C.L. Drake (Eds.), The Geology of Continental Margins. Springer-Verlag: pp. 617-627.

DEFRETIN-LEFRANC, S. 1965. Etude et révision Phyllopodes Conchostracés en provenance d' U.R.S.S. Ann. Soc. Geol. Nord. 85: 15-47.

GHOSH, S.C.; S.C. SHAH 1977. Estheriella taschi sp. nov., a new Triassic conchostracan from the Panchet Formation of East Bokaro Coalfield, Bihar. J. Asia. Soc. 19 (1-2): 14-18.

KOBAYASHI, T. 1973. On the classification of the fossil Conchostraca and the discovery of estheriids in the Cretaceous of Borneo. Geol. Paleontol. SE Asia 13: 42-72.

--- 1975. Upper Triassic estheriids in Thailand and the conchostracan development in the Mesozoic era. Geol. Paleontol. SE Asia 16: 57-87.

KYLE, R.A., A. FASOLA 1978. Triassic palynology of the Beardmore Glacier area of Antarctica. Palinologia num. es. 1: 315-318.

LEGUIZAMON, R.R. 1974. Hallazgo del genero Leaia (Conchostraco) en el Permico Argentino. Acta 1 Congr. Argentino Paleontol. Biostratig. 1: 357-369.

MARLIERE, R. 1950. Ostracoda et Phyllopoda du Système du Karroo au Congo Belge et les régions avoisantes. Ann. Mus. Congo Belge Sci. Géol. 6:11-38.

MARTYNOV, A. 1928. Permian Fossil insects of north-east Europe. Travaux du Musée Géologique, Académie des Science de l'URSS 4: 1-4, 104-118.

MENDES, J.C. 1954. Conchostracos do Sul do Brasil. Paleontol. Paraná. Curitiba, Brazil. pp. 154-164.

MESSALIRA, S. 1974. Contribucão ao Conhecemento da Estragrafía e Paleontol. do arenito Bauru. Inst. Geol. Geog. 51: 120-121

NOVOJILOV, N.I. 1970. Vymershie Limnadoidei (Conchostraca-Limnioidea) Izdat. Akad. Nauka: 1-237.

PINTO, I.D. 1956. Artropódos da Formacão Santa Maria (Triássico Superior) do Rio Grande do Sul, com notícias sobre alguns restos vegetais. Bol. Soc. Bras. Geol. 5(1): 75-94.

SHEN, Y.B. 1978. Leaiid conchostracans from the Middle Devonian of South China, with notes on their origin, classification and evolution. International Symposium on the Devonian System, 1978. Nanking.

TASCH, P. 1969. Antarctic leaiid zone: seasonal events: Gondwana correlations. Gondwana Stratigraphy. Buenos Aires: International Union of Geological Sciences (UNESCO). Pp. 185-194.

--- 1975. Nonmarine Arthropoda of the Tasmanian Triassic. Pap. Proc. R. Soc. Tasmania 109:97-106.

--- 1978. Permian palynomorphs (Coalsack Bluff, Mt. Sirius, Mt. Picciotto) and other studies. Antarc. J. U.S. 13 (4): 19-20.

---; J.M. LAMMONS 1978. Palynology of some lacustrine interbeds of the Antarctic Jurassic. Palinologia num. es. 1: 455-460.

---; P.J. JONES 1979a. Carboniferous and Triassic Conchostraca from the Canning Basin, Western Australia. BMR Bulletin 185: 1-19.

---; ---. 1979b. Lower Triassic Conchostraca from the Bonoparte Gulf Basin, Northwest Australia, with a note on Cyzicus (Euestheria) minuta from the Carnarvon Basin. BMR Bulletin 185: 23-30.

---; P.M. OESTERLEN 1977. New data on the "Phyllopod Beds" (Karroo System) Northern Angola. South Central Geological Society America Annual Meeting (El Paso, Texas) Abstract with Program: 77.

---; W. VOLKHEIMER 1970. Jurassic Conchostraca from Patagonia. Univ. Kansas Paleontol. Contrib. 23 pp.

Permo-Carboniferous palynology of Gondwana:

Progress and problems in the decade of 1980 (Abstract)

ELIZABETH M. TRUSWELL
Bureau of Mineral Resources, Canberra, Australia

The decade 1970-80 has seen the emergence of new regional, palynologically based bio-stratigraphic schemes, or the modification of existing ones, from late Carboniferous through Permian sequences on all of the major continents that formerly constituted Gondwana. Schemes comprising loose-defined assemblage-zones and/or interval-zones have been published for south, east and west Africa, India, South America and Antarctica (Fig. 1 this paper) and for eastern and western Australia (Kemp et al. 1977, *BMR Journal of Australian Geology and Geophysics* 2; 177-208). Inter-regional correlation based on these zones is for the most part of a broad kind, but the beginnings of a widely applicable chronostratigraphy are discernible, and relatively high-resolution correlation is possible between selected intervals in Australia, India and southern Africa.

The biostratigraphic scheme from the northern Karroo Basin of South Africa provides a particularly useful correlative standard for western Gondwana. Stratigraphically useful pteridophyte spores which are common to Australia and South Africa permit detailed correlation between the two areas. It is possible to suggest, on this basis, that the South African Palynozone 4d correlates with the eastern Australian Lower Stage 5 boundary--or, in lithological terms, that microfloral changes occurring at an horizon within the Upper Ecca were synchronous with those occurring within the Aldebaran Sandstone of the Bowen Basin of Queensland.

During the decade there has been some emphasis on the palynology of late Paleozoic glacial deposits, with data becoming available from all the major continents. This wealth of recent data has failed to produce any evidence in support of the idea that glaciation commenced in South America/southern Africa in the Carboniferous and migrated across the supercontinent to India and Australia in the Permian. Facies control on the microfloras from glacial deposits must, however, influence interpretation.

Information has also become available concerning the palynology of sediments transitional between the Permian intervals of coal measure sedimentation and the red-bed sequences which overlie them. Understanding of time relationships in this interval has been made easier by the delineation of the *Protohaploxypinus microcorpus* Zone in Australia, and by the recognition of the probable latest Permian age of this zone through its relationship to the Salt Range sequence. Identifying the zone beyond Australia, India and Pakistan remains problematical--no microfloras as young as *P. microcorpus* are known from South America or southern Africa. Sakamena Group assemblages from Malagasy are probably younger, and the time relationships of late Permian sequences in Gabon are difficult to assess because of provincial differences between that region and Australia.

Palynological information accumulated during the decade points also to the existence of phytogeographic subprovinces within the *Glossopteris*-dominated Permian vegetation of Gondwana. In general, eastern Gondwana microfloras are distinguished by a greater diversity of pteridophyte spores: a greater morphological diversity of gymnosperm pollen is characteristic of the western part of the supercontinent. This distinction is most apparent in the later part of the Permian, when the influence of glaciation had waned.

Fig. 1 -Stratigraphic columns from Africa, Malagasy, India, South America and Antarctica, showing for most cases the relationship of palynostratigraphic units to local lithological units. In some instances, i.e. Zaire and peninsular India, no palynological units independent of lithology have been defined; the lithological sequences have been included here for clarity. The figure should not be interpreted as a correlation chart, although I have attempted to place lithological units that are probably correlative at approximately the same level. The correlation lines joining the Zambian and Tanzanian columns are from Utting (1978: Palynology 2: 53-68).

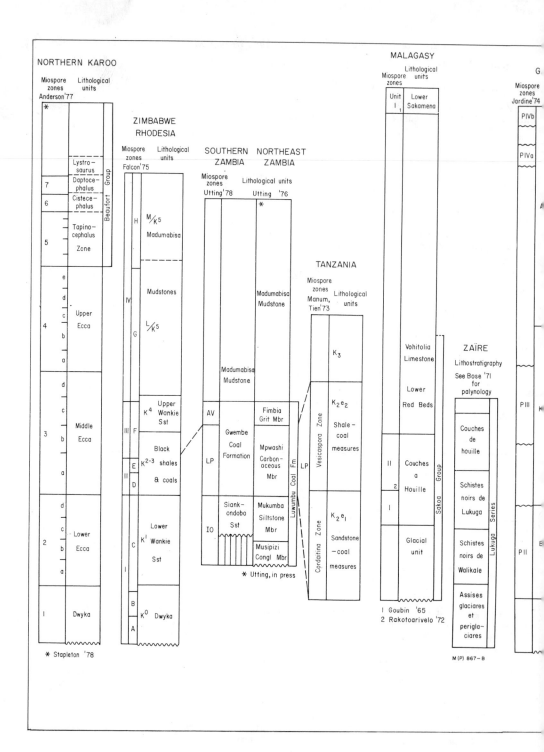

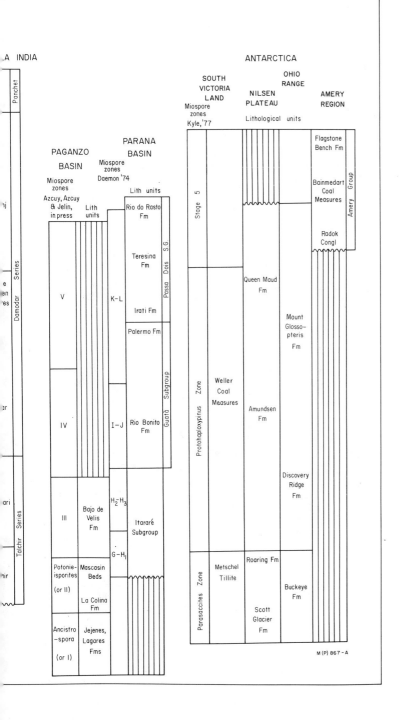

Stratigraphical and geographical distribution of Gondwana megaspores

PIERRE PIERART
Mons University, Belgium

The geographical distributions and stratigraphical ranges of Gondwana megaspores are variable.   Two main groups are distinguishable: one exclusively found with the *Glossopteris* flora, and the other thought to have migrated from the Carboniferous basins in the vicinity of the glaciation limits.   A few other megaspores are characterized by their endemism (e.g. *Proximalicirculates katangaensis* in Zaïre or some species of *Jhariatriletes* and *Singhisporites* in India).

INTRODUCTION

In spite of the small number of *Protolepdodendrales* and *Lepidodendrales* in Gondwana, megaspores are fairly well represented. The stratigraphical ranges of megaspores are not well known because the extraction of these fossils is only possible from carboniferous sediments with more than 25% of volatile matter, and because only a few papers have so far been devoted to the subject.   Nevertheless Indian palynologists have described 33 "species" belonging to 12 "genera" distributed in the Lower Gondwana stages (Bharadwaj & Tiwari 1970; Lele & Chandra 1974).

In the following account, the term "megaspores of the *Glossopteris* flora", which is preferred to the more general term "Gondwana megaspores", is defined as those found exclusively with the *Glossopteridae*.   This group of megaspores is poorly represented in the *Glossopteris* flora and is mostly found in Africa and South America.

We shall use the expression "transgressive megaspores" to distinguish those megaspores whose parent plants have migrated into the area inhabited by the *Glossopteris* flora from floras that were living in adjacent areas during the Carboniferous period.

Transgressive megaspores are represented by some species of the Lower Carboniferous of North Africa (Egypt, Chad, Niger) and of the early Upper Carboniferous of South America (Argentina), and in the Lower Gondwana sediments of South Africa and Brazil.   This group is important in the *Glossopteris* flora.   It is likely that they were mixed with the *Lepidodendropsis* flora (Chaloner & Lacey 1973).   These megaspores (*Sublagenicula brasiliensis, Duosporites tenuis, Setosisporites furcatus*, ...), which had already appeared in the early Carboniferous, seem to characterize a larger Gondwana area which therefore includes North Africa and the western parts of South America (Pierart 1978).

This is in good agreement with the view of Lacey (1975: 132) who suggested that "The "mixed" floras of South Africa and South America resulted from a combination of the migration inwards of "northern" elements from peripheral early Carboniferous stocks and a migration outwards of "*Glossopteris* flora"." This conclusion raises the possibility that the *Lycopodiopsidaceae* originated from the *Sublepidodendraceae*, whose megaspores did not change over a long geological period.

THE GROUP OF *Duosporites congoensis* AND *Trileites endosporitiferus*

The genus *Duosporites* Hoeg, Bose & Manum is characterized by an ornamented mesosporium.   The ornament can not always be recognised in poorly-preserved specimens.   For instance, it is impossible to distinguish *D. congoensis* from *T. endosporitiferus* (Singh) Potonie, characterized by an unornamented mesosporium, when the exine is too black.   Thus when the material is only studied in reflected light the name *T. endosporitiferus* is ambiguous and may mean both *T. endosporitiferus* and *D. congoensis*.   Dettman (1961) proposed *Banksisporites* for megaspores similar to *Duosporites* but lacking the pits on the mesosporium.   In conclusion, we have two nomenclature systems: the first is only useable in reflected light with the genus *Trileites*; the second uses transmitted light and allows the distinction between *Duosporites* (with cushions) and *Banksisporites* (without cushions).

The *Trileites* group (including *Duosporites* and perhaps *Banksisporites*) is well represented in Zaïre, Brazil, Argentina, India, Niger, Chad and Australia (Harris 1969).   Some species from the Chad and South Africa, described by Dijkstra (1971) should be studied in transmitted light to check their taxonomic position in relation to *D. congoensis*.

Trileities tchadiensis Dijkstra and *T. irregularis* Dijkstra from South Africa are very similar to *D. congoensis* and *T. endosporitiferus*. Lachkar (1979) regarded *T. tchadiensis* as a *Pseudovalvisisporites*, but the circular depression between the contact faces and the equator are characteristic of *Duosporites*. *Triletes irregularis* Dijkstra and *T. subnitens* Dijkstra, both from South Africa, are very close to *T. tchadiensis* and belong to *Duosporites*. Some individuals described by Lachkar as *Psudovalvisisporites agadesensis* (p. I, fig. 28, 29) may also be included. This group belongs, probably, to the "transgressive megaspores".

THE GROUP OF *Sublagenicula brasiliensis* (DIJKSTRA) C.I.M.P.

The megaspore species *S. brasiliensis* Dijkstra 1955 has a thick wall, sometimes covered with conate-granulose-spinose elements. The subgula is highly variable in size. This species has been reported from Africa (Pierart 1959). *Lagenoisporites nuda* reported from Anatolia by Yahsiman and Ergönül (1959) is probably *S. brasiliensis*. The species belongs to the group of the "transgressive megaspores"; it is distributed in the Lower Carboniferous of Agades (Lachkar 1979), in the early Upper Carboniferous of Western Argentina (Spinner 1969) and in the Lower Gondwana rocks in general. *Dijkstraea brasiliensis* (Dijkstra) Pant & Srivastava is an unnecessary combination.

Many fossils reported under other names from the Lower Carboniferous of Chad seem to belong to *Triletes mutabilis* Dijkstra described from Egyptian (Dijkstra 1956), Chad (Dijkstra 1971) and Niger specimens (Lachkar 1979).
Triletes bulbatus, with its two forms, seems also to belong to *brasiliensis*. The forma *spinulata* must be designated *bulbata* (first form).
Triletes compactus Dijkstra and *T. dulcis* Dijkstra are probably small forms of *S. brasiliensis*. *Triletes turnaui* Dijkstra and *T. brachytrachelos* Dijkstra are both from South Africa and synonymous with *S. brasiliensis*.
Finally, *Gulatriletes barakarensis* Bharadwaj & Tiwari, in spite of the spines is likewise a variety of *brasiliensis* or a closely related species.

In conclusion, seven forms that are very similar to each other probably belong to *Sublagenicula brasiliensis*. This species has the characteristic distribution of the "transgressive megaspores" (Lower Carboniferous of Chad, Niger and Egypt; Carboniferous of Argentina; Permian of South Africa, Brazil, India, Australia, etc.).

THE GROUP OF *Duosporites tenuis* (DIJKSTRA) PIERART

Duosporites tenuis is known from the Lower Gondwana in Brazil and Zaïre. Spinner (1969) described this species from the early Lower Carboniferous of Western Argentina (La Rioja province). *Triletes plicatus* Dijkstra from the Lower Carboniferous of Egypt and Chad, *Hexalaesuraesporites sinuosus* Lachkar from the Agades Visean Basin are possibly synonyms of *D. tenuis*.
Like *S. brasiliensis*, *Duosporites tenuis* belongs to the group of "transgressive megaspores" (lower Carboniferous of Egypt, Chad, Agades; Carboniferous of Argentina; Lower Gondwana in general).

THE GROUP OF *Biharisporites*

This group, represented by *B. spinosus* (Singh) Pot. emend. Bharadwaj & Tiwari and similar forms known by other possibly synonymous names, *B. distinctus* (Bharadwaj & Tiwari), *B. arcuatus* (Bharadwaj & Tiwari) probably is one of the megaspores of the *Glossopteris* flora defined above. The related genera *Srivastavaesporites* and *Talchirella* are also members of this group. The assemblage is partly represented in the mixed flora of Hazru (Turkey).
The genera *Jhariatriletes*, *Singhisporites*, *Manumisporites* and *Surangeaesporites* belong either to this group exclusively represented in the *Glossopteris* flora or perhaps to the group of endemic species described below. They are not known outside the area of the *Glossopteris* flora. *Jhariatriletes* and *Singhisporites* appear at the base of the Upper Barakar stage (Bharadwaj & Tiwari 1970).

GROUP OF ENDEMIC SPECIES

Proximalicirculates katangaensis from the Shaba (Lukuga Basin) (Pierart 1975) appears to be endemic. A second example of endemism, *Sublagenicula* n.sp., will be published in a forthcoming paper.

CONCLUSIONS

This brief survey of the megaspores found in Gondwana shows the existence of two megaspore groups. The first group, the "megaspores of the *Glossopteris* flora", is found exclusively with the *Glossopteridae*. The second, the "transgressive megaspores" is found mixed with the *Glossopteris* flora and also in lower to upper Carboniferous sediments around the periphery of the *Glossopteris* area. It is interesting to note that the "transgressive megaspores" come from the Lower or Middle Carboniferous situated not far from the limits of glaciation and mix with the Glossopteridae. The presence of these megaspores gives the appearance of an extension of the "*Glossopteris* flora" towards the North of Africa (Pierart 1978) and towards the East in South America (Spinner 1969). This hypothesis seems strong since the *Lepidodendropsis* flora forms a peripheral stock capable of migration and evolution.

If we consider the clockwise rotation of Gondwana during the early Carboniferous (Ziegler *et al.* 1979), we can see that western Argentina has travelled towards the same latitude as that of North Africa, allowing migration of the *Lepidodendropsis* flora (parent plants of the megaspores) from North Africa towards Argentina.

In spite of the high homogeneity between the different basins of Gondwana (the "Assise à couches de houille de la Luena" in Africa and the Rio Grande do Sul coalfield in Brazil for example), there are some endemic species in Africa, India and Australia. In the limnic basin of the Lukuga, we have described *Proximalicirculates katangaensis* and *Sublagenicula micropapillatus*. Indian palynologists have also described many species which could be endemic in India. These megaspores may constitute a third group.

Further research will be necessary to confirm this hypothesis of megaspore origins in Gondwana. The data already given by Dijkstra, Spinner and Lachkar constitute a solid base for this concept of "mixed" flora.

ACKNOWLEDGMENTS

It is a big pleasure for me to thank Professor William S. Lacey, University College of North Wales, for checking the manuscript of this paper.

REFERENCES

BHARADWAJ, D. C.; R. S. TIWARI 1970. Lower Gondwana megaspores. A monograph. Palaeontographica, B: 129 (B): 1-65.

CHALONER, W. G.; W. S. LACEY 1973. The distribution of Late Palaeozoic floras. In N. F. Hughes (Ed.), Organisms and Continents through Time. Paleontological Association, London. Pp. 271-89.

DETTMAN, M. E. 1961. Lower Mesozoic megaspores from Tasmania and South Australia. Micropaleontol. 7: 71-76.

DIJKSTRA, S. J. 1955. Some Brazilian megaspores Lower Permian in age, and their comparison with Lower Gondwana spores from India. Meded. Geol. Sticht. 9: 5-10.

--- 1956. Lower Carboniferous Megaspores. Meded. Geol. Sticht. 10: 5-18.

--- 1971. The megaspores of boring Chad. Meded. Geol. Sticht. 22: 25-35.

--- 1972. Some megaspores from South Africa and Australia. Palaeontol. Africana 14: 1-13.

HARRIS, W. K. 1969. The occurrence and identification of megaspores in Permian Sediments. South Australia. Q. Geol. Notes 31: 1-3.

LACEY, W. S. 1975. Some problems of "mixed" floras in the Permian of Gondwanaland. In K.S.W. Campbell (Ed.), Gondwana Geology. ANU Press: Canberra. Pp. 125-134.

LACHKAR, G. 1979. Le Bassin houiller viséen d'Agades (Niger). II. Palynologie: Megaspores. Palinologia I: 43-54.

LELE, K. M.; CHANDRA ANIL 1974. Studies in the Talchir flora of India. 9. Megaspores from the Talchir formation in the Johilla coalfield, M.P., India. The Palaeobotanist 21(2): 238-247.

PANT, D.D.; G. K. SRIVASTAVA 1961. Structural studies on Lower Gondwana Megaspores. Part. I. Specimens from Talchir coalfield, India. Palaeontographica, B 109 (B): 45-61.

---;--- 1962. Structural studies on Lower Gondwana Megaspores. Part 2. Specimens from Brazil and Mhukuru Coalfield, Tanganyika. Palaeontographica, B 111 (B): 96-111.

PIERART, P. 1959. Contribution à l'étude des spores et pollen de la flore à Glossopteris contenus dans les charbons de la Luena (Katanga). Mem. Acad. R. Sci. Coloniales, Cl. Sci. Nat. Méd. 8: 1-80.

--- 1975. Systématique, distribution stratigraphique et géographique des megaspores du Carbonifère et du Permien. 7me Cong. Int. Stratig. Geol. Carbonifère B IV: 93-102.

--- 1978. Quelques remarques concernant les mégaspores du Gondwana. Ann. Soc. Géol. Nord XCVII: 405-408.

--- in press. Some microfloral Taphocoenoses from the Lower Gondwana of Zaïre (Assise a couches de houille de la Luena). Troisième Congrès du Gondwana

---; S. J. DIJKSTRA 1961. Étude comparée des mégaspores permiennes du Brésil et du Katanga. C-R Cong. Avance. Études Stratig. Carbonifère, Heerlen 2: 541-544.

SPINNER, E. 1969. Preliminary study of the megaspores from the Tupe formation. Quebrada del Tupe, La Rioja, Argentina. Pollen Spores XI 3: 669-685.

YAHSIMAN, K.; Y. ERGÖNÜL 1959. Permian Megaspores from Hazru (Diyarbakir). Bull.

Min. Res. Explor. Inst. Turkey 53: 94-101.
ZIEGLER, A. M.; C. R. SCOTESE; W. S.
McKERROW; M. E. JOHNSON; R. K. BAMBACH 1979.
Paleozoic Paleogeography.   Ann. Rev.
Earth Planet. Sci. 7: 473-502.

Fifth International Gondwana Symposium / Wellington / New Zealand / 11-16 February 1980

Geological background to a Devonian plant fossil discovery,
Ruppert Coast, Marie Byrd Land, West Antarctica

G. W. GRINDLEY & D. Ç. MILDENHALL
N. Z. Geological Survey, Lower Hutt, New Zealand

Plant fossils from dark carbonaceous slaty argillite erratics on the southernmost nunatak of Mt Hartkopf (Milan Rock), head of Land Glacier (76° 01'S, 140° 42'W), are the first fossils discovered in Marie Byrd Land.   Two taxa (*Drepanophycus* n. sp.) and cf. *Haplostigma irregulare* (Schwartz) Seward indicate late Middle or early Upper Devonian age.   ?*Protolepidodendron* sp. and dichotomous axes of the order Psilophytales also occur.   The high rank of the sediments precluded extraction of palynomorphs.

The fossiliferous erratics came from nearby sub-ice outcrops to the SE and are similar to dark carbonaceous slates and quartzo-feldspathic sandstones with carbonized plants conformably underlying the calcalkaline metavolcanic sequence of the Ruppert Coast, formerly considered Jurassic by radiometric dating.   The metavolcanic sequence now seems to be late Devonian (or early Carboniferous), the earliest known calcalkaline volcanism in this sector of the circum-Pacific.

Age revision of the Ruppert Coast sequence accords with recent reconstructions of the Southwest Pacific sector of Gondwana.

INTRODUCTION

Marie Byrd Land, coastal West Antarctica, a link in the circum-Pacific chain of Meso-zoic-Cenozoic folded mountain ranges, has been the subject of much speculation, because of incomplete examination, scarcity of outcrop and lack of fossils to date the folded pre-Cenozoic sequences.   This paper records the first fossil remains dis-covered in this enigmatic terrain and discusses their geologic significance.

The First and Second Byrd Antarctic Expe-ditions concentrated on Western Marie Byrd, showing that the Rockefeller Mountains and Edsel Ford Ranges consist of unfossilife-rous metasediments intruded by granitic plutons.   A United States helicopter-supported survey of Marie Byrd Land and Ellsworth Land continued for three seasons from 1966 (Wade & Wilbanks 1972).   Metamor-phic rocks were discovered in the Fosdick Range (Wilbanks 1972) north of the folded metasedimentary terrane, and the granitic plutons were divided into two groups, late Devonian - early Carboniferous and early Cretaceous respectively, on the basis of preliminary radiometric age determinations (Boudette *et al*. 1966; Halpern 1968; Crad-dock 1972).   Russian scientists accompanied the team during two seasons and provided additional geological, geochemical and ra-diometric data (Klimov 1967; Lopatin & Or-lenko 1972; Krylov *et al*. 1970).

Geological maps of western Marie Byrd Land have also appeared.   Stratigraphic and structural features of the metasedi-mentary terrane (Swanson Group) in western Marie Byrd Land are described by Wade and Long (in press).   Acritarchs from the Swanson Group (Iltchenko 1972), are con-sidered to indicate a late Proterozoic (late Riphean-Vendian) age, perhaps extend-ing up into the earliest Cambrian on the basis of microphytoliths.   K/Ar whole-rock data (375-475 m.y.) indicate folding, cleavage formation and regional metamorphism in the late Ordovician - early Devonian (Krylov *et al*. 1970; Adams, this vol.) Ford Orogeny of Wade (1975).

North of the Fosdick metamorphic complex, the Ruppert Coast sector is underlain by an early Cretaceous granitic batholith, within which are strongly folded, cleaved metavolcanic enclaves, all contact-meta-morphosed by adjoining granite.   The age of this metavolcanic terrane has been con-troversial, with three divergent views:

(1)   Klimov (1967) suggested that the metavolcanics are a eugeosynclinal facies of the metasedimentary Swanson Gp and com-pared them with supposed Early Paleozoic metavolcanics of the Antarctic Peninsula.

(2)   Lopatin & Orlenko (1972), Grikurov & Lopatin (1975), and Grikurov *et al*. (1977) favoured a Late Paleozoic age, based apparently on K/Ar whole-rock dates of 295 and 370 m.y. from similar folded metavolcanics in eastern Marie Byrd Land (Lopatin *et al*. 1974) and comparisons with the Antarctic Peninsula.

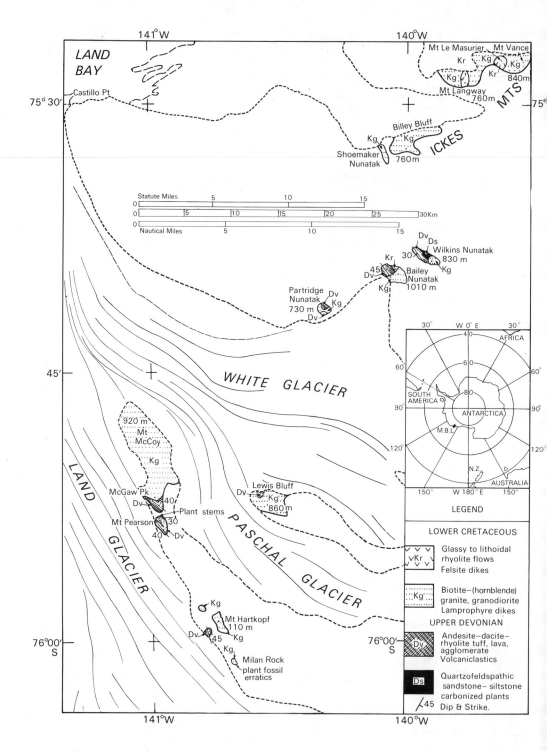

Fig.1--Geological map of the Ruppert Coast, Marie Byrd Land, showing location of plant-fossil-bearing erratics and distribution of Upper Devonian metasediments and metavolcanics.

Fig.2--Metasediments of Wilkins Fm exposed in a tight reclined anticline (axis indicated), conformably overlain by metavolcanics (contacts arrowed) on east face of Wilkins Nunatak. Intrusive early Cretaceous granite at top left.

(3) Wade (1969), Craddock (1972), Spörli & Craddock (this vol.) favour a younger Mesozoic age, partly based on comparisons with Antarctic Peninsula, where Jurassic calcalkaline volcanics are intruded by the early Cretaceous Andean plutonic suite, and partly on preliminary Rb/Sr analyses giving rather unreliable and certainly minimum Middle Jurassic model ages.

Although metasedimentary volcaniclastic rocks were found within the metavolcanics, no fossils were discovered before 1977 (Grindley & Mildenhall 1978; Wade 1978) that would enable these divergent views to be evaluated.

GEOLOGICAL OUTLINE (Fig 1.)

Sparse rock outcrops are on tabular nunataks with steep cliffed margins exposed by recent retreat of the ice sheet. The Land and Paschal Glaciers discharge into a small ice shelf in Land Bay. All the nunataks have flattish upper surfaces produced by recent glacial planing of a mid-Miocene glacial surface, possibly modifying a late Cretaceous erosion surface (Le Masurier & Wade 1975).

Wilkins Fm (Mid-Upper Devonian): The oldest rocks of the Ruppert Coast are shallow-water clastic sediments, exposed in a tight SE-dipping reclined anticline in a steep cliff face, accessible from the windscoop on the north side of Wilkins Nunatak, the type section (Fig. 2). About 100 m of beds are overlain conformably by metavolcanics (Fig. 3) and intruded by early Cretaceous granite. Thin-bedded quartzo-feldspathic sandstone and siltstone contain plane-parallel, locally graded layers, 2-10 mm thick, of light grey sandstone and siltstone (Fig. 4). Small recumbent folds and asymmetric drag folds are common (Fig. 5). Massive, dark grey, non-calcareous carbonaceous siltstone and coarse grey sandstone lenses, up to 3 m, are also present. Black carbonaceous shaly layers, flint-hard due to contact metamorphism, show shiny flecks of graphitised plant material without any recognisable venation. In thin section, the sandstones contain polygonal quartz, albite-oligoclase, iron oxide, and porphyroblastic green-brown biotite (chlorite) and muscovite. Feldspar is essentially

Fig.3--Well-bedded quartzose sediments of Wilkins Fm, partly faulted against overlying andesitic metavolcanics (contacts arrowed) at type section on east face of Wilkins Nunatak. Light-coloured dikes and veins of early Cretaceous microgranite (X) intrude both formations.

Fig.4--Plane-parallel
quartzose sandstone and
siltstone with minor
faults.   Wilkins Fm,
Wilkins Nunatak.

Fig.5--Recumbent fold
in quartzose sandstone
and siltstone, Wilkins
Fm, Wilkins Nunatak.

restricted to coarser layers.

Plant-bearing Erratics: Carbonaceous
argillites and quartzo -feldspathic sandstones comprise a small percentage of
erratic blocks littering glaciated granitic bedrock at Lewis Bluff, Mt Hartkopf
and Milan Rock (Fig. 1).   Most of the metasedimentary rocks are either contact-
altered or coarse-grained, thus unsuitable
for preservation of plants.   At Milan
Rock, during a stop to collect granite for
radiometric dating, one of us (G.W.G.), by
splitting and examining erratics, discovered lycopod impressions in dark, cleaved
carbonaceous argillite, of lower rank but
similar to finer grained beds at Wilkins
Nunatak.   Bedding is identified by thin
sandstone partings at a low angle to

cleavage, and plant fragments are found by
careful splitting of this one lithology.
Other rocks include coarse quartzo-feldspathic sandstone containing dark argillite
flakes, thin-bedded carbonaceous siltstone
containing plant rootlets, conglomerate
with metavolcanic, crenulated phyllite and
quartz sandstone pebbles, and cross-bedded
arkosic sandstone.   Coarser beds are
believed to be deltaic channel deposits,
while finer beds with plant fossils may
have been deposited in a prodelta or overbank situation.   Metasedimentary pebbles
resemble the metagreywacke-phyllite sequence (Swanson Gp) of the Ford Ranges to
the south, from which the Wilkins Fm may
have been derived as a clastic, deltaic
post-orogenic wedge.

Glacial striations on the granitic bedrock have a prevailing SE orientation (Karlen & Melander 1978) and the metasedimentary outcrops are believed to be located less than 5 km to the SE. The fossils include *Drepanophycus* n. sp., cf. *Haplostigma irregulare* (Schwartz) Seward, psilophytalean axes and *?Protolepidodendron* stems, indicating a late Middle or early Late Devonian age (Grindley *et al.* in press). High rank precluded extraction of palynomorphs.

Ruppert Coast Metavolcanics (Upper Devonian-Carboniferous?): Enclaves of calcalkaline metavolcanics are found along the Ruppert Coast between Land and Hull Glaciers. The dominant lithology is dark grey feldsparphyric andesite but ranges from spilitic basalt through andesite and dacite to rhyolite. They are strongly folded about NW-SE axes, with a near-vertical axial-plane cleavage (Fig. 6). Dips generally do not exceed 45°, except adjacent to granite plutons.

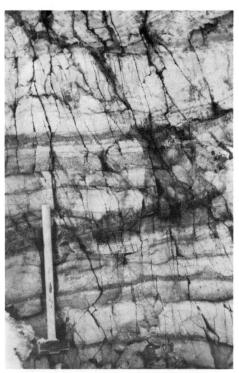

Fig.7--Graded andesitic tuffs at base of Ruppert Coast Metavolcanics, north end of Pearson Peak.

The basal beds are volcaniclastic sediments ranging from pebble conglomerate to stratified sandstones (Fig. 7). Pebbles are mainly feldsparphyric andesite, but pink coarse-grained granite and granodiorite occur locally (e.g. Mt Pearson). Volcaniclastics on the north side of Mt Pearson contain unidentifiable plant stems. On Bailey Nunatak, dark argillite and quartz sandstone clasts in andesitib breccias resemble the underlying metasedimentary sequence.

Massive andesite and dacite flows and agglomerates constitute the major part of the metavolcanic sequence at Wilkins and Bailey Nunataks and at Mts Pearson and Hartkopf. Rhyolites are confined to Partridge Nunatak. Flow banding, volcaniclastic layers, feldspar phenocryst alignments and platy jointing are all subparallel and can be used to determine the regional structure (Fig. 1) of NW-trending open folds, overprinting earlier recumbent folds.

The metavolcanics are metamorphosed regionally to low green-schist facies assemblages (albite-chlorite-epidote-calcite-sphene) and are commonly recrys-

Fig.6--Andesitic metavolcanics dipping gently south and intruded along steep meridional cleavage by Cretaceous lamprophyre dike. View south of Pearson Peak, Ruppert Coast.

tallised near granite contacts (albite-epidote-hornfels facies) producing green to brown biotite, clinozoisitic epidote, muscovite, tremolite-actinolite and rare garnet. Quartz, calcite, epidote, actinolite, chlorite and iron oxides fill veins and vesicles. Thicker flows, less affected by shearing and recrystallisation, commonly preserve fresh andesine (An_{45-50}) phenocrysts but clinopyroxenes are invariably reconstituted.

On the Ruppert Coast, K/Ar dating of the metavolcanics has been largely unsuccessful, providing early Cretaceous (100-110 m.y.) minimum ages because of nearby granites (Krylov et al. 1970; Spörli & Craddock, this vol.). In eastern Marie Byrd Land, K/Ar dating of folded "metadiabase-metarhyolite-dacite porphyries", probably correlative with the Ruppert Coast Metavolcanics (Grikurov et al. 1977: 41) gave latest Devonian-Carboniferous ages (295-370 m.y. Lopatin et al. 1974, Table 1). Rb/Sr analyses of three metavolcanic samples (Spörli & Craddock, this vol.) did not provide a viable isochron.

A late Devonian-Carboniferous age is therefore preferred, based on the inferred age of the conformably underlying Wilkins Fm and the Russian K/Ar ages cited above.

Granitoid Complexes. (early Cretaceous): Early Cretaceous granitoid complexes, widespread throughout western Marie Byrd Land, intrude the Swanson Fm in the Ford Ranges, the Ruppert Coast Metavolcanics (Metcalfe et al. 1978; Spörli & Craddock, this vol.) and an older Paleozoic gabbro-granodiorite complex on the Hobbs Coast.

In the Ruppert Coast sector the granitoid complexes range from biotite and hornblende-bearing adamellite and granodiorite in the west (Mt Hartkopf, Mt McCoy, Lewis Bluff) to more alkalic granites in the east, containing up to 50% perthitic alkali feldspar and minor alkaline amphibole and clinopyroxene (Mt Langway, Mt Le Masurier, Mt Vance). The granitoids are calcalkaline, typical I-type granites, characteristic of many circum-Pacific Mesozoic batholiths (e.g. Separation Point Batholith of New Zealand and the Sierra Nevada Batholith of California).

The granitoids have been dated by K/Ar dating of hornblende (96 m.y.) and Rb/Sr dating of biotite (92 m.y.) at Bailey Nunatak (Spörli & Craddock, this vol.). Felsic and mafic dike rocks, including lamprophyres, accompanied and followed granite intrusion. A quartz porphyry dike at Bailey Nunatak provided a K/Ar whole rock date of 98 m.y. while a mafic dike intruding granite at Billey Bluff (Fig. 1) gave a date of 113 m.y. (Spörli & Craddock, this vol.).

Rhyolitic Volcanics (mid-late Cretaceous?): Small extrusions of pink porphyritic rhyolite, associated with the Cretaceous granites of Bailey Nunatak, Mt Langway and Mt Le Masurier (Fig. 1) are intruded by mafic and felsic dikes accompanying the latest phases of granite intrusion. Some are hydrothermally altered and their original mineralogy is indeterminate. Relatively unaltered flows contain ragged phenocrysts of untwinned alkali feldspar, clouded with epidote and calcite, in a devitrified matrix of quartz and alkali feldspar, sphene, iron oxide, calcite and chlorite. Quartz phenocrysts appear to be absent. Some non-porphyritic, finely crystalline phases may be intrusive microgranites. Chilled margins show evidence (brecciation and recrystallised quartz mosaics) for considerable stress during intrusion.

The Ruppert Coast rhyolites may be earlier phases of the rhyodacite domes at Mt Petras and Mt Galla in central Marie Byrd Land described by Le Masurier and Wade (1976), dated as late Cretaceous (80-88 m.y.) by D.C. Rex.

REGIONAL SIGNIFICANCE

The discovery of Middle-Late Devonian plants in Marie Byrd Land has led to several conclusions.

(1) The occurrence of a typical southern hemisphere lycopod (*Haplostigma*) points to a close Gondwana association with East Antarctica, East Australia, South Africa and South America. *Drepanophycus* is a further link with the South African floras (Plumstead 1967; Grindley et al. in press).

(2) Plant-bearing metasediments of Milan Rock (Wilkins Fm) are associated with conglomerates containing crenulated phyllite clasts, that were probably deposited on a delta by rivers draining the folded metasedimentary Swanson Gp, which according to K/Ar dating was folded and uplifted during the Ordovician-early Devonian.

(3) The calcalkaline Ruppert Coast Metavolcanics (Upper Devonian-Carboniferous?), conformably overlying the Wilkins Fm, are the earliest known in this sector of the circum-Pacific, predating the Lower Permian calcalkaline volcanics of the New Zealand Geosyncline.

(4) Upper Devonian calcalkaline volcanism, unknown from New Zealand, is known from the New England Geosyncline in N.S.W., where the Baldwin Fm (containing *Haplostigma*) is essentially derived from an andesitic volcanic arc to the west (Leitch 1974).

(5) The continuation of a late Paleo-

zoic andesite arc and inferred subduction zone from NE Australia through New Zealand to West Antarctica is implicit in recent continental reconstructions of the SW Pacific (Grindley & Davey, in press; Cooper *et al*. in press).

ACKNOWLEDGMENTS

With the support of Dr M. Turner, (Office of Polar Programmes, N.S.F.), the late Dr F.A. Wade (Texas Technical University) kindly invited one of us (G.W.G.) to participate in the 1977-78 U.S. expedition to Marie Byrd Land, where Drs John Wilbanks (scientific leader) and Wesley Le Masurier (project leader) incorporated him in the field programme. Field work would not have been possible without the support of helicopter pilots from Air Development Squadron VXE6. Thanks are due to colleagues at N.Z. Geological Survey for photography, typing, draughting and comments on the MS. Dr Rosemary Askin (Inst. Polar Studies, Ohio) assembled and despatched the Wilbanks plant collection after the untimely death of Dr J.M. Schopf, who first recognised the new species of *Drepanophycus*.

REFERENCES

ADAMS, C.J.D. Geochronological correlations of Precambrian and Paleozoic Orogeny in North Victoria Land (East Antarctica), Marie Byrd Land (West Antarctica), New Zealand and Tasmania. (This volume)

BOUDETTE, E. L.; R. F. MARVIN; O. E. HEDGE 1966. Biotite, potassium-feldspar and whole rock ages of adamellite, Clark Mountains, West Antarctica. Prof. Pap. U.S. Geol. Surv. 550-D, D190-194.

COOPER, R.A.; C..A. LANDIS; W. E. LE MASURIER; I. G. SPEDEN in press. Geological history and regional patterns in New Zealand and West Antarctica; their paleotectonic and paleogeographic significance. In C. Craddock (Ed.) Antarctic Geoscience, University of Wisconsin: Madison.

CRADDOCK, C. 1972. Geological Map of Antarctica 1 : 5,000,000. Am. Geog. Soc. New York.

GRIKUROV, G. E.; B.G. LOPATIN 1975. Structure and Evolution of the West Antarctic part of the Circum-Pacific Mobile Belt. In K.S.W. Campbell (Ed.) Gondwana Geology. ANU Press: Canberra. Pp. 639-650.

---; G. A. ZNACHKO-YAVORSKY; E. N. KAMENEV; M. G. RAVICH 1977. Explanatory Notes to the Geological Map of Antarctica (scale 1 : 5,000,000). Research Institute of the Geology of the Arctic, Leningrad, 1976.

GRINDLEY, G. W.; F. J. DAVEY in press. The reconstruction of New Zealand, Australia and Antarctica (review). In C. Craddock (Ed.), Antarctic Geoscience, University of Wisconsin: Madison.

---; D. C. MILDENHALL 1978. Discovery of fossils in Marie Byrd Land, Antarctica. N.Z. Geol. Soc. Newsl. 45: 33-34.

---;---; J. M. SCHOPF, in press. A mid-late Devonian flora from the Ruppert Coast, Marie Byrd Land, West Antarctica. J. Roy. Soc. N.Z. 10.

HALPERN, M. 1968. Ages of Antarctic and Argentine rocks bearing on continental drift. Earth Planet. Sci. Lett. 5: 143-47.

ILTCHENKO, L. N. 1972. Late Precambrian Acritarchs of Antarctica. In Adie R.J. (Ed.), Antarctic Geology and Geophysics. Universitetsforlaget: Oslo. Pp. 599-602.

KARLEN, W.; O. MELANDER 1978. Reconnaissance of the glacial geology of Hobbs Coast and Ruppert Coast, Marie Byrd Land. Ant. J. U.S. 13(4): 46-47.

KLIMOV, L. V. 1967. (Some Results of geological investigations in Marie Byrd Land in 1966-67). Inf. Bull. 65, Sov. Antar. Exped. 6(6): 555-59.

KRYLOV, A. Ya.; B. G. LOPATIN; T. I. MAZINA 1970. Age of rocks in the Ford Ranges and on the Ruppert Coast (western part of Byrd Land). Inf. Bull. 80, Sov. Antar. Exped. 8(2): 64-66.

LEITCH, E. C. 1974. The geological development of the southern part of the New England Fold Belt. J. Geol. Soc. Aust. 21(2): 133-56.

LE MASURIER, W. E.; F. A. WADE 1976. Volcanic history in Marie Byrd Land: Implications with regard to Southern Hemisphere tectonic reconstructions. In O. Gonzalez-Ferran (Ed.), Proc. Sym. Andean. Antar. Volcanology Probs., Santiago. IAVCEI: Rome. Pp. 398-424.

LOPATIN, B. G.; A. Ya. KRYLOV; O. A. ALIAPYSHEV 1974. Main tectono-magmatic stages of development of Marie Byrd Land and Eights Coast, West Antarctica according to radioactive data (in Russian). In Antarktika 13: 36-51. Akademia Nauk: Moscow.

---; E. M. ORLENKO 1972. Outline of the Geology of Marie Byrd Land. In R.J. Adie (Ed.) Antarctic Geology and Geophysics. Universitetsforlaget: Oslo. Pp. 245-50.

METCALFE, A. P.; K. B. SPÖRLI; C. CRADDOCK 1978. Plutonic rocks from the Ruppert Coast, West Antarctica. Antar. J. U.S. 13(4): 5-6.

PLUMSTEAD, E. P. 1967. A general review of the Devonian fossil plants found in the Cape System of South Africa. Palaeont. Africana 10: 1-83.

SPÖRLI, K. B.; C. CRADDOCK   Geology of
the Ruppert Coast, Marie Byrd Land, Antarctica.   (This volume)

WADE, F. A. 1969. Geology of Marie Byrd Land
Sheet 18.   In Geologic Map of Antarctica
1: 1,000,000.   Antar. Map Fol. Ser. Pl.
III Fol. 12 - Geology.   Am. Geog. Soc.:
N.Y.

--- 1975.   Swanson Group, Ford Ranges,
Marie Byrd Land.   Antar. J. U.S. 10(5):
244-5.

--- 1978.   Geologic survey of Ruppert -
Hobbs Coast Sector, Marie Byrd Land.
Antar. J. U.S. 13(4): 4-5

---; J. R. WILBANKS 1972.   The Geology of
Marie Byrd Land and Ellsworth Lands,
Antarctica.   In R.J. Adie (Ed.)   Antarctic Geology and Geophysics.Universitets-
forlaget: Oslo.   Pp. 207-214.

---; D. R. LONG in press.   The Swanson
Formation, Ford Ranges, Marie Byrd Land -
evidence for direct relationship with Rob-
ertson Bay Group, northern Victoria Land.
In C. Craddock (Ed.), Antarctic Geoscience,
University of Wisconsin: Madison.

WARNER, L. A. 1945.   Structure and petrography of the southern Edsel Ford Ranges,
Antarctica.   Proc. Am. Phil. Soc. 89(1):
78-122.

WILBANKS, J. R. 1972.   Geology of the
Fosdick Mountains, Marie Byrd Land.
In R.J. Adie (Ed.) Antarctic Geology and
Geophysics.Universitetsforlaget: Oslo.
Pp. 277-86.

Triassic palynology of the Warang Sandstone (Northern Galilee Basin) and its phytogeographic implications

N. J. DE JERSEY & J. L. McKELLAR

Geological Survey of Queensland, Brisbane, Australia

Palynofloras from the Middle and Late Triassic of Australia have been assigned by Dolby and Balme to the Onslow Microflora, recorded from the Carnarvon Basin of Western Australia, and the Ipswich Microflora, recorded from eastern and southern Australia. It is shown that representatives of both the Ipswich and Onslow microfloras can be closely associated geographically and stratigraphically; this association is illustrated by their co-occurrence within the Middle Triassic Warang Sandstone (central Queensland), where miospore assemblages from three stratigraphic boreholes have been investigated.

From the biostratigraphic aspect, the Onslow elements in the microfloras of the Warang Sandstone indicate correlation with the lower part of the *Staurosaccites quadrifidus* Assemblage-zone of the Carnarvon Basin; the Ipswich elements indicate correlation with the lower to middle Moolayember Formation of the southern Bowen Basin. This enables the palynofloral succession of the Carnarvon Basin to be related more closely to that of the Bowen Basin than was previously possible.

From the phytogeographic aspect, comparison with the distribution of the Onslow and Ipswich microfloras shown by Dolby and Balme would indicate an abnormally high paleolatitude for the Onslow elements in the Warang Sandstone. It is suggested that the Triassic orientation of the Australian continent was less clockwise than in the Gondwana reconstruction used by Dolby and Balme.

INTRODUCTION

Until recently, data were inadequate to establish whether the Gondwana Triassic flora exhibits temperature-controlled differentiation from equator to pole, as does the extant flora, or whether it was relatively more uniform than the latter. Dolby and Balme (1976) provided evidence for such differentiation in Middle and Late Triassic palynofloras from Australia, which they classified into the Onslow Microflora, from the Carnarvon Basin of Western Australia, and the Ipswich Microflora from eastern and southern Australia. They attributed the differences between these microfloras to paleolatitude; the Onslow Microflora was believed to represent a temperate rain-forest association between about 30° and 35°S, while the less-diverse Ipswich Microflora was regarded as typical of latitudes about 40°-55°S. Both microfloras have many species in common; the greater diversity of the former is due to the presence of several species, largely of gymnospermous affinity.

More recently, evidence contrary to the pattern recorded for the Onslow and Ipswich microfloras by Dolby and Balme was provided by the discovery of *Staurosaccites quadrifidus* Dolby 1976 in the Warang Sandstone (McKellar 1977: 395). It is a distinctive element of the Onslow Microflora in the Carnarvon Basin; its presence in abundance in assemblages from the northern Galilee Basin, at paleolatitudes comparable with those of other localities for the Ipswich Microflora (Dolby & Balme 1976) was therefore unexpected (Fig. 1). To provide further data on this apparently anomalous occurrence of *S. quadrifidus*, a more detailed investigation has been under-

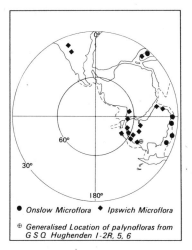

● *Onslow Microflora*   ◆ *Ipswich Microflora*

⊕ *Generalised Location of palynofloras from GSQ Hughenden 1-2R, 5, 6*

Fig. 1--Relation between paleolatitudes of palynofloras from GSQ Hughenden 1-2R, 5, & 6, and those of the Onslow & Ipswich type (Dolby & Balme 1976, fig. 8).

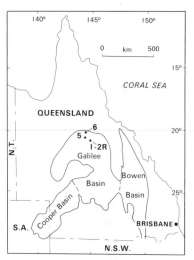

Fig. 2--Locations of GSQ Hughenden 1-2R, 5, and 6 in the northern Galilee Basin.

taken by N.J. de J. on the basis of additional samples from the Warang Sandstone. This investigation has supported and extended the results of McKellar (1977, unpubl.) who recorded *S. quadrifidus* from two stratigraphic wells, GSQ Hughenden 1-2R and GSQ Hughenden 5, in which it is associated with a suite of species characteristic of the lower to middle Moolayember Fm of the Bowen Basin. Study of the additional material has provided further records of

S. quadrifidus, and has also led to the discovery of other distinctive elements of the Onslow Microflora. Here we give an account of the palynoflora of the Warang Sandstone, from both biostratigraphic and phytogeographic aspects, and discuss the significance of the presence of Onslow elements in the assemblages.

PALYNOFLORAL ASSEMBLAGES

Miospore assemblages in 15 productive samples from the Warang Sandstone in three stratigraphic wells (de Jersey & McKellar, unpubl.) include four assemblages (GSQ Hughenden 1-2R, 381.99 m; GSQ Hughenden 5, 378.45 m and 428.16 m and GSQ Hughenden 6, 191.48 m) recognised previously (McKellar, unpubl.). Locations of the three wells are given in Fig. 2; species of biostratigraphic significance recorded in the present study are in Figs. 5 and 6 below.

Apart from *S. quadrifidus*, which is often abundant, other distinctive elements of the Onslow Microflora comprise *Infernopollenites clastratus* Dolby & Balme 1976, *Lunatisporites acutus* Leschik 1955 and cf. *Rimaesporites aquilonalis* Goubin 1965, which are relatively rare and more sporadically distributed (Fig. 3). The remaining species in the assemblages are members of the Ipswich Microflora, all of which have been recorded previously from the Bowen Basin. The additional material not only supports and extends the evidence for

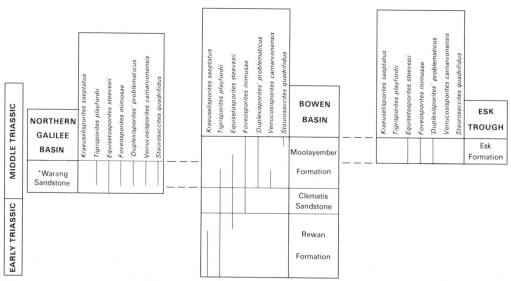

Fig. 3--Relationship of Warang Sandstone to Triassic of Bowen Basin and Esk Fm. * = Warang Sandstone.

association of elements of the Onslow and Ipswich Microfloras provided by McKellar (1977) but also indicates that there is no evidence for any sharp differentiation between the two in the Warang Sandstone. There is, on the contrary, ample evidence for gradation between them. For example, in GSQ Hughenden 1-2R, the proportions of *S. quadrifidus* range from 3.5-2.0% over only 9.71 m. In GSQ Hughenden 6, the species is present in an assemblage from 191.31 m but absent in one from 191.48 m. The other Onslow elements - *I. claustratus*, cf. *R. aquilonalis* and *L. acutus* - also display a pattern of sporadic distribution in the assemblages. These marked variations in the pattern of distribution of Onslow elements in the assemblages are presumably related to fluctuating environmental conditions during the time of deposition.

The association of Onslow and Ipswich Microfloras in the Warang Sandstone has both biostratigraphic and phytogeographic implications. It enables the assemblages from the Warang Sandstone (and therefore their equivalents in the Bowen Basin) to be correlated more precisely with the Carnarvon Basin sequence, than was previously possible. It also indicates that, if the Gondwana reconstruction adopted by Dolby and Balme (1976) is accepted, the Onslow Microflora extended into abnormally high paleolatitudes in eastern Australia during the Middle Triassic.

THE WARANG SANDSTONE AND THE CARNARVON BASIN SEQUENCE

The Onslow content of the Warang Sandstone palynoflora (*S. quadrifidus*, *I. claustratus* and cf. *R. aquilonalis*) indicates significant similarity to the lower part of the *S. quadrifidus* Assemblage-zone of the Carnarvon Basin (Fig. 4). Species which make their first appearance above the basal part of the *S. quadrifidus* Zone in the Carnarvon Basin are absent.

On the basis of the correlation indicated, *"Duplexisporites" problematicus* appears later in the Carnarvon Basin than in the northern Galilee Basin, but this anomaly is readily explained as an environmental effect. *"D". problematicus* is absent from assemblages from GSQ Hughenden 1-2R, which contain *S. quadrifidus*. Conversely, it is present in assemblages from GSQ Hughenden 5 and 6, in which *S. quadrifidus* is relatively rare or absent. It is suggested that assemblages with common *S. quadrifidus* were derived from an environment unfavourable to *"D". problematicus* and that, the latter is thus not represented in the *S. quadrifidus* Assemblage-zone of the Carnarvon Basin. It makes its first appear-

ance at the base of the succeeding *Samaropollenites speciosus* Assemblage-zone, in which *S. quadrifidus* is only sporadically distributed (Dolby & Balme 1976: 151).

Higher in the sequence, in the Upper Triassic, rare specimens of the typical Onslow species, *Minutosaccus crenulatus* Dolby 1976, occur in the Ipswich Coal Measures of eastern Australia (de Jersey & McKellar, unpubl.). In conjunction with the distribution of other species, it favours correlation of the Ipswich Coal Measures with the *Minutosaccus crenulatus* Assemblage-zone of Dolby and Balme (Fig.4).

It must be emphasised that in any biostratigraphic comparison of Triassic sequences from the Carnarvon Basin with those of Queensland, due consideration should be given to the incompleteness of the stratigraphic record in eastern Australia. In Queensland there is evidence for a time break between the Esk Fm in the Esk Trough (and its equivalents in the upper Moolayember Fm of the Bowen Basin) and the Ipswich Coal Measures of the Ipswich Basin (de Jersey 1972). The duration of this interval is uncertain, but portions of the *Staurosaccites quadrifidus* Zone and *Samaropollenites speciosus* Zone of the Carnarvon Basin may be represented.

THE WARANG SANDSTONE AND THE BOWEN BASIN SEQUENCE

Using species of restricted range, the palynoflora of the Warang Sandstone can be correlated with a restricted portion of the Bowen Basin sequence. The stratigraphic distributions of the important species are provided elsewhere (de Jersey & McKellar, unpubl.). The two most significant species are *"Duplexisporites" problematicus* and *Verrucosisporites carnarvonensis*. The former is confined to the Moolayember Fm and the latter to the lower to middle Moolayember Fm, in reference sections previously studied (de Jersey & Hamilton, 1967; de Jersey 1968, 1970a). With one exception, the Warang Sandstone assemblages can be correlated with those of the lower to middle Moolayember Fm. The exception is a low-yield assemblage from 428.16 m in GSQ Hughenden 5. This may be an impoverished Moolayember assemblage lacking key species. Alternatively, from its lack of *"D". problematicus* and *V. carnarvonensis*, and the relatively short interval separating it from higher assemblages containing those species, it may be a correlative of part of the upper Clematis Sandstone. Consequently, the Warang Sandstone (as represented in GSQ Hughenden 1-2R, 5 and 6) is shown (Fig. 3) as equivalent to the upper Clematis Sandstone and lower Moolayember

ASSEMBLAGE ZONES (Carnarvon Basin) / SPECIES

Species (columns): Lunatisporites pellucidus, Lunatisporites acutus, Kraeuselisporites saeptatus, Tigrisporites playfordii, Nevesisporites limatulus, Staurosaccites quadrifidus, cf. Rimaesporites aquilonalis, Infernopollenites claustratus, Samaropollenites speciosus, Aulisporites astigmosus, Annulispora folliculosa, Duplexisporites problematicus, Minutosaccus crenulatus

Assemblage zones (rows):
- Minutosaccus crenulatus
- Samaropollenites speciosus
- Straurosaccites quadrifidus
- Tigrisporites playfordii
- Kraeuselisporites saeptatus

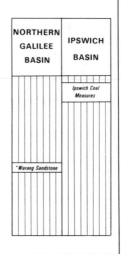

NORTHERN GALILEE BASIN | IPSWICH BASIN

Ipswich Coal Measures

*Warang Sandstone

Fig. 4--Relationship of Warang Sandstone & Ipswich Coal Measures to Assemblage-zones of Carnarvon Basin.
* = Warang Sandstone

Fm in the Bowen Basin. On the basis of mio-spore distribution (de Jersey 1972), the Esk Fm is also correlated with the upper Moolayember Fm of the Bowen Basin. *Staurosaccites quadrifidus* penetrated briefly into the Bowen Basin of the upper Moolayember Fm (McKellar 1977: 395).

The present investigation of the Warang Sandstone is based solely on assemblages from GSQ Hughenden 1-2R, 5 and 6. There is evidence that in other parts of the northern Galilee Basin, the Warang Sandstone may have a longer time range. For example, correlation of the Warang Sandstone in GSQ Hughenden 7 with a lower interval in the Bowen Basin, the lower Rewan Fm, has recently been demonstrated (McKellar 1979: 295, 300).

PHYTOGEOGRAPHIC RELATIONSHIPS

The assemblages recorded from the three Hughenden sections provide evidence for close stratigraphic and geographic association of the Onslow and Ipswich microfloras, and for a gradation between them. Dolby and Balme (1976: fig. 8) plotted on the Gondwana reconstruction of Smith *et al.* (1973) the locations of then-known assemblages assigned to both microfloras. On this paleogeographic reconstruction (Fig. 1), the Onslow Microflora in the Galilee Basin is at a much higher paleolatitude than any previous record, well within the geographic belt to which Dolby and Balme assigned the Ipswich Microflora.

One explanation of the anomaly which has been considered is that the distribution

of the two microfloras was partly influenced by environment of deposition. However, the Warang Sandstone is of continental (fluviatile) origin, like sediments in other areas in eastern Australia from which the Ipswich Microflora has been recorded, and unlike the sediments in the Carnarvon Basin, which are of marine or paralic origin. The occurrence of Onslow elements in the Warang Sandstone cannot be attributed to a near-shore environment of deposition.

A second possible explanation is that local geographic influences, such as prevailing winds and ocean currents, enabled plant communities yielding the Onslow Microflora to extend southwards in eastern Australia into higher paleolatitudes. Although climatic influences may have contributed, it seems unlikely that they constituted the major factor responsible, because the latitude of the northern Galilee Basin was so much greater, in the Smith *et al.* reconstruction, than that of other occurrences of the Onslow Microflora.

In the Gondwana reconstruction (Smith *et al.* 1973), the Australian continent is rotated clockwise relative to its present orientation, so that the eastern Australian localities of the Ipswich Microflora are displaced polewards. With a different reconstruction, in which there is less clockwise rotation of Australia, the palynofloras recorded from the Warang Sandstone, with their inter-gradation of Onslow and Ipswich Microfloras, would lie in intermediate latitudes between those of the Onslow Microflora in the Carnarvon

Basin and the Ipswich Microflora in the Perth Basin. The Ipswich Microflora of the lower Moolayember Fm of the southern Bowen Basin would also conform to this pattern, as it would cover paleolatitudes close to those of the Perth Basin. Accordingly, due consideration should be given to paleolatitudes assigned to these Queensland palynofloras, in assembling data for the reconstruction of Triassic Gondwana.

REFERENCES

DE JERSEY, N. J. 1968. Triassic spores and pollen grains from the Clematis Sandstone. Geol. Surv. Qld. Publ. 338, Palaeontol. Pap. 14. 44 pp.

--- 1970. Early Triassic miospores from the Rewan Formation. Geol. Surv. Qld. Publ. 345, Palaeontol. Pap. 19. 29 pp.

--- 1972. Triassic miospores from the Esk Beds. Geol. Surv. Qld. Publ. 357, Palaeontol. Pap. 32. 40 pp.

---; M. HAMILTON 1967. Triassic spores and pollen grains from the Moolayember Formation. Geol. Surv. Qld. Publ. 336, Palaeontol. Pap. 10. 61 pp.

DOLBY, J. H.; B. E. BALME 1976. Triassic palynology of the Carnarvon Basin, Western Australia. Rev. Palaeobot. Palynol. 22: 105-168.

McKELLAR, J. L. 1977. Palynostratigraphy of core samples from the Hughenden 1 : 250 000 sheet area, northern Galilee and Eromanga Basins. Qld. Govt. Mining J. 78: 393-399.

--- 1979. Palynostratigraphy of core samples from GSQ Hughenden 7. Qld. Govt. Mining J. 80: 295-302.

SMITH, A. G.; J. C. BRIDEN; G. E. DREWRY 1973. Phanerozoic world maps. In N. F. Hughes (Ed.), Organisms and continents through time. Palaeontol. Soc. London Spec. Pap. 12: 1-42.

Fig. 5--1: *Guttatisporites visscherii* de Jersey, 1968. Equatorial focus. 2: *Verrucosisporites carnarvonensis* de Jersey & Hamilton, 1967. Distal-equatorial focus. 3: *"Duplexisporites" problematicus* (Couper) Playford & Dettmann, 1965. Median focus. 4, 5: *Minutosaccus crenulatus* Dolby, 1976. 6: *Equisetosporites steevesii* (Jansonius) de Jersey, 1968. Equatorial focus. 7: *Infernopollenites claustratus* Dolby & Balme, 1976. Proximal focus. 2, 7 = x 562 ; others, x 750.

Fig. 6--1: *Lunatisporites pellucidus* (Goubin) Helby in de Jersey, 1972. Proximal focus. 2: *Duplicisporites granulatus* Leschik emend. Scheuring, 1970. Distal-equatorial focus. 3: *Triadispora crassa* Klaus, 1964. Proximal focus. 4, 5: *Staurosaccites quadrifidus* Dolby, 1976. Median focus. 6, 7: cf. *Rimaesporites aquilonalis* Goubin, 1965. (6 = equatorial, 7 = polar views.) 8: *Lunatisporites acutus* Leschik, 1956. Proximal focus. All, x 750.

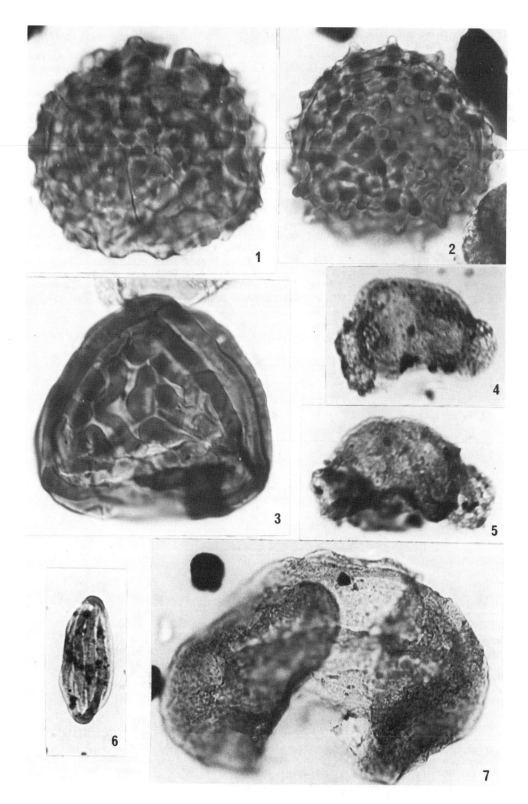

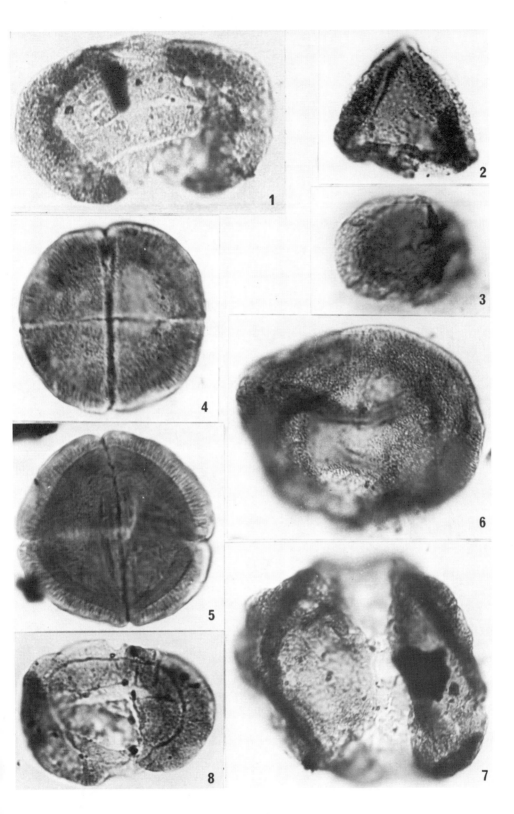

37

The flora from the Permian nonmarine sequences of India and Australia:
A comparison

J. F. RIGBY
Geological Survey of Queensland,
Brisbane, Australia

S. C. SHAH
Geological Survey of India, Calcutta

Peninsular India has extensive nonmarine sequences whereas Australia in general has Permian alternating marine and nonmarine sequences. The nonmarine sequences contain similar but not identical successions of floras. Intercontinental correlation of floral zones may be possible.

In India and eastern Australia, the megaflora is divisible into a number of assemblages containing genera in common, but different at the specific level. This is considered to have been caused by homoplasy rather than by close proximity of India and Australia.

COMPARISON OF THE FLORAS

Some genera are very rare, being known from one or very few localities, although there may be quite a number of specimens. These will not be considered as they are of no value for international correlation. They include *Belemnopteris, Buriadia, Palaeovittaria, Pecopteris, Rubidgea, Walkomiella* and several others.

SPHENOPSIDS

Trizygia speciosa Royle is common in the Barakar and Raniganj Fmns. It is very rare in eastern Australia.

Phyllotheca australis Brongniart from Australia differs from the various *Phyllotheca* species common in India. *P. indica* Bunbury from the Raniganj Fmn has been differentiated from *P. australis* (Kulkarni 1970).

Schizoneura gondwanensis Feistmantel occurs throughout much of the Indian succession, above the Talchir Fmn. It is limited to the Bandanna Fmn in Queensland and has been found at an unnamed horizon in New South Wales.

Raniganjia is quite common in the Raniganj Fmn and Bandanna Fmn equivalents. Specimens have also been found in the Blair Athol Coal Measures and the Barakar Fmn. Pant and Nautiyal (1967) demonstrated that both Indian species were variants of a single species and suggested that Australian specimens belonged in the same species. Examination of actual specimens from India and Australia suggests they might be distinct species.

Lelstotheca robusta (Feistmantel) Maheshwari is limited to the Barakar Fmn in India, whereas in Australia it is more widely distributed, being known from both the Greta and Newcastle Coal Measures in New South Wales and equivalents of the Bandanna Fmn in Queensland.

FERNS AND FERN-LIKE FOLIAGE

Maithy (1974) has shown that the Australian and Indian species of *Neomariopteris* differ but their ranges are similar. *N. lobifolia* (Morris) Maithy occurs in the Greta Coal Measures and later in Australia. The widespread *N. polymorpha* (Feistmantel) Maithy occurs throughout the Damuda Gp along with some less common species.

Botrychiopsis is an index genus for the Early Permian. It occurs either with typical Lower Gondwana plants or alone in the Joe Joe Fmn and the Karharbari Fmn. So far it is not certain whether the Australian and Indian species are the same or distinct. We consider this genus to be indicative of a cooler climate.

Glossopteris AND *Gangamopteris*

The genus *Glossopteris* includes approximately 100 species and typifies the Lower Gondwana or Permian floras of Gondwana. Its peak of variety in species occurs in the Raniganj Coalfield where 40 species are known (Shaila Chandra & Surange, in prep.). These authors recognise 62 species in the Permian of India, of which only one is represented in Australia.

Rigby *et al.* (in prep.) are at present revising the Australian species of *Glossopteris* using the same parameters for the definition of species as used in India. The only species unquestionably in common between eastern Australia and Peninsular India is the rather simple form, *G. communis*. Some forms previously given the same specific name in India and Australia are now known to be misidentified, for example, *Glossopteris browniana, G. ampla* and *G. linearis* in India, and *G. indica* and *G. angustifolia* in Australia.

No critical review of species of *Gangamopteris* in Australia has been made although the type species, *G. angustifolia*, is from the Sydney Basin.

Surange (1973) has given data for 15 Indian species of *Gangamopteris*. Only one species occurs in the Barakar Fmn or above.

Three of these species have been recorded in Australia. The nature of the *Gangamopteris* leaf, which lacks a midrib, makes its separation into a large number of species inappropriate.

FRUCTIFICATIONS

A number of different genera, most with more than one species, have been proposed for female gymnospermous fructifications from the Lower Gondwana sequence. Many have been found intimately associated with *Glossopteris* or *Gangamopteris* leaves. There has been a tendency to place specimens from various parts of Gondwana into the same species. In many cases this has been encouraged by indifferent preservation and lack of scientific communication.

The first genus to be recognised as a *Glossopteris* fructification was *Scutum* in 1952. *Scutum* species in India are small in the Early Permian, become larger and more varied until Raniganj time when they reached their peak of development (Surange & Shaila Chandra 1979). In Australia, the biggest and most highly developed species of *Scutum* occurs in the Early Permian Blair Athol Coal Measures. The only other genus in common between Australia and India is *Senotheca* which is known by a number of specimens from one locality in India (Banerjee 1969) and by two specimens from Australia (Rigby 1978). Originally placed in the same species, they are now known to be distinct. *Ottokaria, Jambadostrobus, Dictyopteridium, Plumsteadiostrobus* occur only in India, and *Isodictyopteridium, Plumsteadia* occur only in Australia. Species are quite rare in the Early Permian in India, but become relatively common in the Raniganj Fmn.

CORDAITALEAN LEAVES

Many localities in both India and Australia have yielded abundant leaves known commonly as *Noeggerathiopsis* but more correctly as *Cordaites* (Meyen 1969). In Australia they occur throughout the Permian, becoming most common in the Newcastle Coal Measures. They reach their peak in number of specimens in the Karharbari Fmn and disappear within the overlying Barakar Fmn. Similar leaves in later horizons in India are probably not *Cordaites* (Meyen 1969).

WHY ARE THERE DIFFERENCES BETWEEN THE INDIAN AND AUSTRALIAN PERMIAN FLORAL SUCCESSIONS?

Previously the Permian flora of Gondwana was considered to be uniform throughout (for example, Arber 1905). While this may be true for genera, it is now known to be untrue for species.

We conclude that the Permian floras of Gondwana had common ancestors, and diverged during or immediately after the Late Paleozoic glaciation. Chaloner and Lacey (197 plotted the floras of about Carboniferous/ Permian transition time showing the Indian and eastern Australian occurrences lying between latitudes 40° and 60°. The plant newly occupying the deglaciated region probably were an impoverished flora. Species proliferated throughout the Permian but the number of genera always remained small.

In India, the climate at the start of the Permian was frigid (Shah & Sastry 1974), t gradually warmed until it became temperate by Raniganj time, based on evidence from invertebrates (Shah 1976), megafloras (Lele 1976) and palynomorphs (Kar 1976). Climates behaved differently in Australia where faunas indicate a return to cool conditions in the Mid-Permian (Dickins 1978). If climate had been the only control, then one would have expected some mingling between the Raniganj flora and the eastern Australian Late Permian floras but mixing did not happen until post Permian time.

Sea penetrated deep into Gondwana during the Permian. It entered between India and Australia and epicontinental seas passed into the heart of Peninsular India (Shah & Sastry 1974). It may have penetrated along an early rifting of the Australia/Antarctic suture (McGowran 1973). To us, it seems likely that at least a transient sea flooded Carnarvon Basin to the Officer Basin during Sakmarian time, however, little is known of the Sakmarian sediments in these basins. Kemp (1979) described the paynoflora as typical of a fresh or possibly brackish environment. Whether marine, fresh or brackish, this belt of depositional activity formed another barrier to migration of floras into or from eastern Australia.

Because the floral successions in Peninsular India and eastern Australia differ, close biostratigraphical correlation will not be possible between the regions. On the other hand, there is mounting evidence that biostratigraphical division based on megafloras within individual regions in the Permian of Gondwana is possible.

The work of Shaila Chandra and Surange

(1980) indicates that even finer subdivision of the assemblage zones and sub zones proposed by Shah *et al.* (1971) for India may be possible.

In so far as they are known biostratigraphic zones in eastern Australia do not conform to the Indian zones. From our observations it may not be possible to erect zones for the whole of Gondwana that will allow fine biostratigraphic correlation. The base of the *Botrychiopsis-Gangamopteris-Glossopteris* flora, and the *Lepidopteris* or *Dicroidium* flora may be the only horizons that can be precisely correlated.

ACKNOWLEDGMENTS

This paper has been printed with the kind permission of the Under Secretary, Department of Mines, Queensland, and the Director, Geological Survey of India.

REFERENCES

ARBER, E. A. N. 1905. Catalogue of the fossil plants of the Glossopteris Flora in the Department of Geology, British Museum (Natural History). British Museum, London lxxiv + 255 pp.

BANERJEE, M. 1969. Senotheca murulidihensis, a new glossopteridean fructification from India associated with Glossopteris taeniopteroides Feist. In H. SANTAPAU et al. (Eds.), J. Sen Memorial Volume, J. Sen Memorial Committee, and Botanical Society of Bengal, Calcutta: 359-368.

CHALONER, W. G.; W. S. LACEY 1973. The distribution of Late Palaeozoic floras. In N.F. HUGHES (Ed.), Organisms and continents through time. Spec. Pap. Palaeontol. 12: 271-290.

DICKINS, J. M. 1978. Climate of the Permian in Australia: invertebrate faunas. Palaeogeog., Palaeoclimatol., Palaeoecol., 23: 33-46.

KAR, R. K. 1976. Microfloristic evidences for climatic vicissitude in India during Gondwana. Geophytology 6(2): 230-244.

KEMP, E. M. 1976. Palynological observations in the Officer Basin Western Australia. Bull. Bur. Mineral Res., Australia 160: 23-39.

KULKARNI, S. 1970. Studies in the Glossopteris flora of India 40.Sphenopteris polymorpha Feistm. (1881) emend. from the Barakar Stage of South Karanpura Coalfield, Bihar, India. Palaeobotanist 18(2): 208-211.

LELE, K. M. 1976. Palaeoclimatic implications of Gondwana floras. Geophytology 6(2): 207-229.

McGOWRAN, B. 1973. Rifting and drift of Australia and the migration of mammals. Science 180: 759-761.

MAITHY, P. K. 1974. A revision of the Lower Gondwana Sphenopteris from India. Palaeobotanist 21(1): 70-80.

MEYEN, S. V. 1969. New data on relationship between Angara and Gondwana Late Palaeozoic floras. Earth Sci. 2: 141-157.

PANT, D. D.; D. D. NAUTIYAL 1967. On the structure of Raniganjia bengalensis (Feistmantel) Rigby with a discussion on its affinities. Palaeontographica 121B: 52-64.

RIGBY, J. F. 1978. Permian glossopterid and other cycadopsid fructifications from Queensland. Publ. Geol. Surv. Queensland 367; Palaeontol. Pap. 41: 1-21.

---; SHAILA CHANDRA & SURANGE, K. R. (in prep.). Monographic study of the Australian species of Glossopteris. Geol. Surv. Qld., Brisbane.

SHAH, S. C. 1976. Climates during Gondwana era in Peninsular India: faunal evidences. Geophytology 6(2): 186-206.

---; M. V. A. SASTRY 1974. Significance of Early Permian marine faunas of Peninsular India. In K. S. W. CAMPBELL (Ed.), Gondwana Geology, Australian National University Press, Canberra: 391-395.

---; G. SINGH; M. V. A. SASTRY 1971. Biostratigraphic classification of Indian Gondwanas. Ann. Geol. Dept., Aligarh Muslim Univ. 5-6: 306-326.

SHAILA CHANDRA; K. R. SURANGE 1980. Revision of the Indian species of Glossopteris. Birbal Sahni Inst. Palaeobot., Lucknow, Mon. 2. 291 pp.

SURANGE, K. R. 1973. Indian Lower Gondwana floras: a review. In K. S. W. CAMPBELL (Ed.). Gondwana Geology. Australian National University Press, Canberra: 135-147.

---; SHAILA CHANDRA 1979. Morphology and affinities of Glossopteris. Palaeobotanist 25: 509-524.

Permian foraminifera of the Sydney Basin

VIERA SCHEIBNEROVÁ
Geological & Mining Museum, Sydney, Australia

Sixty species of smaller foraminifera have been distinguished in marine Permian sediments of the Sydney Basin. Three-quarters of them are agglutinated forms. The agglutinated species are dominated by species of the genera *Hyperammina* (10 species), *Reophax* (7 species), *Ammobaculites* and *Trochammina* (4 species). Remaining genera are represented by one or two species each. Many of these species were recently recorded from the Sydney Basin for the first time, although originally they were described from Western Australia.

No fusulinid foraminifera have been found either in the Sydney Basin or any other Permian deposits in Australia.

INTRODUCTION

Permian foraminifera of the Sydney Basin attracted the attention of micropaleontologists as early as 1905 when Chapman and Howchin described thrity-five species of agglutinated and calcareous foraminifera from the Pokolbin limestone. Crespin and Parr (1941) described one new genus and four new species of agglutinated foraminifera from the Sydney Basin. Crespin (1958) described several agglutinated and calcareous species from the Dalwood and Maitland Gps in the Hunter Valley, northern Sydney Basin.

The smaller foraminiferal faunas of the Sydney Basin are dominated by agglutinated forms, with one sporadic calcareous species. They were classified (Scheibnerová, in press a) within four groups:
1. Long-ranging forms occurring throughout the Permian.
2. Species occurring only in the Late Permian.
3. Species which occur in short time intervals within more than one lithostratigraphic unit but not throughout the Permian.
4. Species which occur in one formation only.

Foraminiferal species of groups 2, 3 and 4 have definite limits in their vertical distribution and many may be of considerable stratigraphic value.

I have also found foraminifera in marine intercalations within the Illawarra, Tomago and Singleton Coal Measures and in the Wombarra claystone of the Narrabeen Gp, in addition to the well-known marine sediments of the Shoalhaven, Dalwood and Maitland Gps.

STRATIGRAPHIC VALUE

1. Long-ranging forms represented by the following agglutinated and calcareous taxa: *Hyperammina fletcheri*, *H. hebdenensis*, *Ammodiscus multicinctus*, *A. nitidus*, *Reophax asper*, *Textularia bookeri*, *Trochammina laevis*, *Digitina recurvata*, *Nodosaria tereta* and *Ammobaculites wandageensis*. These species were found to occur throughout the whole of the marine Permian sequence.

2. Agglutinated and calcareous species which occur only in the Maitland Gp and time-equivalent strata: *Reophax subasper*, *Trochammina pulvilla*, *Tolypammina undulata*, *Dentalina grayi*, *Frondicularia hillae*, *Hyperammina expansa*, *Lingulina antiqua*, *Hyperammina callytharraensis*, *Nodosaria raggatti*, *Frondicularia aulax*, *Hyperammina elegans*.

3. Species with discontinuous ranges within more than one lithostratigraphic unit but not occurring throughout the Permian section: *Trepeilopsis australiensis*--Dalwood Gp, Mulbring Siltstone; *Geinitzina triangularis*--Dalwood Gp, Branxton Fm, Mulbring Sst.; *Ammobaculites eccentrica*--Allandale and Branxton Fms.; *Ammobaculites woolnoughi*--Dalwood and Maitland Gps.; *Thuramminoides teicherti*--Maitland Gp and base of Singleton Coal Measures; *Thuramminoides phialaeformis*--Dalwood and Maitland Gps.; *Frondicularia woodwardi*--Dalwood and Maitland Gps.; *Hemigordius harltoni*--Branxton Fm, Mulbring Sst, and Newcastle Coal Measures; *Reophax tricameratus*--Maitland Gp. and Singleton Coal Measures; *Thuramminoides sphaeroidalis*--Maitland Gp. and Singleton Coal Measures; *Reophax minutissimus*--Mulbring Sst and Singleton Coal Measures; *Hemigordius schlumbergi*--Mulbring Sst and Singleton Coal Measures; *Rectoglandulina serocoldensis*--Maitland Gp and Singleton Coal Measures; *Textularia improcera*--Mulbring Siltstone and Singleton Coal

Measures; *Trochammina subobtusa*--Ruther-
ford and Farley Fms and Singleton Coal
Measures; *Hyperammina fusta*--Branxton Fm
and Singleton Coal Measures; *Trochammina
laevis*--Branxton Fm, Mulbring Sst and
Singleton Coal Measures; *Pelosina ampulla*--
Allandale, Rutherford, Farley and Branxton
Fms, Singleton Coal Measures; *Hyperammina
fletcheri*--Dalwood Gp, Branxton Fm and
Mulbring Sst; *Haplophragmoides pokolbin-
ensis*--Dalwood Gp and Branxton Fm.
4. Species which occur only in one formation: *Calcitornella elongata*--Pokolbin
limestone; *Ammobaculites crescendo*-- Allan-
dale Fm; *Ammodiscus oonahensis*--Allan-
dale Fm (middle part); *Tritaxia* sp.--
Rutherford Fm; *Frondicularia impolita*--
Rutherford Fm; *Calcitornella heathi*--
Rutherford Fm; *Hyperammina elegantissima*--
Mulbring Sst; *Hyperammina vagans*--Mulbring
Sst; *Glomospirella nyei*--Branxton Fm;
Lingulina davidi--Branxton Fm; *Hippocrep-
inella biaperta*--Branxton Fm; *Geinitzina
caseyi*--Mulbring Sst; *Reophax audax*--
Mulbring Sst; *Brachysiphon rudis*--Mulbring
Sst.

Comparing the stratigraphical occurrences
of 51 species in both Western Australia and
the Sydney Basin shows that:
1. 49 species persisted into younger Permian sequences in the Sydney Basin while
becoming extinct in Western Australia.
2. Only two species occurred earlier in
the Sydney Basin than in Western Australia.

This summary shows a definite trend indicating that, as a rule, species of foraminifera persisted later into the Permian
in eastern Australia that in western Australia. This observation agrees with the
stratigraphical occurrences of Permian
invertebrate genera such as *Keeneia,
Eurydesma, Deltopecten* and others noted by
Clarke and Banks (1975 : 465) and
Dickins (1978 : 33). Dickins (1978)
considered this to have been the result of
a warming in Western Australia while eastern Australia remained relatively cool.

PALEOGEOGRAPHICAL VALUE

As generally known (Crowell & Frakes 1971;
and others) large parts of Australia were
covered by glaciers in the Late Paleozoic.
Evidence for terrestrial glaciation in the
Sydney Basin is confined to Late Carboniferous sediments, but numerous isolated
megaclasts and dropstones occurring
throughout Permian marine sediments indicate that the climate was seasonally cold,
allowing icebergs to be transported by
marine currents, probably from a source
farther to the south (Herbert, in press).

All Permian foraminiferal assemblages
of Australia and especially those of the

Sydney Basin show the following features
of cold-water assemblages:

Absence of fusulinid foraminifera. Th
most important larger foraminifera of the
Permian are Fusulinacea. However, the
Fusulinacea were sensitive to the physica
properties of their environment and are
mostly restricted to certain lithological
units of the Pennsylvanian and Permian.
They appear to have been restricted to of
shore marine environments and are now fou
in limestones, highly calcareous shales
and less commonly in sandstones.

Fusulinaceans are almost world-wide in
their distribution, having been recorded
all continents except Australia and
Antarctica. The most southerly occurrenc
known are in Patagonia off southern Chile
and on the South Island of New Zealand.
It can be postulated that the lack of fusu
linaceans in Australia is a result of cold
temperatures.

Abundance of agglutinated foraminifera.
Foraminiferal assemblages of the marine
Permian of the Sydney Basin and other part
of Australia are impoverished. Although
many calcareous taxa are known to occur,
assemblages are dominated by agglutinated
taxa.

Permian to recent agglutinated genera
such as *Ammodiscus, Hyperammina, Thurammin
Ammobaculites, Hippocrepinella, Proteoina,
Psammosphaera, Reophax* and *Spiroplectammin*
are known to favour cold waters. They
often occur in sediments closely associate
with glacigene sediments. Crespin (1958)
mentioned *Hippocrepinella* in the Quamby
Mudstone in Tasmania, which was deposited
above the Stockton Tillite, and also mentioned the presence of agglutinated forms
in the glacial deposits of the Nangetty
Fm of the Irwin Sub-basin, in the upper
part of the Lyons Gp of the Carnarvon Basi
and in the Grant Fm of the Canning Basin.

*Prominent intermittent occurrence of the
calcareous forminifera Nodosariidae and
Miliolacea.* The calcareous imperforate
genera *Hemigordius, Calcitornella,
Trepeilopsis, Plummerinella, Orthovertella
Streblospira* and *Flectospira* are all
restricted to the Late Paleozoic, and sinc
they do not have any recent survivors we
known only indirectly about their ecological requirements. Crespin (1958: 16, 18)
recorded *Hemigordius* and *Calcitornella* fro
glacigene sediments as well as *Nodosaria
tereta* from the Grant Fm (Crespin 1958).

*North-south differentiation of foramini-
feral faunas.* Crespin (1958) described
some 30 species of foraminifera from the
Permian of the Bowen Basin (Springsure
area) in Queensland. Of thirty species
only some five are agglutinated. V.
Palmiera (pers. comm.) is at present

44

studying the Permian foraminifera of Queensland and has confirmed the numerical predominance of calcareous foraminifera in the Permian assemblages in the area. This contrasts with the fauna in the Sydney Basin, where of some 60 species determined by the present authors only ten are calcareous. As suggested by the observation that during the Phanerozoic calcareous foraminifera have predominated in warmer waters, this north-south differentiation of Permian forminiferal faunas supports the geographic orientation of Australia during the Permian envisaged by Irving (1964).

Despite intensive sampling no conodonts have ever been found in Permian marine sediments of Australia. According to Nicoll (1976), the lack of conodonts in the Permian in comparison with their relative abundance in the Carboniferous and Triassic, is a result of a cold climate associated with extensive glaciation.

There is no sedimentological evidence for a Permian marine environment deeper than the inner and outer shelf in the Sydney Basin. According to the known ecology of recent *Nubecularia* it can be assumed that Permian *Nubecularia* and associated foraminifera probably lived in shallow quiet water with slow sedimentation rates. Textulariidae are known to favour fine-grained clastic sediments and shallow water, grazing on algae. The abundance of *Nubecularia* and other attached forms such as *Calcitornella* and Textulariidae in the Sydney Basin indicate shallow depth, probably within the photic zone.

Permission to publish this paper was granted by the Under Secretary, New South Wales Department of Mineral Resources and Development.

REFERENCES

CLARKE, M. J.; M. R. BANKS 1975. The stratigraphy of the Lower (Permo-Carboniferous) parts of the Parmeener Super-Group, Tasmania. In K.S.W. Campbell (Ed.), Gondwana Geology. 454-467.

CRESPIN, I. 1958. Permian Foraminifera of Australia. Bull. 48, Bur. Miner. Resour. Geol. Geophys. 1-207.

---; W. J. PARR 1941. Arenaceous foraminifera from the Permian of New South Wales. J. Proc. Soc. N.S.W. 74: 300-311.

CROWELL, J. C.; L. A. FRAKES 1971. Late Palaeozoic glaciation of Australia. J. Geol. Soc. Aust. 17 (2): 115-155.

DICKINS, J. M. 1978. Climate of the Permian in Australia: the invertebrate faunas. Palaeogeo., Palaeoclimatol., Paleoecol. 23: 33-46.

HERBERT, C. in press. Evidence for glaciation in the Sydney Basin and Tamworth synclinorial zone. Geol. Surv. N.S.W. Bull. No. 26, (C. Herbert & R. Helby, Eds.)

IRVING, E. 1964. Paleomagnetism and its application to geological and geophysical problems. Wiley: New York.

NICOLL, R. S. 1976. The effect of late Carboniferous - Early Permian glaciation on the distribution of conodonts in Australia. Geol. Assoc. Canada, Spec. Pap. 15: 273-278.

SCHEIBNEROVÁ, V. in press a. Permian foraminifera of the Sydney Basin. Mem. Geol. Surv. N.S.W. Palaeontol. 19.

--- in press b. Permian foraminifera of the Sydney Basin. Geol. Surv. N.S.W. Bull. 26, (C. Herbert & R. Helby, Eds.)

A geo-paleontological synthesis of the Gondwana formations of Uruguay

ALVARO MONES
Museo Nacional de Historia Natural/
Faculdad de Humanidades y Ciencias,
Montevideo, Uruguay

ALFREDO FIGUEIRAS
Faculdad de Humanidades y Ciencias,
Montevideo, Uruguay

The Gondwana formations of Uruguay occupy two basins. The most extensive one occupies the northern departments of the country and is subdivided into two parts: one in the NE, filled mainly with Eogondwanic rocks (Carboniferous-Permian), and one in the NW, filled mainly with Neogondwanic rocks (Jurassic-Cretaceous). The other basin lies in SE Uruguay and contains Jurassic-Cretaceous rocks without equivalents in Argentina and Brazil. Brief lithological descriptions and fossil lists for each formation are given. The oldest known Gondwana rocks in Uruguay are the San Gregorio and Tres Islas Fms (Upper Carboniferous to Lower Permian). They are followed by the Fraile Muerto Fm (Lower Permian), Mangrullo Fm (Middle Permian), Paso Aguiar and Yaguarí Fms (Upper Permian), Puerto Gómez, Tacuarembó, and Arapey Fms (Upper Jurassic to Lower Cretaceous), and Valle Chico, and Arequita Fms (Lower Cretaceous). A correlation table with Argentina and Brazil is given.

SAN GREGORIO FM (UPPER CARBONIFEROUS OR LOWER PERMIAN)

History: Guillemain (1912: 241), Caorsi & Goñi (1958: 32-34).

Lithology: Diamictites, shales, sandstones and rhythmites, with fossiliferous phosphoritic-calcareous concretions.

Boundaries: Unconformably overlying crystalline basement and/or Lower Devonian beds and overlain conformably by Tres Islas Fm.

Distribution: Outcrops in the Depts. of Cerro Largo, Rivera, Tacuarembó, Durazno, Soriano, and Río Negro, and subsurface reports from Artigas, Salto, and Paysandú. Type locality: San Gregorio, Tacuarembó. Thickness: 286 m.

Origin: Sediments of glacial and fluvio-glacial origin.

Biostratigraphy: Assemblage Zone of *Eoasianites (Glaphyrites) rionegrensis* (Francis 1975: 548-549). PORIFERA, *Itararella gracilis* Kling & Reif; MOLLUSCA, Nautiloidea, *Dolorthoceras chubutense* Closs; Ammonoidea, *Eoasianites (Glaphyrites) rionegrensis* Closs; OSTEICHTHYES, *Mesonichthys antipodeus* Beltan, *Carbonilepis uruguayensis* Beltan; *Gondwanichthys maximus* Beltan; *Elonichthys macropercularis* Beltan; *Rhadinichthys rioniger* Beltan; *Itararichthys microphthalmus* Beltan; *Daphnaechelus formosus* Beltan; MICROSPOROMORPHI 33 different types (Martínez-Macchiavello 1963); POLLEN & SPORES: A prolific flora has been described (Marquez-Toigo 1976).

Age: The goniatite indicates a Uralian (or Stephanian) age (Upper Carboniferous) but the pollens and spores indicate a Lower Permian Autunian (Sakmarian) or Artinskian age. The palynological evidence is preferred, and the age of the San Gregorio Fm is considered to be lower Permian.

Correlations: According to Delaney & Goñi (1963: 9), the San Gregorio Fm can be correlated with the Suspiro Fm (or Facies) of the Itarare Gp of Southern Brazil and with the lower part of the Charata Fm of Argentina (Padula 1972: 216-218). The presence of *Eoasianites (Glaphyrites)* in the Dwyka Series of SW Africa, as well as in the upper part of the Tepuel System of Argentina, allows correlation of these beds with the San Gregorio Fm.

Note: See Tres Islas Fm.

TRES ISLAS FM (UPPER CARBONIFEROUS OR LOWER PERMIAN)

History: Walther (1911: 594), Caorsi & Goñi (1958: 34-35).

Lithology: Grey, brown or yellowish cross-bedded sandstones, with conglomeratic beds associated with grey to purple shales, siltites, and thin local coal beds.

Boundaries: Conformable with San Gregorio and Fraile Muerto Fms.

Distribution: Depts. of Cerro Largo, Tacuarembó, and Durazno. Type locality: Tres Islas, Cerro Largo. Thickness: 100 m.

Origin: Continental sediments deposited in shallow water under cold weather conditions; partial contemporaneity with San Gregorio sediments.

Biostratigraphy: Microspores of herbaceous plants with dominant Arthrophyta are indeterminate (Martínez-Macchiavello 1976: 28).

Age: Like the San Gregorio Fm, the Tres Islas Fm may be Upper Carboniferous but is more probably Lower Permian.

Correlations: The Rio Bonito Fm of Brazil and the middle part of the Charata Fm of Argentina (Padula 1972: 216-218) are correlatives.

Notes: Following Walther (1935: 266), Bossi (1966: 110-133) maintains that it is difficult to distinguish because of the interbedding of sandstones and glacial deposits that does not permit mapping as separate units. The Rio Bonito Fm of Brazil is characterized by the coal beds that, in Uruguay, are of insignificant thickness and lateral extent. Bossi recognizes two members of one formation which he informally named "San Gregorio-Tres Islas Fm". We more conservatively recognize two different formations.

FRAILE MUERTO FM (LOWER PERMIAN)

History: Falconer (1931: 11-12), Caorsi & Goñi (1958: 35-36).

Lithology: Grey to bluish-grey sandy shales and sandstones.

Boundaries: Conformable with Tres Islas and Mangrullo Fms.

Distribution: Depts. of Cerro Largo, Tacuarembó, and Rivera. Type locality: Fraile Muerto, Cerro Largo. Thickness: 200 m.

Origin: The sediments were deposited in marine (perhaps brackish) shallow water environments, under unstable platform conditions.

Biostratigraphy: Zone of Palaeonisciformes. MOLLUSCA, ?"*Nuculana*" sp.; OSTEICHTHYES, Palaeonisciformes, ?*Acrolepis* sp.; ?*Elonichthys* sp.

Age: A Lower Permian age is generally accepted, based mainly on stratigraphical relations. We do not agree with the Upper Carboniferous age assigned by Padula (1972: 218). Age is not determinable from the fossils.

Correlations: The Fraile Muerto Fm is correlated with the Palermo Fm of Brazil and with the upper part of the Charata Fm in Argentina (fide Padula, op. cit.).

Notes: The original biostratigraphic name, 'zone of Palaeoniscidae', (Francis 1975: 549), is changed because the family Palaeoniscidae is not represented in the fossils. Bossi (1966: 133-134) included Fraile,Muerto, Mangrullo, Paso Aguiar, and Yaguarí Fms, which comprise the Caraguatá Gp, while Elizalde *et al.* (1970: 55) included the three former formations under the informal name "grey pelitic sediments". "*Nuculana*" was reported by Harrington (1945), but the biochron of this genus does not agree with the age assigned to the sediments.

MANGRULLO FM (MIDDLE PERMIAN)

History: Guillemain (1912: 255-257), Walther (1919: 93-109), Caorsi & Goñi (1958: 36-37).

Lithology: Black, sometimes bituminous, shales, and grey limestones.

Boundaries: Conformable with Fraile Muerto and Paso Aguiar Fms. According to Martínez-Macchiavello (1977: 41-42) unconformable with the former.

Distribution: Outcrops in the Depts. of Cerro Largo, Tacuarembó, and Rivera, and in subsurface in Artigas, Salto, and Paysandú. Type locality: Mangrullo, Cerro Largo. Thickness: 70 m.

Origin: Warm brackish water environment of low salinity and low pH.

Biostratigraphy: Assemblage Zone of *Mesosaurus brasiliensis* (Francis 1975: 549-550). REPTILIA, *Mesosaurus brasiliensis* McGregor.

Age: Its stratigraphical relations, as well as paleoentomological (Pinto 1972: 253, Rio Grande do Sul) and palynological (Menendez 1976: 27, Sao Paulo) studies in the Iratí Fm indicate a Kazanian Age (Middle Permian) rather than the Lower Permian proposed by Padula (1972: 220). From systematic study of the Mesosauria (Araujo 1977: 115) only a Permian s.l. age can be concluded (see Tres Islas Fm).

Correlations: The Mangrullo Fm is correlated with the Iratí Fm of Brazil and with the upper part of the Charata Fm of Argentine (Padula & Mingramm 1969: 1043-1045).

Notes: The poor preservation of the mesosaur remains does not allow reliable generic identification, but stratigraphical and geographical considerations favour *Mesosaurus*. Rey Vercesi (1933: 13) reports fish-scales from the Yaguarí-well between 169 and 204 m, but no later discoveries have been reported.

PASO AGUIAR FM (UPPER PERMIAN)

History: Walther (1923: 66), Falconer (1931: 15-17), Caorsi & Goñi (1958: 37-38).

Lithology: Grey sandy shales and sandstones.

Boundaries: Conformable with Mangrullo and Yaguarí Fms. Yaguarí Fm overlies unconformably in the Quebracho I, Salto I, and Gaspar I wells (Padula & Mingramm 1968: 304-305).

Distribution: Depts. of Cerro Largo, Rivera, Salto, and Paysandú. Type locality: Paso Aguiar on the Negro River (Depts of Tacuarembó and Cerro Largo). Thickness: 50 m.

Origin: Open lake environment of warm brackish water with normal pH.

| PÉRIOD | AGE | URUGUAY | | BRAZIL | ARGENTINA |
		NORTH	SOUTHEAST		
Upper Cretaceous	Senonian				
Lower Cretaceous	Wealdian	Arapey	Valle Chico	Serra Geral	Serra Geral
			Migues | Arequita		
Upper Jurassic	?	Tacuarembó	Puerto Gómez	Botucatú	Tacuarembó
?Lower Triassic		Buena Vista		Armada	Buena Vista
Permian — Upper	?	Yaguarí (Buena Vista / Yaguarí)		Estrada Nova	Chacabuco
		Paso Aguiar		Caveira	
Middle	Kazanian	Mangrullo		Iratí	
Lower	?	Fraile Muerto		Palermo	
Lower Permian and/or Upper Carboniferous	Autunian and/or Uralian	Tres Islas		Rio Bonito	Charata
		San Gregorio		Itararé	
Lower Devonian	Coblencian				

Table 1. Correlation of the Gondwanic formations of Uruguay, Brazil, and Argentina.

Biostratigraphy : Assemblage Zone of *Ferrazia cardinalis* (Francis 1975: 550). MOLLUSCA, *Ferrazia cardinalis* Reed, *Pyramus anceps* (Reed); GYMNOSPERMAE, *?Dadoxylon* sp.

Age and Correlations: We agree with the Upper Permian age proposed for the Estrada Nova Fm of Brazil (Runnegar & Newell 1971: 16-17). The Caveiras Facies of the Estrada Nova Fm. is correlated with Paso Aguiar Fm (see Yaguarí Fm). According to Padula & Mingramm (1969: 1044) the Paso Aguiar Fm can be correlated with the upper part of the Chacabuco Fm of Argentina.

YAGUARI FM (UPPER PERMIAN)

History: Walther (1919: 123-124), Caorsi & Goñi (1958: 38-41).

Lithology: Two members are distinguished: Yaguarí s.str. consisting of red, purple, grey, and greenish shales, overlain by the Buena Vista Member of brick-red friable sandstones and thin intercalations of dark-red clays.

Boundaries: Conformable with Paso Aguiar except in subsurface in NW Uruguay, and overlain unconformably by Tacuarembó Fm.

Distribution: Depts. of Cerro Largo, Tacuarembó, and Rivera, and in subsurface in Artigas, Salto, and Paysandú. Type locality: Yaguarí River, Rivera. Thickness: Yaguarí s. str., 177 m; Buena Vista Member, 348 m.

Origin: Yaguarí s.str. is of lacustrine fresh, clear, and smooth water origin with a somewhat higher pH than normal. Buena Vista is fluvially reworked sand. Both Members were deposited in warm humid conditions (Delany & Goñi 1963: 14-16).

Biostratigraphy: Assemblage Zone of *Terraia altissima* (Francis 1975: 550-551). MOLLUSCA, *Terraia altissima* Cox; *?Pyramus falconeri* (Cox); *Leptoterraia aegra* (Cox); ARTHROPODA, Conchostraca, *?Euestheria* sp.; silicified woods. All fossil records are from Yaguarí s.str.

Age and Correlations: Due to its lithological similarity with the Santa María or Rosário do Sul Fm in Rio Grande do Sul, Brazil, the Yaguarí Fm is currently regarded as Upper Triassic, but we agree with Figueiredo Filho (1972: 225-230) that Yaguarí Fm could be correlated with the Armada Facies of Estrada Nova Fm in Rio Grande do Sul; in this case a correlation could be established between Paso Aguiar Fm and the Caveiras Facies of Estrada Nova. Correlation with the Rio do Rasto Fm of Santa Catarina, Brazil, can no longer be sustained. In Argentina, Padula & Mingramm (1968: 321-323) applied the name "Buena Vista" for the red sediments that they separate from their Alhuampa Fm. The correlation seems reasonable but the Upper Triassic age assigned does not. A Permo-Triassic boundary age for the Buena

49

Vista Member as a whole or for its upper
beds cannot be excluded. Nearly all correlations are based on lithological similarities.

TACUAREMBO FM (UPPER JURASSIC TO LOWER CRETACEOUS)

History: Walther (1911: 575-609),
Caorsi & Goñi (1958: 42-44).

Lithology: Cross-bedded, friable to
hard compact fine to very fine sandstones,
the lower part with a higher clay content.
The colour varies from white through grey,
yellowish-green, brownish, and reddish to
pink.

Boundaries: Unconformable on Yaguarí
Fm. Overlain by and in part interdigi-
tating with the Arapey Fm.

Distribution: Depts. of Tacuarembó,
Rivera, Artigas, Salto, and Paysandú.
Type locality: near Paso Santander on the
Tranqueras Creek, Tacuarembó. Thickness:
850 m.

Origin: The lower part predominantly of
fluvio-lacustrine origin; the upper part
eolian (fossil dunes).

Biostratigraphy: Assemblage Zone of Semi-
onotidae (Francis, 1975: 551). MOLLUSCA,
Gastropoda: Viviparidae?; OSTEICHTHYES,
Semionotidae: *?Lepidotes* sp.; REPTILIA,
Crocodilia, *Meridiosaurus* n.sp. Mones (in
press).

Age: Absolute dates of the lower lava
flows of the Arapey Fm allow us to assign an
Upper Jurassic age to the lower part of the
Fm and a Lower Cretaceous age to the upper
part. The fossils occur in the lower part
but are of no value for dating.

Correlations: Correlated definitely with
the Botucatu Fm. of Brazil, and possibly
with the Tacuarembó Fm of Argentina (Padula
& Mingramm 1968: 306-316).

ARAPEY FORMATION (UPPER JURASSIC TO LOWER CRETACEOUS)

History: Walther (1911: 575-609), Caorsi & Goñi (1958: 44-47).

Lithology: Compact and vesicular tholeiitic basalts.

Boundaries: Overlying and in part interbedded with the Tacuarembó Fm. The overlying rocks (Upper Cretaceous or Cenozoic)
are unconformable.

Distribution: Depts. of Artigas, Salto,
Paysandú, Río Negro, Durazno, Tacuarembó,
and Rivera. Type locality: Arapey River
near the railroad bridge. Thickness:
more than 1,000 m.

Origin: Numerous individual lava flows
resulting from quiescent volcanic activity
fed by great tension cracks originated from
diverse local foci.

Age: Most samples from Brazil, Argenti
and Uruguay give a Lower Cretaceous age
(radiometric ages of about 120-130 m.y.)
Some flows can be assigned to the Upper
Jurassic (147-149 m.y.) (Cordani & Vando
1967: 216-218; Cortelezzi & Cazeneuve
1967).

Correlations: Puerto Gómez Fm possib
only represents a lateral facies of the
Arapey Fm. Lava flows are known to occu
in a vast area in South America, includir
Brazil, Argentina and Paraguay, where the
are known as Serra Geral Fm. Similar ba
saltic rocks are found in South Africa,
India, Antarctica, Australia, and Tasman

PUERTO GOMEZ FM (UPPER JURASSIC)

History: Walther (1927: 363-364), Ser
(1944: 32-33), Caorsi & Goñi (1958: 47-
48).

Lithology: Microlithic, porphyritic an
amygdaloid dark lavas.

Boundaries: The basaltic flows came up
through the most recent granitic intrusic
(550-500 m.y.). They are overlain unconform
ably by the Arequita Fm.

Distribution: Dept. of Treinta y Tres.
Type locality: Puerto Gómez, Treinta y
Tres. Thickness: more than 1,000 m.

Origin: Submarine lava flows associate
with the SE tectonic framework of the
country.

Age: The lowest lava flows have radiometric ages of 140 m.y. (Upper Jurassic,
Umpierre in Bossi *et al.* 1975: 16).

Correlations: Probably related to the
Puerto Gómez Fm are the "Canelones lavas"
(Jones 1956: 38-40, 101-103), and the
"Lascano lavas" (Walther 1923: 57, et
auct.). The lava flows of the Depts. of
Canelones, Maldonado, Rocha, and Lavallej
are correlated with this formation, and
probably correspond to an early lateral
facies of the Arapey Fm. Apparently the
are no equivalent rocks in Argentina and
Brazil.

MIGUES FM (LOWER CRETACEOUS)

History: Jones (1956: 49-56).

Lithology: Marls, silts, clays, sandstones, and frequently ferruginous conglomerates, with lower bituminous beds.
It shows vertical as well as horizontal
changes of facies.

Boundaries: Probably conformable on the
Puerto Gómez Fm; unconformable with Upper
Cretaceous or Cenozoic sediments.

Distribution: Depts. of Canelones,
Lavalleja, and Treinta y Tres. Type
locality: Migues, Canelones. Thickness:
more than 1,000 m.

Origin: The Santa Lucia graben area was
filled with sediments proceeding from the

decomposition of the underlying lavas.

Biostratigraphy: Assemblage Zone of Pollen (Francis 1975: 562-563). POLLEN: *Monocolpites medius*-type, cf. Cycadeae, *Tricolpites troedsoni*-type, *Triporites ivereseni*-type, *Striatiletes* sp., *Psilatriletes guadensis*, *Echitriletes* sp., *Verrutriletes* sp., cf. *Retitriletes*, *Ephedra*-type, *Echimonocolpites* or *Echimonoletes* sp., *Verrumoletes* sp.

Age: A Wealdian Age (Lower Cretaceous) seems to be most probable (Hammen in Goñi & Hoffstetter 1964: 144-145).

Correlations: The "Tala sandstones" (or "Tala facies") of Jones (op. cit.) are included in this formation by Goñi & Hoffstetter (1964: 154-155), but Jones considers them to be part of the "areniscas con dinosaurios" (Asencio Fm, Upper Cretaceous). The Migues Fm is supposed to be the same age as the eolian facies of the Tacuarembó Fm and the Arapey Fm. Possibly the Río Salado Fm of Argentina is the same age as the Migues Fm.

Notes: Zambrano (1975: 461) distingui-shes two different formations: the Castellanos Fm which includes the bituminous beds and has provided the pollen remains, and lying upon it, the Migues Fm. As extension and lithology of the two proposed formations are not well known, we consider all these sediemtns as one stratigraphic unit.

AREQUITA FM (LOWER CRETACEOUS)

History: Walther (1923: 57), Rossi (1966: 201-210).

Lithology: Rhyolites, trachytes, syenites, microsyenites, and micropegmatites.

Boundaries: Lies unconformably on the Puerto Gómez Fm.

Distribution: Depts. of Lavalleja, Maldonado, and Rocha. Type locality: Cerro Arequita, Lavalleja. Thickness: unknown.

Origin: Final products of differentiation of basaltic magmas.

Age: Radiometric dates determined by Umpierre (in Bossi 1969: 188) indicate a Lower Cretaceous age (120-130 m.y.).

Correlations: Equivalent rocks are not known in Argentina or Brazil.

VALLE CHICO FM (LOWER CRETACEOUS)

History: Bossi *et al.* (1975: 17-18).

Lithology: Trachytes, syenites, and porphyritic trachytes, and syenites.

Boundaries: unknown.

Distribution: Dept. of Lavalleja. Type of locality: between Mariscala and Colon, Lavalleja. Thickness: unknown.

Origin: Intrusive complex.

Age: (Umpierre *in* Bossi *et al.* 1975: 18) absolute dates indicate a Lower Cretaceous age (120 m.y.).

Correlations: It is a little-known formation, probably related to the Arequita Fm and all the Jurassic-Cretaceous extrusive activities.

REFERENCES

ARAUJO, D. C. 1977. Taxonomia e relações dos Proganosauria da Bacia do Parana. An. Acad. Brasil. Cien. 48: 91-116.

BOSSI, J. 1966. Geología del Uruguay. Montevideo: Universidad de la República. 469 pp. (1969. Idem, 2d. ed.).

---, et al. 1975. Carta geológica del Uruguay. Montevideo: Dirección de Suelos y Fertilizantes. 32 pp.

CAORSI, J. H.; GOÑI, J. C. 1958. Geología uruguaya. Bol. Inst. Geol. Uruguay 37: 1-73.

CORDANI, U. G.; P. VANDOROS 1967. Basaltic rocks of the Paraná basin. In J.J. Bigarella et al. (Eds.), Problems in Brazilian Gondwana Geology. Curitiba. Pp. 207-231.

CORTELEZZI, C. R.; H. CAZENEUVE 1967. Estudio geocronológico de los basaltos de Nogoyá (Prov. de Entre Ríos) y su relación con las rocas efusivas del Sur de Brasil y Uruguay. Rev. Mus. La Plata (Geol.) 6: 19-32.

DELANEY, P. J.; J. GOÑI 1963. Correlação preliminar entre las formações Gondwânicas do Uruguay e Rio Grande do Sul, Brazil. Bol. Parana. Geo.g. 8-9: 3-21.

ELIZALDE, G., et al. 1970. Carta geológica del Uruguay. 3 Segmento Aceguá, sector XXX. Montevideo: Universidad de la República. 126 pp.

FALCONER, J. D. 1931. Memoria explicativa del mapa geológico de la región sedimentaria del departamento de Cerro Largo (Formación de Gondwana). Bol. Inst. Geol. Perfor. 12: 1-22.

FIGUEIREDO-FILHO, P.M. 1972. A faciologia do Grupo Passa Dois no Rio Grande do Sul. Rev. Brasil. Geol. 2: 216-235.

FRANCIS, J. C. 1975. Esquema bioestratigráfico regional de la República Oriental del Uruguay. Actas 1 Cong. Argent., Paleontol., Biostratig. 2: 539-568.

GOÑI, J.; R. HOFFSTETTER 1964. Uruguay. Lexique Stratigraphique International, 5 Amérique Latine (9a): 1-202.

GUILLEMAIN, C. 1912. Beiträge zur Geologie Uruguays. N. Jahrb. Mineral., Geol., Paläontol. 33: 208-264.

HARRINGTON, H. J. 1945. Algunas observaciones sobre el Sistema de Gondwana en el Uruguay. Primera Reunión de Comunicacio-

nes del IPIMIGEO (Sección Argentina).
pp. 5-19.

JONES, G. H. 1956. Memoria explicativa y mapa geológico de la región oriental del departamento de Canelones. Bol. Inst. Geol. Uruguay 34: 1-193.

MARQUEZ-TOIGO, M. 1976. Some new species of spores and pollens of Lower Permian age of San Gregorio Fromation of Uruguay, Paraná Basin. An. Acad. Brasil. Cien. 46: 601-616.

MARTINEZ-MACCHIAVELLO, J. C. 1963. Microesporomorfos tipos contenidos en el glacial de la base del Sistema de Gondwana del Uruguay (Cuchilla de Zamora, Dpto. de Tacuarembó). Bol. Univ. Parana (Geol.) 10: 1-14.

--- 1967-77. Resumen geológico de los sub-bituminosos de la región NE de la República O. del Uruguay. Rev. Minera 33: 28-32; 34: 41-46.

MENENDEZ, C. A. 1976. Contenido palinológico de estratos pérmicos con Mesosaurus de Rio Claro, São Paulo, Brasil. Rev. Mus. Argent. Cien. Nat. (Paleontol.) 2: 1-30.

PADULA, E. L. 1972. Subsuelo de la mesopotamia y regiones adyacentes. In A. Leanza (Ed.), Geología Regional Argentina, Córdoba: Academia Nacional de Ciencias. Pp. 213-235.

PADULA, E.; A. MINGRAMM 1968. Estratigrafia distribución y cuadro geotectónico-sedimentario del "Triásico" en el subsuelo de la llanura chaco-paranense. Actas Terc. Jorn. Geol. Argent. 1: 291-331.

---;--- 1969. Permian subsurface beds of the chaco-mesopotamian region, Argentina and their relatives in Uruguay, Paraguay and Brazil. In UNESCO (Ed.), Gondwana Geology. Paris. Pp. 1041-1051.

PINTO, I. D. 1972. Late Paleozoic insects and crustaceans from Parana basin and their bearing on chronology and continental drift. An. Acad. Brasil. Cien. 44: 247-260.

REY-VERCESI, D. 1933. Terrenos gondwánicos del departamento de Rivera. Bol. Inst. Geol. Perfor. 20: 1-15.

RUNNEGAR, N.; N. D. NEWELL 1971. Caspian-like relict molluscan fauna in the South American Permian. Bull. Am. Mus. Nat. Hist. 146: 1-66.

SERRA, N. 1944. Memoria explicativa del mapa geológico del departamento de Treinta y Tres. Bol. Inst. Geol. Uruguay 31: 1-43.

WALTHER, K. 1911. Ueber permotriassische Sandsteine und Eruptivdecken aus del Norden der Republik Uruguay. N. Jahrb. Mineral., Geol., Paläontol. 31: 575-609.

--- 1919. Líneas fundamentales de la estructura geológica de la República Oriental del Uruguay. Rev. Inst. Nac. Agron. (2) 3: 1-186.

--- 1923. Estudios geomorfológicos y geológicos. Rev. Inst. Hist. Geog. Uruguay 3: 1-348.

--- 1927. Consideraciones sobre los restos de un elemento estructural, aún desconocido del Uruguay y el Brasil más meridional. Bol. Acad. Nac. Cien., Córdoba 30: 349-382.

--- 1935. Contribucion a la petrología y estratigrafía del glacial gondwánico uruguayo y apuntes a propósito de un croquis sinóptico del Gondwana brasileño. Bol. Inst. Geol., Perfor. 22: 257-275.

ZAMBRANO, J. J. 1975. Cuencas sedimentarias en el subsuelo de la Provincia de Buenos Aires y zonas adyacentes. Rev. Asoc. Geol. Argentina 29: 443-469.

The genus *Saurichthys* (Pisces, Actinopterigii) during the Gondwana Period

LAURENCE BELTAN
Museum National d'Histoire Naturelle,
Paris, France

ANDREA TINTORI
Università degli studi di Milano, Italy

Saurichthys Agassiz, 1834 had a world-wide distribution during the Triassic Period and is represented in many Triassic formations. It displays mixed archaic and advanced characters, and the sequence of fossils does not indicate orthogenesis. The fossils occur in marine and (less frequently) in freshwater deposits, and they provide evidence of the beginning of the opening of the Mozambique Channel, between Malagasy and Africa, in Eotriassic times. The occurrence of the genus in Gondwana and in northern continents indicates that the Tethys and Arctic oceans were connected.

INTRODUCTION

Many Triassic fish assemblages including especially Chondrostei and Holostei, Coelacanthini have been found in different stages of several geological provinces between which relationships have been often established. Chondrostei consist of many orders and families, and among them is the family Saurichthyidae. Among the genera belonging to this family, *Saurichthys* is without doubt the most important. Indeed, it is represented by numerous species, generally coming from marine deposits, and is now well known owing to several authors' investigations. We here point out the complexity of the genus, which has a great paleozoogeographic distribution.

NORTHERN HEMISPHERE DISTRIBUTION

Canada. Specimens referred to *Saurichthys* have been found in the Lower Triassic Sulphur Mountain Formation of the Wapiti Lake area, British Columbia (Fig. 1A). Although descriptions of these remains especially concern the dermal skull, they have been referred to new or known species (Schaeffer & Mangus 1976).

Greenland. In the Eotriassic of East Greenland there are very well-preserved saurichthids. They are interesting, especially because of the endocranium (Nielsen 1936).

Spitsbergen. The Saurichthids are well known in SW Spitsbergen, especially in the middle Lower Triassic. They have been very well studied in nearly all anatomical respects, and several species have been erected: *S. ornatus, S. wimani, S. elongatus, S. hamiltoni* (Fig. 1B) (Stensio, 1925-1932).

Germany. The type-species *Saurichthys apicalis* Agassiz has been found in the Muschelkalk and the Lower Keuper of Germany.

Austria. *Saurichthys krambergeri*

Schlosser (Griffith 1962) is known from the Upper Triassic near Salzburg (Fig. 1C); *S. calcaratus* (Griffith 1977) occurs in the Carnian (Upper Triassic) among the Triassic fishes from Polzsberg near Lunz. This genus has been described under the specific name of *S. striolatus* (Bronn), from the Upper Triassic of Raibl (Griffith 1959).

France. The species *S. daubrei* Firtion occurs in the Lower Triassic (Upper Buntsanstein, Volzia Sandstones) near Walsselone (Bas-Rhin) (Lehman 1974).

Italy - Middle Triassic of Lombardy (A.T.) In the Lombardy basin, *Saurichthys* are recorded from the "Scisti bitumosi di Bergamo" Formation, of Middle Triassic age, the most important outcrops of which are between Besano (Varese-I), and the San Giorgio Mountain (Tessin-C.H.). The vertebrate level is 10 m thick with alternating beds of bituminous black-shales and grey to brown dolomite. They were first described in the last century (Bellotti 1857; Bassani 1886). Some years ago Griffith (1959) described two specimens of *S. curioni* (Bellotti), with special reference to the skull roof. From the same locality, Bassani (1886) described two species of *Saurichthys*, only one of which was accompanied with some remarks on the skull (*S. stoppani*); the other one, *S. intermedius*, was known only by the anal and dorsal fins. Bellotti created *S. robustus* (Bassani 1886) for specimens from Perledo, from a unit a little younger than the one at Besano. Previous authors considered *S. curioni* and *S. robustus* very similar but not conspecific. Unfortunately all the types were destroyed during the last world war, and only undescribed specimens, all collected in recent years at Besano, are stored in the Civic Museum of the Natural History of Milan. Of this material specimens V450/V451 belong to *S. stoppani*,

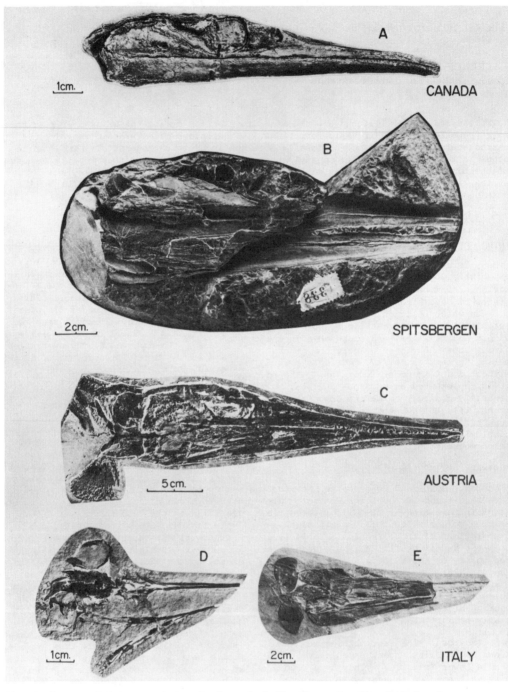

Fig. 1--A. *Saurichthys* sp. Lateral view of skull. B. *S. hamiltoni*. Lateral view of the skull, and the visceral skeleton. C. *S. krambergeri*. Skull in lateral view. D. *S. stoppani*. Skull roofing bones and lower jaws. E. *S.* sp. Skull in dorsal and lateral view.

V449/V452 to *S.curioni*. Some remarks about the skull of *S.stoppani* are given below.

Saurichthys stoppani (Bassani, 1886). A small (20-25 cm) *Saurichthys*, of which about 1/3 is the skull. The skull is, in general, well preserved (Fig. 1D). As usual, dermopterotico-extrascapular bones are large, with posteriorly small medial extrascapulars and suprascapulars, and then a rather big supracleithral bone. The parietal region is not clear; the frontal bones are elongated and triangular. The dermosphenotic is also visible and supports an infraorbital sensory canal extending backward along the dermopterotico-extra-scapular. Anterior to the frontal, postrostals and rostropremaxillaries constitute the upper part of the very slender beak. A detached maxillary allows a good reconstruction of this bone: trapezoidal, the greatest height is in the posterior region from whence it gradually becomes reduced anteriorly. The mandibles are long and gently increase their height backward. Ornamentation is always fine; skull roof bones have a pitted surface with radial ridges; on the other side longitudinal striae, and sometimes thinner transversal ones, are on the beak bones.

Norian of Lombardy (A.T.) New localities for marine vertebrates have recently been found. Shallow sheltered lagoons (Zorzino Limestone), with anoxic bottom conditions along a large carbonate tidal flat "Dolomia principale", allow a very good preservation of the rich fauna. In all places, *Saurichthys* is present with many specimens, usually about half a metre long. Here I will give only the general features of a specimen (n° ET13z,b) I collected in 1978.

Saurichthys sp.a (Fig. 1E) Standard length = 69.5 cm; 20.5 cm (30% s.l.) belonging to the head (opercular included). The body is very slender with a height of about 4% of the s.l.

The right side and the roof of the skull are visible; the top of the beak is preserved on the counterpart (left side visible).

The skull roof has a pair of powerful trapezoidal dermopterotico-extrascapulars with a coarse irregular ornamentation. The parietal region seems to be occupied by a single, rather big bone. Frontals are elongated and, posteriorly, contact briefly with the dermopterotico extrascapulars.

The posterior portion of the maxillary is very high and its posterior margin almost vertical. Close to it the vertical branch of the preopercular is visible. The premaxillary begins anterior to the orbit and increases gradually in height.

The mandible is very long and forms the whole lower jaw. All bones bear fine vertical ridges in ornamentation. Tubercles also are on the most anterior part of the mandible. Teeth are triangular, well-compressed laterally, with alternating large and small teeth. The premaxillary teeth are bigger than the corresponding ones on the mandible. From the top of the beak the main teeth increase their size till the 13th, and the distance between the teeth increases backward.

The opercular is a semilunar bone, higher than long. Ornamentation is composed of gentle radial ridges, more pronounced in the anterior region, and by growth lines. The internal surface is smooth.

The pectoral fins are just behind the opercular. The pelvic ones begin slightly anterior to the midpoint between the end of the skull and the caudal notch. The dorsal and the anal fins, opposite each other, are halfway between the pelvic fins and the caudal notch. Pectoral and pelvic fins are compounded by unjointed lepidotrichia; on the other hand, dorsal, anal and caudal fins have jointed lepidotrichia with a long article.

Four longitudinal scale-rows constitute the whole squamation, but only the dorsal one seems to be complete, from the caudal fin to the posterior end of the head.

Spain. *Saurichthys* sp. is known in two NE provinces--the Muschelkalk near Barrenco de Oden, Lerida (Lehman 1964), and in Catalonia near Tarragona (Fig. 2A) (Beltan 1972).

Turkey. Recently, a new marine fish and a placodont reptile fauna of Ladinian age have been discovered in SW Turkey (Beltan *et al.* 1979). The outcrop is in the region of Seydischir (Western Taurides) and constitutes limestone. Some fish remains belong to *Saurichthys* (Fig. 2B). It is the first time that this genus has been found in Turkey.

Nepal. An endocranium of a new *Saurichthys*, *S. nepalensis* Beltan & Janvier, 1978, (Fig. 2C) was recently discovered in the Lower Triassic (Scythian) part of Thini Gaon region in the NW Annapurnas in Nepal. This was also the first report of this genus in this part of the world.

SOUTHERN HEMISPHERE DISTRIBUTION

South Africa. An elongate jaw with teeth was discovered in the *Cynognathus* Zone corresponding to the Upper Beaufort Series (Lower Triassic, Olenikian). It belongs to the genus *Saurichthys* and is the first record of this genus found on this continent; it is noteworthy that it comes from a freshwater deposit (Griffith 1978).

Malagasy. Triassic fishes occur in two important outcrops in the western part of Malagasy. The NW deposit is mainly marine and is Eotriassic. This horizon, near

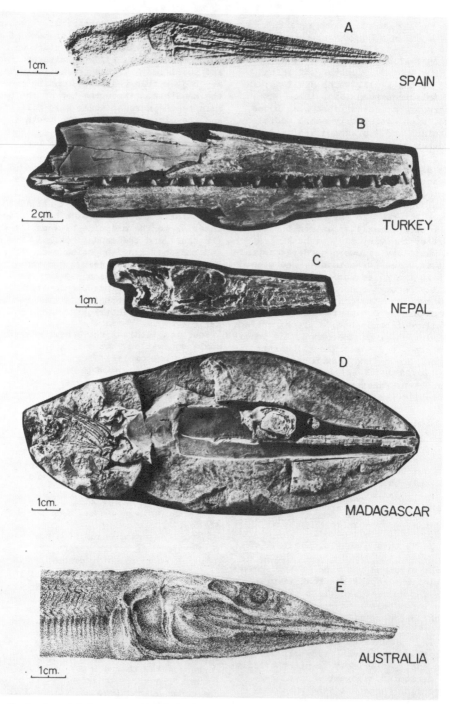

A SPAIN

B TURKEY

C NEPAL

D MADAGASCAR

E AUSTRALIA

Fig. 2--A. *Saurichthys* sp. Skull in lateral view. B. *S.* sp. Part of rostrum in lateral view. C. *S. nepalensis*. Anterior part of the endocranium in sagittal view (natural) and posterior part of rostrum in lateral view. D. *S. madagascariensis*. Skull in lateral view, and small part of the endocranium. E. *S. gigas*. Lateral view of skull.

Ambilobe, Antsaba, has yielded an abundant ichthyofauna in a good state of preservation. The fossils are contained in nodules which when broken show anatomical imprints in negative relief. Among the Actinopterigians, the family Saurichthyidae is well represented: *Saurichthys madagascariensis* Piveteau, 1944 (Fig. 2D), *S. stensiöi* Lehman, 1952, *S. piveteaui* Beltan, 1968, and many *Saurichthys* spp.

The SW deposits (Permotriassic middle Sakamena Fm) have yielded many fishes almost identical to those of the north. Fishes in nodules have been found near the localities of Beroroha and Mandronarivo in the Tulear area. The deposits are lacustrine with marine intercalations (Lehman *et al.* 1959). *Saurichthys* has been identified among the Actinopterygians.

Australia. Triassic fishes flourished in the lakes in New South Wales. Specimens of *Saurichthys* occur in the Middle Triassic of Brookvale: *Saurichthys parvidens* Wade, 1935; and in the Lower Triassic of Gosford: *S. gigas* (A. Smith-Woodward) 1890, *S. gracilis* (A. S-W.) 1890.

ANATOMICAL REMARKS

Scales. One notices the absence of orthogenesis in the reduction of the number of scales. Indeed, *Saurichthys* from the Lower Eotriassic of Greenland are completely covered with scales, while those of Madagascar from the same level have only four longitudinal scale rows. The number is again reduced in the Spitzbergen specimen from the Upper Eotrias. In *Saurichthys* sp., discussed above, four scale rows constitute the whole squamation.

Dermal skull. A great variability has been observed regarding the number and shape of parietal bones. Specimens from Spitsbergen have generally four, rarely more; *S. curioni* and *S.* sp. from Norian Lombardy have only one parietal, and *S. piveteaui* has 33 of varied shape and size. Moreover, in *S. madagascariensis* the number of these bones varies between four and one, and in some specimens the parietal zone is indicated by the presence of pit lines. It seems that this variability is connected with the growth. The Triassic genera *Australosomus* Piveteau (British Columbia, Greeland, Tanzania, Madagascar) and *Macroaethes* Wade (Australia), both Pholidopleuridae, have the parietal area sub-divided into several small bones. The rostral zone may present many ossicles, as in *S. stensioi* and *S. piveteaui.* One may conclude that the skull roof was very variable in *Saurichthys.*

In *S. krambergeri* the angular of the jaw is small, while in specimens from Spitsbergen and Madagascar this bone occupies a large part of the lower jaw. The supraangular and the quadratojugal exist in Spitsbergen forms and *S. krambergeri,* but are often absent in *Saurichthys* from Madagascar. These bones (primitive features) disappear during evolution; this observation shows that it is not a case of geological age.

Neurocranium. This is well known. With a long rostrum and a long oto-occipital region, it is generally well ossified, but in the orbitotemporal region some differences are noticed.

The orbit is wide in *Saurichthys* from Spitsbergen and Madagascar, small in *S. nepalensis.* The same is the case for the dorsal fontanelle. The fenestra optica is small in *S. ornatus,* wide in *S. wimani* of the same fish horizon, small in *S. nepalensis* and wide in *S. madagascariensis.* The anterior myodomes are well developed in the latter two, but not in *S. ornatus.* The geological provenance being considered, one notices, once more, that there is no orthogenesis of the regression in the bony tissues through geological time.

Teeth. Many authors (Stensiö 1925; Lehman 1952; Griffith 1959; Beltan & Janvier 1978; Beltan *et al.* 1979) have described teeth of *Saurichthys.* In *S. ornatus, S. madagascariensis* for instance, conical teeth of various size originate in medial lamellae of rostropremaxillaries and of dentaries, vomeres and parasphenoids. In *S. ornatus* the median part of the anterior half of the parasphenoid, the dermopalatine, and ectopterygoid, the horizontal lamella of the entopterygoid, and the most anterior part of the mixicoronoid have very small teeth. Small teeth are also observed in *S. parvidens. S. nepalensis* shows large teeth in the posterior part of the vomeres. It is noteworthy that *Saurichthys* sp. from the Ladinian of Turkey has a rostrum with a great number of teeth, and there are three types of teeth according to their shape and position. On the labial edge of the rostropremaxillary and the corresponding part of the dentary, there are series of small spheroidal teeth which decrease in size laterally. Medially to these labial teeth the medial lamellae of the restropremaxillary and of the dentosplenial bear a series of very large and stout teeth which represent the main tooth rows of the jaws. These teeth are conical in shape and their average height is 1 cm; they have a small enameloid cap, á large pulp cavity easily seen in an X-ray. In *S. krambergeri* the pulp cavity reaches up into the apical cap; this is not the case for *S. wimani*. Finally, in the genus *Saurichthys* variously developed teeth cover all visceral bones. Their shape ranges from very small, spheroidal, tubercle-like to conical.

PALEOBIOGEOGRAPHICAL REMARKS

The genus *Saurichthys* exists in all Triassic stages and has world-wide distribution.

In northwestern Madagascar, the fish horizon is marine. Indeed, deposits show ammonites and decapod crustaceans, but this out-crop was doubtless near the shore because terrestrial stegocephalians and one reptile have been recorded. On the contrary, the southwestern fish zone is deltaic (the presence of dipnoans is noticed) but with marine intercalations. In South Africa *Saurichthys* has been found in freshwater deposits (Griffith.1978). In the same out-crop at Vaalbank near Burghersdorp, in the *Cynognathus* Zone of the Beaufort Series, dipnoans have been found (Gardiner and Jubb 1975). The *Saurichthys* record, however, was definitely in place.

These data indicate the rifting between Africa and Madagascar during the Lower Triassic age, or perhaps even before, during the Permian and the beginning of the opening of the Mozambique channel which was perhaps wider in the north than in the south. Moreover, the occurrence of this genus in Gondwana and in the Northern Hemisphere suggests that the Tethys Sea (where *Saurichthys* existed) and the Arctic Ocean were connected throughout Western Europe in a NE direction by sea. The Eotriassic SE fish horizon of Greenland and that of the SW of Spitsbergen seem to bear out this hypothesis.

CONCLUSION

The genus *Saurichthys* is according to this outline a heterogeneous, ubiquitous fish adapted to both marine and fresh waters. Several species characterized it, and its evolution is not orthogenetic. Its existence in marine deposits (e.g., Greenland, Spitsbergen, Western Europe) and lacustrine, freshwater and deltaic outcrops (South Africa, Southern Madagascar, Australia) suggests two hypotheses:

Saurichthys was a potomodromous fish which might swim up river to spawn.

Saurichthys might live either in seas or in rivers and lakes. One notices this phenomenon in the Recent teleostean marine genus *Salmo (Salmo salar)* which goes up rivers to spawn, and the same species can live in large lakes of Quebec, for instance, and never return to the sea.

ACKNOWLEDGMENTS

The authors are grateful to Prof. G. Pir (Milan) for the loan of specimens coming from Besano and thank Mr G. Chiodi (Milan) Miss J. Crapart and Mr D. Serrette (Paris) for the photographs.

REFERENCES

BASSANI, F. 1886. Sui fossili e sull'eta degli schisti bitumosi Triasici di Besano Lombardia. Atti. Soc. Ital. Sci. Nat. 29: 15-72.

BELLOTTI, C. 1857. Descriptione de alcune nuove specie di pesci fossili di Perledo e altre localita lombarde. In A. Stoppani, Stud. geol. paleontol. Lombardia. : 419-438

BELTAN, L. 1968. La faune ichthyologique l'Eotrias du N.W. de Madagascar: Le neuroc Cah. Paléontol., CNRS 135 p.

--- 1972. La faune ichthyologique du Muschelkalk de la Catalogne. Mem. R. Acad. Artes Barcelona, Tercera epoca No. 760-41/1 291-325.

BELTAN, L.; Ph. JANVIER 1978. Un nouveau Saurichthyidae (Pisces, Actinopterigii) Saurichthys nepalensis n.sp. du Trias inférieur des Annapurnas (Thakkhola, Népal) et sa signification paléobiogéographique. Cybium 3ème série 4: 17-28.

BELTAN, L.; J.M. DUTUIT 1978. Recapitula-tion des affinités gondwaniennes anté-jurassiques de Madagascar (Poissons, Amphi-biens, Reptiles). Apports récents à la Géologie du Gondwana. Ann. Soc. Géol. Nor 97: 357-362.

BELTAN, L. Ph. JANVIER; O. MONOD; F. WEST-PHAL 1979. A new marine fish and placo-dont reptile fauna of the Ladinian age from Southwestern Turkey. Neues Jahrb. Geol. Palaeontol. 5: 257-267.

GARDINER, B. G.; R. A. JUBB 1975. A pre-liminary catalogue of identifiable fossil fish material from Southern Africa. Ann. S. Afr. Mus. 67/11: 381-440.

GRIFFITH, J. 1959. On the anatomy of Sau-richthyid fishes, Saurichthys striolatus (Bronn) and S.curioni (Bellotti). Proc. Zool. Soc. London 132/4: 587-606.

--- 1962. The Triassic Fish Saurichthys krambergeri Schlosser. Paleontology 5/2: 344-354.

--- 1977. The Upper Triassic fishes from Polzberg bei Lunz, Austria. Zool. J. Linn. Soc. 60/1: 1-93.

--- 1978.   A fragmentary specimen of
Saurichthys sp. from the Upper Beaufort
series of South Africa.   Ann. S. Afr. Mus.
76/8:   299-307.
LEHMAN, J. P. 1952.   Etude complémentaire
des Poissons de l'Eotrias de Madagascar.
K. Sven. Vetenskapsakad. Handl. 2/6: 192.
--- 1964.   Etude d'un Saurichthyide de la
région d'Oden (Espagne).   Ann. Paléontol.
Vertébr. 50/1:   25-30.
LEHMAN, J. T.; C. CHATEAU; M. LAURAIN;
M. NAUCHE 1959.   Les Poissons de la Saka-
mena moyenne Paléontologie de Madagascar 28.
Ann. Paléontol. Vertébr. 45:   177-219.
NIELSEN, E. 1936.   Some few preliminary
remarks on Triassic Fishes from East Green-
land.   Medd. Groenl. 112/3:   55p.
PIVETEAU, J. 1944-45.   Les Poissons du Trias
inferieur.   La famille des Saurichthydes,
Paléontologie de Madagascar 25.   Ann.
Paléontol. 31:   79-88.
SCHAEFFER, B.; M. MANGUS 1976.   An early
Triassic fish assemblage from British
Columbia.   Bull. Am. Mus. Nat. Hist. 156/5:
517-563.
STENSIÖ, E. 1925.   Triassic fishes from
Spitzbergen.   K. Sven. Vetenskapsakad. Handl.
2/1:   261 p.
--- 1932.   Triassic Fishes from East Green-
land.   Medd. Groenl. 83/3:   298 p.
WADE, R. T. 1935.   Triassic Fishes of Brook-
vale New South Wales.   British Museum (NH):
London.

The stratigraphic distribution of the Dicynodontia of Africa reviewed in a Gondwana context

A. W. KEYSER

Geological Survey of South Africa, Pretoria

In a new biostratigraphic subdivision of the Beaufort Group of South Africa, considerable reliance was placed on the Dicynodontia. Seven assemblage-zones are recognised, with ages ranging from Middle Permian to Middle Triassic.

The only Permian dicynodont that occurs on three continents is the genus *Endothiodon*, which occurs in at least three formations in Africa and formations in Brazil and India. Small endothiodonts resembling *Pristerodon* are now known from Africa and India.

The Triassic genus *Lystrosaurus* has a wide distribution in Gondwana and also occurs in the northern hemisphere. The genus *Kannemeyeria* also has a wide distribution in Gondwana and is also known from China and Russia.

The relationships of the newly-recognised assemblage-zones of the Beaufort Group to other formations on the Gondwana continents are discussed.

The Beaufort Gp consists of a succession of mudstones, siltstones and sandstones deposited on a vast floodplain during Upper Permian and Lower Triassic times. The sandstones are mostly river channel fill deposits, and the mudstones and siltstones represent overbank silts and muds. Fossil reptiles are extremely common in these overbank sediments and can be used for biostratigraphic subdivision. The discovery of uranium in the Beaufort Gp led to an urgent demand for a more detailed lithostratigraphic subdivision of it. As is to be expected, there are few recognisable marker beds in such a flood plain deposit, and therefore a detailed biostratigraphic subidivsion was needed. Keyser and Smith (1979) proposed a new subdivision to assist with the correlation and recognition of the lithostratigraphic units.

In the new biostratigraphic subdivision of the Beaufort Gp considerable reliance was placed on the Dicynodontia. Seven assemblage-zones are recognised. This subdivision has since been accepted by the South African Committee on Stratigraphy. These assemblage-zones are, in the usual sequence:

Kannemeyeria-Diademodon Assemblage-zone
Lystrosaurus-Thrinaxodon Assemblage-zone
Dicynodon lacerticeps-Whaitsia Assemblage-zone
Aulacephalodon-Cistecephalus Assemblage-zone
Tropidostoma-Endothiodon Assemblage-zone
Diictodon-Pristerognathus Assemblage-zone
Dinocephalian Assemblage-zone

The new assemblage-zones can now for the first time be linked to the lithostratigraphic units of the Beaufort Gp.

The Beaufort Gp is highly fossiliferous. In fact, vertebrate fossils are so common that they can be used for practical biozonation on a 1:50 000 scale map.

Fossils that can be assigned to the Dinocephalian Assemblage-zone are only known from the Beaufort Gp of the Republic of South Africa and the Lower Madumabisa Mudstone of the Bingwe district of Zimbabwe-Rhodesia, and as far as could be ascertained do not occur in Russia, with *Moschops* being the only common genus. It is possible that the differences between the other genera are being over-emphasized.

The *Diictodon-Pristerognathus* Assemblage-zone has no exact equivalent outside South Africa. The zone is characterised by the absence of both dinocephalians and large dicynodonts other than *Endothiodon*, which is occasionally encountered. The importance of this zone is that 90% of the known uranium occurrences of the Beaufort Gp are found in it.

The Dinocephalian Assemblage-zone coincides with the former Lower and Middle *Tapinocephalus* Zone, while the *Diictodon-Pristerognathus* Assemblage-zone coincides with the former Upper *Tapinocephalus* Zone.

The *Tropidostoma-Endothiodon* Assemblage-zone is contained in a stratigraphic unit which includes very little sandstone. It is equivalent to Broom's (1905) *Endothiodon* Zone and Kitching's (1977) Lower *Cistecephalus* Zone. It coincides with the range-zone of *Tropidostoma* and the acme-zone of *Endothiodon*, the latter occurring occasionally above and below this assemblage-zone. The genus *Tropidostoma* is known only from the Beaufort Gp of South Africa and the Madumabisa Mudstone Fmn at Charisa in Zimbabwe-Rhodesia, where it also occurs

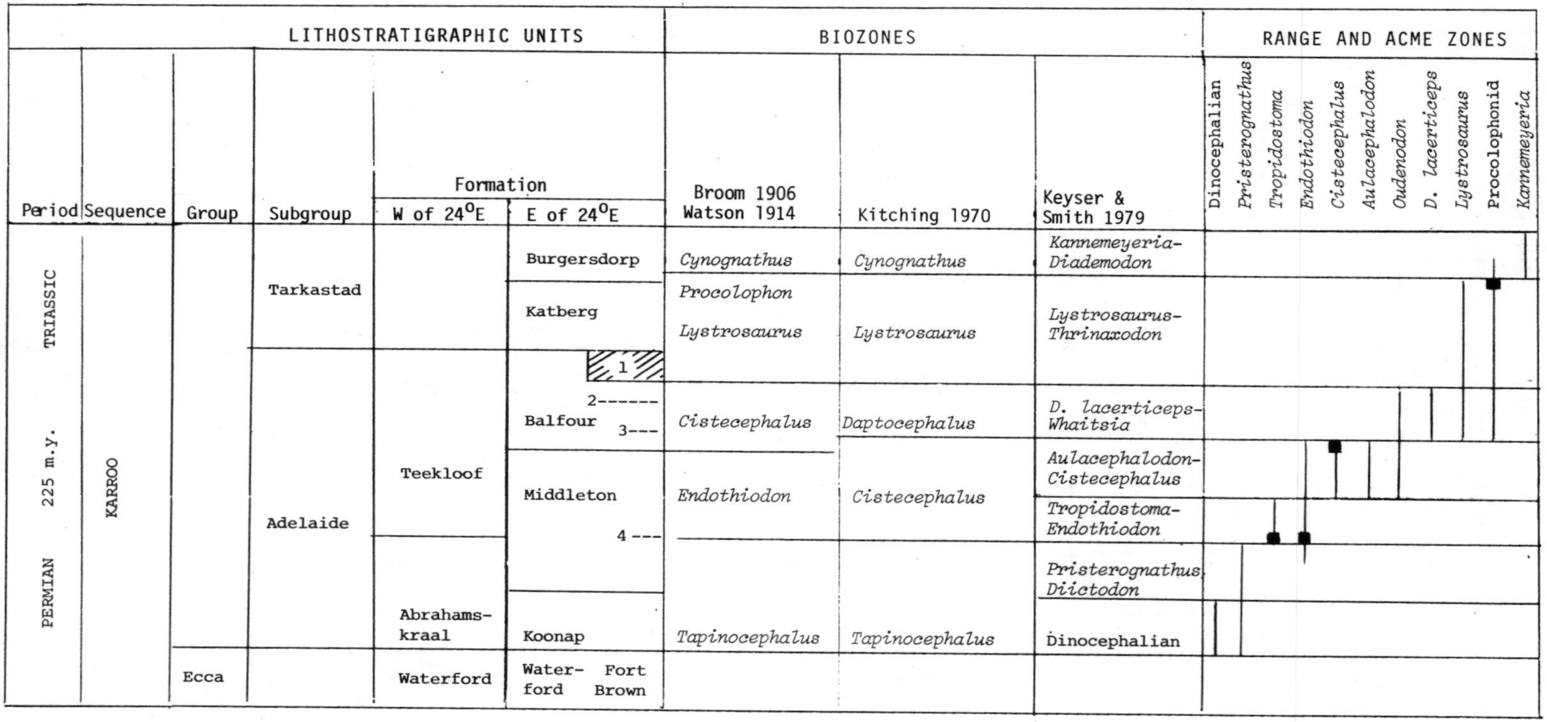

RANGE AND ACME ZONES: *Kannemeyeria*, Procolophonid, *Lystrosaurus*, *D. lacerticeps*, *Oudenodon*, *Aulacephalodon*, *Cistecephalus*, *Endothiodon*, *Tropidostoma*, *Pristerognathus*, Dinocephalian

BIOZONES

LITHOSTRATIGRAPHIC UNITS				Broom 1906 Watson 1914	Kitching 1970	Keyser & Smith 1979
Period	Sequence	Group	Subgroup	Formation W of 24°E	Formation E of 24°E			

Formation: W of 24°E / E of 24°E

Subgroup	W of 24°E	E of 24°E	Broom 1906 Watson 1914	Kitching 1970	Keyser & Smith 1979
Tarkastad		Burgersdorp	*Cynognathus*	*Cynognathus*	*Kannemeyeria–Diademodon*
		Katberg	*Procolophon*		
			Lystrosaurus	*Lystrosaurus*	*Lystrosaurus–Thrinaxodon*
		Balfour 3 --- 2 ----- 1	*Cistecephalus*	*Daptocephalus*	*D. lacerticeps–Whaitsia*
Adelaide	Teekloof	Middleton 4 ---	*Endothiodon*	*Cistecephalus*	*Aulacephalodon–Cistecephalus*
					Tropidostoma–Endothiodon
	Abrahams-kraal	Koonap	*Tapinocephalus*	*Tapinocephalus*	*Pristerognathus–Diictodon*
	Waterford	Water-ford Fort Brown			Dinocephalian

Ecca

PERMIAN   225 m.y.   TRIASSIC

KARROO

1 - Parlingkloof Member; 2 - Barbaerskrans Member; 3 - Oudeberg Member; 4 - Poortjie Member   ALL INFORMAL

Fig.1--Relationship between bio- and lithostratigraphy of the Beaufort Group.

62

together with *Endothiodon*. There can be no doubt that the latter locality is a time equivalent of the *Tropidostoma-Endothiodon* Assemblage-zone. The genus *Tropidostoma* also occurs in the Madumabisa Mudstone Fmn of the Luangwa Valley in Zambia, the Tete district of Mozambique and the Ruhuhu depression of Tanzania. *Endothiodon* is the only therapsid genus other than *Lystrosaurus* known to occur on three Gondwana continents. It is known from the Pranhita-Godavari Valley of India (Kutty 1972) and from the Rio do Rasto Fmn of the Parana Basin of Brazil (Barbarena, pers. comm.). The most likely correlation of other formations that contain *Endothiodon* is with the *Tropidostoma-Endothiodon* Assemblage-zone of South Africa. It must always be kept in mind, however, that the range of *Endothiodon* extends into both the underlying and overlying assemblage-zones.

The *Aulacephalodon-Cistecephalus* Assemblage-zone coincides with the range zone of both these zone fossils. The latter genus is, however, very rare towards the bottom of its range. At the top of the assemblage-zone a very distinct *Cistecephalus* acme-zone is developed. The genus occurs in the Madumabisa Mudstone Fmn of the Luangwa Valley in Zambia. A closely related genus *Kawingasaurus* occurs in the Kawinga series of the Ruhuhu depression of Tanzania. Specimens that closely resemble *Cistecephalus* are known from the Pranhita-Godavari Valley of India. The genus *Aulacephalodon* is a more plentiful zone fossil, but it is as yet not known from deposits outside the Beaufort Gp of South Africa.

The *Dicynodon lacerticeps-Whaitsia* Assemblage-zone replaces Kitching's (1977) *Daptocephalus* Zone, the range being identical. The change is necessitated by the fact that *Daptocephalus leoniceps* Owen is due to be declared a junior synonym of *Dicynodon lacerticeps* Owen by Cluver (in preparation.

Dicynodon lacerticeps is also known from the Ruhuhu Beds of Tanzania from where von Huene described it under the name *Platypodosaurus*. Casts of a specimen that closely resemble *Dicynodon lacerticeps* have been sent from the USSR to the British Museum (Natural History) and the Palaeontological Institute of the University of Tübingen. These casts were referred to *Dicynodon annae*. As far as I could ascertain, no description of such a species has yet appeared in the literature. The homonym *D. annae* has been used for a species of *Diictodon* in South Africa. It is likely, however, that a time equivalent of these beds exists in Russia.

What was formerly known as the Middle Beaufort Stage, is now called the Katberg Sandstone Fmn. The therapsid genera *Lystrosaurus* and *Thrinaxodon* characterize this zone. These genera are known from the Fremouw Fmn of Antarctica and the Panchet Beds of India. They are as yet not known from any of the other Gondwana continents. The genus *Lystrosaurus* is also known from the Jiucaiyuan Fmn of China and from the Dvina V Beds of Russia.

The Burgersdop Fmn, formerly known as the Upper Beaufort Stage and referred to as the *Cynognathus* Zone, is now called the *Kannemeyeria-Diademodon* Assemblage-zone. The genera *Kannemeyeria* and *Cynognathus* are also known from the Puesto Viejo Fmn of Argentina. The genus *Kannemeyeria* and other forms very closely related to it have a wide distribution across Gondwana and the northern hemisphere. *Kannemeyeria* occurs in the Omingonde Fmn of SW Africa, the Upper Madumabisa Mudstone Fmn of Zambia, and the Manda Beds of Tanzania. Closely related forms are known from the Er Ma Ying Fmn of the People's Republic of China and from Cisuralian Russia.

The Dicynodontia of the Upper Triassic of South America are not known from Africa, although there is some resemblance with the forms from the northern hemisphere. A species that resembles *Ischigaulastia* from the Santa Maria and Ischigaulasto Fmns of South America has been found in the Manda Beds of Tanzania. This specimen still awaits description.

The Dicynodontia are the most important fossils for stratigraphic correlation in the Beaufort Gp of South Africa. It is hoped that they will fulfil a similar function in other parts of Gondwana in future.

REFERENCES

BROOM, R. 1905. Notes on the localities of some type specimens of the Karroo fossil reptiles. Rec. Albany Mus. 1: 275-278.
KEYSER, A. W.; R. M. S. Smith 1979. Vertebrate biozonation of the Beaufort Group with special reference to the western Karroo Basin. Ann. Geol. Surv. S. Afr. 12: 1-35.
KITCHING, J. W. 1977. The distribution of the Karroo vertebrate fauna. Mem No.1, Bernard Price Inst. Palaeontol. Res., Johannesburg.
KUTTY, T. S. 1972. Permian reptilian fauna from India. Nature 237: 462-463.

Permo-Triassic continental deposits and vertebrate faunas of China

CHENG ZHENGWU
Chinese Academy of Geological Sciences, Beijing

This paper deals chiefly with the Permo-Triassic continental succession in China, with six vertebrate horizons or faunas and their relationship with corresponding faunas in the other parts of the world, discussion on the ages of the various faunas, the Permo-Triassic boundary and the various boundaries within the Triassic, and the derivation of kannemeyerids from dicynodontids by the late Permian or early Triassic and their middle Triassic age.

INTRODUCTION

The Permo-Triassic deposits in China may be roughly separated into northern continental and southern marine regimes by a line joining the Kunlun-Qinling-Dabieshan ranges.

There are two exceptions to these general remarks: the Middle Triassic Badong Fmn in Sangzhi, Hunan Province, within the southern marine regime contains continental intercalations yielding *Lotosaurus*; in Qishan and Lingyou of Shaanxi within the primarily northern continental regime there occur such marine bivalves as *Eumorphotis*, *Leptochondria*, etc. testifying to the presence of early Triassic seas (Yang Zunyi *et al*. 1979), but some geologists are sceptical about the exact horizon of these bivalves.

The continental Permo-Triassic is remarkable for its wide distribution, great thickness (3-4 km) and abundant fossils. North China is one of the important regions in the world for the study of terrestrial deposits and vertebrates.

STRATIGRAPHIC PROVINCES, STRATIGRAPHIC SEQUENCES, AND VERTEBRATE FOSSILS

On both lithologic and faunistic features the Northern Permo-Triassic regime may further be subdivided into the Northwest China and North China regimes.

In NW China represented by Xinjiang we find the following succession (in ascending order):

The Upper Permian consists of the Upper Jijichao Gp or Taodonggou Gp in the lower part and the lower Cangfanggou Gp in the upper, containing fishes and reptiles respectively.

The Upper Jijichao Gp includes the Lucaogou and Hongyanchi Fmns in the Junggar basin, represented by black shale, oil shale, dolomite and greyish-green sandstone, mudstone, marl, etc. of lacustrine-paralic facies. In Turfan Basin, the equivalent Taodonggou Gp is composed of continental dark red conglomerate, sandstone, and mudstone intercalated with greyish-green, greyish-white arenaceous mudstone and marl. Both groups are 604-1837 m thick and contain fossil fishes, such as *Turfania* Liu & Wang, *Sinoniscus* Liu & Wang, *Tienshaniscus* Liu & Wang.

The Lower Cangfanggou Gp (from bottom to top, Quanzijie and Wutonggou Fmns) is first predominantly dark purplish-red, yellowish-green sandy mustone intercalated with conglomerate, 372-853 m thick, yielding dicynodonts such as *Striodon* Sun, *Jimusaria* Sun, *Dicynodon* Owen, *Turfanodon* Sun, *Kunpania* Sun.

The Lower Triassic consists of the Jiucaiyuan Fmn in the lower part and the Shaofanggou Fmn in the upper, both lacustro-fluvial deposits.

The Jiucaiyuan Fmn is chiefly purplish and dark red mudstone, sandstone with limy concretions and greyish-green sandstone, 170-370 m thick, yielding vertebrate fossils: *Lystrosaurus* Cope, *Chasmatosaurus* Haughton, *Prolacertoides* Young, *Santaisaurus* Koh, Labyrinthodontia.

The Shaofanggou Fmn is made up of mudstone containing calcareous concretions, purplish-red massive sandstone, conglomerate, intercalated with greyish-green sandstone and mudstone, 109-352 m. So far no definite vertebrate fossils have been reported. On palynological evidence from the green intercalations it is early Triassic.

The Middle Triassic: On the basis of stratigraphic succession, lithology and vertebrates, the lower part of the Kelamayi Fmn, which contains *Parakannemeyeria*, is definitely Middle Triassic, and the upper is Upper Triassic. This differs from North China, but for the time being the entire Kelamayi Fmn is put in the Middle Triassic as advocated in the Triassic correlation adopted by the second All-China Stratigraphic Congress.

The Lower Kelamayi Fmn consists of purplish-red, dark purplish-red sandstone, argillaceous siltstone with calcarous concretions intercalated with greyish-green

Fig. 1--The distribution of the Permo-Triassic vertebrate fossils in China.
⊗ = fossil locality   1 = Yaomoshan; 2 = Fukang; 3 = Turpan; 4 = Fugu; 5 = Wubu;
6 = Tougchuan; 7 = Lishi; 8 = Ninwu; 9 = Wuxiang; 10 = Hongdon; 12 = Pingquan;
13 = Sangzhi.

sandstone, 121 m thick, with such verte-brates as *Sinosemionotus* Yuan & Koh, *Parotosaurus* Young, *Turfanosaurus* Young, *Vjushkovia* Huene, *Parakannemeyeria* Sun.

The Upper Triassic from the bottom up-wards is composed of Upper Kelamayi, Huang-shanjia and Haujiagou Fmns. The Upper Ke-lamayi Fmn chiefly consists of yellowish-green, greyish-black sandstone, mudstone and shale 107-380 m thick, with fishes--*Fukangichthys* Su, *Bogdania* Young and *Fukangolepis* Young--and a late Triassic flora.

The Huangshanjie and Haojiagou Fmns in-clude mainly a series of greyish-yellow, greyish-green mudstone, sandstone with marly lenses and thin coal seams, reaching a thickness of 184-829 m. They are rich in late Triassic flora, but so far no verte-brate fossils have been reported.

In North China the Permo-Triassic is fairly well exposed in the Shaanxi-Gansu-Ningxia basin (the Ordos) and Shanxi.

The Upper Permian contains the Upper Shi-hezi and Sunjiagou Fmns. The first is chiefly dark purple, purplish-red and purple mudstone, siltstone, alternating with grey, greyish-green, greyish-white sandstone, av-eraging 100-200 m. In Jiyuan, Henan Pro-vince, this formation contains *Bystrowiana* Young, *Honania* Young, *Tsiyuania* Young,

Taihangshania Young, *Wangwusaurus* Young, *Hwanghocynodon* Young.

The Sunjiagou Fmn is chiefly dark and pur-plish-red mudstone and siltstone intercala-ted with purplish-grey or greyish-green and greyish-white arkose, 100-300 m thick, containing in its middle and upper parts *Shihtienfenia* Young & Ye, *Shansisaurus* Cheng, Tapinocephalidae and Dicynodontia.

The Lower Triassic comprises the Liujia-gou and Heshanggou Fmns. The former is a series of purplish-red and greyish-purple arkose intercalated with a small amount of purplish-red siltstone, conglomerate, and cross-bedded sandstone, 349-633 m thick, yielding only fragmentary unidentified bones. In Shanxi plant fossils have been collected by Wang Lixin and Wang Zichiang from the upper part--*Pleuromeia jiaocheng-ensis* Wang, *P. rossica* Newberry, *Neocala-mites* sp. According to Qu Lifan the sporo-pollen analysis gives a subzone of *Lund-blatispora-Taeniaesporites-Cycadopites*, which indicates an early Triassic age.

The Heshanggou Fmn consists of brick-red, purplish-red mudstone, sandy mudstone intercalated with purplish-grey, greyish-green arkose, 103-280 m thick, containing in its upper part *Ceratodus heshanggou-ensis* Cheng, Benthosuchidae, Procolophonid, *Fugusuchus* Cheng, Thoriodontia.

The Middle Triassic Ermaying Fmn is the most widespread unit of the Permo-Triassic in N. China and the most rich in fossils. It is chiefly dark purplish-red mudstone and silty mudstone alternating with greyish-green to greyish-yellow sandstone with numerous calcareous concretions in the mudstone, 414-600 m. thick. It may be subdivided into the lower, with *Paoteodon* Chow & Sun, Procolophonidae, Thecodontia, *Ordosiodon* Young, *Ordosia* Hou, *Shaanbeikannemeyeria* Cheng, *Parakannemeyeria xingxianensis* Cheng; and the upper, with Capitosauridae, *Neuprocolophon* Young, *Shansisuchus* Young, *Fenhosuchus* Young, *Wangisuchus* Young, *Shansiodon* Ye, *Sinokannemeyeria* Young, *Parakannemeyeria* Sun.

The Upper Triassic consists of the Tongchuan (lower) and Yanchang (upper) Fmns, both forming a gradational (from coarse to fine) series of detrital rocks, yellowish-green to greenish-grey sandstone, mudstone, and shale, topped by coal beds. In the middle and upper parts of the former, there occur *Hybodus youngi* Liu and Plesiosauria, the first assigned to Late Triassic or probably Middle Triassic age (Liu Xianting 1962). The associated plant fossils are, however, considered to be of late Middle Triassic (Huang Zhigao & Zhou Huiqing, in press).

Therefore, six faunas may be recognised (see Table 1).

DISCUSSION

The Permo-Triassic vertebrates collected in the last 20 years from Xinjiang and N. China may roughly be grouped into six faunas (from old to new), though some of the material needs further study.

The *Turfania* fauna in the Upper Jijicao and Tondonggou Gps south and north of the Tianshan mountains is Late Permian (Liu Xianting). Probably corresponding to them are the upper Shihezi vertebrates of Jiyuan, Henan, which are comparable to those of the *Endothiodon* Zone of South Africa and Zone III of Soviet Russia, and are referred to the late Permian (Young 1979). This fauna still needs further study, as the fossils are fragmentary and its exact horizon is uncertain.

The Dicynodonts-Pareiasaures fauna is a combination of the known materials from NW and N. China. Fragmentary dicynodont fossils lately discovered in the Sunjiagou Fmn in Pingchuan, Hebei, of N. China have been incorporated with the vertebrate fossils of Xinjiang. The dicynodonts of Xingjiang studied by Young (1934) and Sun Ailing (1973, 1978) are comparable to the S. African material and assigned to the Late Permian. Pareiasaures in N. China (Young 1963; Cheng Zhenwu, in press) is correlated with the *Cistecephalus* Zone of S. Africa and Zone IV of Soviet Russia, assigned to the uppermost Permian.

The *Lystrosaurus* fauna has so far been found only in the Xinjiang region (Young & Yuan 1934a; Sun Ailing 1973), but it is known in various continents and is recognised as an Early Triassic fauna. Our study from the Heshangou Fmn of Shaanxi-Kansu-Ningxia basin suggests that it appeared in the earliest Triassic.

The Labyrinthodontes-Procolophonids fauna was recently discovered in the Shaan-Gan-Ning basin (Table 1). Among them, Benthosuchidae represents a cosmopolitan Early Triassic amphibian; *Ceratodus heshanggcuensis* Cheng is similar to *Ceratodus donensis* subsp. Vorobyeva & Minich of the Early Triassic Baskunchak series; *Fugusuchus* Cheng is intermediate in skull structure between Early Triassic *Chasmatosaurus* and Middle Triassic *Shansisuchus*, *Erythrosuchus*. *Procolophonia* was lately discovered from the top of the Heshanggou Fmn in N. Shaanxi by the Institute of Vertebrate Paleontology and Paleoanthropology, and according to a preliminary report by Sun Ailing *et al.* on its skull is an advanced type and similar to the late Early Triassic *Procolophon* of S. Africa. In short, the present fauna should be late Early Triassic.

The kannemeyerid fauna is abundant throughout the Ermaying Fmn of N. China--chiefly *Shaanbeikannemeyeria* in the basal part, *Sinokannemeyeria* in the upper, and *Parakannemeyeria* throughout both. This has convinced us that the Ermaying Fmn should be assigned as a whole to Middle Triassic. (According to Sun Ailing, the base of the Ermaying is of latest Early Triassic age.)

Shaanbeikannemeyeria Cheng, recently discovered in the basal part of the Ermaying Fmn in Shaanxi, is fairly similar in skull characters to *Kannemeyeria erithrea* and the middle Triassic *Ularokannemeyeria* of Soviet Russia, and is also middle Triassic. The upper part of the Ermaying containing *Sinokannemeyeria* belongs undoubtedly to the middle Triassic. The whole kannemeyerid fauna can be assigned to the middle Triassic and is equivalent to the *Cynognathus* Zone of S. Africa and Zone VI of Soviet Russia (Table 2).

The Fukang fauna is named in place of the *Fukangichthys* fauna of Liu Xianting because in our opinion, the fishes are very few and altogether too fragmentary; besides, amphibians and reptiles are included. According to Young (1978) and Su Tetsao (1978), this fauna contains *Bogdania*, *Fukanglepis*, and *Fukangichthys*, corresponding respectively

Table 1--The continental Permo-Triassic sequence and vertebrate fauna in China.

Age	NORTHWEST CHINA — XINJIANG (Formation)	NORTHWEST CHINA — XINJIANG (Fauna)	SHAANGANNING BASIN (Formation)	SHAANGANNING BASIN (Fauna)	NORTH CHINA — SHANXI	HEIBEI PINGQUAN	HENAN JIYUAN	VERTEBRATE FAUNA
UPPER TRIAS.	Haojiagou F, Huangshanjie F	UPPER: Fukangichthys, Bogdania, Fukangolepis	Yanchang F; Tongchuan F	Hybodus, Pleiosauridae				Fukang fauna (VI)
MIDDLE TRIASSIC	Kelamayi F	LOWER: Sinoseminotus, Parotosaurus, Turfanosaurus, Vjushkovia, Parakannemeyeria	Ermaying F	UPPER: Capitosauridae, Shansisuchus, Shansiodon, Sinokannemeyeria, Parakannemeyeria; LOWER: Paoteodon, Ordosiodon, Ordosia, Shaambeikannemeyeria, Parakannemeyeria, Thecodontia, Procolophonidae	Capitosauridae, Neoprocolophon, Sinognathus, Shansisuchus, Fenhosuchus, Wangisuchus, Shansiodon, Sinokannemeyeria, Parakannemeyeria		Traversodontoides	Kannemeyerid fauna (V)
LOWER TRIASSIC	Shaofanggou F	Labyrinthodontia, Prolacertoides, Santaisaurus, Chasmatosaurus, Lystrosaurus, Urumchia	Heshanggou F	Ceratodus, Capitosauridae, Benthsuchidae, Procolophonia, Codontosauria, Fuguisuchus				Labyrinthodontes Procolophonid fauna (IV)
LOWER TRIASSIC	Jiucaiyuan F	Jimusaria, Dicynodon, Turfanodon, Striodon, Kunpania	Liujiagou F					Lystrosaurus (III)
UPPER PERMIAN	Lower Cangfanggou G		Sunjiagou F	Shihtienfenia, Shansisaurus, Tapinocephalidae		Dicynodontia		Dicynodonts–Pareiasaures fauna (II)
UPPER PERMIAN	Upper Jijicao G	Labyrinthodontia, Chichia, Stenotisaurus, Turfania, Tienshanicus	Upper Shihezi F				Labyrinthodontia, Bystrowiana, Tsiyuania, Honania, Tathangsharia, Wangnesaurus, Huangohocynodon	Turfania fauna / Jiyuan fauna (I)

Region / Series	China	South Africa	India	U.S.S.R.	Europe	South America	North America
UPPER TRIASSIC	Fukang fauna (VI)	Molteno Beds	Maleri	Zone VII	Keuper	Colorados F (Arg) Ischigualasto F (Arg)	Chinle Dockum
MIDDLE TRIASSIC	Kannemeyerid fauna (V)	*Cynognathus* Zone	Yerrapalli F	Zone VI		Santa Maria (Brazil) Chanares F (Arg) Cacheuta Gp	
LOWER TRIASSIC	Labyrinthodontes-Procolophonid fauna (IV) *Lystrosaurus* fauna (III)	*Procolophon* Zone *Lystrosaurus* Zone	Panchet F	Zone V	Bunter		Moenkopi
UPPER PERMIAN	Dicynodonts-Pareiasures fauna (II) *Turfania* (I) Jiyuan fauna fauna	*Cistecephalus* Zone *Endothiodon* Zone *Tapinocephalus* Zone	Bijori Beds	Zone IV Zone III Zone II Zone I			

Table 2--The correlation of the Permo-Triassic vertebrate fauna of China with the other parts of the world.

to related genera in the Dockum and Chinle of N. America, all Late Triassic. In N. China corresponding to the Fukang is the Tongchuan Fmn, which yields *Hybodus youngi* Liu (a middle or late Triassic fish) and Plesiosauria. Associated plants are identified as late Middle Triassic (Huang & Zhou, in press). I am inclined to assign the Tongchuan Fmn to latest Middle to early Late Triassic.

PERMO-TRIASSIC BOUNDARY (Upper Permian and Lower Triassic)

The *Striodon* skull discovered in the top of the Late Permian Wutonggou Fmn in Xiaolongkou, Zimusar, Xinjiang looks like *Dicynodon osborni* of S. Africa (Sun Ailing) and is latest Permian to earliest Triassic in age. This large dicynodon has lately been assigned by Sun to the latest Permian, in agreement with associated plant and invertebrate fossils. The basal part of the overlying Jiucaiyuan Fmn contains typical Early Triassic *Lystrosaurus*. These two formations are conformable and separated only by a thin arkose layer. The Wutonggou Fmn is composed of yellowish-green beds topped occasionally by variegated strata and overlain conformably by purplish-red Jiucaiyuan Fmn. The Permo-Triassic boundary should be located between the two.

LOWER AND MIDDLE TRIASSIC BOUNDARY

This is more distinctly marked in the Shaanxi-Kansu-Ningxia basin where the Heshanggou Fmn contains Early Triassic vertebrates, but the advanced *Fugusuchus* skull is late Early Triassic. The procolophonid

found at the top of the Heshanggou Fmn, being close to the S. African *Procolophon*, further indicates that this formation belongs to late Early Triassic. In Xilougou, Fugu County, also the procolophonid fauna occurs in the top of Heshanggou Fmn. Above the latter 30-40 m of sandstone underlies the Ermaying Fmn, with the Middle Triassic *Shaanbeikannemeyeria* fauna in its basal mudstone. These two faunas differ strikingly, though some fossils of each may occupy reciprocal horizons. Kannemeyerids never appear in the underlying Heshanggou Fmn. They most probably were derived by the Early Triassic from dicynodontids. Though the two formations are conformable, they are lithologically different, and the procolophonids above the boundary differ sharply from those below. Hence it is held here that the Lower-Middle Triassic boundary should be drawn between the Heshanggou and the Ermaying Fmns.

MIDDLE AND UPPER TRIASSIC BOUNDARY

This boundary is debatable: e.g., whether the Tongchuan Fmn belongs to late middle or early late Triassic is not yet settled.

Judging from evidence in Xinjiang, the lower Kelamayi Fmn contains the middle Triassic *Kannemeyeria* fauna, whereas the upper carries the late Triassic Fukang fauna; therefore the middle-upper Triassic boundary should be fixed between the middle and upper Kelamayi Fmn. However, much more work remains to be done before this boundary is well established.

CONCLUSIONS

Six Permo-Triassic vertebrate fossil horizons or faunas have been recognised in China during the past twenty years.

Our knowledge of Permo-Triassic vertebrates permits correlation with equivalent units in other parts of the world, especially with zones of the Karroo System. This is essential to the deciphering of the relationship between the southern and northern continents, and the time of their fragmentation.

Since the early sixties we have been working on the Permo-Triassic and its vertebrate faunas in the Shaanxi-Kansu-Ningxia basin and have discovered a new fossil horizon--the Heshanggou vertebrates. In recent years, excavations there by the Institute of Vertebrate Paleontology and Paleoanthropology have recovered numerous fossils. For the time being we propose to call them the Labyrinthodontes-Procolophonids fauna. The discovery of benthosuchid amphibians has widened the possibility of correlation of our faunas with those in Europe and other continents.

The kannemeyerids form an independent Triassic fauna and are of the same cosmopolitan distribution as the *Lystrosaurus*. They flourished in Middle Triassic times but were extinct by early Late Triassic.

REFERENCES

BONAPARTE, J. F. 1970. Annotated list of the South American Triassic tetrapods. Proc. 2nd Gondwana Symp., S. Africa. CSIRO: Pretoria.

CHENG ZHENGWU in press. Mesozoic stratigraphy and paleontology of the Shaan-Gan-Ning (Ordos) Basin. II--Part Vertebrata.

COLBERT, E. H. 1969. Gondwanaland and the distribution of Triassic tetrapods. In A. J. Amos (Ed.), Gondwana Stratigraphy. UNESCO: Paris. Pp. 355-374.

HUANG ZHIGAO; ZHOU HUIQUING in press. Mesozoic stratigraphy and paleontology of the Shaan-Gan-Ning (Ordos) Basin. I--Part Plant.

KEYSER, A.W. 1979. A new Dicynodont genus and its bearing on the origin of the Gondwana Triassic. 4th Gondwana Symp., India: Sec. III: 185-188.

KITCHING, A. W. 1977. The distribution of the Karroo vertebrate fauna. Mem. 1, Bernard Price Inst. Paleontol. Res. Univ. Witwatersrand, Johannesburg. Pp. 1-131.

LIU HSIENTING 1962. Two new species of Hybodus from N. Shensi. Vert. Palasiatica 6(2): 150-156.

---; WANG NIENCHUNG 1978. The Upper Permian fish fauna Z of Dzungari. Mem. Inst. Vert. Paleontol. Paleoanthropol. Acad. Sinica 13: 1-18.

ROMER, A. S. 1975. Intercontinental correlations of Triassic Gondwana vertebrate faunas. In K.S.W. Campbell (Ed.), Gondwana Geology. ANU Press: Canberra. Pp. 469-473.

SU TETSAO 1978. A new Triassic palaeoniscoid fish from Fukang, Sinkiang. Mem. Inst. Vert. Paleontol. Paleoanthropol. Acad. Sinica 13: 55-59.

SUN AILIN 1973. Permo-Triassic dicynodonts from Turfan, Sinkiang. Mem. Inst. Vert. Paleontol. Paleoanthropol. Acad. Sinica 10: 53-68.

--- 1978. Two new genera of Dicynodontidae. Mem. Inst. Vert. Paleontol. Paleoanthropol. Acad. Sinica 13: 19-25.

YANG ZUNYI; YIN HONGFU; LING HEMAO 1979. Marine Triassic fauna from Shihchienfeng Gp in the northern Weihe River Basin, Shaanxi Province. Acta Paleontol. Sinica 18(5): 465-4?

YIN HONGFU; LIN HEMAO 1979. Triassic marine sediments from Northern Shaanxi with note of the age of the Shihchienfeng Gp. Acta Stratig. Sinica 3(4): 233-241.

YOUNG, C. C. 1978. A Late Triassic vertebrate fauna from Fukang, Sinkiang. Mem. Inst. Vert. Paleontol. Paleoanthropol. Acad. Sinica 13: 60-67.

--- 1979. A new late Permian fauna of Jiyuan, Henan. Vert. Palasiatica 17(2): 99-113.

---; H. K. YEH 1963. On a new Pareiasaurus from the Upper Permian of Shansi, China. Vert. Palasiatica 7(3): 195-312.

---; P. L. YUAN 1934. On the occurrence of Lystrosaurus in Sinkiang. Bull. Geol. Soc. China 13: 575-580.

---; --- 1934. On the discovery of a new Dicynodon in Sinkiang. Bull. Geol. Soc. China 8: 564-574.

ZHANG FA KU 1975. A new thecodont lotosaur from Middle Triassic of Hunan. Vert. Palasiatica 13(3): 144-147.

Permo-Triassic boundary in the marine regime of South China

YANG ZUNYI
Beijing Graduate School of Wuhan College
of Geology, China

WU SHUNBAO & YANG FENGQING
Wuhan College of Geology, Hubei, China

During Late Permian to early Middle Triassic, marine and continental conditions of sedimentation existed in South China and North China, respectively. The Permo-Triassic boundary east of the Kangdian oldland in South China is shown in (1) maps of lithofacies for latest Permian (Changxinian) and earliest Triassic (Dayeian) times, and (2) maps showing the distribution of types and nature of lithologic contexts, for the same period. Four categories and 13 types of litho-contacts are given. The contacts vary from conformable to unconformable. On the basis of the history of transgression and regression, paleontology, especially the presence of a "mixed" Permo-Triassic fauna, and absence of any weathered crust, we hold that there is an extensive area of conformable Permo-Triassic sequences where the Permo-Triassic boundary is represented.

INTRODUCTION

The Permo-Triassic boundary problem has attracted particular attention for over a decade (Kummel 1972; Kummel & Teichert 1970; Logan & Hills 1973; Nakazawa *et al.* 1970, 1975; Newell 1973, 1978; Zhao *et al.* 1978; Chen 1978; Liao 1979), as it is the boundary between the Paleozoic and Mesozoic and is marked by significant organic extinctions and evolution.

In South China marine Permian and Triassic are extensive and in many places form continuous sequences. The marine regime of South China is separated from the northern continental regime roughly by the line joining Lianyun Harbour, Huanshi, Xiangfan, Tiansui and Xiling. This study is based chiefly on our field work in the area east of the Kangdian oldland, although reference is also made to the area west of it.

MARINE PERMIAN AND TRIASSIC

In order to facilitate the discussion, a general review of the development of systems concerned is given in Table 1.

LATE PERMIAN CHANGXINIAN DEPOSITS (Fig. 1)

The latest Permian deposits investigated here fall into three categories and seven types. The first category is chiefly clastics, mainly distributed on the eastern flank of the Kangdian oldland, forming the Xuanwei Fm, but also on the western flank of Cathaysia, making up the informally named "Dalong Fm."

It includes three different combinations: terrestrial clastics (Xuanwei, Yunnan); paralic clastics (Laowuji, Panxian of Guizhou and Daniaoshan, Zhenxiong of Yunnan);

marine clastics (Geding, Zhangping and Yanshi, Longyan of Fujian).

The second category is composed mainly of limestone facies (= Changxin Fm) and distributed in the area of eastern Sichuan, northern Guizhou, western Hubei and southern Shaanxi; the Jiangnan oldland; and a NE strip penetrating the siliceous facies.

The third category is dominated by siliceous rocks, siliceous limestone, and siliceous clastics (Dalaong Fm) and extends mainly through Guangxi, Hunan, southern Guizhou and northern Guangdong as well as in parts of Jiangxi, Hubei, Jiangsu and Anhuei. It includes chiefly siliceous rock (Lingling and Laiyang of Hunan); a combination of siliceous rock, cherty limestone and clastics (Laibing, Guangxi, and Mingyuexia, Guangyuan of Sichuan); and siliceous clastics with tuff intercalations (Liuzhou and Datang of Guangxi).

A seventh, combined, lithology occurs west of the Kangdian oldland, where one finds a great thickness of alternating clastics, carbonate rock, and slate together with much volcanic rock, indicating deposition in a mobile zone.

EARLY TRIASSIC DAYE DEPOSITS (Fig. 2)

Two categories of Early Triassic fall within our study. The first is dominated by clastic deposits divided into marine, argillaceous and silty rocks on both flanks of the Kangdian oldland--the Feixianguan or Kayitou Fm, on the west flank of Cathaysia--Changping, Datian (= Qikou Fm); the west flank of Yunkai oldland--Laibing, and Wuxuan (= Majiaoling Fm), and marine or paralic sandstone, siltstone and

Series	Stage	Chief units and facies

MARINE TRIASSIC GENERAL FEATURES

Series	Stage	Chief units and facies
UPPER TRIASSIC	Rhaetian / Norian / Carnian	E: Anyuan Group: continental coal-bearing with marine intercalations W: Lower part (Maantang Fm, Xiaotangzi Fm) mainly marine clastics; upper part (Xujiahe Fm) continental coal-bearing
MIDDLE TRIASSIC	Ladinian	Large-scale regression, leaving no deposits in most areas. In W: marine represented by Tianjingshan Fm
	Anisian	Luikoubo Fm: Marine calcareous strata Badong Fm: Marine calcareous and clastic strata Huangmaqing Fm (lower part): terrestrial clastics
LOWER TRIASSIC	Olenikian	Jialingjiang Fm, Yongningchen Fm, Upper Qinglong Fm: all marine calcareous strata
	Induan	Daye Fm: marine calcareous rocks Feixianguan Fm: marine clastics Yelang Fm: calcareous and clastic transitional facies

PERMIAN GENERAL FEATURES

Series	Stage	Chief units and facies
UPPER PERMIAN	Changxin	Changxin Fm: mostly marine carbonates Dalong Fm: marine siliceous rock, siliceous limestone and siliceous clastics Xuanwei Fm: alternating marine and terrestrial deposits or terrestrial clastics
	Longtan	Longtan Fm: paralic coal-bearing Wujiaping Fm: transitional strata and marine carbonates
LOWER PERMIAN	Maokou	Maokou Fm: marine carbonates Danchong Fm: marine siliceous rocks, siliceous clastics with transitional deposits
	Qixia	Qixia Fm: marine calcareous rock; locally, coal-bearing clastics at base = Liangshan Fm

Table 1--General features of the Permian and Marine Triassic in South China.

mudstone, transgressing on terrestrial beds of diverse ages (= Xia Huanchang Fm) of S. Qilianshan and Liuchiakou and Heshangkou Fms of Linyou, Shaanxi).

The second chiefly contains widely-distributed calcareous mudstone and argillaceous limestone. It comprises calcareous mudstone, marl and argillaceous limestone-- often transitional with clastics (Nanchang, Zigong, Luzhou of Sichuan and the area west of the Xuefeng oldland (Yelang Fm); argillaceous limestone, limestone intercalated with calcareous mudstone with wide distribution (the Lower Qinglong Fm of Jiangsu, Anhui, and the Daye Fm of Hubei, Sichuan, Hunan, and Guangdong); mainly argillaceous limestone and limestone (southern Guangxi (=Lolou Fm)); and calcareous

mudstone, argillaceous limestone and limestone intercalated with tuff (NE of Guilin)

West of the Kangdian oldland corresponding deposits consist chiefly of sandstone, slate and limestone of great thickness, which are flysch deposits of the mobile zone.

LITHOLOGIC MARINE PERMIAN AND TRIASSIC CONTACTS (Fig. 3)

Here, four categories and 13 types may be discerned (see Table 2). The most important are the contact between the siliceous facies (Dalong Fm) and the argillo-calcareous facies (Daye Fm), and the contact between the chert-nodular limestone facies (Changxin Fm) and the argillo-calcareous facies (Daye Fm). The former

Fig.1 我国南方海相晚二叠世晚期(长兴期)岩相略图
Sketch map of lithofacies for the marine Late Permian
(Changxin age) of South China

Fig. 1--Lithofacies for the marine Late Permian (Changxin age) of South China. A = Cathaysia; B = Kandiania; C = Yunkainia; D = Jiannia; E = North China land area.

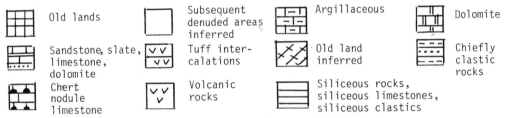

Old lands

Sandstone, slate, limestone, dolomite

Chert nodule limestone

Subsequent denuded areas inferred

Tuff inter-calations

Volcanic rocks

Argillaceous

Old land inferred

Siliceous rocks, siliceous limestones, siliceous clastics

Dolomite

Chiefly clastic rocks

occurs mainly in the central portion of the study area, and the latter is in eastern Sichuan, western Hubei, northern Guizhou and southern Shaanxi as well as around the Jiangnan oldland.

The contact between the Permian and the Triassic has generally been held to be disconformable or at most paraconformable. However, after recent research, both lithological and paleontological, the Permo-Triassic contact, in our opinion, is conformable over much of the area under investigation. It may be noticed that in Table 2, the category II of contact is generally conformable, and the category III is conformable except for a few cases of disconformity (e.g., Xiangxi of western Hubei, Zhenan of northern Sichuan and central Giangxi).

Unconformities exist chiefly around the Kangdian oldland and the southern margin of the Shaan-Gan-Ning oldland, where the lower Triassic overlaps terrestrial Per-

mian. Major unconformities between the two systems occur in the mobile belt west of the Kangdian oldland (Table 2).

EVIDENCE FOR CONTINUOUS SEDIMENTATION

Late Permian marine transgression. The early part of the Early Permian was marked by an extensive transgression, and the early Late Permian (or, locally, late Early Permian) witnessed a regression when the greater part of S. China became land and coal-bearing strata were deposited. The area of epicontinental seas of the world "decreases from approximately 43% in early Permian to 13% in latest Permian and then increases to 34% in the early Triassic" (Schopf 1974), but in S. China the latest Permian sea still held sway and was succeeded by an equal or greater Triassic transgression, which is very helpful to the solution of the Permo-Triassic boundary in S. China.

During the latest Permian there was a

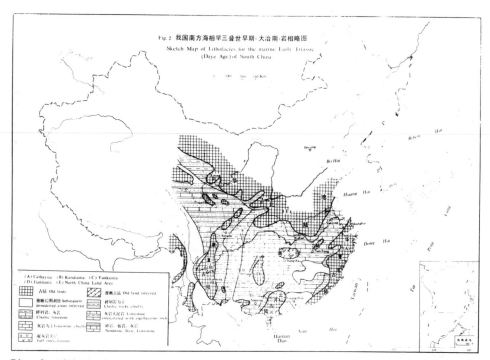

Fig. 2 我国南方海相早三叠世早期（大治期）岩相略图
Sketch Map of Lithofacies for the marine Early Triassic
(Daye Age) of South China

Fig. 2--Lithofacies for the marine Early Triassic (Daye age) of South China.
A - E as in Fig. 1.

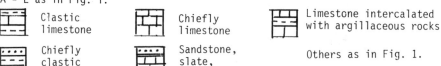

Clastic limestone

Chiefly limestone

Limestone intercalated with argillaceous rocks

Chiefly clastic rocks

Sandstone, slate, limestone

Others as in Fig. 1.

distinctive and fairly stable sedimentation pattern in S. China. A carbonate platform chiefly occupied eastern Sichuan, western Hubei, northern Guizhou, southern Shaanxi and the surrounding districts of the Jiangnan oldland, whereas a relatively stagnant sea--where siliceous rock, siliceous mudstone, and siliceous limestone were formed--mainly covered the central portion of our area.

A clastic facies is distributed chiefly around the oldlands. The same sedimentation pattern persisted with little change until the end of the Permian. The distribution of fossil assemblages is fairly regular. Benthic invertebrates such as fusulinids, brachiopods, and bivalves are typical of the carbonate platform, while such pelagic forms as ammonoids, pectens, and small brachiopods are typical of the relatively stagnant sea. Ammonoids occur occasionally in carbonate rocks. Note that both ammonoids (*Rotodiscoceras, Pseudotirolites*) and fusulinids (*Palaeofusulina*) marking the highest Permian horizon, are distributed over almost

the whole area concerned, indicating no large-scale regression at the end of the Permian.

A significant siliceous component in the sediments throughout the Permian is probably related to volcanic activity. On the carbonate platform the amount of silica relative to lime was sufficient only to form nodules in limestones. On the other hand, in the relatively stagnant sea, inhabited by few benthic invertebrates and supplied with little continental carbonate, siliceous material was the chief constituent of the deposits. Further, as the supply of siliceous material was restricted, it formed only thin massive or thin-bedded layers (only 2-10 m thick at Nandan in Guangxi and 10-30 m at Lianyuan and Shaoyang in Hunan). The thickness increases where clastics and carbonates are mixed with the silica. It is generally agreed from geologic mapping that the calcareous facies (Changxin Fm) and the siliceous facies (Dalong Fm) are of the same age.

The sedimentation pattern was altered by

74

Fig. 3. 我国南方海相二叠—三叠系接触关系略图
Sketch map showing marine Permo-Triassic Contacts.

Fig. 3--Marine Permo-Triassic contacts. A - E as in Fig. 1.

Areas without Periassic deposits | Mainly conformable | Conformable or unconformable | Mainly un-conformable

I Type of lithologic contact

further transgression at the beginning of the Triassic. Deposition became more uniform with predominant marl facies (Daye Fm), and the fauna assumed a new aspect.

Paleontologic evidence. The highest Permian horizon is marked by ammonoids (*Rotodiscoceras, Pseudotirolites, Pleurondoceras*) in the Dalong Fm (siliceous facies) (Zhao *et al.* 1978) and fusulinids (*Palaeofusulina sinensis*) in the Changxin Fm (calcareous facies). These taxa are so widespread over S. China that they prove the complete development of the uppermost Permian here.

The ammonoid *Otoceras woodwardi* zone conventionally marks the very beginning of the Triassic. In S. China it is rare, known only at two locations (Longtan, Jiangsu, and Changxin, Zhejiang). Conversely, *Ophiceras, Claraia wangi,* and *C. griesbachi* are fairly abundant. The argillo-calcareous rocks contain mostly ammonoids, and the clastic facies bivalves. As these earliest fossils occur extensively, we must conclude that a great part of the area under study was covered by the earliest Triassic deposits.

The so-called "transitional beds" have been reported from a number of places in S. China. Beds with typical early Triassic *Ophiceras* and *Claraia* often have relict Permian brachiopods--eight genera and 11 species from 14 localities in 11 provinces (Liao 1979). They have been spotted by our team at several other places, represented mainly by *Crurithyris, Fusichonetes* and various other Paleozoic relict genera such as *Bellerophon* (gastropod) and *Towapteria* (bivalve) in the Xiajaichao section. Others reported are *Pteria* (bivalve) in the Huopu section, Panxian, and *Hollinella* (ostracod) in western Guizhou. The earliest Triassic *Otoceras* itself may be regarded as a Paleozoic ammonoid relict. The fact that a number of Paleozoic elements persisted (owing to their adaptability as late as Early Triassic) serves to verify the transition between the Permian and the Triassic within the greater part of the area under investigation. However, as there are marked differences between the fossil assemblages of the two systems, the boundary between them is fairly conspicuous.

75

If there was practically continuous deposition between the Permian and Triassic, how can we explain the marked differences in biotic composition between the two? What caused such Permo-Triassic exinctions as, e.g., trilobites, fusulinids, and rugose corals? Among various hypotheses proposed, we are inclined to accept sea-floor spreading, which controls, in general, marine transgression and regression, and thus exerts great influence on the waxing and waning of marine life. Though the area of shallow marine seas in S. China did not decrease conspicuously in the Late Permian, the variety of habitats decreased sharply in the rest of the world, causing great stress to various groups of organisms. The Early Triassic transgression changed Late Permian marine environments to such an extent that a number of new taxa appeared in abundance. Some Permian forms which could adapt to the new environments held over.

Lack of diastrophism between the Permian and Triassic. Except for the mobile zone lying west of the Kangdian oldland where volcanism was occurring, the vast platform east of the same oldland was comparatively quiet during the Permian, and not until the Middle Triassic (Ladinian) did extensive regression occur.

No trace of weathering, either fossil soil or erosional surface, has been found in the Permo-Triassic sections we regard as being conformable.

In many sections one or more clay-layer averaging 10 cm each occur near the Permo-Triassic boundary. They have been regarded as products of weathering, and the Permo-Triassic contact as a disconformity.

Category & type		Latest Permian Changxinian lithology	Earliest Triassic Daye lithology	Nature of contact	Type localities
I	I₁	Terrestrial clastics	Marine clastics	Conformable or disconformable	Fuyuan, Xuanwei of Yunnan Houpu, Panxian of Guizhou
	I₂	Paralic clastics	Marine clastics	Conformable or disconformable	Daniaoshan, Zhenxiong of Yunnan Laowuji, Panxian of Guizhou
	I₃	Marine clastics	Marine clastics	Conformable	Changding, Longyan of Fujian
II	II₁	Siliceous rock chiefly	Mudstone, argillaceous limestone, limestone	Conformable	Laiyang, Lingling of Hunan
	II₂	Siliceous rock, clastics, limestone	Mudstone, argillaceous limestone, limestone	Conformable	Laiping, Guangxi, Mingyuexia, Guanyuan of Sichuan, Sanguanying, Panxian of Guizhou
	II₃	Siliceous rock and limestone	Clastics chiefly	Conformable	Qingzhen, Guiyang &c. of Guizhou
	II₄	Siliceous rock, siliceous limestone	Mudstone, argillaceous limestone, limestone	Conformable	Chaoxian, Anhui; extensively distributed in Guangxi, Hunan, Guizhou and Guangdong
	II₅	Siliceous rock, siliceous mudstone intercalated with tuff	Mudstone, argillaceous limestone, limestone	Conformable	Liuzhou, Heshan of Guangxi
III	III₁	Chert-nodular limestone chiefly	Argillaceous limestone, limestone, mudstone	Conformable; locally disconformable	Xiejiachao, Chuxian of Sichuan; Changxin, Zhejiang, and extensively distributed in E. Sichuan N. Guizhou, W. Hubei, S. Shaanxi
	III₂	Chert, nodular limestone, argillaceous mudstone	Mudstone, argillaceous limestone	Conformable	Xixian, Shaanxi
IV	IV₁	Terrestrial clastics or lacking	Overlapping marine clastics or marine intercalations	Disconformable	Tianjun, Qinghai and Linyou, Shaanxi
	IV₂	Extrusives	Sandstone, slate + limestone	Unconformable	Markang, W. Sichuan and parts of Yunnan
	IV₃	Dolomite, limestone sandstone slate		Unconformable Conformable	Western Qinling

Table 2--Nature and types of marine Permian-Triassic contacts in South China.

Recent examination of them disclosed as many as 10 layers, each less than 10 cm thick, in the section at Chaoxian, Auhui, where they are intercalated within siliceous rocks or siliceous clastics. The clay layers are uniformly bedded with distinct bedding planes and homogeneous composition, and without unevenly weathered surfaces or residual soil. Petrographically they are products of marine sedimentation as seen in one of the Guizhou sections, which contains well-preserved cephalopods. Judging from these features the clay layers are normally-deposited beds resulting from intermittent supply of clay particles under particular conditions and are not products of weathering.

CONCLUSIONS

This discussion shows that in S. China the Permian and Triassic form continuous sequences, although there is a series of different lithologic contacts between the two. Many sections are normal stratigraphic sequences with prolific fossils, which provide abundant data for biostratigraphic research. Consequently, S. China could provide a suitable section as a candidate for the Permo-Triassic boundary stratotype. Our working group on the Permian-Triassic boundary has been studying toward this end.

ACKNOWLEDGMENTS

Participants in the study of the Permo-Triassic boundary problem include Li Zhiming, Yin Hongfu, Niezetung, Ting Meihua and Liu Jinhua.

REFERENCES

CHEN CHUZHEN 1978. The lower boundary of the Triassic in Southwest China. Acta Stratig. 2(2): 160-162.
KUMMEL, BERNHARD 1972. The lower Triassic (Scythian) ammonoid Otoceras. Bull. Mus. Comp. Zoo. 143(6): 365-417.
---; C. TEICHERT (Eds.) 1970. Stratigraphic boundary problems: Permian and Triassic of West Pakistan. Univ. Kansas Press: Lawrence. 474 pp.
LIAO ZHOTING 1979. Brachiopod assemblage zones of Changxin stage and brachiopods of the mixed fauna in S. China. Acta Stratig. 3(3).
LOGAN, A.; L. V. HILLS (Eds.) 1973. The Permian and Triassic systems and their mutual boundary. Mem. 2, Canad. Soc. Pet. Geol. Calgary. 766 pp.
NAKAZAWA, I.; I. ISHII; Y. BANDO; T. MAEGOYA; D. SHIMIZU; Y. NOGAMI; T. TOKUOKA; S. NOHIDA. 1970. Preliminary report on the Permo-Trias of Kashmir. Mem. Fac. Sci. Kyoto Univ., Ser. Geol. Mineral. 37: 163-72.
NAKAZAWA, K.; H. M. KAPOOR; K. ISHII; Y. BANDO; Y. OKIMURA; T. TOKUOKA; et al. 1975. The Upper Permian and the Lower Triassic in Kashmir, India. Mem. Fac. Sci. Kyoto Univ., Ser. Geol. Mineral. 42(1): 1-106.
NEWELL, N. D. 1962. Paleontological gaps and geochronology. J. Paleontol. 36: 592-610.
--- 1973. The very last moment of the Paleozoic Era. In A. Logan, L. V. Hills (Eds.).
--- 1978. The search for a Paleozoic-Mesozoic boundary stratotype. Oesterreich. Akad. Wiss. Schrift. Erdwiss. Komm. 4: 9-19.
SCHOPF, T.J.M. 1974. Permo-Triassic extinctions: Effects of area on biotic equilibrium. J. Geol. 82: 129-143.
ZHAO JINKO; LIANG XILAO; ZHENG ZHUOGUAN 1978. Late Permian cephalopods of S. China. Palaeontol. Sinica 154, ser. B, 12: 166-194.

Permian paleogeography of Peninsular and Himalayan India and the relationship with the Tethyan region

J. M. DICKINS
Bureau of Mineral Resources, Canberra, Australia

S. C. SHAH
Geological Survey of India, Calcutta

The *Glossopteris* flora characteristic of the non-marine sequences of Peninsular India is found in the Lesser and Tethyan Himalayas and in southern Tibet north of the Himalayas.

The *Eurydesma* and younger Early Permian faunas of Peninsular India are found in the Lesser Himalayas (including the Salt Range), in the Tethyan Himalayas and on the northern slopes of the Himalayas (southern Tibet). They are cold and temperate water faunas. In the Lesser and Tethyan Himalayas and in southern Tibet they are overlain by a warm water Late Permian fauna.

During the Permian, Peninsular India, the Lesser and Himalayan India and southern Tibet are considered to have formed a single block or plate.

INTRODUCTION

A comparison of the marine faunas of India and Western Australia (Dickins & Shah 1979) gave paleogeographic conclusions for the Himalayas and adjacent regions during the Permian which are the subject of the present paper.

The Himalayas have been divided into four subparallel belts from south to north (Gansser 1964), the Sub-Himalayas, Lesser Himalayas, Great Himalayas, and Tethyan Himalayas. Late Paleozoic rocks have not been identified in the Sub-Himalayas or the Great Himalayas.

In Gondwanan reconstructions, Peninsular India is generally considered to have been separated by a wide ocean from the northern part of the present Himalayas in the Late Paleozoic and Early Mesozoic, but recently some doubt about this has arisen. Hsü (1978) postulated a boundary between the Indian and Eurasian Plates in southern Tibet. Crawford (1974) placed a plate boundary north of Tibet. Pal and Chaloner (1979) discussed a boundary along the Tien Shan-Mongol Mountain Belts to Vladivostok.

NON-MARINE FOSSILS

Peninsular India is the typical area for the continental deposits of Gondwana. Its sequence of fossil floras is well known and has been reviewed recently by Shah *et al.* (1971).

Continental Permian with a *Glossopteris* flora like that of the Raniganj Fmn occurs in the Eastern Himalaya at Darjeeling, Bhutan and Arunachal Pradesh (Acharyya *et al.* 1975). The Lesser Hima-

laya on the western side have non-marine beds only at the Salt Range in the Early Permian *Glossopteris* beds.

In Kashmir in the Tethyan Himalayas plant fossils occur at five early Artinskian to early Kungurian horizons, including *Gangamopteris, Glossopteris, Vertebraria, Psygmophyllum, Rhaebdotaenia, Lepidostrobus* and lycopods (Kapoor and Shah 1979). The flora has mixed affinities but is most closely related to the floras of Peninsular India and the Salt Range. A *Gangamopteris* flora has been recorded in the Aghil Range (Norin, in Gansser 1964). Farther to the north *Lystrosaurus* fauna similar to that of South Africa and Peninsular India occurs in the foothills óf the Tien Shan Range (Sun 1973).

In Tibet the Qubu Fmn has yielded *Glossopteris communis, G. indica, G. angustifolia, Sphenophyllum speciosum, Raniganjia qubuensis* and *Dichompteris qubuensis*, a flora similar to that of the Raniganj Formation of Peninsular India (Hsü 1978).

The 700 m thick conformably underlying Jilong Fmn has yielded a *Stepanoviella* fauna of late Sakmarian to early Artinskian age--comparable with the Umaria marine beds or the *Eurydesma* fauna of Peninsular India (Guo 1976; Ching Yukan *et al.* 1977).

MARINE BIOTA

In the Gondwana Basins in Peninsular India marine fossils are found at Manendragarh, Rajhara, Umaria and Badhaura (Fig. 1). These faunas were fully reviewed by Dickins and Shah (1979).

In the Lesser Himalayas the best Lower Permian faunas so far known are from the

Fig.1. Peninsular India and
Himalayas. Freshwater sediment:
cross-hatched; marine - solid.
A. Early Permian marine trans-
gression. NB. South of the
Himalayas the sea has generally
been considered to have pene-
trated along arms leaving much
of northern India above sea.
B. Late Permian transgression.
C. Transgression of latest
Permian.

05/93

Salt Range, the Garhwal Himalayas,
Arunachal Pradesh and Sikkim.

In the Salt Range the oldest, the
Eurydesma fauna, occurs in the Boulder
Beds and the Olive Beds (Waagen 1891; Reid
1936) which apparently correspond to the
Dandot Fmn of the Nilawahan Gp (Fatmi
1974). The fauna needs revision but
closely resembles the equivalent faunas of
Peninsular India and is of Asselian to
early Sakmarian (Tastubian) age, with
Eurydesma mytiloides (Reed) and species of
Ambikella and *Cyrtella* (or *Pseudosyrinx*).
Brachiopods are particularly well repre-
sented in the Amb Fmn (Lower Productus
Limestone) and are closely related to those
from Badhaura, Peninsular India, and from
the Agglomeratic Slate of Kashmir. The
Wargal Fmn (Middle Productus Limestone) is
Late Permian (not older than Kazanian), and
overlies the Amb Fmn with a significant
stratigraphic hiatus. The same hiatus can
be recognised at places in the Lesser and
Tethyan Himalaya regions and probably in
Iran (Sestini 1966) and southern Arabia
(Dickins, pers. observ.).

The large, Early Permian, Subansiri,
Arunachal Pradesh, fauna including brachiopods identified as *Stepanoviella,
Subansiria* and *Cyrtella,* pelecypods, gastropods, bryozoans, conulariids and crinoids (Dickins & Shah 1979) is closely related to the faunas from Badhaura, Peninsular India and the Himalayas including the
Agglomeratic Slate of Kashmir.

The better-known South Sikkim fauna
(Sahni & Srivastava 1956) contains *Euryde*
mytiloides (Reed), *Ambikella* and *Keeneia*
and is closely related to the Asselian-Ta
tubian faunas of Peninsular India.

An early Permian fauna from the Garhwal
Himalaya has strong relationships with the
Peninsular fauna and with other Himalayan
faunas of the same age (Waterhouse & Gupta
1978, 1979).

Distinctive Upper Permian faunas in par
representing the *Lamnimargus himalayensis*
zone of Waterhouse and Gupta (1979) and
Waterhouse (1978), occur in the Lesser
Himalayas and in the Tethyan Himalayas.
Elements of it are found in Iran, Malaysia
Indonesia, and Japan. Abundant fossils
in the Salt Range Wargal and Chhidru Fmns
include characteristic species of the
brachiopods *Aulosteges, Callispirina,
Cleiothyridina, Hemiptychina, Linoproduc-
tus, Lyttonia, Neospirifer, Richthofenia,
Waagenoconcha, Whitspakia* and many more
(Grant 1970). Some other invertebrate
groups are well represented. Ammonoids
and fusulinids are sporadic. The age
can be taken as Guadalupian (in part
Kazanian) and may extend into the Dzhul-
fian. Warm water is indicated by the
high diversity, the genera present,
colonial corals, the types of algae,
fusulinids, and other criteria. It is a
part of the Tethyan faunal province
stretching from Europe across the Middle
East and middle and southern Asia, China,

80

	STAGES OF WORLD SCALE	PENINSULAR INDIA	LESSER HIMALAYAS	SALT RANGE PAKISTAN	KASHMIR	TETHYAN HIMALAYAS AND TIBET
UPPER PERMIAN	DORASHAMIAN	PANCHET FM		KATHWAI MBR OF MIANWALI FM —?—	KHUNAMUH FM (EI)	—?—
	DZHULFIAN	NIDHPUR BEDS	—?—	—?— CHHIDRU FM	ZEWAN FM (DIV IV)	NEPAL, LADAKH CHAMBA, SPITI
	'KHACHIK'	RANIGANJ FM	GARHWAL	WARGAL FM (KALABAGH MBR AT TOP)	ZEWAN FM (DIV I-III)	KUMAON, NORTH SIKKIM
	'GNISHIK' (≡ KAZANIAN) —?—		BHUTAN			
LOWER PERMIAN	KUNGURIAN	KULTI SH				
	BAIGENDZHINIAN			AMB FORMATION		
	ARTINSKIAN) AKTASTINIAN				—?—	
	STERLITAMAKIAN SAKMARIAN) TASTUBIAN	BARAKAR FM KARHARBARI FM TALCHIR FM UMARIA BADHAURA	SUBANSIRI GARHWAL SOUTH SIKKIM	DANDOT FM	AGGLOMERATIC SLATE	MT EVEREST CHAMBA, LADAKH AND SPITI
	ASSELIAN	RAJHARA & MANENDRAGARH				

05/94

Fig.2.   Correlation chart.

Japan, Indonesia and probably to northern Australia (Thomas 1957).

In the Tethyan Himalayas, the *Eurydesma* fauna is well represented in the Agglomeratic Slate of Kashmir (Kapoor & Shah 1979) and is also present in the Chamba Synclinorium (Kapoor 1973), Ladakh (Gupta 1978; Acharyya & Shah 1975) and Spiti (Srikantia *et al*. 1978), with *Eurydesma*, *Deltopecten*, the brachiopods *Trigonotreta* (or *Brachythyrinella*), *Ambikella*, *Cyrtella* and *Taeniothaerus*. An Asselian to late Sakmarian age is indicated.   The faunas have close relationships with Peninsular India and the Lesser Himalayas, and indicate cold to temperate water conditions in contrast to overlying warmer water faunas.

The Upper Permian fauna in the Zewan Formation of Kashmir, in the Tethyan Himalayas, contains many of the characteristic Salt Range species.   Amongst them are the foraminifer *Coloniella*,   the brachiopods *Leptodus*, *Costiferina*, *Cancrinella*, *Waagenoconcha*, *Spiriferellina*, *Spirigerella*, *Stenocisma*, *Whitspakia*, *Marginifera* (or *Lammimargus*), *Neospirifer*, *Cleothyridina* and *Spinomarginifera* and the ammonoid *Cyclolobus* (Nakazawa *et al*. 1975).

The fauna in Nepal is similar to that in Kashmir and the Salt Range as are less well-known faunas in Ladakh, the Chamba Synclinorium, Spiti, Kumaon Himalaya and North Sikkim with *Lammimargus* present at all localities (Dickins & Shah 1979; Waterhouse 1978).   Although the Salt Range element can be recognised the nature of the Permo-Triassic transition is not altogether clear.

Diener (1903) described Upper Permian fossils from Chitichun No.1 in southern Tibet similar to other Himalayan fossils elsewhere, especially the Salt Range. Recent Chinese work in southern Tibet has recorded not only Upper Permian faunas (Zhang S. Xin & Chin Yukang 1976; Mu ·et *al*. 1973) but also a cold to temperate fauna probably entirely Sakmarian (Ching Yukan *et al*. 1977) (see also Waterhouse & Gupta 1979) from the northern slopes of Mount Everest (Mount Qomolangma).   This fauna is close to that of the Umaria beds of Peninsular India.

CONCLUSIONS

Paleontological studies now show that the Early Permian cold water (*Eurydesma* fauna) and temperate faunas of Peninsular India occur in two structural units of the Himalayas as well.   At the species level the faunas have taxonomic elements in common which distinguish them from faunas of similar age in other Gondwana countries. The Lower and Upper Permian of the two regions contain matching changes of faunas related to changes in water temperature. This suggests that the sequences which contain the faunas were laid down in a single or several connected basins with similar water temperature, which allowed

81

free and rapid migration. The same was also concluded for the Early Permian by Waterhouse & Gupta (1979).

In the Permian the Himalayas and S. Tibet apparently lay against the peninsula and flanked the southern margin of the Tethys, the sea episodically transgressing far over the Indian Shield (or Plate) during major transgressions. The sea penetrated farther during the Early Permian. During the Late Permian it penetrated only as far as the Lesser Himalayas (Fig. 1).

Early Permian faunal data from Iran (Sestini 1966) and possibly Afghanistan and southern China (Waterhouse & Gupta 1977, 1979) suggest that these areas were also part of the same sea. The Late Permian data show the same relationship even more strongly.

The fossil flora and fauna both suggest that Peninsular India, Lesser and Tethyan Himalayas were parts of the same block during the Permian and that the boundary of this block lay north of the Tethyan Himalayas and southern Tibet (Hsü 1978). Structural work also points to the proximity of India and Tibet (Crawford 1974; Ray & Acharyya 1976), as do geophysical data on seismicity and deep seismic soundings in the Himalayan region (Kaila & Hari Narain 1976) and paleomagnetic data (Pal & Bhimsankaran 1977, p. 394). The paleontological information, however, is even more far reaching in indicating not only that India and Tibet were adjacent in the Permian but also that they were near to other parts of southern and central Asia. Paleomagnetic and floral data from the Lower Carboniferous point to the same conclusion (Pal & Chaloner 1978).

ACKNOWLEDGMENTS

We are grateful to the Geological Survey of India and the Bureau of Mineral Resources for the arrangements which have made this joint work possible. We are also grateful to many colleagues in a number of countries whose work has contributed to the understanding of the problems discussed. Publication is by permission of the Director-General of the Geological Survey of India and the Director of the Bureau of Mineral Resources. The figures were drafted by Gwen Bates and Rosa Fabbo and the manuscript was typed by Pat Porter.

REFERENCES

ACHARYYA, S. K.; S. C. GHOSH; R. N. GHOSH; S. C. SHAH 1975. The continental Gondwana Group and associated marine sequences of Arunachal Pradesh. Himalayan Geol. 5: 60-82.

ACHARYYA, S. K.; S. C. SHAH 1975. Biostratigraphy of the marine fauna associated with the diamictites of the Himalayas. Bull. Indian Geol. Assoc. 8(2): 9-23.

CHING YUKAN; LIANG XILUO; WEN SHIHHSUAN 1977. Additional material of animal fossils from the Permian deposits on the northern slope of Mount Qomolangma Feng. Sci. Geol. Sin. 7(3): 236-249.

CRAWFORD, A. R. 1974. The Indus suture line, the Himalayas, Tibet and Gondwanaland. Geol. Mag. 111(5): 369-380.

DICKINS, J. M. in press. Late Palaeozoic climate - with special reference to the invertebrate faunas. Comptes Rendus Ninth International Congress of Carboniferous Stratigraphy and Geology, Washington and Urbana, 1979.

---; S. C. SHAH 1979. Correlation of the marine Permian sequence of India and Western Australia. Fourth International Gondwana Symposium, Calcutta, 1977.

DIENER, C. 1903. Permian fossils of the Central Himalayas. Palaeontol. Indica, ser. 15, 1(2).

FATMI, A. N. 1974. Lithostratigraphic units of Kohat-Potwar Province, Indus Basin Pakistan. Mem. Geol. Soc. Pakistan 10.

GANSSER, A. 1964. Geology of the Himalayas. New York Interscience Publishers John Wiley and Sons.

GRANT, R. E. 1970. Brachiopods from Permian-Triassic boundary beds and age of Chhidru Formation, West Pakistan. In B. Kummel C. Teichert (Eds.), Stratigraphic boundary problems. Univ. Press, Kansas. Pp. 117-1

GUO SHUANGXING 1976. Plant fossils from Rikaze Group of the Jolmo Lungma Region. Report of the Scientific Expedition in the Mount Jolmo Lungma Region 1966-1968, Palaeontology: 411-424

GUPTA, V. J. 1973. Indian Palaeozoic stratigraphy. Delhi: Hindustan Publishing Corporation (India).

HSÜ, J. 1978. On the botanical evidence for continental drift and Himalayan uplift Palaeobotanist 25: 131-145.

KAILA, K. L.; HARI NARAIN 1976. Evolution of the Himalayas based on seismotectonics and deep seismic soundings. Preprints Section II; Structure, Tectonics, Seismicity and Evolution. Himalayan Geology Seminar, New Delhi 1-30.

KAPOOR, H. M. 1973. On the stratigraphy of Bhadarwah and Bhallesh, J. & K. J. Palaeontol. Soc. India: 55-66.

---; S. C. SHAH 1979. Lower Permian in Kashmir Himalaya; a discussion. Section 1: Geology, Stratigraphy and Palaeontology, Himalayan Geology Seminar. Geol. Surv. India, Misc. Publ. 41.

MU, AN-TZE; WEN SHIH-HSUAN; WANG YI-KANG; CHANG PING-KAO 1973. Stratigraphy of the Mount Jolmo Lungma Region in Southern Tibet. Sci. Sin. 16(1): 1-111.

NAKAZAWA, K.; H.M. KAPOOR; K. ISHII; Y. BANDO; T. OKIMURA; T. TOKUOKA. 1975. The Upper Permian and the Lower Triassic in Kashmir, India. Mem. Fac. Sci., Kyoto, Geol. Mineral. Ser. 42(1).

PAL, A. K.; W. G. CHALONER 1979. A Lower Carboniferous Lepidodendropsis flora in Kashmir. Nature 281: 295-297.

PAL, P. C.; F. L. S. BHIMSANKARAN 1977. Origin of the Himalaya - an appraisal of the palaeomagnetic data. Himalayan Geol. 7: 379-398.

RAY, K. K.; S. K. ACHARYYA 1976. Conceal-ed Mesozoic-Cenozoic Alpine Himalayan Geosyncline and its petroleum possibilities. AAPG Bull. 60(5): 794-808.

REED, F. R. C. 1936. Some fossils from the Eurydesma and Conularia Beds (Punjabian) of the Salt Range. Palaeontol. India, n.s. 23(1).

SAHNI, M. R.; J. P. SRIVASTAVA 1956. Dis-covery of Eurydesma and Conularia in the eastern Himalayas and description of asso-ciated faunas. J. Palaeontol. Soc. India 1(1): 202-214.

SESTINI, N. F. 1966. The geology of the Upper Djadjerud and Lars Valleys (North Iran) II. Palaeontology, brachiopods from Geirud Formation Member D (Lower Permian). Riv. Ital. Paleontol. Stratigr. 72(1): 9-50.

SHAH, S. C.; COPAL SINGH; M.V.A. SASTRY 1971. Biostratigaphic classification of Indian Gondwana. Int. Symp. Stratig. Min. Res. Gondwana System, Aligarh. Pp. 306-321.

SRIKANTIA, S. V.; O. N. BHARGAVA; H. M. KAPOOR 1978. A note on the occurrence of the Eurydesma and Deltopecten assemblage from the Kulip formation (Permian) Baralacha Baralacha Ben Area, Lahaul Valley, Himachal Himalayas. J. Geol. Soc. India 19: 73-75.

SUN AI-LIN 1973. Permo-Triassic reptiles of Sinkiang. Sci. Sin. 16(1): 152-156.

THOMAS, G. A. 1957. Odlhaminid brachio-pods in the Permian of Northern Australia. J. Palaeontol. Soc. India 2: 174-182.

WAAGEN, W. 1891. Salt Range fossils. Palaeontol. India, Ser. 13, 4(2): 89-242.

WATERHOUSE, J. B. 1978. Permian Brachio-poda and Mollusca from North-west Nepal. Palaeontogr., Abt A: 160 (1-6): 1-175.

---; V. J. GUPTA 1977. Permian faunal zones and correlations of the Himalayas. Bull. Indian Geol. Assoc. 10(2): 1-19.

---; --- 1978. Early Permian fossils from the Bijini Tectonic Unit, Garhwal Himalaya. Rec. Res. Geol. 4: 410-437.

---; --- 1979. Early Permian fossils from southern Tibet, like faunas from Peninsu-lar India and Himalayas of Garhwal. J. Geol. Soc. India 20: 461-464.

ZHANG, S. XIN; CHING YUKAN 1976. Upper Palaeozoic brachiopods from Mount Jolmo Lungma region. Report of the Scientific Expedition in the Mount Jolmo Lungma (Qomolungma) Region (1966-1968), Palaeon-tology 2: 159-270.

Mesozoic stratigraphy of India and its paleogeographic implications (Abstract)

S. C. SHAH & M. V. A. SASTRY
Geological Survey of India, Calcutta

A review of the Mesozoic stratigraphy of India indicates that during Triassic and the Lower Jurassic, the sea was restricted to the Himalayan region stretching from Kashmir in the west to Bhutan in the east. The sea then extended southwards from the Indus Basin (which was the westward continuation of the Himalayan Basin from pre-Jurassic times) to the western Rajasthan and Cutch. During Lower Cretaceous times marine conditions continued on the western coast.

The subsurface data from the Cauvery Basin in South India indicate that this part of the country first experienced marine oscillations of the Mesozoic Era in late Jurassic.

The marine oscillations of late Jurassic covered more areas on the east coast, but there were temporary regressions both on the west and east coasts where non-marine sedimentation also took place. In the late Lower Cretaceous, the sea transgressed the Barmer and Jaisalmer Basins in Rajasthan, Cutch, and Katiawar, and this continued till the Middle Cretaceous. Then the sea withdrew completely from those areas.

The marine transgression from the west entered along the Narbada Valley and reached the interior as far as Jabalpur, where the Lametas--which are considered partly marine-- were deposited. Marine conditions continued for the rest of Mesozoic times.

On the east coast, the marine transgression covered the coastal tracts from Assam to the Cauvery Basin during Upper Cretaceous. The Andaman Group of islands were also covered by sea during late Cretaceous.

During Jurassic and Lower Cretaceous times, the Indus Basin, Rajasthan and Cutch were together as they are found today. The presence of Tethyan elements in the Jurassic of Cutch indicates that the sea washing the shores of the west coast was part of the Tethyan sea that also covered the Himalayan region.

Rajasthan, Cutch, Narbada Valley and southern India were part of the Indian Peninsula, which remained a single land mass from Precambrian times.

Recent advances in the study of the Gondwana stratigraphy of India (Abstract)

N.D.MITRA & C.S.RAJA RAO
Geological Survey of India, Calcutta

The intracratonic Gondwana basins of peninsular India do not bear record of synchronous deposition of similar sequences. Periodic uplift along rift shoulders of these fault-bounded basins took place at varying rates and times as recorded by local unconformities Accordingly, there has been disagreement regarding definition, nomenclature and group-ing of the Gondwana sediments on a regional scale. The following Gondwana basins occurring in broad belts, however, show similar characteristics and have more or less similar tectono-sedimentological histories: (i) Rajmahal-Malda-Purnea-Galsi master basin: thick Permian coal measures and Lower Cretaceous lava flows (ii) Damodar-Koel-eastern Son valley basins: uninterrupted sequence of Permian and Lower Triassic sediments followed unconformably by Upper Triassic sediments (iii) Mahanadi-southeastern Son valley basins: sediments of the Permian period (iv) Godavari basin: development of a continuous sequence of Permian and Triassic beds succeeded by Lower Jurassic and Lower Cretaceous sediments with distinctive unconformities (v) Rewa basins: Permian and Triassic sediments with a break at the base of the Rhaetian (vi) Wardha-Kamptee basins: a local unconformity within the Permian succession and sedimentation ending in the Lower Triassic period (vii) Satpura basins: Permian, Lower and Middle Triassic and Cre-taceous sediments with a hiatus during the entire Jurassic period and (viii) Athgarh-Krishna-Godavari-Palar-Cauveri and Umia basins: mainly continental and marine Lower Cretaceous sediments deposited along emerging coastlines. A synthesis of the strati-graphic data shows that the Gondwana sedimentation came to a halt in the Lower Permian in a few basins, Lower Triassic in some and Upper Triassic in most of the basins, excepting those along the coast.

Correlation and classification of the Basal and Lower Permian Coal Measures of India (Abstract)

T. N. BASU

Central Mine Planning & Design Institute Ltd., Ranchi, India

The Lower Gondwana coals of India occur in three distinct formations--the Karharbaris (Basal Coal Measures), the Barakars (Lower Coal Measures) and the Raniganj (Upper Coal Measures).

The depositional history and conditions of sedimentation are reflected in the qualitative characteristics and petrographic composition of the coals. This has helped considerably in identifying and correlating the formations and coal seams in the various Gondwana coalfields of the country.

The Karharbaris (30 - 300 m), a product of a high-energy environment, contain better quality coals (up to 20% ash, rarely over 25%). The washability results show that for reduction of 1% ash, the average yield goes down by as much as 10% (on coals crushed to 3") for obtaining cleans at 16% ash level. Shallow water conditions, under which such coals were formed, were extremely favourable for "oxidative dehydrogenation" by aerobic microbial populations leading to the formation and concentration of sub-hydrous macerals. Thus, the inertinite content of these coals varies from 35 to 55%, and the coals have a low hydrogen content (4 - 5%). As a result of the high inertinite content, the coals are dull with a faintly developed laminated structure.

The Barakars--Lower, Middle, and Upper--have coal seams with distinctive qualitative and petrographic composition. The Lower Barakars contain thick, highly-interbanded seams in which the dirt is highly intergrown. As a result, the seams are of inferior quality (25 - 35% ash), with an erratic phosphorus content as high as 0.572%. Washability characteristics are slightly better than those of Karharbari coals, as the loss in yield is about 5 - 6.2% for each 1% reduction in ash on washing (coals crushed to 3") for obtaining cleans at 16% ash level. The inertinite content varies from 20 to 35%, with corresponding increase in the vitrinite content (60 - 75%). As a result, the coals have a fairly well-developed laminated structure and are usually orthohydrous.

The Middle Barakar coals are of comparatively better quality (15 - 25% ash), show better washability characteristics (for every 1% reduction in ash, loss is yield is 3 - 5%), and exhibit a well-developed laminated structure. The coals are orthohydrous in character. It has, however, not yet been possible to differentiate the horizon petrographically.

The Upper Barakar coals exhibit quality and cleaning characteristics more or less similar to those of the Middle Barakars. The loss in yield is 3% for every 1% reduction in ash (on coals crushed to 3") for obtaining cleans at 16% ash level. The coals are rich in vitrinite and show well-developed laminated structure and are orthohydrous.

Compared with the Basal and Lower Permian coals, the Upper Permian coals (Raniganj Formation) have high vitrinite content (70 - 84%), with a corresponding decrease in inertinite content (12 - 20%).

Gondwana coal basins of India and their resource potential (Abstract)

N. D. MITRA, C. LAHA & U. K. BASU
Geological Survey of India, Calcutta

Gondwana coal basins occupy well-defined grabens on the peninsular craton of India. The reserves of coal from these basins constitute about 1% of the world resources and account for 4% of the world's production. The bulk of the accessible Gondwana coal in India is of inferior quality, due to well-dispersed detrital mineral matter. Coals with less than 17% ash content form only 3% of the total. These include the coal seams of the Damodar Valley, which are of higher rank as they have been exposed during coalification to a higher geothermal gradient. There is also a degree of correlation between the coal rank and the lamprophyre intrusions from the subcrustal levels. The overburden does not seem to have favoured much chemical reaction which would lead to an increase in rank. This is evident from the subhydrous rank of Godavari coal despite a cover of 2 km of strata over the coal measures.

The economic resources of Gondwana coals of India amount to 85.6×10^9 tonnes, and subeconomic resources contribute an additional 25.4×10^9 tonnes. The recoverable reserves would be approximately 50% of the economic resources.

Application of borehole geophysics for exploration of coal
in Gondwana coalfields of eastern India (Abstract)

J. R. KAYAL
Geological Survey of India, Calcutta

Borehole geophysical logging was developed for petroleum exploration, but nowadays is successfully used for coal identification. In coring, the usual procedure for coal exploration, structural conditions and core loss often pose a problem in correlation and determining lateral variation of coal seams and estimating the reserves. Well logging is then found to be a valuable technique for the economic evaluation of the coal deposits.

Single electrode resistance, self-potential, multielectrode short (16 inch) normal and 6 foot lateral resistivity, gamma-ray and continuous temperature logging have been carried out in the West Bokaro, North Karanpura and Jharia Coalfields. The geophysical logs are used for locating the seams, determining their accurate depths and thicknesses and approximate quality.

Gondwana coals of India are mostly sub-bituminous to bituminous, and they are characterised by very high resistance/resistivity compared to interbedded sandstone and shale. Low or high self-potential values for coal depend on the borehole fluid conditions. Single electrode resistance and self-potential logs are mostly used, as they give a detailed picture of the formations penetrated through the borehole. Relative competence of the roof and floor conditions and the overburden lithology can be determined from the logs. Multielectrode short normal resistivity curves almost resemble the single electrode resistance log and is used for correlation. A 6-foot lateral device having higher current penetration is found to be more useful in differentiating coal and highly resistive sandstone beds than either the single point resistance or short normal resistivity logging devices. Physicochemical properties of coal are responsive to electrical logs, and 'finger prints' in the log indicate variations of coal quality.

Burnt coals (locally known as Jhama) which are caused by dolerite and micaperidotite intrusions are characteristically highly conductive beds. Very low resistivity and self-potential values are recorded against Jhama sections of the coal beds. Self-potential values indicate their effective porosity.

Natural gamma-ray radiation is maximum for shale and minimum for coal and has thus been effectively used with electrical logs for correlation and identification of coal beds. In cased and dry boreholes gamma-ray logs alone have been used to detect coalseams. Since gamma-counts increase with shale content and 99% of ash content in coal is due to the shale contamination in it, it is possible to make a rough estimation of ash content from the gamma-ray log. A linear relationship of gamma-counts with ash content of known coal samples has been drawn by the least-squares fitting method, the correlation coefficient being 0.97 to 0.99 and the standard error of ash-content ±2.51 to ±3.78%.

Temperature logs define geothermal gradients between $38^{\circ}C$ and $49^{\circ}C/km$, in the coalfields. Thermal conductivity of coal being low compared to shale and sandstone, high electrical resistivity together with high thermal resistivity differentiates coal seams from other formations (like sandstone) having high electrical resistivity but low thermal resistivity. It is observed that thermal resistivity increases with decreasing grade of coal.

A dinosaur from New Zealand

R. E. MOLNAR
Queensland Museum, Fortitude Vy, Australia

A vertebra from the Mata Series (Upper Cretaceous) of North Island, New Zealand, is identified as a theropod dinosaur caudal. The amphiplatyan centrum with elevated transverse processes and other features exclude identification as marine reptile. Terrestrial tetrapods known or inferred to have lived in New Zealand during the Upper Cretaceous were forms probably widespread throughout Gondwana before New Zealand's isolation. A brief survey of other known Upper Cretaceous insular dinosaur faunas indicates that these faunas comprised some dwarf and relict forms.

INTRODUCTION

Concretionary boulders in the valley of the Mangahouanga Stream, a tributary of the Te Hoe River, North Island, New Zealand, have produced an abundant fauna of marine vertebrates (Keyes 1977; Speden 1973). This fauna of plesiosaurs, mosasaurs, teleosts, elasmobranchs, and invertebrates, has been assiduously collected and prepared by Mr and Mrs M. Wiffen.

Among this material is a single vertebra (N.Z. Geological Survey CDl) that represents, as will be here shown, a dinosaur. The alleged absence of dinosaurs from New Zealand has been considered important for the interpretation of its faunal history (Fleming 1962, and especially Darlington 1965), and thus this discovery warrants treatment at some length. In particular it supports the suggestion of Cracraft (1974) that some forms not now represented in New Zealand in fact reached the islands and later became extinct.

STRATIGRAPHY

The concretionary boulders from which these fossils have been recovered derive from the Upper Cretaceous Mata Series. This series is of Piripauan-Haumurian age, which corresponds to the Campanian and Maastrichtian ages (Wellman 1959).

DESCRIPTION

The vertebra consists of the centrum with much of the arch: both transverse processes, the neural spine and the left prezygopophyseal process are missing (Fig. 1). A ventral sulcus is present on the centrum, with marked chevron facets demonstrating that it is a caudal vertebra. The transverse processes are well elevated and project laterally from the lamina. The neural spine, although robust at the level of the break, is very strongly inclined,

rising behind rather than above the posterior face of the centrum. The spine continues anteriorly across the plane of the lamina as a sharp ridge descending abruptly between the prezygopophyseal processes. The prezygopophyseal facet is inclined at about 30° to the vertical.

TAXONOMIC IDENTIFICATION

As this vertebra occurred in a marine fauna it needs to be established that it does not derive from some marine reptile. The vertebra is not procoelous as are those of mosasaurs and symoliophids (Nopcsa 1925), and mosasaur caudals lack elevated transverse processes (Russell 1967), thus this caudal does not derive from either (Fig. 2). Ichthyosaur caudals are considerably more anteroposteriorly compressed, strongly amphicoelous, and without elevated transverse processes. Plesiosaur caudal centra are usually strongly depressed (Romer 1956), showing paired ventral foramina. The neural spines are little inclined to the vertical, and the transverse processes are not elevated. Chelonian caudal centra are usually not amphiplatyan, and lack elevated transverse processes (Hoffstetter & Gasc 1969; Romer 1956). The neural spines are little developed (Romer 1956).

Eusuchian crocodilians are equally out of the question because of their procoelous caudals. Some mesosuchians, dyrosaurids in particular, have similar caudals, but they lack elevated transverse processes (Thevinin 1911, Pl. 3) and have centra with quadrangular or subquadrangular central articular facets (Buffetaut 1976). Neural spines are high and not strongly inclined (Swinton 1937).

The vertebra most resembles those of ornithopod and theropod dinosaurs. Several ornithopods (e.g. *Iguanodon*, *Thescelosaurus*) show strongly inclined,

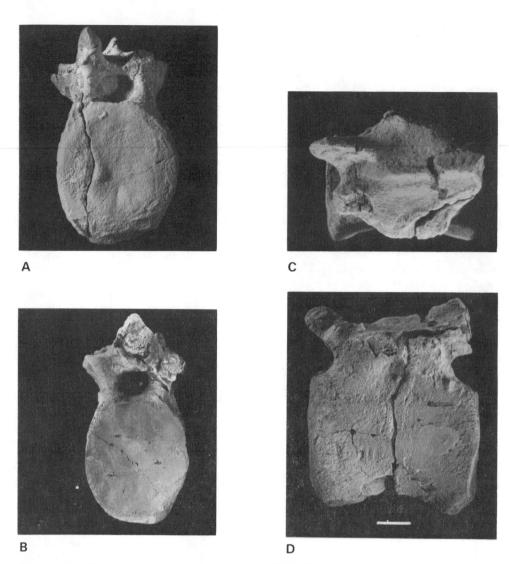

A

C

B

D

Fig. 1--The Mangahouanga dinosaur vertebra (NZGS CD1) in anterior (A), posterior (B), dorsal (C), and R lateral (D) aspects. Bar = 1 cm.

posteriorly overhanging neural spines. But I know no ornithopod caudals with both transverse processes and strongly inclined neural spines, as the former structure is found on only the proximal caudals, and the latter condition in intermediate and distal caudals.

Proximal caudals of several theropods and some prosauropods resemble the Mangahouanga caudal in the elevated position of the transverse processes, e.g. *Allosaurus fragilis* (Madsen 1976, Pl. 32) and *Elaphrosaurus bambergi* (Janensch 1925, Taf. IV), but most do

not show the strongly inclined, over-hanging neural spine. The proximal caudals of *Poikilopleuron bucklandii* however, do closely approach this, with elevated transverse processes and an inclined spine located far posteriorly (Eudes-Deslongschamps 1837, Pl. 2). The Mangahouanga vertebra would represent a spine placed farther back than the *P. bucklandii* caudals, but otherwise are very similar having even the ventral sulcus. Because of this I consider it most likely that the Mangahouanga caudal derive from a theropod. However, it is obvious

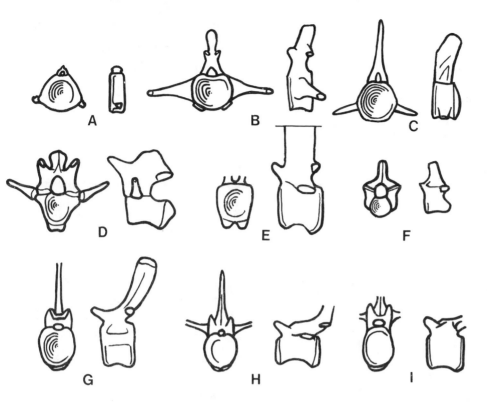

Fig. 2--Caudal vertebrae, anterior and R lateral aspects. A, ichthyosaur (*Ophthal-mosaurus*); B, plesiosaur (*Cryptocleidus*); C, mosasaur (*Platecarpus*, two different vertebrae); D, marine turtle (modern, unidentified); E, dyrosaurid crocodile (anterior aspect): unidentified specimen, R lateral aspect: *Dollosuchus*); F, symoliophid (*Symoliophis*); G, ornithopod (*Iguanodon*); H, theropod (*Poikilopleuron*: for G and H the two aspects represent two different vertebrae); and I, the caudal from Mangahouanga Stream. Vertebrae A - G represent mid-caudals, H an anterior caudal. Not to scale. (From various authors; D and I original.)

that a single, incomplete vertebra is hardly a substantial basis for an identification, and although I think it unlikely, it is possible that the vertebra is of an ornithopod.

DISCUSSION

The discovery of an undoubted terrestrial reptile in the Upper Cretaceous of New Zealand raises several points of interest, with regard to New Zealand and to insular dinosaur faunas in general. These are: the composition of the Upper Cretaceous terrestrial fauna of New Zealand; the origin of these forms and the route by which they reached New Zealand; and, the relation of dinosaur body size to the area of New Zealand. This leads to some more general considerations: the existence and composition of other insular dinosaur faunas and the existence of other insular effects

on faunal elements (Carlquist 1965; Sondaar 1977).

To the hypothetical (but likely) Upper Cretaceous terrestrial fauna of New Zealand suggested by Fleming (1962), of 'ratites' (i.e. ancestors of the kiwi and moas), a sphenodont and a leiopelmatid, can be added a (probably) theropod dinosaur. Presumably some prey population ('ratites' or other saurians) was also present.

Although some reconstructions show New Zealand still on the Australian continental shelf as late as the Paleocene (e.g. Smith & Briden 1977), there is evidence for the opening of the Tasman Sea before the end of the Cretaceous (Griffiths & Varne 1972), about 80 m.y. ago (Pitman *et al.* 1968). Some authors (e.g. Suggate 1972) suggest an even earlier date; Fleming (1975) summarises this literature. So, it is possible that

New Zealand was well separated from Australia (and Antarctica) by the time of the
Mangahouanga dinosaur, especially as these
geographical reconstructions place New
Zealand at the edge of the continental shelf,
rather than near the (present) shoreline.

Possibly excepting the sphenodontids,
the faunal elements of Upper Cretaceous
New Zealand, hypothetical and actual,
are all from groups known from Gondwana
before New Zealand's isolation. The
frog *Leiopelma* appears related to the
Jurassic South American frog, *Notobatrachus*
(Estes & Reig 1973), so the leiopel-
matids may be indigenous to Gondwana.
Theropod (and ornithopod) dinosaurs are
known from all continents except Antarctica. Birds presumably originated in the
Upper Jurassic of Europe, although Ellenberger (1974) reports evidence of feather-
ed bipeds in the Upper Triassic of South
Africa, and had already reached Victoria
(Australia) by the Lower Cretaceous
(Talent *et al*. 1966). Sphenodontids
were seemingly most diverse in Europe and
North America, with one genus from the
Triassic of South Africa (Romer 1966).
Cretaceous terrestrial tetrapod faunas of
the southern hemisphere, however, are
poorly known, and sphenodontids may have
been widespread in Gondwana. The absence
of sphenodontids from Upper Cretaceous
micro-vertebrate sites in North America
and Asia suggests that they were already
scarce, at least in the north, at this
time.

Thus birds and theropods were very likely
to have already been in the New Zealand
region before it split off from the Austro-
Antarctic land mass. Sphendontids, if not
already present, probably spread south
through South America or Africa, across the
tip of South America to West Antarctica
and· to New Zealand.

The work of Harestad and Bunnell (1979),
based on McNab (1967) relates body mass to
home range size for mammals, birds and
reptiles. The existence of an insular
dinosaur prompts the estimation of the
home range size and comparison with the
size of New Zealand. They give formulae
for home ranges of herbivores, omnivores
and carnivores. The Mangahouanga dinosaur
was probably about 4 m long: extrapolation
from the estimated dinosaur weights of
Colbert (1962) gives a weight of about 0.4
tonnes (irrespective of whether it was a
theropod or an ornithopod). The formulae
of Harestad and Bunnell give home ranges
of about 4 km^2 for an herbivore, 34 km^2
for an omnivore, and 18,500 km^2 for a
carnivore. The data of Schaller for lions
(1972) weighing 0.1 to 0.2 tonnes and with

ranges to 4,700 km^2 indicate that this is
not incredible. If Upper Cretaceous New
Zealand had been approximately the same
size as now (about 270,000 km^2) this would
imply a very small population, fewer than
100, even if my weight estimate is off by
75%. However Schaller's data also indicat
that ranges of thousands of km^2 are at the
large end of a spectrum of sizes, and this
may be assumed for the Mangahouanga dinosaur as well.

Other insular dinosaur faunas are known,
all from the Upper Cretaceous: from Trans-
sylvania, Hungary and Austria (Nopcsa
(1923), Madagascar, and India (albeit
India was a large island). These faunas
are usually ignored in the literature, so
some comments are in order here. Unusual
phenomena attend the evolution of insular
endemic animals (e.g. Carlquist 1965;
Mertens 1934; Sondaar 1977); two of these,
dwarfism and relicts, can be seen in the
Upper Cretaceous.

Several mechanisms of insular endemic
dwarfing have been recently proposed
(Soulé 1966; Case 1978; Heaney 1979;
Wassersug *et al*. 1979) and criticised
(Dunham *et al*. 1978), and instances of
this phenomenon are usually reported from
the Pleistocene and Recent only (e.g.
Carlquist 1965; Sondaar 1977), although
Nopcsa (1923) pointed out an instance
(the Transylvanian insular dinosaurs).
Three--*Telmatosaurus, Struthiosaurus*
and *Magyarosaurus*--are noticeably smaller
than their contemporaries, and the latter
two are probably the smallest members of
their families. There is no indication
from the literature that any of the
specimens are juveniles. This phenomenon
has not yet been observed in the other
Cretaceous insular faunas. Much new
material has been recovered from Madagasca
(Russell *et al*. 1976; Obata & Kanie 1977)
that is now under study, so it is prematur
to state what may be found in the Malagasy
fauna.

Nopcsa (1923) has also claimed that the
Transylvanian insular tetrapods included
relict forms, and *Telmatosaurus* and the
nodosaurs do seem to be relicts. A
striking relict has been reported from
India. Stegosaurids appeared to have
become extinct at the end of the Jurassic,
except in China where a single genus
survived into the Lower Cretaceous (Dong
1973). In India this family survived
into the Maastrichtian (Yadagiri &
Ayyasami 1979).

Finally, it may be noted that future
work on fossil tetrapods from insular,
isolated land masses may bear on the
question of dinosaur extinction. The

94

catastrophic models (e.g. Russell 1977) would predict the extinction of dinosaurs everywhere nearly simultaneously (geologically speaking) while the ecological replacement models (e.g. Sloan 1976) do not. Any evidence for later extinction on an island, such as New Zealand or India (or even Australia) would weigh against the catastrophic theories.

SUMMARY

A caudal vertebra from the Upper Cretaceous Mata Series of North Island, New Zealand, probably derives from a theropod dinosaur. The form of the centrum and arch exclude reference to mosasaurs, ichthyosaurs, symoliophids, mesosuchians, plesiosaurs and chelonians. The oval central articular surfaces exclude reference to the dyrosaurids. Similarities to the caudals of *Poikilopleuron bucklandii* suggest that this caudal derives from a theropod, although similarities to ornithopod caudals can be seen and indicate that this identification is tentative.

To the three inferred members of the New Zealand Upper Cretaceous fauna (a leiopelmatid, a sphenodontid, and one or more 'ratites') can be added an actual dinosaur. Of these, there is good reason to suspect that all but (perhaps) the sphenodontids were widespread in Gondwana at the time of the separation of New Zealand.

Estimation of the weights of the Mangahouanga dinosaur and hence of the home range indicates that, if it was carnivorous, the home range may have been as much as an order of magnitude less than the area of New Zealand. Presumably this was not an equilibrium situation.

Other insular dinosaur faunas are known, and some of these had elements that were dwarfed or relictual. Dwarfed forms are known from Transylvania and relictual forms from both Transylvania and India.

ACKNOWLEDGMENTS

Mrs Joan Wiffen enthusiastically brought this specimen to my attention and kindly granted permission to study it. Assistance and encouragement was provided by Dr E. Buffetaut, Mr G. Czechura, Sir Charles Fleming, Dr P. Kott, Dr D.A. Russell and Dr M. Wade. Miss Jennifer Hamilton tolerantly typed the paper, largely at the last minute.

REFERENCES

BUFFETAUT, E. 1976. Une nouvelle definition de la famille de Dryosauridae De Stefano, 1903 (Crocodylia, Mesosuchia) et ses consequences: inclusion des genres Hyposaurus et Sokotosuchus dans le Dyrosauridae. Geobios 9: 333-336.

CARLQUIST, S. 1965. Island Life. New York: Natural History Press, 451 pp.

CASE, T.J. 1978. A general explanation for insular body size trends in terrestrial vertebrates. Ecology 59: 1-18.

COLBERT, E.H. 1962. The weights of dinosaurs. Am. Mus. Nov. 2076: 1-16.

CRACRAFT, J. 1974. Continental drift and vertebrate distribution. Ann. Rev. Ecol. Syst. 5: 215-261.

DARLINGTON, Jr., P.J. 1965. Biogeography of the Southern End of the World. Cambridge: Harvard University Press, 236 pp.

DONG, Z. 1973. Dinosaurs from Wuerho. Mem. Inst. Vert. Paleontol. Paleoanthro. Acad. Sinica 11: 45-52. (In Chinese).

DUNHAM, A.E.; D.W. TINKLE: J.W. GIBBONS 1978. Body size in island lizards: a cautionary tale. Ecology 59: 1230-1238.

ELLENBERGER, P. 1974. Contribution à la classification des Pistes de Vertébrès du Trias: Les types du Stormberg d'Afrique du Sud (IIeme Partie: Le Stormberg superieur-I. Le biome de la zone B/1 on niveau de Moyeni: ses biocénoses). Palaeovert., Mem. Extraord. 1974: 1-147.

ESTES, R.; O.A. REIG 1973. The early fossil record of frogs, a review of the evidence. In, J.L. Vial, ed., Evolutionary Biology of the Anurans. Columbia: University of Missouri press, pp. 11-63.

EUDES-DESLONGCHAMPS, M. 1837. Mémoire sur le Poikilopleuron bucklandii, grande saurien fossile, intermédiaire entre les crocodiles et les lézards. Mém. Soc. Linn. 6: 5-114.

FLEMING, C.A. 1962. New Zealand biogeography, a paleontologist's approach. Tuatara 10: 53-108.

--- 1975. The geological history of New Zealand and its biota. In G. Kuschel (ed.) Biogeography and Ecology in New Zealand. The Hague: W. Junk, pp. 1-86.

GRIFFITHS, J.R.; R. VARNE 1972. Evolution of the Tasman Sea, Macquarie Ridge and Alpine Fault. Nature Phys. Sci. 235: 83-86.

HARESTAD, A.S.: F.L. BUNNELL 1979. Home range and body weight - a re-evaluation. Ecology 60: 389-402.

HEANEY, L.R. 1978. Island area and body size of insular mammals; evidence from the tri-colored squirrel (Callosciurus prevesti) of southwest Asia. Evolution 32: 29-44.

HOFFSTETTER, R.; J.-P. GASC 1969. Vertebrae and ribs of modern reptiles. In C. Gans (Ed.) Biology of the Reptilia, Volume 1, Morphology A. London: Academic Press. 201-310.

JANENSCH, W. 1925. Die Coelurosaurier und Theropoden der Tendaguru-Schichten Deutsch-Ostafrikas. Palaeontographica,

Suppl. 7, R.l, T.1: 1-99.

KEYES, I.W. 1977. Records of the northern hemisphere Cretaceous sawfish genus Onchopristis (order Batoidea) from New Zealand. N.Z. J. Geol. Geophys. 20: 263-272.

MADSEN, J.H. Jr., 1976. Allosaurus fragilis: a revised osteology. Utah Geol. Miner. Surv. Bull. 109: 1-163.

McNAB, B.K. 1963. Bioenergetics and the determination of home range size. Am. Nat. 97: 133-140.

MERTENS, R. 1934. Die Insel Reptilien. Zoologica 84: 1-205.

NOPSCA, F. 1923. On the geological importance of the primitive reptilian fauna in the uppermost Cretaceous of Hungary; with a description of a new tortoise (Kallokibotion). Quart. J. Geol. Soc. London 79: 100-116.

--- 1925. Ergebnisse der Forschungsreisen Prof. E. Stromers in den Wüsten Ägyptens. II. Wirbeltier-Reste der Baharîje-Stufe (unterstes Cenoman). 5. Die Symoliophis-Reste. Abh. Bayer. Akad. Wiss., Math.-naturwiss. Abt. 20(4): 1-27.

OBATA, I.; Y. KANIE 1977. Upper Cretaceous dinosaur-bearing sediments in Majunga region, northwestern Madagascar. Bull. Nation. Sci. Mus. Tokyo C, 3: 161-173.

PITMAN, W.C.: E.M. HERRON: J.R. HEIRTZLER 1968. Magnetic anomalies in the Pacific and sea floor spreading. J. Geophys. Res. 73: 2069-2085.

ROMER, A.S. 1956. Osteology of the Reptiles. Chicago: University of Chicago Press, 772 pp.

--- 1966. Vertebrate paleontology (3rd ed.), Chicago: University of Chicago Press. 468 pp.

RUSSELL, D.A. 1967. Systematics and morphology of American mosasaurs. Peabody Mus. Nat. Hist. Bull. 23:1-237.

--- 1977. The biotic crisis at the end of the Cretaceous Period. Syllogeus 12: 11-23.

RUSSELL, D.D.; P. TAQUET; H. THOMAS 1976. Nouvelles récoltes de Vertébrés dans les terrains continentaux du Crétacé supérieur de la région de Majunga (Madagascar). Bull. Soc. géol. France, Suppl. 5: 204-208.

SCHALLER, G. 1972. The Serengeti Lion. Chicago: University of Chicago Press. 479 pp.

SLOAN, R.E. 1976. The ecology of dinosaur extinction. In C.S. Churcher (Ed.) Athlon. Toronto: The Royal Ontario Museum, 134-154.

SMITH, A.C.; J.C. BRIDEN 1977. Mesozoic and Cenozoic Paleocontinental Maps. Cambridge: Cambridge University Press.

SONDAAR, P.Y. 1977. Insularity and its effect on mammal evolution. In M.K. Hecht et al. (Eds.) Major Patterns in Vertebrate Evolution. New York: Plenum Press. pp. 671-707.

SOULÉ, M. 1966. Trends in the insular radiation of a lizard. Am. Nat. 100: 47-6

SPEDEN, I.G. 1973. Distribution, stratigraphy and stratigraphic relationships of Cretaceous sediments, western Raukumar Peninsula, New Zealand. N.Z. J. Geol. Geophys. 16: 243-268.

SUGGATE, R.P. 1972. Mesozoic-Cenozoic development of the New Zealand Region. Pacific Geol. 4: 113-120.

SWINTON, W.E. 1937. The crocodile of Maransart (Dollosuchus dixoni (Owen)). Mem. Mus. Roy. d'Hist. Natur. Belg. 80: 1

TALENT, J.A.; P.M. DUNCAN; P.L. HANDBY 19 Early Cretaceous feathers from Victoria. The Emu 66: 81-86.

THEVININ, A. 1911. Le Dyrosaurus des Phosphates de Tunisie. Ann. Paléont. 6: 95-108.

WASSERSUG, R.J.; H. YANG; J.J. SEPKOSKI, D.M. RAUP 1979. The evolution of body size on islands: a computer simulation. A Nat. 114: 287-295.

WELLMAN, H.W. 1959. Divisions of the New Zealand Cretaceous. Trans. R. Soc. N.Z. 87: 99-163.

YADAGIRI, P.; K. AYYASAMI. 1979. A new stegosaurian dinosaur from the Upper Cretaceous sediments of south India. J. Geol. Soc. India 20: 521-530.

The continental margin of Gondwana, principally in central western Argentina:
Jurassic and Lower Cretaceous palynomorphs and calcareous microfossils

WOLFGANG VOLKHEIMER
Museo Argentino de Ciencias Naturales,
Buenos Aires

EDUARDO MUSACCHIO
Centro de Investigaciones en Recursos
Geologicos, Buenos Aires

The Andean border of Gondwana in western Argentina contains a nearly complete sequence of Sinemurian to Aptian-Albian strata. The well-known ammonite zones and intertonguing with marine and continental beds enable the dating of terrestrial and marine plant microfossils, foraminifers and ostracods. The Neuquén Basin provides nine assemblages of palynomorphs and six of calcareous microfossils.

In some zones relations with Australian and Indian microfloras have been recognized, and in others, cosmopolitan and South African, as well as northern Hemisphere, affinities in calcareous microfossils are recognised.

Palynomorphs and calcareous microfossils in the San Luis Basin, Río Chubut Embayment, Deseado "Massiv", and Austral Basin are also described.

INTRODUCTION

Marine and non-marine sedimentary rocks of Jurassic and Lower Cretaceous age are abundant and well distributed in several sedimentary basins of southern South America (Sadras *et al.* 1973: 8; Zambrano & Urien 1970: 1387, Fig. 8).

Palynostratigraphic studies have been published from the Neuquén Basin (Lower Cretaceous), San Luis Basin (Lower Cretaceous), Río Chubut Embayment (Upper Jurassic and Lower Cretaceous) and Deseado "Massiv" (Lower Cretaceous). Previous information about calcareous microfossils exists from Neuquén Basin, Golfo de San Jorge Basin and Austral Basin (Fig. 1).

NEUQUÉN BASIN

The most complete palynological and micropaleontological record is that of the Neuquén Basin, where Sinemurian (?), Pliensbachian, Toarcian, Aalenian, Bajocian, Bathonian-Lower Callovian, Middle Callovian, Oxfordian, Tithonian, Hauterivian-Barremian and Aptian to Albian microfloras have been studied.

The Lower Jurassic palynomorph assemblages were summarized by Volkheimer (1976). Two palynostratigraphic units were distinguished within the Lower Jurassic. The lower one (Unit 1, Table 1) corresponds to the Sinemurian (?), Pliensbachian and Lower Toarcian and is characterized by *Nevesisporites vallatus*, *Cadargasporites verrucosus*, *Verrucosisporites varians*, and high frequencies of *Classopollis simplex* and other species of *Classopollis*. The presence of

Todisporites minor indicates that the lowermost strata are not older than Liassic. At the southernmost end of the Neuquén Basin (Alicurá, Río Limay area) the Lower Jurassic Nestares Fm contains *Skarbysporites elsendoornii*, known from the Pliensbachian and Lower Toarcian of the Vicentinian Alps, and a large number of species of the genera *Deltoidospora*, *Concavisporites*, *Auritulinasporites*, *Biretisporites*, *Dictyophyllidites*, *Neoraistrickia*, *Duplexisporites*, *Contignisporites*, *Lycopodiumsporites*, *Marattisporites*, bisaccate and monosulcate grains, *Classopollis* and acritarchs (*Leiosphaeridia* spp.).

The Lower limit of the upper unit (Unit 2) is marked by the incoming of *Callialasporites dampieri*, *C. segmentatus* and *Inaperturopollenites turbatus*. This unit embraces the Upper Toarcian and Aalenian. The age of the microfloras is gauged by associated ammonites (Volkheimer 1973: 116-7).

The first appearance of *Microcachryidites antarcticus* marks the lower limit of Unit 3 (Bajocian). In continental beds deposited near the coast, this unit is characterized by high frequencies of *Classopollis* spp. (±40%) and *Araucariacites australis* (up to 42%). Other important forms present are *Ischyosporites marburgensis* and *Uvaesporites minimus*. The upper limit of Unit 3 is characterized by the first appearance of *Microcachryidites castellanosii*.

Unit 4, mainly Bathonian to Middle Callovian, yields well-preserved microfloras in the southern part of the Neuquén Basin (Volkheimer 1972; Volkheimer & Rosenfeld, in prep.). Characteristic forms are

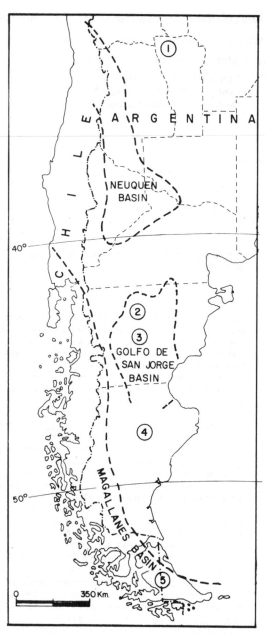

Fig.1--Location of the sedimentary basins mentioned in the text: 1: San Luis Basin; 2: Rio Chubut Embayment; 3: Golfo de San Jorge Basin; 4: Deseado "Massiv"; 5: Austral Basin.

Microcachryidites castellanosii, Equiseto-sporites menendezii, Cycadopites punctatus Osmundacidites diazii, Ischyosporites volkheimeri and *Verrucosisporites varians Classopollis* is generally less frequent th in the Bajocian and in the Liassic, but it frequency varies strongly with environment and has been observed to oscillate within a few metres of sediment from 0.5 to 43.5% Unit 4 may be subdivided into two sub-unit 4a (lower and middle part of the Lajas Fm) and 4b (Red beds of the Lajas Fm). In Unit 4b of the Middle Callovian, *Maratti-sporites scabratus* is more frequent than i any other Jurassic or Cretaceous formation of this basin and locally reaches more tha 50% of the spectrum.

Unit 5. In the Puente Arroyo Picún Leufú area marine pelites of the *Reineckei* spp. bearing Lotena Fm (mostly Middle Call vian) contain a mainly marine association *Nannoceratopsis*, many other dinoflagellate and acritarchs, that alternates with asso-ciations of marine (*Leiosphaeridia hyalina, L. staplinii, L.* spp.) and terrestrial for like *Classopollis* spp., *Callialasporites trilobatus, C. segmentatus, C. microvelatu Araucariacites australis, Microcachryidite antarcticus* and *Staplinisporites caminus.*

Unit 6. The Upper Oxfordian Auquinco Fi at the southern end of Sierra de la Vaca Muerta contains a distinctive species of *Gemmatriletes*, not observed in other forma-tions of the same basin, and a distinctive *Classopollis* with a generally heptangular distal porus. Several other species of *Classopollis, Araucariacites australis*, monosulcate grains, *Eucommiidites, Trisac-cites microsaccatus, Microcachryidites, Podocarpidites, Alisporites, Vitreisporite pallidus*, the *Taurocusporites - Duplexispo-rites - Polycingulatisporites* - complex and *Callialasporites turbatus, C. trilobatus* an *C. microvelatus* (Volkheimer & Moroni, in prep.).

Unit 7. The base of Unit 7 is defined by the incoming of *Equisetosporites caichigüensis*; the top by the first appearance of *Cyclusphaera psilata*. The unit's assemblage is best represented in the Vaca Muerta Fm. Characteristic spe-cies are: *Equisetosporites caichigüensis, Trisaccites microsaccatus, Interulobites variabilis, Classopollis simplex* and *Leiosphaeridia hyalina.*

Unit 8 is characterized by *Cyclusphaera psilata*. The base is marked by its first appearance, the top, provisionally, by the incoming of tri- to hexacolpate pollen grains. Characteristic species are: *Cyclusphaera psilata, Inaperturopollenites limbatus, Classopollis classoides, Callia-lasporites trilobatus, C. microvelatus,*

98

Taurocusporites segmentatus. Two sub-units are recognised, the lower (8a) characterized by one species of *Cyclusphaera (C. psilata)*, and the upper (8b), by co-existing thick-walled *C. psilata* and *C.* sp. A, distinguished by a thinner equatrrial wall (cf. Volkheimer *et al.* 1977). In Unit 8 *Bisaccate* sp. A, characterized by smooth, non-reticulate bladders, appears for the first time.

South of the Dorsal, Unit 8a occurs in the Mulichinco Fm and the lower and middle portion of the Agrio Fm, and Unit 8b in the upper part of the Agrio Fm and in the La Amarga Gp.

Unit 8 represents the Valanginian, Hauterivian, Barremian and probably part of the Aptian.

Unit 9 is well represented in the higher part of the Huitrín Fm in its type locality. The base is marked by the first appearance of *Huitrinipollenites transitorius* and *Stephanocolpites mastandreai*. The top cannot yet be defined. Characteristic species are *Huitrinipollenites transitorius, Stephanocolpites mastandreai, Clavatipollenites* spp., *Classopollis classoides, Callialasporites trilobatus, Cicatricosisporites australiensis, Tricolpites* sp. cf. *T. sagax, Retitricolpites* sp. cf. *vulgaris, Taurocusporites segmentatus.*

Calcareous microfossils. In the same basin, three well-preserved diverse marine foraminifera and ostracod assemblages of Lower-Middle Callovian and Hauterivian ages and one non-marine ostracod and charophyte assemblage of Barremian age are known. Several other assemblages in marine Tithonian and Berriasian beds and one in non-marine Aptian-Albian strata represent local environmental peculiarities and are of little paleobiogeographic value.

Callovian foraminifers and marine ostracods (Musacchio 1978, 1979c; Musacchio in Dellapé *et al.* 1979). An outer shelf benthonic foraminifera and ostracod assemblage with *Reineckia* spp. is associated with palynomorphs Units 4b. The foraminifera (52 spp.) include Nodosariidae (28), Lituolacea (13), Miliolacea (4) and the Epistominidae (2). More than 75% of the species are cosmopolitan, and the fauna can be compared with those of similar age in North Europe. Ten species of ostracods belong to seven genera, six of which are known in the Northern Hemisphere. Affinities with others from East Africa are less pronounced.

Hauterivian foraminifera and marine ostracods (Musacchio 1978, 1979b, 1979c). Two Hauterivian normal marine to hyposaline foraminifera and ostracod assemblages occur in the southern outcrops of the Agrio Fm. The lower, associated with the *Holcoptychites*

neuquensis horizon, roughly corresponds with the palynomorph Unit 8a. The higher, associated with the *Crioceratites* spp. horizon, corresponds only partly with palynomorph Unit 8b. In both assemblages the Nodosariidae (36 species) are dominant amongst the Foraminiferida (54 species). The Polymorphinidae (8 species) are more frequent in hyposaline facies, and agglutinated species are uncommon. Most of the species are cosmopolitan (more than 75%). On the contrary, many of the ostracods are unknown outside the basin. Many of the species have affinities at higher taxonomic levels with species of similar age, principally in South Africa. The Andean ostracodal assemblage has a strong southern hemisphere character.

Barremian non-marine ostracods and charophythes (Musacchio 1970, 1971a, 1971b). A Barremian continental ostracod (14 species) and charophyte (4 species) assemblage replaces the latest Hauterivian marine assemblages in the La Amarga Fm. Five species of *Cypridea* together with *Looneyellopsis chinamuertensis* Musacchio, *Huillicythere grambasti* Musacchio, *Atopochara trivolvis triquetra* Grambast and *Triclypella* sp. are the most typical elements.

Other calcareous microfossils. Next to the Jurassic-Cretaceous boundary some stratigraphical units which retain a strong diachronism across the Neuquén Basin carry several assemblages of calcareous microfossils. The assemblages are controlled by lithology rather than age.

The Vaca Muerta Fm--the oldest marine part of the "Andico" composed by black bituminous pelites--includes a low-diversity foraminiferal fauna with radiolarians and scarce ostracods.

In the overlying epineritic limestones of the less well distributed Picún Leufú Fm, a badly preserved ostracod fauna with *Amphicytherura* was found in Tithonian levels. At the top of the Picún Leufú Fm, several regressive epiclastic Berriasian beds include marine and brackish ostracods (with occasional *Cypridea* and charophytes) and foraminifera (Musacchio 1979b).

In the Ranquiles Fm, a Lower Cretaceous (presumably Aptian-Albian) non-marine (oligohaline?) assemblage with *Cypridea* (2 species), *Manteliana (?) ulianai* Musacchio, *Rayosoana quilimalensis* Musacchio and *Flabellochara* cf. *harrisi* (Peck) and other non-marine ostracods and charophytes and small foraminifers are known (Musacchio & Palamarczuk 1975).

SAN LUIS BASIN

Yrigoyen (1975) published palynological data of L. Stover (Esso Production Research

PALYNOMORPHS

PALYNOMORPHS	1	2	3	4a	4b	5	6	7	8	9
Skarbysporites elsendoornii Van Erve 1977	—									
Nevesisporites vallatus de Jersey & P. 1964	—									
Cadargasporites verrucosus Reis & W. 1969	—									
Todisporites minor Couper 1958	—	—								
Dictyophyllidites mortoni (de J.) Playf. & Dettm. 1965	—	—								
Verrucosisporites varians Volkh. 1972	—	—								
Callialasporites dampieri (Balme) Dev 1961		—	—	—	—	—	—	—		
Callialasporites segmentatus (Balme) Srivast. 1963		—	—	—	—	—	—	—		
Inaperturopollenites turbatus Balme 1957		—	—	—	—	—	—	—		
Ischyosporites marburgensis de Jers. 1963		—	—	—	—	—	—	—		
Microcachryidites antarcticus Cooks 1947				- - -	- - -	- - -	- - -	- - -		
Microcachryidites castellanosii Menéndez 1968				—	—					
Equisetosporites menendezii (Volkh. 1972)				—						
Ischyosporites volkheimeri Filatoff 1975				—	—	- - -	- - -	- - -	- - -	- - -
Nannoceratopsis nov. sp. Volkh. (in prepar)						—				
Gemmatriletes nov. sp. Volkh. & Moroni (in prepar)							—			
Equisetosporites caichiguensis Volkh. & Qu. 1975								—		
Interulobites variabilis Volkh. & Quattr. 1975								—		
Cyclusphaera psilata Volkh. & Sep. 1976									—	
Inaperturopollenites limbatus Balme 1957									—	
Bisacado sp. A (in: Volkh., Caccav. & Sep. 1977)									—	
Stephanocolpites mastandreai Volkh. & Salas 1975									—	—
Huitrinipollenites transitorius Volkh. & Salas 1975									—	—
Retitricolpites sp.										—

CALCAREOUS MICROFOSSILS

CALCAREOUS MICROFOSSILS	A	B	C	D1	D2	E	F	G
Progonocythere neuquenensis Musacchio 1979	—							
Eucytherura? leufuensis Musacchio 1979	—							
Gen. et sp. indet. (Mus. in Dellapé et al. 1979)	—							
Paracytheridea? sp. 1 (in Musacchio 1979 b)		—						
Paracytheridea? sp. 2 (in Musacchio 1979 b)		—						
Progonocythere cf. *reticulata* Dingle & al. 1972			—					
Cytherelloidea andica Musacchio 1979				—				
Amphicytherura (S.) *theloides* Dingle 1969				—	—	—		
'Paranotacythere' *maruchoensis* Musacchio 1979				—	—	—		
Procytherura kroemmelbeini Musacchio 1979					—	—		
Atopochara trivolvis triquetra Grambast 1969						—		
Cypridea ludica Musacchio 1971						—		
Looneyellopsis chinamuertensis (Musacchio 1970)						—		
Huillicythere grambasti Musacchio 1979						—		
Triclypella sp. (in Musacchio 1971)						—		
Cypridea diminuta Vanderpool 1928							—	
Cypridea craigi Musacchio 1975							—	
Cypridea amerikana Musacchio 1975							—	
Flabellochara aff. *harrisi* (Peck 1941)							—	- -
Rayosoana spp.								- -
Mantelliana? ulianai Musacchio & Pal. 1975								- - -

Table 1--Stratigraphic distribution of selected forms in the Jurassic and Lower Cretaceous of the Neuquén Basin.   Palynomorphs: 1: Sinemurian, Pliensbachian & Lower Toarcian; 2: Upper Toarcian & Aalenian; 3: Bajocian; 4a: Lower Callovian; 4b: Middle Callovian; 5: Middle Callovian; 6: Oxfordian; 7: Tithonian & Berriasian; 8: Valanginian, Hauterivian & Barremian; 9: Aptian & Albian.   Calcareous microfossils: A: Middle Callovian; B: Berriasian; C: Valanginian; D1: Lower Hauterivian; D2: Upper Hauterivian; E: Barremian; F: Aptian; G: Albian.

Co.) from the La Cantera Fm. Amongst 17 genera and 29 species, the following important forms were listed: *Appendicisporites tricornitatus, Classopollis classoides, Taurocusporites segmentatus, Cicatricosisporites* sp., similar *C. australiensis.* Stover considered the following nine species as diagnostic for the La Cantera Fm (locality El Toscal): *Lycopodiacidites erraticus, Taurocusporites cureatus, Ephedripites bilateralis, E. martinalis, Monosulcites cuyoensis, Classopollis hyalinus, Phyllocladidites ovalis, Pityosporites choreatus, Pityosporites doris.* As stated by Stover, the flora is Lower Cretaceous.

RIO CHUBUT EMBAYMENT

The following species have been identified in poorly-preserved palynomorphs from lacustrine beds of the Callovian to Upper Jurassic Cañadon Asfalto Fm (Cerro Cóndor area, Chubut Province) (Volkheimer 1970): *Classopollis classoides, Callialasporites trilobatus, Callialasporites dampieri, Araucaricites australis, Lycopodiumsporites austroclavatidites.*

GOLFO DE SAN JORGE BASIN

A Lower Cretaceous (presumably Aptian) continental ostracod (12 species) and charophyte (2 species) assemblage is the only well-preserved one yet found in the Chubut Gp (Musacchio & Chebli 1975). It is the southernmost record of the "Wealden facies" (or facies with *Cypridea*) in the world. The species of *Cypridea: C. diminuta* Vanderpool, *C. craigi* Musacchio, and *C. amerikana* Musacchio, and the charophytes: *Flabellochara* aff. *harrisi* (Peck) and *Stellatochara* aff. *mundula* Peck, strongly resemble others from the Rocky Mountains and nearby parts of the United States. A dispersion through barriers could explain this "polarity".
Two other assemblages (unpublished) have been found in the Chubut Gp. A Lower Cretaceous one includes *Cypridea* spp., *Huillicythere* sp. amongst other ostracods and charophytes, and a Late Cretaceous one with *Ilyocypris (Ilyocypris)* sp. and *I. (Neuquenocypris)* sp. amongst other ostracods and charophythes (Musacchio 1974).

DESEADO MASSIV

The microflora of the Baqueró Fm, Lower Cretaceous of Santa Cruz Province (Archangelsky & Gamerro 1965, 1966 a, b, c, d, 1967) is among the best preserved ones of this age in southern South America. It is characterized by
1- Gondwanic elements, common with Australia (cf. Archangelsky & Gamerro 1967):

Cicatricosisporites hughesii, Alisporites grandis, Podocarpites ellipticus, Trisaccites microsaccatus, Microcachryidites antarcticus, Inaperturopollenites limbatus.
2- Stratigraphically important elements also present in non-gondwanic areas: *Sestrosporites pseudoalveolatus, Rouseisporites reticulatus, Taurocusporites segmentatus, Trilobosporites apiverrucatus, Trilobosporites pulverulentus, Trilobosporites trioreticulosus, Foraminisporites dailyi, Contignisporites cooksonii, Densoisporites velatus, Aequitriradites spinulosus, Aequitriradites verrucosus, Aequitriradites baculatus, Schizosporis reticulatus, Clavatipollenites hughesii, Classopollis torosus.*

AUSTRAL BASIN

From the subsoil of Tierra del Fuego some benthonic foraminiferal faunas from Chile and Argentina were recognized. Sigal *et al.* 1970 distinguished gondwanian, indianian and cosmopolitan affinities in two assemblages from Chile of Late Jurassic (8 species) and Tithonian-Neocomian (16 species) ages. Together with latest Neocomian foraminifers, the ostracod *Majungaella* cf. *nematis* Grekoff, was found.
In Argentina, Malumian & Masiuk 1975, described a Lower Cretaceous assemblage (28 species) from the Pampa Rincon Fm. These authors emphasized the malgachean affinities of the fauna.
Additional information about calcareous microfossils in the Austral Basin can be found in Natland *et al.* 1972 and Rossi de Garcia 1978.

CONCLUSIONS

Nine marine and non-marine palynological and six calcareous microfossil biostratigraphic units of Early Jurassic to Lower Cretaceous ages are recognised in the Andean Region, principally from the Neuquén Basin. Important species for stratigraphic correlation are shown in Table 1, in which international stages define chronological units.

REFERENCES

ARCHANGELSKY, S. J. C. GAMERRO 1965. Estudio palinológico de la Formación Baqueró (Cretácico), provincia de Santa Cruz: I. Ameghiniana 4 (5): 159-170. 1966a-c--II: 4(6): 201-209; III: 4(7): 229-236; IV: 363-372.
---;--- 1966d. Spore and Pollen types of the Lower Cretaceous in Patagonia (Argentina). Rev. Paleobot. Palynol. 1: 211-217.

DELLAPÉ, D.; G. PANDO; M. ULIANA; E. MUSACCHIO 1979. Microfosiles marinos jurásicos y consideraciones estratigráficas sobre la Formación Lotena. Actas VII Congr. Geol. Arg. (in press), Buenos Aires.

MALUMIÁN, N.; V. MASIUK 1975. Foraminíferos de la Formación Pampa Rincón (Cretácico Inferior) Tierra del Fuego, Argentina. Rev. Esp. Microp. VII (3): 579-600.

MUSACCHIO, E. 1970. Ostrácodos de las superfamilias Cytheracea y Darwinulacea de la Formación La Amarga (Cretácico Inferior) en la Provincia de Neuquén, República Argentina. Ameghiniana VII (4): 301-316.

--- 1971a. Charophytas de la Formación La Amarga (Cretácico Inferior) en la Provincia de Neuquén, Argentina. Rev. Mus. La Plata (n. ser.) Pal. VI (37): 19-38.

--- 1971b. Hallazgo del género Cypridea en Argentina y consideraciones estratigráficas sobre la Formación La Amarga (Cretacico Inferior) en la Provincia de Neuquén. Ameghiniana VIII (2): 105-125.

--- 1974. Microfósiles del Grupo Chubut. Yacimientos Petrolíferos Fiscales (unpublished), Buenos Aires.

--- 1978. Microfauna del Cretácico Inferior y el Jurásico. Actas VII Congr. Geol. Arg. Relatorio, 147-161.

--- 1979a. Algunos microfósiles calcáreos marinos y continentales del Jurásico y el Cretácico Inferior de la República Argentina. Actas II° Congreso Arg. Paleont. y Estrat. (in press).

--- 1979b. Ostrácodos del Cretácico Inferior en el Grupo Mendoza Provincia de Neuquén, Argentina. Actas VII Congr. Geol. Arg. (in press). Buenos Aires.

--- 1979c. Datos Paleobiogeográficos de algunas asociaciones de foraminíferos, ostrácodos y carofitas del Jurásico Medio y el Cretácico Inferior de Argentina. Ameghiniana (in press).

---; G. CHEBLI 1975. Ostrácodos no marinos y carofitas del Cretácico Inferior en las Provincias de Chubut y Neuquén, Argentina. Ameghiniana XII (1): 70-96.

---; S. C. PALAMARCZUK 1975. Microfósiles calcáreos de la Formación Ranquiles (Cretácico Inferior) en la Provincia de Neuquén, Argentina. Ameghiniana XII (4): 306-314.

NATLAND, M. L.; P. GONZÁLEZ; A. CAÑON; M. ERNST 1974. A System of Stages for correlation of Magallanes Basin Sediments. Geol. Soc. Am. Mem. 138, 125.

ROSSI DE GARCÍA, E. 1977. El género Novocythere (Ostracoda) (Perforacion SC-1). Santa Cruz, República Argentina. Ameghiniana XIV (1-4): 117-121.

SADRAS, W. 1973. Evaluación de formaciones en Argentina. Public Cia Schlumberger S.A., 215 pag., Buenos Aires.

TASCH, P.; W. VOLKHEIMER 1970. Jurassic conchostracans from Patagonia. Univ. Kansas Paleontol. Contrib. Pap. 50: 1-23.

VOLKHEIMER, W. 1968. Esporas y granos de polen del Jurásico de Neuquén (República Argentina). I. Descripciones sistemáticas. Ameghiniana 5 (9): 333-370.

--- 1970. Jurassic microfloras and paleoclimates in Argentina. II Intern. Gondwana Symp., Proc. Papers: 543-549, Pretoria.

--- 1972. Estudio palinológico de un carbón caloviano de Neuquén y consideraciones sobre los paleoclimas jurásicos de la Argentina. Rev. Mus. La Plata (n. s.), Pal. 6, 101-157.

--- 1973. Palinología estratigráfica del Jurásico de la Sierra de Chacai-Có y adyacencias (Cuenca Neuquina, Republica Argentina). I Ameghiniana 10 (2): 105-131.

--- 1976. Liassic microfloras of the Neuquen Basin (Argentina). Relations with other gondwanic areas. IV Internat. Palynol. Conf. Lucknow (in press).

---; M. CACCAVARI DE FELICE; E. SEPULVEDA 1977. Datos palinologicos de la Formacion Ortíz (Grupo La Amarga), Cretácico Inferior de la Cuenca Neuqina (República Argentina). Ameghiniana XIV (1-4): 59-74.

---; U. ROSENFELD (in prep.). Mikropaläobotanische Charakteristik von Paläoenvironments der mitteljurassischen Lajas-Folge (Neuquén-Becken, Argentinien).

YRIGOYEN, M. R. 1975. La edad cretacica del Grupo Gigante (San Luis) y su relacion con cuencas circunvecinas. Actas I Congr. Argent. Paleont. Biostr. 2: 29-56.

ZAMBRANO, J. J.; C. M. URIEN 1970. Geological outline of the basins in southern Argentina and their continuation off the Atlantic shore. J. Geophys. Res. 75 (8): 1363-1396.

The Gondwana group along the Himalayan zone in northeastern India and its paleogeographic significance (Abstract)

J. ROY CHOWDHURY & C.S. RAJA RAO
Geological Survey of India, Calcutta

A narrow, 300-km long belt of Gondwana formations in front of the Himalayas comprises a continental and marine facies. Sedimentation commenced with diamictite, followed by a marine sequence of black shale, siltstone, sandstone, shale, and estuarine thin coal seams with intercalations of basic volcanics. The sequence is thicker in the west, and late-stage volcanics are dominant in the east.

The sequence is abruptly terminated against the NW-SE-trending Mismi thrust. The eastern Himalayan Gondwana trough was a part of the Indian plate and was close to the Tethyan Shelf Zone. The eastern end of the belt was probably contiguous with the sequence in western Australia, and the two have been separated by the Mismi thrust. Tectonic modifications of physiography during the late Paleozoic led to independent Mesozoic histories of the two areas.

The Indian Gondwana margin extends northwards at least to the Lachi Series in Sikkim. Serpentinite associated with the thrusts of the Lohit Himalayas and those near the Mismi massif (SE extension of Tibetan Cathaysia) may mark the junction of the Greater Indian Gondwana plate with the Cathaysian-Laurasian land mass.

Silurian glaciation in central South America

JOHN C. CROWELL
University of California, Santa Barbara, USA

A. C. ROCHA-CAMPOS
Universidade de Sao Paulo, Brazil

RAMIRO SUÁREZ-SORUCO
Yacimientos Petrolíferos Fiscales Bolivianos,
Santa Cruz, Bolivia

The Cancañiri Formation consisting in part of diamictite containing rare striated and faceted boulders is interpreted as deposited during the late Ordovician-Silurian refrigeration recorded in the Amazon basin, central Sahara region, South Africa, and western Europe. Paleomagnetic data suggest that the Bolivian portion of the Gondwana supercontinent moved across the south pole during Early Silurian time to account in part for the glaciation. Tentative paleographic reconstructions indicate that seas in the Cancañiri depositional basin and in the Amazon basin provided evaporation sources for nearby ice caps or sheets on the Pampean massif and on the Guiana and Brazilian shields.

INTRODUCTION

Continental glaciation near the transition between the Ordovician and Silurian Periods left a marked imprint on Gondwana continents. Our purpose here is to describe briefly the record of this ancient ice age in Bolivia and adjoining parts of southeastern Peru and northern Argentina (Fig. 1). Cancañiri Fmn strata preserve a faint record of this and are in a tectonically complex belt in the high eastern Andes. They crop out locally and are known from drill holes in a belt about 1500 km long and up to 400 km wide. Although many stratigraphic sections have been described and measured, in particular by geologists of the Yacimientos Petrolíferos Fiscales Bolivianos, studies aimed at reconstructing the basin of deposition have just begun. We here review what is now known about this mid-Paleozoic glaciation and speculate on its relation to Silurian geography.

STRATIGRAPHY

Lower Silurian strata in Bolivia were termed the Cancañiri Greywacke as early as 1919 but mentioned first in publication in 1935 (Turneaure), and continental glaciation was first discussed by Keidel (1940). The name Zapla Glacial Horizon was applied to beds of about the same age in Argentina by Schlagintweit (1943). Rivas and Carrasco (1968) formally established the Cancañiri Fmn in Bolivia, although the name Sacta is used in places

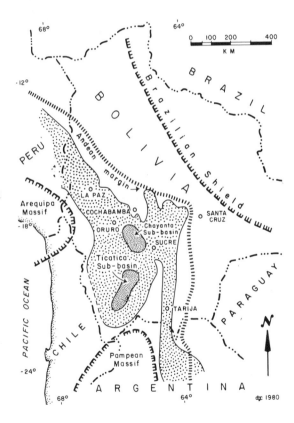

Fig. 1--Map showing distribution of Silurian Cancañiri Fm in Bolivia and adjoining Peru and Argentina.

by some (Branisa *et al.* 1972). In Argentina the term Zapla has been replaced by Mecoyita Fmn (Turner 1964), in Peru locally by San Gaban Fmn (Suárez-Soruco MS, 1977), and in northern Chile, part of the Gualchagua Fmn may correlate with the Cancañiri. It contains iron, tin, antimony, and other metals in the Sucre region of Bolivia and in northern-most Argentina. In outcrop the strata are indurated and locally cleaved but are not significantly metamorphosed.

The Cancañiri Fmn lies mainly discon-formably upon older beds, assigned to the Ordovician System and locally with an angular discordance of a few degrees. Near the border between Argentina and Bolivia, it rests upon grey sandstone layers of the Santa Rosita Fmn, which, at places, may be as old as Tremadocian. Northward, the Cancañiri lies upon successively younger stages and near Tarija overlies Caradocian beds. Whether Ashgillian beds are present either below or within the formation is not yet known, so it is assigned entirely to the Silurian System as discussed below. In Bolivia the San Benito Fmn (white quartzitic sandstone with occasionally interbedded thin dark silt and shale) underlies it, and the Cancañiri is overlain conformably by the Kirusillas Fmn, also largely consisting of dark grey shale with inter-bedded sandstone layers. In northern Argentina it is overlain conformably by the Lipeon Fmn of shale beds with flaser structure.

The thickness of the Cancañiri Fmn in Bolivia measured by geologists of the Yacimentos Petrolíferos Fiscales Bolivianos, ranges from 21 - 680 m. In southeastern Peru, about 200 m have been identified (Dávila and Ponce de León 1971). Nearly 1400 m may occur in the Ticatica Sub-basin and over 1000 m in the Chayanta Sub-basin (Fig. 1). Although details of thickness changes are not yet known it is likely that the formation is transgressive to the south and west.

LITHOLOGY

The Cancañiri Fmn consists of diamictite with interbedded laminated dark shale, bedded sandstone and conglomerate (Fig. 2). Stratigraphic sections so far studied consist of those made up almost entirely of diamictite to those of about half diamictite. Most exposures at the base of the formation consist of stone-rich diamictite lying unconformably upon Ordovician beds although at places the base is conglomeratic and sandy. Glacial striations at the unconformable base

of the formation have not been confidentl identified. The top of the formation is defined as the top of the highest diamic-tite layer which in Bolivia is in sharp contrast with sandstone and shale of the Kirusillas Fmn. Many stratigraphic secti are incomplete because of faulting.

The diamictites, dark brown in fresh outcrops but tan with distinctive spheroi weathering in most exposures, range from massive to bedded, from sandy to argill-aceous, and from sparse-stoned to crowded Sandstone lenses, many of which are defor and rolled up by deformation shortly afte deposition, are characteristic. Many of these, averaging 2 m thick and 10 m long, are pod- or cigar-shaped and erode out resistantly in the softer massive diamic-tite. Detached folds and hooks and wisps of sandstone within massive diamictites attest to dismemberment of sandstone lens followed by their deformation and kneadin while the mass was still soft and pliable.

Some diamictite layers occur in beds only a few centimetres thick with wavy or gradational contacts, and isolated stones thicker than the layers of diamictite occur locally. Indisputable dropstones have not been identified. Other diamictite beds grade into clay grits or are overlain by calcareous horizons, and a biostrome has been re-ported (Bransia *et al.* 1972). Inter-bedded current-bedded and ripple-marked sandstone beds contrast sharply with poorly sorted layers gradational into true diamictites. Rare interbedded sandstone layers display grading and sole marks, including flute casts, groove casts, and irregular load forms.

Laminated dark green shale layers, some with ferruginous concretions, are inter-calcated within the Cancañiri diamictites. Laminations are thin and continuous, but no lonestones were seen within them. However, large stones up to 150 cm in diameter occur within diamictite layers as much as 40 cm thick, interbedded in sandstone and shale units.

Megaclasts within the Cancañiri Fmn are usually no larger than 15 cm in diameter although locally they reach 150 cm. The stones are generally rounded and subrounded with sphericities of 0.5 to 0.7 (Rodrigo *et al.* 1977) and include quartz, grey granitic gneiss, quartzite, pink granite, grey pebble conglomerate, and slate. At places blocks and pieces of sandstone and shale, from both within the formation and units below it, occur within massive diamictite. Some large stones are faceted, and a few of the facets are glacially striated.

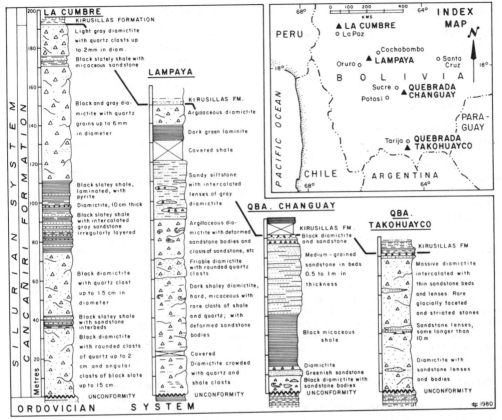

Fig. 2--Four stratigraphic columns of the Cancañiri Fm; locations shown on index map (insert). The sections were measured and described by geologists of the Yacimentos Petrolíferos Fiscales Bolivianos (J. Aguilar, G. Jordán, J. M. López, E. Rodriguez, D. Rojas, and R. Suárez-Soruco).

Some pink granites are similar to those of the Precambrian massifs of Santa Victoria, Argentina, and the Arequipa region of southern Peru. Lohmann (1965) has inferred that much of the debris came from the Pampean massif on the south and the Arequipa massif or Altiplano block on the west, but until more is known about the petrology of both potential source areas and derived clasts, this correlation is tentative. On the northeast, igneous and metamorphic rock types suggest derivation from the Brazilian shield as well. Incomplete data suggest stones are smaller and there are more deformed sand bodies in the middle of the depositional region.

Compositional variations in the diamictite matrices are notable. On the south, the matrix is sandier and in the centre of the depositional region it is argillaceous. Interbedded sandstones consist of lithic subfeldspathic wackes and quartz-wackes with argillaceous

matrices (Rodrigo *et al.* 1977). Detrital mica and authigenic chlorite and pyrite are widespread and the matrix clay is mostly illite.

AGE

Stratigraphic position places the Cancañiri Fmn within the interval Ashgillian-Wenlockian. The youngest strata underlying the formation are assigned to the Caradocian Stage and are zoned on the basis of graptolites, trilobites, and brachiopods (Suárez-Soruco 1976, Fig. 28). Fossils or strata of the Ashgillian Stage have not been recognized, and the highest zones identified are the *Orthograptus truncatus*, *Homalonotus bistrami*, and *Bistramia elegans* zones, all assigned to the Caradocian Stage. Beds overlying the diamictites are assigned to the *Neoveryhachium carminae* and *Pristiograptus colonus* zones of the Ludlovian Stage.

Unfortunately, the Cancañiri Fmn contains

very few invertebrate fossils, and the few brachiopods collected were not diagnostic of age. Palynological investigations have been more helpful and indicate a Wenlockian age (*Duvernaysphaera jelinii* zone) based on acritarchs and chitinozoans. Inasmuch as the Ordovician strata have been slightly folded, uplifted, eroded, and depressed after Caradocian time, it is likely that these events took place during the Ashgillian, and during part or all of the Llandoverian. In summary, the Cancañiri Fmn is considered as primarily Wenlockian, but deposition may have commenced in the Llandoverian. This is somewhat younger than the Llandoverian age given by Berry and Boucot (1972, Fig. 2).

PALEOGEOGRAPHY

On the basis of rare marine fossils, primarily brachiopods, the Cancañiri Fmn is inferred to be almost entirely marine.

This interpretation is supported by bioturbated horizons within the formation and stratigraphically just above it. The widespread distribution of the formation indicates deposition in a large body of water, more than 400 by 1500 km, arguing for a sea rather than a huge lake. In addition, the occurrence of the bioturbated layers and rare fossils suggest that the Cancañiri Sea (?) was not always frozen over.

Diamictites with deformed sandstone bodies suggest deposition upon unstable slopes during rapid clastic sedimentation in quiet water with little current or wave reworking. Large quantities of fine-grained detritus were carried in, and much moved downslope, mixing and churning as it moved. Beds and lenses of sandstone were therefore deformed shortly after they were deposited, and kneaded into the moving diamicton mass. Tentative interpretations on the sources of the detritus, largely based on inferences from clast types, suggest that they lay largely to the west and south but included a contribution from the Brazilian shield on the east. Data from thickness and facies changes, from paleocurrents, and from stone petrology matched to provenance areas are as yet incomplete so that the documentation for these source regions is insecure. Moreover, tectonic movements are complicated, and it may be that crustal blocks have been moved about in the Andean region since the early Paleozoic in ways not yet recognized.

It is likely that some glaciers occupied the region bordering the Cancañiri basin on the west and south in view of the occurrence of faceted and glacially striated exotic blocks. It is not known, however, whether the inferred glaciers were alpine or were ice caps or piedmont glaciers reaching the sea and extending out upon the water. That ice shelves did exist, however, is reasonable in view of the huge quantity of mud and silt now present in the diamictites. One way of forming these rocks is by the subaqueous wasting of ice shelves along with detritus from subglacial streams, and more understanding of such an environment is needed (see, e.g. Kurtz and Anderson 1979). Present knowledge, however, does not permit sound judgment of the role of glaciation here nor of the proportionate role of downslope sliding and mixing.

We therefore infer that the paleoclimate was cold in regions marginal to the Cancañiri basin, and that these regions included highlands if not true mountains. It may be that ice caps lay upon parts of the Brazilian shield to the east as well if the identification of debris in the diamictites from such a source is confirmed. Glaciers on the Brazilian and Guiana shields might also have contributed to diamictites in the Trombetas Fmn accumulating to the north in the middle Amazon basin (Caputo *et al.* 1972; Rocha-Campos *in press*).

Continental glaciation at the end of the Ordovician Period is especially well documented in the central Saharan region of northern Africa (Beuf *et al.* 1971). In addition, evidence of glacier activity has been reported from many other regions in the Northern Hemisphere, and as far south as the Cape Ranges of South Africa (Harland 1972; Frakes 1979). Here, about 200 km NNE of Cape Town, the Pakhuis Conglomerate Fmn has morainal material associated with striated surfaces and stones as well as complex soft sediment folds and faults believed formed when glacial ice reached the basin (Rust *in press*). Evidence of glaciation at approximately this time is also reported from West Africa: in the Taoudeni Basin of Mauritania; in the Tindouf Basin, near the common boundaries between Morocco, Algeria, and Mauritania (Deynoux *et al.* 1972) and from Sierra Leone (Tucker and Reid 1973). These West African glaciations are primarily Ashgillian but may include Caradocian and Llandoverian times.

Across the present Atlantic Ocean in Brazil, the Lower Silurian Trombetas Fmn in the middle Amazon Basin displays possible evidence of glaciation (Caputo *et al.* 1973; Rocha-Campos *in press*). The formation outcrops in two E-W-trending belts along the margins of the basin and

has also been penetrated in wells within
the basin. The Nhamundà Member of the
formation contains diamictites, but at
present sedimentological and other data
indicating convincing glacial origin are
lacking. Shale beds of the Pitinga
Member above the Nhamundà member have
abundant marine fossils indicating a
Llandoverian age. If these widespread
marine diamictites are shown to document
glaciation, they will provide a connection across South America between the
African and Bolivian records of the
Ordovician-Silurian ice age.

Although as yet no paleomagnetic
measurements have been made on rocks of
the Cancañiri Fmn, other paleomagnetic
investigations show that the Bolivian
region lay at high latitudes. Reinter-
pretation of Gondwanan poles by Schmidt
and Morris (1977) and by Morel and
Irving (1978) suggest that apparent polar
wander paths may include complex loops
(Fig. 3). The assembled Gondwana supercontinent may have crossed the pole in
such a way that the apparent polar
wander path moved across South America
from the Saharan region to the Paleo-
Pacific Ocean during the Ordovician and
Silurian, and then back to now-central
Africa by the Devonian. The elongate
Amazon basin would therefore have been
an antarctic intracratonic sea, perhaps
at times ice-covered and connected to
open ocean near the juncture between

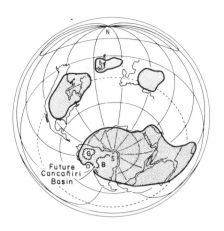

Fig. 4--Late Ordovician map showing posi-
tioning of Gondwana supercontinent upon
the S. Pole, redrawn from Morel & Irving
1978, Fig. 8). The Cancañiri basin as
shown is inferred to develop at the beginning of the Silurian Period after the
time interval of this map, as the APWP
followed that shown in Fig. 3. G = Guiana
shield; B = Brazilian shield.

northwestern Africa and northern South
America (Fig. 4). Perhaps as well ice
caps or sheets lay upon both the Guiana
and Brazilian shields. This polar siting
and the existence of near-polar seas to
provide accessible evaporative moisture
to precipitate as snow on nearby shield
regions would reasonably explain the
distribution of continental glaciers at
the beginning of the Silurian Period. The
fitting of the Pampaean and Arequipa
massifs into this scheme, however, is
still not clear, primarily because of the
difficulties in removing the overprinting
of later complex tectonic events in the
Andean region.

ACKNOWLEDGMENTS

The first-named author acknowledges with
thanks courtesies extended by Yacimientos
Petrolíferos Fiscales Bolivianos during
short visits to Bolivia in 1975 and 1976,
mainly through arrangements made by the
Continuing Education Program of the
American Association of Petroleum Geologists. Some research support was also
provided by United States NSF Grant EAR
77-06008. Critical comments on the manuscript by Mario Vicente Caputo, Nicholas
Christie-Blick, and Julia M.G. Miller are
appreciated.

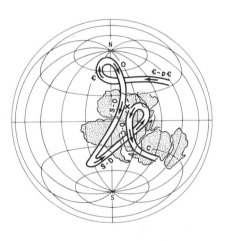

Fig. 3--Apparent polar wander path for the
Lower Paleozoic, redrawn from Morel & Irving (1978, Fig. 1). X = previously-pub-
lished connecting path in Africa (McElhinny et al. 1974). pϾ = Precambrian; O =
Ordovician; Ͼ = Cambrian; S = Silurian;
D = Devonian; C = Carboniferous.

REFERENCES

BERRY, W.B.N.; A.J. BOUCOT (Eds.) 1972.
Correlation of the South American
Silurian Rocks. Geol. Soc. Am. Spec.
Paper 133. 59 pp.

BEUF, S.; B. BIJU-DUVAL; O. DeCHARPAL;
P. ROGNON; O. GARIEL; A. BENNACEF 1971.
Les gres du Paleozoique inferieur au
Sahara--Sedimentation et discontinuities,
evolution structurale d'un craton.
Inst. Fr. Pétrole-Sci. Tech. Pétrole 18
464 pp.

BRANISA, L.; G.A. CHAMOT; W.B.N. BERRY;
A.J. BOUCOT 1972. Silurian of Bolivia.
Geol. Soc. Am. Spec. Paper 133: 21-31.

CAPUTO, M.V.; R. RODRIGUES; D.N.N.
de VASCONCELLOS 1972. Nomenclatura
estratigráfica da bacia do Amazonas.
An. XXVI Congr. Bras. Geol., Soc. Bras.
Geol. 3: 35-46.

CAROZZI, A.V.; H.R.P. PAMPLONA; J.C.
deCASTRO; C.J.A. CONTREIRAS 1973.
Ambientes deposicionais e evoluçao tecto-
sedimentar da seção clástica palezóica
da bacia do Medio Amagonas. An. XXVII
Cong. Bras., Geol., Soc. Bras. Geol. 3:
279-314.

DÁVILA, J.; V. PONCE deLEÓN 1971. La
sección del río Inambari en la faja
Subandina del Peru y la presencia de
sedimentitas de la Formación Cancañiri
(Zapla). Rev. téc. Yacimientos Petrol.
Fisc. Bolivianos 1 (1): 67-85.

DEYNOUX, M.; O. DIA; J. SOUGY; R. TROMPETTE
1972. La glaciation "Fini-Ordovicienne"
en Afrique de l'ouest. Bull. Soc. géol.
minéral. Bretagne, Ser. C 4 (1): 9-16.

FRAKES, L.A. 1979. Climates throughout
geologic time. Elsevier Amsterdam. 310 pp.

HARLAND, W.B. 1972. The Ordovician ice
age. Geol. Mag. 109: 451-456.

KEIDEL, J. 1941. Paleozoic glaciation in
South America. Proc. Eighth Am. Sci. Cong.
4: 89-108.

KURTZ, D.D.; J.B. ANDERSON 1979. Recognition and sedimentologic description of
Recent debris flow deposits from the Ross
and Weddell Seas, Antarctica. J. Sediment.
Petrol. 49: 1159-1170.

LOHMANN, H.H. 1965. Paläozoische Vereisun-
gen in Bolivien. Geol. Runds. 54: 161-165.

McELHINNY, M.W.; B.J.J. EMBLETON 1974.
Australian palaeomagnetism and the
Phanerozoic plate tectonics of eastern
Gondwana. Tectonophys. 22: 1-29.

---; J.W. GIDDINGS; B.J.J. EMBLETON 1974
Palaeomagnetic results and late Precambrian
glaciations. Nature 248. 557-561.

MOREL, P.; E. IRVING 1978. Tentative
paleocontinental maps for the early
Phanerozoic and Proterozoic. J. Geol.
86: 535-561.

RIVAS, S.; R. CARRASCO 1968. Geología
y yacimientos minerales de la región de
Potosí, Tomo 1. Parte geol. Bol. Serv.
Geol. Bolivia 11. 95 pp.

ROCHA-CAMPOS, A.C. (in press). The Late
Ordovician (?)-Early Silurian Trombetas,
Formation, Amazon Basin, Brazil. In
M.J. Hambrey and W.B. Harland (Eds),
Earth's Pre-Pleistocene Glacial Record.
Cambridge Univ. Press.

RODRIGO, L.A.; A. CASTAÑOS; R. CARRASCO
1977. La formacion Cancañiri, sedimen-
tologia y paleogeográfia. Rev. Geoci.
Univ. Mayor San Andrés, La Paz, Bolivia,
1 1: 1-22.

RUST, I.C. (in press). Lower Palaeozoic
Pakhuis Tillite, South Africa. In M.J.
Hambrey and W.B. Harland (Eds), Earth's
Pre-Pleistocene Glacial Record. Cambridge
Univ. Press.

SCHLADINTWEIT, O. 1943. La posición estratigráfia del yacimento de hierro de
Zapla y la difusión de horizonte glacial
de Zapla en la Argentina y en Bolivia. Re
Min., Geol., Mineral. 13(4): 115-127.

SCHMIDT, P.W.; W.A. MORRIS 1977. An
alternative view of the Gondwana Paleozoic
apparent polar wander path. Canad. J.
Earth Sci. 14: 2674-2678.

SUÁREZ-SORUCO, R. 1976. El Sistema
Ordovício en Bolivia. Rev Tec. Yacimiento
Petrol. Fisc. Bolivianos 5: 111-223.

---; 1977. Bosquejo de la estratigráfia y
paleogeográfia de la formación Cancañiri
(Silurico) en Bolivia. Mimeographed
typescript. 16 pp, 20 figures.

TUCKER, M.E.; P.O. REID 1973. The sedimentology and context of Late Ordovician
glacial marine sediments from Sierra
Leone, West Africa. Palaeogeog.,
Palaeoclimat., Palaeoecol. 13: 289-307.

TURNEAURE, F. S. 1935. The tin deposits
of Llallagua, Bolivia. Econ. Geo. 30:
170-190.

TURNER, J.C.M. 1964. Descripción
geológica de la Hoja 2c, Santa Victoria.
Inst. Nac. Geol. Min., Buenos Aires, Bol.
102.

Late Carboniferous glacial and fluvioglacial deposits
in the Tshipise Basin, South Africa

J. N. J. VISSER
University of the Orange Free State,
Bloemfontein, South Africa

H. J. VAN DEN BERG
South African Iron & Steel Industrial Corp.

In the Tshipise Basin glacial strata more than 80 m thick which consist of three lithofacies form the base of the Karroo sequence and are conformably overlain by coal measures. The diamictite facies consists of subangular to rounded clasts, up to 2 m in diameter, of predominantly extrabasinal rocks in a sandy matrix and contains upward-fining cycles and banding showing deformational structures. Sandstone facies consists of upward-fining cycles of coarse to medium-grained, horizontally to cross-bedded sandstone grading into siltstone and mudstone. Mudstone facies consists predominantly of brownish mudstone showing distorted bedding and containing scattered sand and grit-size grains and organic fragments.

The sequence fines upwards and westwards. The diamictite accumulated subglacially but grades westwards into fluvioglacial sediments consisting of alternating diamictite and sandstone beds deposited at the ice margin. The fining-up sandstones were laid down by braided streams in a proglacial environment, while the mudstone facies represents marsh deposits. Indirect evidence indicates paleo-ice-flow from the east while sedimentation took place during glacier retreat.

INTRODUCTION

The Tshipise Basin contains scattered outliers of Karroo strata on Early Precambrian basement along an E-NE-trending belt about 250 km long and 25 km wide close to the northern boundary of South Africa. The weakly consolidated Karroo rocks were preserved on the southern down-thrown sides of post-Karroo faults. Surface mapping of the area was carried out periodically since the start of the century, but no detailed investigation of the Karroo strata was undertaken and no indisputable glacial deposits were identified. The presence of the Dwyka Fm or its equivalent was considered doubtful (Van Zyl 1950: 27; Haughton 1969: 377; Frakes & Crowell 1970: 2276).

Recent coal exploration in the Tshipise Basin yielded a wealth of borehole information about the basal beds of the Karroo sequence.

STRATIGRAPHY AND LITHOLOGY

The stratigraphy for the basin is given in Table 1. The Beaufort Gp of the Main Karroo Basin apparently is not represented in the Tshipise Basin, but Haughton (1969: 378) mentions that the dinosaurs in this basin are identical with those of the Elliot Fm near Lesotho. On stratigraphical similarities and paleontological evidence one can assume that depositional processes in the Main Karroo and Tshipise basins followed the same sequence of events.

The Glacial zone (Fig. 1) rests unconformably on Precambrian basement and its upper contact is defined at the base of the carbonaceous beds of the coal facies, but this contact is in places transitional with intervening coarse-grained rocks which contain no evidence of glacial action and show affinities with the overlying coal facies. The Glacial zone thins and finally pinches out against an E-NE-trending high in the basin floor but thickens north and southwards from this high.

Most of the glacigene rocks are represented by borehole cores in which sedimentary structures and the geometry of sedimentary bodies cannot be studied. Fortunately,

TABLE 1   Karroo sequence in the Tshipise Basin

Max. thickness (m)	Lithology	Correlation with Main Karroo Basin
(Top) 1200	Lava (limburgite and olivine basalt)	Drakensberg Volcanic Group
120	Fine to medium-grained sandstone (Bushveld sandstone)	Clarens Formation
400	Alternating greenish grey siltstone and sandstone and reddish mottled mudstone; carbonate concretions; dinosaur remains	Elliot Formation
130	Cross-bedded feldspathic sandstone and grit; grey mudstone	Molteno Formation
	U N C O N F O R M I T Y	
350	Carbonaceous shale; intercalated coal seams and sandstone lenses	Ecca Group
(Base) 80+	Diamictite; sandstone; mudstone (Glacial zone)	Dwyka Formation

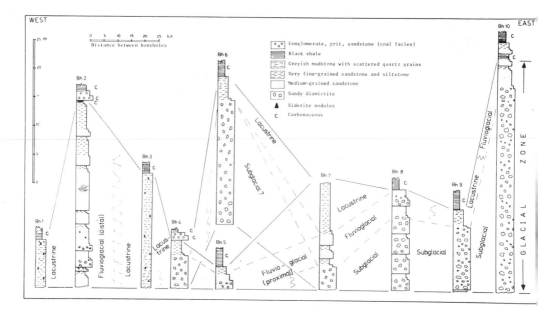

Fig.1--Stratigraphic sections and facies interpretation based on borehole data from the Tshipise Basin.

four small outcrops yielded valuable information. Three lithofacies recognised in the Glacial zone show a pronounced facies change from east to west except for the area around Boreholes 2 and 6, which appear to be anomalous (Fig. 1).

Sandy Diamictite Facies. This has a maximum thickness of 51 m and is well developed towards the east and NE of the basin but absent in the west. The facies consists of upward-fining cycles (averaging 2-3 m thick) defined by an upward decrease in clast size or a gradation from diamictite to sandstone (Fig. 1).

The light-coloured rock which consists of unsorted scattered clasts in a sandy matrix, usually has a massive appearance. The clasts are matrix supported, but local gradations to an orthoconglomerate or pebbly sandstone occur. Distorted fine-grained laminae (1-5 mm thick) partly fold around clasts. Sometimes these laminae form the upper parts of thin beds of unsorted coarse material at the base grading upwards into a small flat-pebble breccia smoothed at the top (Fig. 2). The moderately sorted matrix consists mostly of coarse sand containing moderately rounded quartz and quartzite grains as well as very angular fragments, with very small amounts of white clayey material trapped between the grains. Where the clasts are composed of argillaceous rock types and lava the matrix is markedly more fine-grained and darker coloured.

The megaclasts vary from 40 mm up to 2 m in diameter with the most common <100 mm. The largest were recorded in the east, as well as at the base of diamictite beds, and tended to be more angular than the smaller ones. A large percentage of the clasts lie with their long axes parallel to the bedding. The extra-basinal clasts are composed of reddish quartzite, glassy quartzite, white quartzite, reddish shale, greenish schist, greenish lava, granite, an

Fig.2--Smoothed surface in diamictite

vein quartz. Fragments of coarse-grained greyish sandstone and mudstone represent intrabasinal material. In the northern boreholes clasts of granite and vein-quartz dominate, but these disappear southwards where quartzite and lava fragments are the most abundant. Clasts exhibiting faceted surfaces and in two instances striations were found.

Sandstone-siltstone Facies. This either forms the upper part of a diamictite cycle or upward-fining cycles grading from coarse-grained moderately to poorly sorted sandstone at the base, through well-sorted fine-grained sandstone to siltstone at the top (Fig. 1). The cycles vary in thickness from about 1 to 7 m. The facies has a maximum thickness of 94 m, but there is no distinct pattern in its thickness distribution, although there appears to be a sympathetic relationship between the diamictite and the sandstone development.

In the east the whitish sandstone is coarser grained and contains appreciable amounts of feldspar. Shale fragments as well as rounded vein-quartz pebbles are sometimes present close to the base of the upward-fining cycles. Moderately rounded quartz and feldspar grains and to a lesser extent rock fragments are the dominant constituents with very subordinate light-coloured clayey matrix. Sedimentary structures are scarce. Cross-bedding, apparently of a low angle, is occasionally seen, and parallel bedding is well developed in the silty sections.

Mudstone Facies. Greyish mudstone forms the tops of upward-fining cycles, and towards the west the entire Glacial zone consists of mudstone which reaches a maximum thickness of about 20 m. It is a very homogeneous rock containing scattered sand-size quartz grains as well as occasional subrounded clasts up to 2 cm and thin grit layers. Rootlet beds are found at the tops of thick mudstone units, and fragments of carbonaceous material are abundant throughout. Small siderite nodules are disseminated throughout the mudstone and sometimes are also present in the underlying siltstone. The amount of siderite tends to increase towards the west. A prominent feature of the thick mudstone units is slumping and distorted bedding.

DEPOSITIONAL HISTORY

Basin Configuration. As the area was severely affected by pre-Karroo as well as post-Karroo faulting (Van Zyl 1959: 11-18) it is difficult to establish the outline of the basin with the limited outcrops available. Pre-Karroo and syndepositional faulting probably controlled the shape, as

well as subsidence in the basin as isopachs of the formations tend to follow a linear pattern similar in strike to the major faulting. Indications are that the basin was of considerably larger extent than the present outcrop belt, and that only remnants of the basin fill have been preserved by the post-Karroo step faulting.

Structural contours of the basement show an E-NE trend, and an east-dipping nose subdivided the basin into northerly and southerly sloping parts which probably converged towards the east. Borehole data show the nose to consist of lava, while the southern and northern parts of the basin have hard resistant quartzite and granitic basements respectively. This is difficult to explain by normal erosional processes, even if block faulting had occurred, but if weathering had taken place under extreme cold conditions frost shattering would have disintegrated the quartzite before the lava.

Depositional Environment. Sedimentation in a glacial environment is indicated by the texture and composition of the rock, the presence of faceted and striated stones, and the close relationship between clast and matrix composition in different parts of the basin. The stratigraphic position of the zone, below a coal facies, corresponds with glacial beds in neighbouring Zimbabwe-Rhodesia (Chappell & Humphreys 1970: 501-505) and in southern Transvaal (De Jager 1976: 305-306). There are thus reasonable grounds to assume that the basal sequence in the Tshipise Basin includes glacial and proglacial deposits (Fig. 1).

The characteristics of the diamictite, including the crude bedding and fine banding, indicate either subglacial (Edwards 1975: 64) or subaqueous outwash (Dreimanis 1976: 37; Rust & Romanelli 1975: 188-191) deposits. A subaqueous deposit can be ruled out because associated sediments indicate very shallow water or subareal processes. Considering the lateral distribution of the diamictite and its association with sandstone and mudstone the deposits represent more than one glacial subenvironment. The apparently massive diamictite showing clast grading and containing angular to rounded material was probably deposited subglacially from an englacial or basal tractional position (Boulton 1978: 776) in the presence of abundant meltwater. The disturbed bedding could indicate stress underneath the ice, while the long-pebble axes lying nearly in the plane of bedding could point to reworking by subglacial streams (Pessl 1971: 103) or embedding in the lodgement till (Boulton 1978: 782-785). The diamictite grading upwards into sandstone was deposited at the snout of the glacier where it was reworked by meltwater streams during rapid melting of the ice, or by possible "jökulhlaups" characteristic of certain glaciers in Iceland (John 1977: 82). Its fluvioglacial origin is evidenced by the lenticular sedimentary bodies, the abrupt changes in grain size and the poorly rounded to subrounded clasts. The smoothed stratigraphic surfaces in the diamictite could have been caused by periodic over-riding of the deposits by the ice (Fig. 2).

The thick sandstone sequences showing upward-fining cycles represent a different environment than the sandstones closely associated with the diamictite, which were interpreted as proximal outwash deposits. The largely horizontal bedding and only occasional cross-bedding, the lack of overbank deposits, and the association with grit favour deposition in a high kinetic energy environment. The sequences best fit braided stream deposits, which are frequently associated with glacial environments (Miall 1977: 51). Periodic floods during melting of the ice, which are common in the proglacial areas, are capable of depositing coarse-grained sand bodies in a more distal proglacial environment.

The presence of rootlet beds in the mudstone ruled out deposition at any great water depth, but for the preservation of organic fragments in the rock under such shallow conditions the oxidation-reduction interface must have laid in the sediment itself. Deposition of clay and silt probably took place in a swamp environment with periodic influx of coarser sand and grit which became mixed with the mud *in situ*. The deformation of the sediment could be ascribed to compaction and bioturbation.

As a whole the sequence represents glacial retreat sedimentation with sub-glacially deposited material being overlain by a proximal outwash facies close to the ice margin. As the ice retreated further a distal arenaceous outwash facies was deposited, while later on fine-grained sediments rich in organic matter were laid down under very shallow lacustrine conditions (Fig. 1). This closely resembles the glacial and fluvioglacial sequences described by Chappell and Humphreys (1970: 505) for the Dwyka in Zimbabwe-Rhodesia.

Ice Paleo-flow Directions. Evidence of paleo-flow directions of the ice is based on facies distribution, clast composition and basement topography. The sediments become finer grained towards the west. The basement contours show approximately east-west trending linear structures which makes

paleo-ice flow perpendicular to that trend very unlikely. The amount of siderite increases from east to west, which could be a function of the lithofacies but could just as well indicate an increase in paleo-temperature westwards (De Villiers & Visser 1977: 7). The close relationship between clast and bedrock composition also favours a parallel east-to-west flow.

It can be concluded that the most likely direction of ice retreat was towards the east, and it is postulated that the ice also came from that direction. The glaciers flowing from the east were probably diverted by the high ridge in the floor with the result that very little sediment was deposited on top of it. This obstruction could also have caused the dumping of material by the glaciers giving rise to anomalous thicknesses adjacent to it.

This comparatively small basin yields valuable data on glacial processes as well as ice flow directions in a region where they were previously unknown. The westerly ice flow (Fig. 3) fits with the glacial model for Zimbabwe-Rhodesia (Frakes & Crowell 1970: Fig. 3) and Northern Natal (Matthews 1970: Fig. 5; Von Brunn 1977: 129) but is in contrast with the Main Karroo Basin where the largest volume of till was derived from the S and SE (Visser 1979: 73). The difference in ice flow direction between south and north must be considered in constructing glacial models for this part of Gondwana.

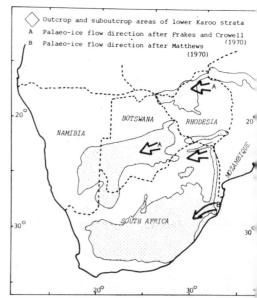

Fig. 3--Ice paleo-flow directions for the eastern portion of southern Africa during the late Carboniferous.

ACKNOWLEDGMENTS

The authors wish to thank the South African Iron and Steel Industrial Corporation Limited (ISCOR) for making data available, in particular Mr M. J. Deats, Divisional General Manager (Mining) and Mr R.A.G. Cuipers, Exploration Manager.

A grant from the Council for Scientific and Industrial Research (CSIR) to the senior author to attend the symposium is gratefully acknowledged. Dr Kevin Hall of the Institute of Environmental Studies of the University of the O.F.S. is also thanked for his helpful suggestions during the investigation.

REFERENCES

BOULTON, G. S. 1978. Boulder shapes and grain-size distributions of debris as indicators of transport paths through a glacier and till genesis. Sedimentol. 25: 773-779.

CHAPPELL, J.; M. HUMPHREYS 1970. Glacial sedimentation in the Lower Karroo, Mid-Zambezi Valley, Rhodesia. Proc. Pap. 2nd Gondwana Symp., S. Africa: 501-506.

DE JAGER, F. S. J. 1976. Coal. In C.B. Coetzee (Ed.), Mineral Resources of the Republic of South Africa. Handbook of the Geological Survey of South Africa 7: 289-330.

DE VILLIERS, P. R.; J. N. J. VISSER 1977. The glacial beds of the Griqualand West Supergroup as revealed by four deep boreholes between Postmasburg and Sishen. Trans. Geol. Soc. S. Africa 80: 1-8.

DREIMANIS, A. 1976. Tills: their origin and properties. In R. F. Legget (Ed.), Glacial Till. Spec. Pub. R. Soc. Canada 12: 11-49.

EDWARDS, M. 1975. Late Precambrian subglacial tillites, North Norway. Theme 1, 9th Int. Sedimentol. Cong., Nice: 61-66.

FRAKES, L. A.; J. C. CROWELL 1970. Late Paleozoic Glaciation: 2, Africa exclusive of the Karroo Basin: Geol. Soc. Amer. Bull. 81: 2261-2286.

HAUGHTON, S. H. 1969. Geological History of South Africa. Johannesburg: The Geological Society of South Africa. 535 pp.

JOHN, B. S. 1977. The Ice Age - Past and Present. London: Collins. 254 pp.

MATTHEWS, P. E. 1970. Paleorelief and the Dwyka Glaciation in the eastern region of South Africa. Proc. Pap. 2nd Gondwana Symp., S. Africa: 491-499.

MIALL, A. 1977. A review of the braided river depositional environment. Earth-Sci. Rev. 13: 1-62.

PESSL, F. 1971. Till fabrics and till stratigraphy in Western Connecticut. In R. P. Goldthwait (Ed.), Till - A Symposium. Ohio State University Press: 92-105.

RUST, B.R.; R. ROMANELLI 1975. Late Quaternary sub-aqueous outwash deposits near Ottawa, Canada. In A. V. Jopling and B. C. McDonald (Eds.), Glaciofluvial and Glaciolacustrine Sedimentation. Spec. Publ. Soc. Econ. Paleontol. Mineral. 23: 177-192.

VAN ZYL, J. S. 1950. Aspects of the geology of the northern Soutpansberg area. Ann. Univ. Stellenbosch 26A: 1-95.

VISSER, J.N.J. 1979. Changes in sediment transport direction in the Cape-Karroo Basin (Silurian-Triassic) in South Africa. S. Africa J. Sci. 75: 72-75.

VON BRUNN, V. 1977. A furrowed intratillite pavement in the Dwyka Group of Northern Natal. Trans. Geol. Soc. S. Africa 80: 125-130.

Sedimentary facies related to Late Paleozoic (Dwyka) deglaciation
in the eastern Karroo Basin, South Africa

VICTOR VON BRUNN
University of Natal, Pietermaritzburg, South Africa

Glaciogene sediments of the Dwyka Formation, deposited on the NW flank of a regional downwarp, record the closing stages of the Late Paleozoic glaciation in the eastern Karroo Basin. The deposits were derived from a major ice sheet which had flowed down the paleoslope before reaching a fjord-indented shoreline.

Multiple sequences comprise tillites, diamictites, conglomerates, sandstones, and silt-stones. Tillite mainly constitutes the base of the Dwyka succession and represents subglacial lodgement from active ice. Diamictite is the product of downslope flowage of meltwater-soaked supraglacial till from a wasting stationary or retreating glacial margin. Deposition in the ice-contact environment is apparent from localised collapse structures. Accumulation of tills occurred under subaerial and subaqueous conditions.

Lithological variations and sedimentary structures associated with fluvioglacial con-glomerate and sandstone sequences indicate deposition in proximal to distal braided reaches. The occurrence of sharp-based silty sandstone units is attributed to sub-aqueous gravity flow in a glaciolacustrine deltaic setting. Horizontally-laminated siltstones were deposited in a low-energy glaciolacustrine environment.

Cyclic repetition of lithologies is ascribed to an oscillatory retreat of continental ice coupled with isostatic rebound during the period of deglaciation.

INTRODUCTION

The Late Paleozoic (Dwyka) sediments discussed in this account were deposited on a gently-inclined paleoslope forming the NW flank of a linear pre-Dwyka downwarp in the eastern Karroo Basin (Fig. 1). This trough, with its axial gradient to the S-SW (Matthews 1970), and trending parallel to the present-day coastline of the African subcontinent, is thought to have marked a line of incipient crustal rupture consti-tuting a local precursor of Gondwana frag-mentation (Hobday & von Brunn 1979). Stri-ated pavements occurring over a wide area attest to a SE flow of continental ice from a major dispersal centre situated on elevated ground to the north. Ice-move-ment and glacial deposition were locally influenced by an irregular topography. In contrast to the massive, uniform tillite exceeding 300 m in thickness in the remnants of the regional trough farther south, the glaciogene deposits on the NW flank display a vertical succession of different facies. The purpose of this contribution is to examine the occurrence and genesis of sedi-ments that record the closing stages of the Dwyka glacial episode.

STRATIGRAPHIC FRAMEWORK

In the region under discussion Dwyka rests unconformably on an uneven, striated Precambrian surface. Due to the bedrock relief thicknesses are extremely variable, ranging from some 300 m in SE-trending paleo-valleys to zero where Dwyka abuts against Precambrian inliers. Lower Ecca shales, generally resting disconformably on the Dwyka, are followed by a coal-bearing arkosic succession. The Permian Ecca de-posits were laid down under fluvio-deltaic conditions.

SEDIMENTARY FACIES

The individual facies representing the Dwyka Fm are distinguished from one another in terms of their respective lithologies and internal sedimentary features.

Tillite Facies. The tillite is essentially massive, hard and bluish-black in colour. Clasts, which are subangular to subrounded and make up between 2% and 20% of the rock, are set in a dark, extremely fine-grained matrix containing crushed crystal and lithic fragments. Most clasts are derived from crystalline basement to the north, but locally-derived angular bedrock material is

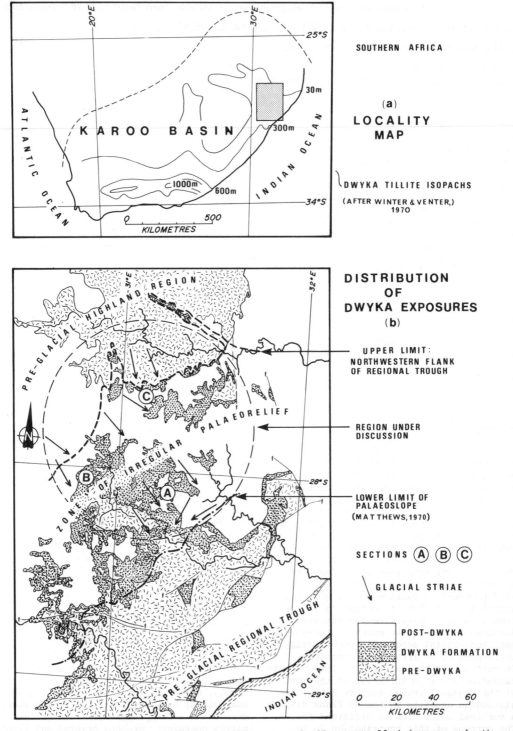

Fig.1--(a) Locality map. Isopachs depict tectonically-controlled downwarps in the eastern and southern Karroo Basin. (b) Distribution of Dwyka exposures on the paleo-slope bordering the western remnant of the former regional downwarp.

118

also abundant. Striated and faceted stones are occasionally noted. Tillite always forms the base of the Dwyka succession; subordinate tillite units are also interstratified with sediments of the succeeding lithofacies.

The tillite is interpreted as a subglacial lodgement deposit, having been released from the debris-laden base of a slowly-moving ice sheet. Unevenness of the subglacial terraine, promoting pressure melting, is likely to have had a significant influence on the lodgement process. Subglacial till accumulations are also attributable to frictional drag (Boulton & Paul 1976) between a till substrate and debris particles in traction. Appreciable volumes of lodgement till accumulated beneath the active ice, particularly in subglacial depressions where subsequent gravitational mass-movement of unconsolidated material was induced by steep slopes.

Diamictite Facies. The term "diamictite" is applied here to describe pale-brown tillite-like units interbedded with other clastic sediments in the Dwyka succession. The rock is generally massive but is often crudely-bedded; discontinuous wavy internal structures are sometimes developed. The matrix is characteristically sandy, and irregular bodies of coarse-grained sandstone and conglomerate occur sporadically. Clasts, the distribution of which is very irregular, constitute up to 60% of some diamictites. Striated stones are virtually absent. The thickness of individual diamictite units attains 10 m but is variable and has a tendency to change over short distances.

The diamictites represent former supraglacial melt-out tills deposited along the margin of a stationary or retreating ice sheet at the time of deglaciation. Emplacement occurred by downslope flow, following till accumulation on the glacier surface during downwasting, as described by Boulton (1972). Such deposition accounts for the well-defined basal contacts typical of diamictite units. The lower portions of diamictite bodies are usually massive and probably represent the basal, parautochthonous element (Boulton & Paul 1976) of former flow tills which had not moved far from their source. Greater mobility, due to the presence of meltwater, and involving the upper, allochthonous element, is apparent where silty wisps, streaks of sandstone (Fig. 2) and fluidal structures in the matrix of clast interstices are prominently displayed in the diamictite. During the slurry-like flow process depletion of the finer fractions by winnowing has led to aggregation of clasts and to lateral facies changes where diamictites pass into either conglomerate or fine-grained sandstone units.

Interdigitation of diamictite with fluvioglacial outwash indicates simultaneous deposition along the ice front at a time when large volumes of meltwater were being released. The conspicuous roundness of clasts in diamictite can be ascribed to prolonged subglacial and englacial transport, and subsequent supraglacial stream activity. Deformation structures, encountered in some diamictites and accentuated by associated stratified sediments, reflect an ice-contact environment. Volume changes, resulting from the decay of buried stagnant ice have caused subsidence or collapse of the till overburden. Where diamictite rests on an undulatory tillite surface it is apparent that supraglacial or subglacial melt-out from debris-laden stagnant ice was let down onto an unconsolidated substrate during the ablation process.

Conglomerate Facies. Conglomeratic rocks are not widely distributed. The most common type is akin to diamictite into which it passes imperceptibly. The matrix, supporting irregularly distributed clasts, is coarse-grained but streaks of fine-grained sandstone are also present. Sorting is poor. Some of the diamictite-conglomerates clearly represent lag concentrates resulting from depletion of fines by winnowing or selective removal by percolating meltwater. In some cases deformed siltstone bands and undulose matrix structures in the rock suggest re-transportation by debris flow which could have been either subaerial or subaqueous (Fig. 3). Units with sharp lower contacts resemble the "Trollheim type" of debris flow of Miall (1978).

Some conglomerates represent diamictites that have been reworked by fluvioglacial processes. Outwash conglomerates are usually clast-supported and consist of sub-angular to rounded cobbles and boulders. From several of the exposures it is apparent that multistorey longitudinal gravel bars developed in an area of high, fluctuating discharge of meltwater from a nearby glacial terminus. Crude, horizontal bedding and clast imbrication suggest migration by downstream accretion of planar gravel sheets under upper flow regime conditions. Intercalated plane-bedded sandstone lenses reflect periodic waning of flood conditions. In these upward-fining sequences, attributed to gradual recession of the ice front, sandstone is structured by trough and planar cross-bedding and ripple-drift cross-lamination. The assemblages represent a proximal to distal transition in proglacial outwash fans. The lower conglomeratic part is comparable with the "Scott type" deposi-

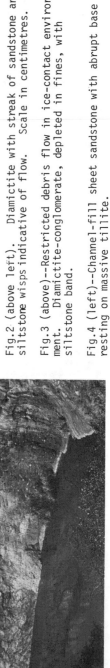

Fig.2.(above left). Diamictite with streak of sandstone and
siltstone wisps indicative of flow. Scale in centimetres.

Fig.3 (above)--Restricted debris flow in ice-contact environment. Diamictite-conglomerate, depleted in fines, with
siltstone band.

Fig.4 (left)--Channel-fill sheet sandstone with abrupt base
resting on massive tillite.

tional profile which merges vertically with the "Platte type" sequence of Miall (1978) that represents sediments deposited in broad, sandy braided reaches. The fluvioglacial accumulations were laid down under conditions similar to those of the "upper fan" and "midfan" (Boothroyd & Ashley 1975) in outwash extending from the Scott and Malaspina glacier margins to the Gulf of Alaska. Upward-coarsening sequences, also observed in the Dwyka fluvioglacials, suggest a systematic change from distal to proximal conditions brought about by glacial readvance.

Sandstone-siltstone Facies. Fine-grained sandstones and siltstones form sheetlike bodies several hundred metres in extent. They attain thicknesses up to 15 m and are recurrent in the succession. The sediments are either sharp-based or constitute part of an upward-fining sequence. They have been found to grade laterally into diamictite. The sheet sandstones are generally massive but internal sedimentary structures are common. Horizontally-laminated siltstones support rafted clasts and are sometimes rippled.

Sediments of this facies were deposited in distal fluvioglacial and glaciolacustrine environments. Deposition from suspension played an important role as is apparent from the common occurrence of different types of ripple-drift cross-lamination (Jopling & Walker 1968) which indicate fluctuations in discharge, sediment load and settling rate.

The fluvioglacial sediments were deposited on the "lower fan" of Boothroyd and Ashley (1975) and form a continuum with the more proximal conglomeratic outwash described above. Planar cross-bedded sets indicate that transverse bars had migrated downstream by avalanche face accretion. Periodic falling stages of sediment-charged flow, when climbing ripples developed on bar surfaces is apparent from the ubiquitous ripple-drift cross-lamination.

Massive silty sandstone units suggest subaqueous mass-flow deposition in a glaciolacustrine delta front setting. Such deltas, built by distal outwash discharging into proglacial lakes, occur as overlapping lobes of sediment deposited in a low-energy environment (Gustavson *et al.* 1975) by hyperpycnal inflow entering a lake as density underflow. Sub-aqueous gravity flow is evident from the sharp and sometimes erosively-based lower contacts of these sandstones which were deposited rapidly from relatively powerful currents with a high sediment load. These deposits sometimes constitute broad channel-fills (Fig. 4) comparable with those in the

Quaternary subaqueous outwash of Canada (Rust 1977). Rapid fallout from turbid underflows, under reduced energy conditions, has produced uniform ripple-drift cross-lamination which is commonly developed in the upper parts of the massive units. Some of the subaqueous outwash could have been deposited from subglacial streams issuing from beyond the grounded shelf zone of wet-based glaciers debouching onto a body of water.

Asymmetric ripples on horizontally-laminated siltsontes are likely to have been formed by underflow currents in the proximal deltaic region of glacial lakes. Laminae in glaciolacustrine siltstones are rarely varved, indicating deposition from fluctuating discharge. Vendl (1978) demonstrates that laminations in a glacier-fed lake can be the product of short-term inflow variations rather than seasonal accumulation. Dropstones, indicative of ice-rafting, are common in the glaciolacustrine laminites and are occasionally noted in the sheet sandstones. Laminated siltstones, draped over boulders, embedded in diamictite, represent deposition in temporary meltwater ponds in ice-contact glacial debris.

STRATIGRAPHIC SECTIONS

Representative sections A, B and C (Fig. 5) display the lithological interrelationships within the Dwyka succession in the region under consideration (Fig. 1b). Common to these examples are multiple sequences. Periodic glacial readvance is apparent where the tillite abruptly overlies diamictites or arenites. The NW-SE orientation of soft-sediment glacial grooves corresponds to that of striations on Precambrian bedrock pavements in the region (von Brunn 1977).

The rose diagrams illustrate paleocurrent trends inferred from sedimentary structures in fluvioglacial sediments. In coarsening-up fluvial sequences collapse- and other ice-deformation structures suggest stagnation and in-place downwasting of debris-laden ice. Upward-fining in diamictite units and the development of current stratification indicates a greater mobility in the upper parts of till flows due to the presence of increased amounts of meltwater.

PALEOENVIRONMENTAL SYNTHESIS

During the Late Paleozoic glaciation continental ice flowed SE from a polar highland in the northern Karroo Basin, across a rugged paleoslope, before reaching a fjord-indented shoreline where ice shelves extended out into a body of water

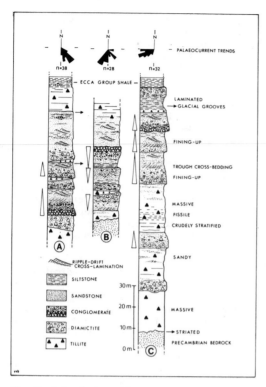

Fig.5--Selected sections
illustrating the lithological
interrelationships within the
Dwyka succession.

occupying a regional downwarp. Lodgement
till was plastered onto the subglacial
terrain, thereby partly masking the irregular topography.

Climatic amelioration led to the com-
mencement of deglaciation which coincided
with progressive thinning and recession
of the ice sheet. Disintegration and
subsequent stagnation of ice along a
digitate marginal zone was controlled by
localised subsurface irregularities. An
accelerated rate of sedimentation by
increasing volumes of meltwater followed
gradual downwasting of the glacial mass.

Complex sedimentary assemblages developed in the proglacial environment where,
due to the uneven configuration of the
subsurface, deposition of glacial and
associated sediments occurred under both
subaerial and subaqueous conditions.
Mass-transport contributed significantly
to the depositional process. Outwash
plains with broad channel patterns, extending from aprons of ice-contact debris,
merged with glaciolacustrine deltas of
marginal lakes. Some subaqueous outwash
accumulated beyond the buoyancy line of

grounded shelf ice. Intermittent isostatic rebound, in response to glacial
unloading, and restricted glacial read-
vances account for the cyclic repetition
of lithologies.

The thermal nature of the ice yielding
the glacial sediments cannot readily be
ascertained. By analogy with modern,
cold polar glaciers, the inner zone (Boulton 1975) could have been temperate thus
permitting the deposition of subglacial
till by lodgement. The occurrence of diamictite, interpreted as a supraglacial
melt-out and representing debris originall[y]
entrained by basal freezing (Boulton 1972)
suggests that the terminal basal zone of
the ice was below melting point. Winnowing and selective removal of fines from
meltwater-soaked supraglacial flow till
could have contributed significantly to
the accumulation of the extensive fine-
grained arenite bodies.

Denudation of the isostatically-reju-
venated highlying region to the north at
the end of the glacial period resulted in
a southward progradation of fluvio-deltaic
complexes into a region that was becoming
progressively inundated. Lower Ecca
argillites, overlying the Dwyka deposits,
represent prodeltaic sediments which had
built across a broad, shallow platform.

ACKNOWLEDGMENTS

A grant from the Council for Scientific
and Industrial Research in Pretoria is
gratefully acknowledged. I am indebted
to Prof. Henno Martin, who kindly reviewed
this manuscript.

REFERENCES

BOOTHROYD, J. C.; G. M. ASHLEY 1975. Processes, bar morphology, and sedimentary
structures on braided outwash fans, north-
eastern Gulf of Alaska. In A.V. Jopling,
McDonald (Eds.), Soc. Econ. Paleontol.
Mineral., Spec. Pub. 23: 193-222.
BOULTON, G.S. 1972. Modern Arctic glaciers
and depositional models for former ice
sheets. J. Geol. Soc. London 128: 361-
393.
--- 1975. Processes and patterns of subglacial sedimentation: a theoretical
approach. In A.E. Wright and F. Moseley
(Eds.), Ice Ages Ancient and Modern.
Liverpool: Seal House Press. Pp. 7-42.
---; M. A. PAUL 1976. The influence of
genetic processes on some geotechnical
properties of glacial tills. Q. J.
Engineer. Geol. 9: 157-194.

GUSTAVSON, T.C.; G. M. ASHLEY; J. C. BOOTHROYD, 1975. Depositional sequences in glaciolacustrine deltas. In A.V. Jopling; B.C. McDonald (Eds.), Soc. Econ. Paleontol. Mineral., Spec. Pub. 23: 264-280.

HOBDAY, D. K.; V. VON BRUNN 1979. Fluvial sedimentation and palaeogeography of an Early Palaeozoic failed rift, southwestern margin of Africa. Palaeogeo., Palaeoclim., Palaeoecol. 28: 169-184.

JOPLING, A. V.; R. G. WALKER 1968. Morphology and origin of ripple-drift cross-lamination with examples from the Pleistocene of Massachusetts. J. Sediment. Petrol. 38: 971-984.

MATTHEWS, P. E. 1970. Paleorelief and Dwyka glaciation in the eastern region of South Africa. 2nd Gondwana Symp. Proc. Pap. Pretoria: CSIR. Pp. 491-506.

MIALL, A. D. 1978. Lithofacies types and vertical profile models in braided river deposits: a summary. In A.D. Miall (Ed.), Canad. Soc. Petrol. Geol. Mem. 5: 597-604.

RUST, B.R. 1977. Mass flow deposits in a Quaternary succession near Ottawa, Canada: diagnostic criteria for subaqueous outwash. Canad. J. Earth Sci. 14: 175-184.

VENDL, M. A. 1978. Sedimentation in glacier-fed Peyto Lake, Alberta. Unpublished M.Sc. thesis, University of Illinois at Chicago Circle, Chicago, U.S.A. 96 pp.

VON BRUNN, V. 1977. A furrowed intra-tillite pavement in the Dwyka Group of northern Natal. Trans. Geol. Soc. S. Africa 80: 125-130.

WINTER, H. DE LA R.; J. J. VENTER 1970. Lithostratigraphic correlation of recent deep boreholes in the Karroo-Cape Province. In 2nd Gondwana Symp. Proc. Pap. Pretoria: CSIR. Pp. 395-408.

Lithofacies analysis of the Late Permian Raniganj coal measures (Mahuda Basin) and their paleogeographic implications

S. M. CASSHYAP
Aligarh Muslim University, India

Lithofacies analysis of the Raniganj coal measures in the Mahuda basin shows four facies; borehole data from this area are analysed. Results of paleocurrent study suggest that the Mahuda basin was drained largely by sinuous streams which flowed dominantly from the southeast to the northwest and locally from southwest to northeast.

It is argued that the alluvial plain wherein Mahuda sedimentation took place, was areally more extensive at the time of deposition (Late Permian) including the area south of the boundary fault. The basin was truncated in the south later possibly in the Middle to Late Mesozoic, after the Gondwana strata were deformed.

INTRODUCTION

The SW part of Jharia coalfield to the north of Damodar river has been practically undescribed until now. This study is designed to provide useful information on the basin framework and tectonic setting at the time of deposition and after.

GEOLOGICAL AND STRUCTURAL SETTING

The Jharia coalfield contains some 1900 m of Gondwana sediments ranging in age from the Permo-Carboniferous to Permian resting above the "unclassified Early Precambrian" rocks (? Archean) unconformably along the northern margin and faulted against them to the south.

The Mahuda basin is underlain largely by the Raniganj coal measures. On lithological grounds and on similarities and differences in sedimentary structures noted in the two strata, the contact between the Raniganj and Barren Measures has been shifted farther south in this study (Casshyap 1978), as shown in Fig. 1. The structure represented by the Mahuda basin is that of a plunging syncline (Fig. 1). The structure is not a local feature; it extends farther eastward over to the E and SE parts of the coalfield. The structure is represented here by a syncline, a continuation of the Mahuda syncline, and a complementary anticline and syncline, all plunging westward (Fig. 1). The sickle-like form of Jharia coalfield so produced may, therefore, well be structurally controlled rather than a primary feature as visualised by Fox (1930).

LITHOFACIES AND SEDIMENTARY CHARACTERS

Coarse to medium sandstone is next to shale-siltstone in abundance. It occurs commonly in lenticular or channel-shaped bodies, 2-30 m wide and up to 15 m thick, which become less frequent upwards and towards the east. Some large scale cross-bedding foresets exhibit back-flow and co-flow structures.

Silty sandstone is below the interbedded shale or above sandstone, rarely >2 m thick, and is from 5-10 m to several tens of metres wide. Fine-grained sandstone exhibits parallel to wavy laminations, climbing ripple laminations, and rib-and-furrow structure. A series of crescentic bedforms resembling "scroll-like" features may also occur locally (Fig. 2).

Interbedded sequence of shale and siltstone exceeds all others by volume and lies, commonly with a sharp contact, either above silty sandstone or directly above medium to coarse sandstone. It is locally carbonaceous and may be overlain or underlain by coal. Enclosed in the facies are small (1-6 m long), lenticular to pod-shaped, cross-laminated or laminated, channel sand bodies (Fig. 3), elongated SE-NW and closely spaced in places.

Exploratory drilling in this basin has revealed about 35 coal horizons.

SUBSURFACE STUDY OF LITHOFACIES

The panel diagram of Mahuda basin (Fig. 4) shows a three-dimensional distribution and shape of persistent and impersistent lithofacies referred to above. Sandstone bodies enclosed in fine clastics are lensoid and essentially channel-shaped. Some are laterally continuous, sheet-like bodies, especially in the southern part; the same sandstone bodies become lensoid when followed northward. Sandstone bodies are on an average greater in number in the lower and middle Raniganj particularly in the southern and western parts of the area. Coal seams recorded as No. III and No. VII

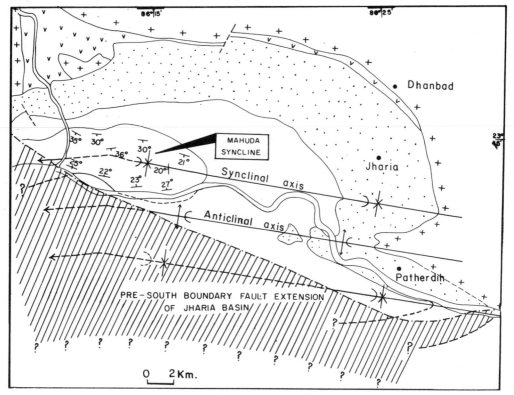

Fig.1--Geological map of Jharia coalfield, showing the structure of Mahuda basin and of the area around. Also shown is likely extension of the above structure farther south before the development of south boundary fault. Symbols refer to stratigraphy: plus (+), Archean; VV, Talchir Fm.; dots (··), Barakar Fm.; blank, Barren M.; Mahuda basin is underlain by the Raniganj Fm.

Fig.2--Elliptical bedforms yielding scroll-like topography (? point-bar) in fine sandstone facies of the Raniganj Formation. The bedforms are profusely cross-laminated and cross-bedded. South bank of river Damodar; about 1 km southeast of Murlidih colliery; looking west.

126

Fig.3--A typical
outcrop of inter-
bedded sequence of
shale-siltstone and
fine sandstone;
the latter occurs
as lenticular and
pod-shapped channel
bodies.   North
bank of river Dam-
odar near Bhatdih
colliery.

in the official logs occur almost continuously across the basin; No. IX seam is
also fairly widespread.   Others do not
appear to be so, due either to non-deposition or to pinching due to sub-stratal compaction during subsidence.

The lithofacies in the panel diagram
tend to close on the eastern side.   The
same cannot be said for the western side
with certainty unless sufficient borehole
logs are available from the area to the west
of the Jamunia river.

Lithofacies Analysis.   Major subsurface
analysis of lithofacies was carried out on
a thicker stratum of ~600 m sequence,
examined for thickness of strata, number of
sandstone beds, sand-shale ratio, and its
relationship with coal.

Isopach lines exhibit a systematic increase in thickness to the central part (JK-
19) defining a basin broader in the west and
narrower and bifurcated in the east (Fig. 5).
There is a progressive thinning in the area
now truncated by the south boundary fault.
The slope indicated by spacing of form-lines
coincides fairly well with the structural
strike except locally in the western and
eastern parts of the basin.

The number of sandstone beds thicker than
3 m are fewer in the northern half than
the southern half (14 to 16) (Fig. 6).
The greater number of sandstone beds in
the south oriented SE-NW may correspond
to stream channels at the time of deposition, and compares well with the paleoflow
direction discussed later.

Distribution of total sand is likewise
interesting.   Sandstone is shown to be

thickest (>240 m) in the south-central
through western part but progressively
thins towards northern and eastern parts of
the basin, where total sand is barely 60 m
thick.   By contrast, fine clastics including shale and siltstone abound (300-400
m) in much of the northern and eastern
parts.   Sand-shale ratio confirms the above
contention (Fig. 7).   Notably, the coal-
rich area does not coincide with the thickest part of the basin, but with that of
fine clastics.   Near the southern boundary
and in the western part, which abound in
sandstone, the total coal is less than 8 m
(Fig. 8).

PALEOFLOW AND GEOLOGIC IMPLICATIONS

Paleocurrent results show that the
channel sand bodies were deposited by
streams which flowed dominantly towards the
NW and N, and locally towards the NE (Fig.9).
Ripple marks and channel axes show similar
orientation, implying their genetic relationship with master streams.   Grand mean
orientation of paleostreams computed for
the entire area is 328 ± 58°, and total
variance is 3480.   This direction coincides well with the northwesterly component of radiating paleodrainage reconstructed for rest of the Jharia coalfield outside the Mahuda basin (Casshyap 1973,
1979b).

PETROGRAPHY

20 thin sections from as many specimens
were subjected to petrographic examination.
Table 1 records the average modal composition.   Quartz is subangular to subrounded,

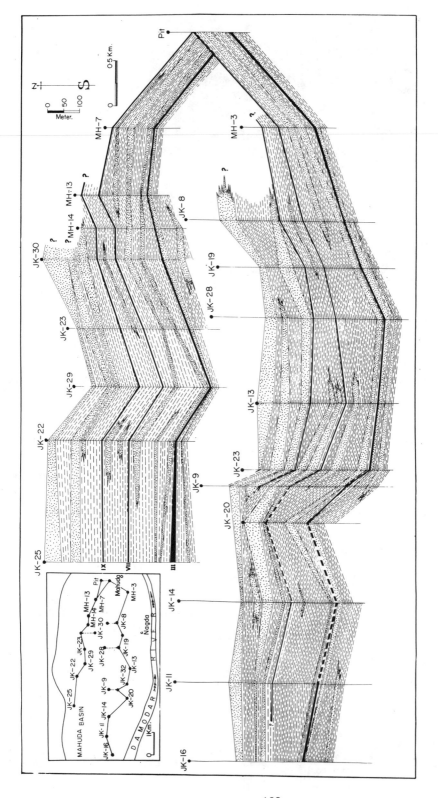

Fig.4--A panel diagram showing subsurface distribution of sandstone bodies enclosed in fine clastic "matrix", including three coal seams namely No. III, No. VII and No. IX; inset shows the location of borehole logs.

128

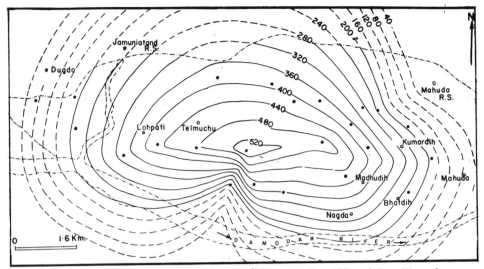

Fig. 5--Isopach map of the stratum (∿600 m) between No. III and No. IX coal seams. Solid dots correspond to borehole localities as shown in inset of Fig. 4. The oval-shaped pattern of form lines and their continuation beyond the present limit of Mahuda basin is noteworthy.

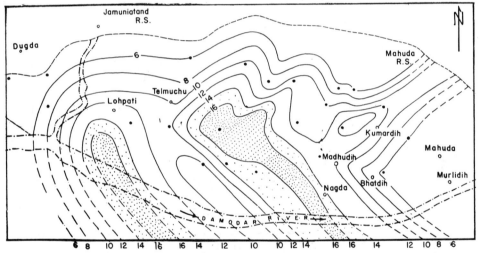

Fig.6--Map showing subsurface distribution of number-of-sandstone beds thicker than 3m in the stratum referred to in Fig. 5. 14 to 16 sandstone beds are confined to portions oriented southeast-northwest.

exhibits moderate to good sorting, and includes both mono- and poly-crystalline varieties; the former being more common than the latter. The feldspar grains are subrounded to subangular or rounded. Each species shows a variable admixture of fresh and altered varieties. Muscovite flakes are commonly torn apart along their cleavage planes or bent around harder minerals. Sand sized fragments of older rocks represent schist, phyllite, quartzite and granite gneiss etc. Detrital matrix is by and large scarce. Chalcedonic silica as cementing material occurs in varying amount (1-11%), replacing the initial matrix. Limonite and carbonate cements are very local and particularly confined to yellow pod-shaped channel sandstone. The Raniganj sandstones are mostly sub-arkose, arkose to lithic arkose (McBride 1963; Casshyap 1967).

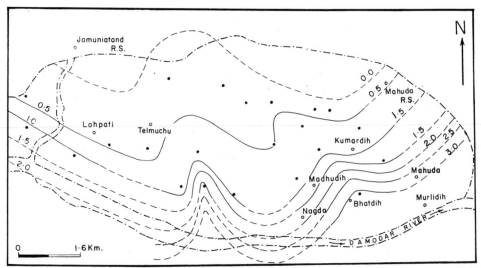

Fig.7--Sand-shale ratio map for the stratum between No. III and IX coal seams.   Note excess of shale over sand in the central and eastern parts.

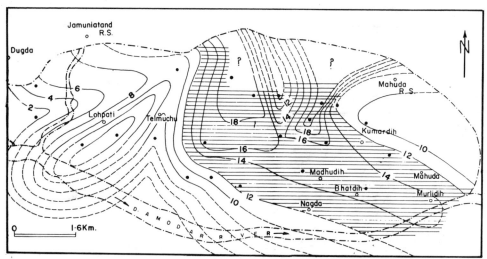

Fig.8--Coal isolith map showing subsurface distribution of total coal (in m). Hatched portion is likely to hold greater potential for workable coal.

BASIN FRAMEWORK AND CONDITIONS OF DEPOSITION

Fig. 9 illustrates a model for Raniganj sedimentation in the Mahuda Basin.   Indeed, the strata in the basin correspond to sediment types deposited in various interrelated subenvironments of an alluvial flood plain, fed by streams flowing SE to NW.   The outcrop and subsurface study reveals that the SW half of the basin was possibly occupied by a network of sinuous streams.   The sand bodies were deposited in stream channels by vertical and locally lateral accretion.   The NE half of the basin may represent largely overbank and backswamp environments that favoured deposition of silt and clay and development of coal-forming swamps.   Small ephemeral channels produced lenticular pod-shaped channel sand bodies, now interbedded with the shale-siltstone sequence.   Local coal-forming swamps may have developed also in interchannel areas in the SW half of the basin favouring deposition of interbedded shale and siltstone and thin coal. At times the basin was converted into extensive coal-forming swamps, perhaps in

130

Table 1--Modal composition (by volume) of Raniganj Sandstone of Mahuda Basin.

Constituents/ Specimen No.	JR/16	JR/2	JR/1	JR/13	JR/14	JR/10	JR/20B	JR/22	JR/20	JR/28
Quartz Resistates										
Plutonic quartz	42.1	52.8	46.5	37.8	32.5	34.0	30.2	39.8	50.9	25.8
Metamorphic quartz	7.8	8.6	16.4	9.3	12.1	13.4	14.1	8.5	7.9	10.5
Sedimentary quartz	5.1	2.5	3.4	3.3	0.2	1.5	2.5	3.6	4.8	1.5
Feldspar										
Microcline	3.6	5.1	5.7	12.3	19.6	20.2	12.2	6.2	8.5	5.4
Orthoclase	2.5	5.4	4.9	5.0	3.1	2.9	8.0	7.5	4.0	1.5
Plagioclase	3.8	2.9	2.0	3.1	2.0	4.4	6.1	3.5	4.6	2.1
Labile Rock Fragments										
Quartzite	7.1	4.4	3.7	5.8	1.7	1.4	0.9	1.0	2.0	1.0
Phyllite & Schist	9.5	6.8	6.5	4.8	8.8	6.1	10.1	10.5	2.8	-
Granitoid rock fragments	-	-	-	-	2.0	-	-	-	-	-
Mica	3.7	2.1	2.2	1.0	0.5	0.8	1.1	1.0	1.0	0.2
Cement										
Cryptocrystalline	4.7	3.2	2.9	5.5	1.0	1.8	7.3	10.8	5.4	-
Carbonate	-	-	-	-	11.1	12.4	-	-	-	30.1
Limonitic	-	-	-	-	-	-	-	-	4.0	3.5
Matrix										
Primary	2.7	1.1	2.1	2.5	1.2	0.2	1.4	2.0	1.1	0.5
Secondary	7.0	4.5	3.3	8.5	4.1	1.1	6.1	5.5	2.8	2.5
Petrified Plant Fragments	-	-	-	-	-	-	-	-	-	15.4

response to greater subsidence, which accounts for the occurrence of persistent coal seams. Stream channels may have shifted either to the west or to the east of the main basin about this time.

LIMIT OF DEPOSITION

The Raniganj strata in the Mahuda basin show a normal depositional contact with the underlying Barren Measures on three sides, to the west, north and east, but lie faulted against the basement Archeans to the south, which may imply that the original basin (s) of deposition was wider than the existing areal limit (Casshyap 1979b). Some early workers put forward a view that the boundary fault to the south of the basin developed pari passu with sedimentation (Fox 1930; Pascoe 1959; Ghosh & Mitra 1970). Others believe that the boundary fault developed much later (Ahmad 1960, 1967; Verma & Singh 1977). Certain features which may have relevance to this problem are:

(1) Form-lines of isopach and other lithofacies maps, when extrapolated, indicate that the original basin may have been oval.

(2) The Mahuda basin represents the structure of a synclinal fold; along its south flank the Raniganj and Barren Measures are truncated by the south boundary falut, so that the underlying Barakar and Talchir rocks are not exposed along the southern boundary of the basin.

(3) A NW paleoflow conforms with the radiating northerly drainage, with no evidence of centripetal drainage even though the original basin possibly was oval.

(4) There is no evidence of coarsening of sand size or occurrence of conglomerate beds in the Raniganj or Barren Measures near the south boundary fault, in outcrops or in the available borehole logs. Also, these sandstones are by no means different texturally and compositionally from those exposed farther N and NW in the downcurrent direction.

(5) A likely site of provenance would be the Precambrian highlands of eastern Chotanagpur plateau, some 70 km SE of the existing south boundary fault.

This suggests that after the deposition of Barren Measures the alluvial plain became more localised during the Raniganj sedimentation, shifting SW to an area apparently more extensive than the present Mahuda basin. The basin, fed by NW-flowing sinuous streams, may well have extended south in the upcurrent direction beyond the existing southern limit at the time of deposition (Fig. 1). After deformation the strata were truncated by the south boundary fault. This boundary fault is post-depositional and post-folding; its creation may have coincided with the major tectonic disturbance that affected this part of peninsular India in the Late Jurassic or Early Cretaceous (Krishnan 1968; Ahmad 1967; Casshyap 1979a).

131

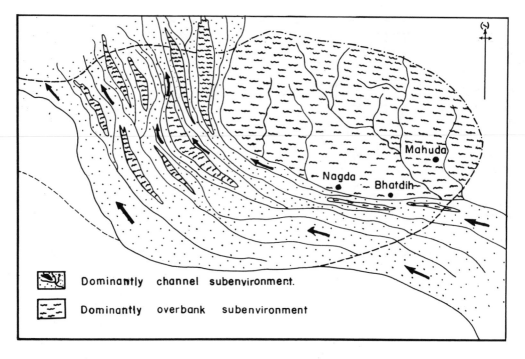

Fig.9--A schematic interpretative map showing dominant sedimentary subenvironments as inferred in the Mahuda basin.

ACKNOWLEDGMENTS

This project was funded by the Council of Scientific and Industrial Research, New Delhi. Messrs N.K. Srivastava, Prabhat Kumar, and Z.A. Khan assisted at different stages of the project. I am grateful to Shree S.K. Pande, Suptd. Geologist, B.C.C.L., Shree Ajit K. Gulathi, Project Officer, Monidih Colliery and Mr R.K. Bhatnagar, Manager Kargali Colliery for providing necessary facilities in the course of field work. Thanks are due to Prof. S.H. Rasul for kindly providing laboratory facilities in the department, and Prof. V.K. Srivasta-va, Dr Zahid A. Khan and Mr R.C. Tewari for stimulating discussions.

REFERENCES

References not included in this list are in the bibliography of Casshyap 1979a.

CASSHYAP, S. M. 1967. On the classifica-tion of argillaceous sandstones. Ann. Geol. Dept. AMU, Aligarh 3: 48-50.
--- 1978. Quantitative lithofacies anal-ysis and sedimentation trends of the coal bearing Raniganj Formation in Mahuda basin of Jharia coalfield, Bihar, India. Tech. Rept., C.S.I.R., New Delhi: 1-34.
--- 1979a. Patterns of sedimentation in Gondwana basins (key paper). IV Inter. Gondwana Symp., Calcutta. 1-34.
--- 1979b. Paleocurrents and basin framework - an example from Jharia coalfield, Bihar, India. IV Inter. Gondwana Symp., Calcutta.
KRUMBEIN, W. C.; L. L. SLOSS 1963. Stra-tigraphy and sedimentation. 2nd ed. Freeman, San Francisco, 1-660.
McBRIDE, E. F. 1963. A classification of sandstone. J. Sediment. Petrol. 33: 664-669.
MEHTA, D. R. S.; B. R. NARAYANA MURTHY 195 A revision of the geology and coal resourc es of the Jharia coalfield. Mem. Geol. Surv. India. 84: 1-142.
SEN, N. 1970. Correlation between the geological structure and geothermal regime in Jharia coalfield and its bearing on Gon dwana tectonics. Ann. Dept. Geol., AMU 5 & 6: 101-147.
VERMA, R. P.; V. K. SINGH 1977. A chrono logy of tectonics and igneous activity in Damodar valley coalfields. IV Inter. Gondwana Symp., Calcutta (pre-print).

Paleochannel analysis of Early Permian streams in East Bokaro Basin,

India – A summary (Abstract)

S. M. CASSHYAP & Z. A. KHAN
Aligarh Muslim University, India

The lithologic assemblage of Karharbari and Barakar Formations in the Gondwana Basin of East Bokaro comprises four distinct facies: namely, conglomeratic and pebbly sandstone; coarse to medium sandstone; shale and siltstone; and coal. The total clastic assemblage (900 m), by and large, shows a progressive decrease in grain-size and scale of cross-bedding from Karharbari up through lower and uper Barakar, illustrating a fining upward megacycle. A paleoflow analysis reveals that the system of streams which transported sediments into the basin flowed dominantly north-northwest. Empirical relationships from modern streams derived by fluvial geomorphologists are used to estimate various channel morphological and hydraulic attributes of ancient Karharbari and Barakar streams. Results indicate that:

The Karharbari streams were moderately deep (3.2 m) and wide (156 m); their sinuosity was low (P = 1.25), for they flowed generally on a steep slope of about 0.00036 with a high flow velocity of 163 cm/sec. Grain size analyses suggest that bed-load constituted about 21% of the total load. The estimated paleochannel parameters and the dominance of coarse clastics in the Karharbari Formation indicate that the channel pattern may correspond with the bed-load of transitional streams synonymous with braided streams.

The lower Barakar streams were shallower (2.85 m) and slightly narrower (135.5 m) than the Karharbari streams, and flowed on a gentler slope, 0.00028 with a flow velocity of 132 cm/sec. These changes in the channel parameters, together with a decrease in bed-load (12%) and increase in fine clastics may account for an increase in channel sinuosity (P = 1.45).

The Upper Barakar streams which still flowed north-northwest became yet narrower (96 m) and shallower (2.1 m), and flowed over a slope which was yet more gentle (0.00018), and at a lower velocity (88 cm/sec). Paleochannel sinuosity increased (P = 2.05). Bed-load material decreased to only about 3% of the total load. Thus the Early Permian Gondwana rivers of eastern India underwent evolutionary changes in channel morphology and hydrodynamic conditions as sedimentation progressed from Karharbari through upper to lower Barakar, following the retreat of the Talchir ice-sheet.

Depositional patterns in limestones of the Kota Formation (Upper Gondwana),
Andhra Pradesh, India

NITYANANDA BHATTACHARYA
Oil & Natural Gas Commission, Dehra Dun, India

Limestones of the Kota Formation are characterized by good bedding, gentle dips, fine-grained texture and content of high Mg-calcite, chert, dolomite, montmorillonite, kaolinite, ostracods and coccoliths. Detailed petrographic, mineralogic and geochemical analyses of these carbonates indicate deposition in marine inter-tidal flats where abundant terrigenous material was brought in by a river estuary. High tides caused mixing of marine and continental sediments. The sea encroached and widened from the Indo-Australian Gulf in early Jurassic times. The formation is classified here as a distinct marine unit in the Upper Gondwana formations of peninsular India.

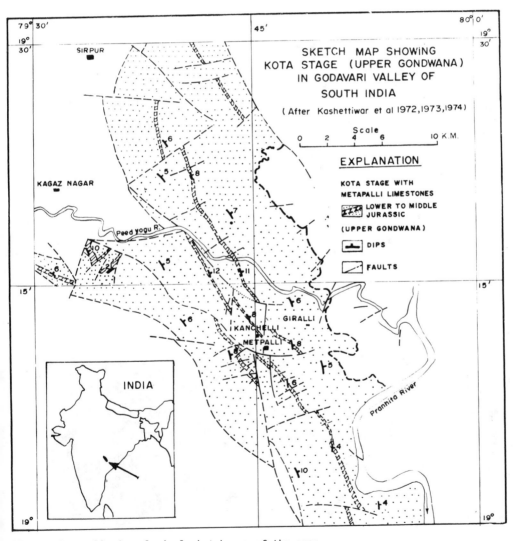

Fig. 1--Generalised geological sketch map of the area.

INTRODUCTION

The Kota Fm, part of the Upper Gondwana Gp in the Pranhita-Godavari valley of peninsular India, has 1200-1600 m of limestones, marls, clays, and conglomerates resting unconformably over the older Maleri beds (Krishnan 1960) (Fig. 1).

Kashettiwar and Balakrishnan (1972) and Kashettiwar *et al.* (1973, 1974) mapped the area and named the carbonate rocks as Metapalli Limestones (15 to 30 m) overlain by a bed of red clay, and characterized by lithographic texture, good bedding and gentle dips of about 5° to 10° to the E and NE. Abundant fish remains have been described (Pascoe 1959; Jain 1973; Yadgiri & Prasad 1977). Govindan (1975) considered the environment of deposition lacustrine on the evidence of ostracod fossils, and Datta *et al.* (1978) regarded it continental on the basis of micromammals. This study proposes a marine environment and tries to answer the following questions: whether limestones of the Kota Fm are marine and/or continental; how land and sea were distributed at the time of formation; and to provide a sedimentation model.

Ninety-three representative samples of limestones and clays from various localities in the Pranhita-Godavari valley were megascopically examined and the limestones subjected to thin-section petrographic analysis. Carbonates and clays were also investigated by differential thermal analysis and X-ray diffraction. Calcium and magnesium in selected limestones were determined by titration; boron and gallium from clays and insoluble residues of limestones were estimated colorimetrically. A few limestones were examined under the scanning electron microscope for texture and coccoliths.

LIMESTONES

Petrography. Limestones of the Kota Fm are light brown, extremely fine-grained, sub-conchoidal, hard, micro-nodular, often with chert lenses and grey laminations. They are micritic with abundant ooze mater occasionally associated with quartz. In clasts of carbonates and broken ostracod shells are present (Fig. 2) in various stages of preservation partially or comple filled with sparry calcite. Pellets with abundant inter-pelletoidal voids occur nea Kanchelli. Microcrystalline ooze matrix tends to form micrite, and terrigenous mat rial like fine-grained quartz and clay is less common. Rare specimens show zonatio of authigenic silica development: an oute rim of cryptocrystalline silica succeeded inwards by agate with fibrous quartz, an intermediate zone of chert, and finally a nucleus of crystallized quartz.

Mineralogy. Limestones at Kazipet, Alkapalli and Kanchelli are dominantly calcite rich; those at Metapalli are dolomite rich, but most have both calcite and dolomite varying widely in relative abundance. Distribution of calcite and dolomite is supported by X-ray diffraction and differential thermal analysis. Scanning electron micrographs indicate that the car bonates have developed from microcrystalline ooze. The calcite grains have very little crystalline form and have many voids, which vary from a few to 20 μm across.

Chert. Chert occurs near Metapalli and Kanchelli and is vertically restricted in zones of flattened lenses in limestones approximately parallel with bedding. The are composed of medium to dark brown micro crystalline silica whose contact with the

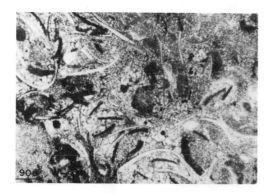
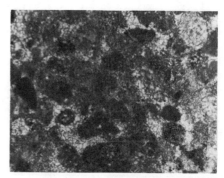

Fig. 2--Thin-sections of limestone.

carbonate portion is quite regular in thin-section.   Some specimens have crystal
overgrowths of calcite around a nucleus of
chert filling spaces within the limestone.

Clay.   The clay bands associated with
limestones at Metapalli are montmorillonite
but also include kaolinite and minor quantities of degraded illite (Fig. 3A).   Content of degraded illite and mixed-layers
increases towards the south.   Montmorillonite structure is very well organized,
but illite is always highly degraded.
Kaolinite is similar to montmorillonite both
in crystallinity and relative abundance.

Distribution of clay minerals in insoluble
residues of the limestone is similar to
that of the clays described above, but with
more illite.   It is highly degraded and
associated with abundant expandable mixed
layers (Fig. 3B).   Kaolinite is minor,
and montmorillonite is poorly organized in
structure.

Chemical Analysis.   Data for 14 samples
indicate high dolomite and/or magnesian cal-
ite (Table 1).   Calcium content varies
between 25% - 30%, and magnesium from 8% -
12% in most cases.   Fig. 4 shows that most
samples have 200 to 400 ppm of boron and 10
to 15 ppm of gallium.

Coccoliths.   Coccoliths were found almost
in all specimens from the region, but they
are more abundant in marls than hard limestones.   Some better-preserved specimens
could be photographed (Fig. 5).   The common
forms have a central hole or depression with
or without repeated layerings and are conical.   The component crystals in layers tend

Table 1--Partial chemical and mineral analyses of specimens of Kota Fm limestone.

Specimen	%Ca	*%Mg	Calcite/Dolomite **Ratio
9	36.87	1.46	3.0
34	26.45	8.75	0.6
36	23.05	13.98	0.3
38	32.04	9.00	1.1
41	29.26	8.15	0.9
42	26.85	12.40	0.5
50	25.85	9.61	0.6
83	39.68	0.00	only calcite
84	23.45	3.53	?
85	29.66	9.85	0.8
86	29.26	2.19	3.1
87	28.66	1.82	3.0
88	33.07	8.51	3.0
89	40.08	3.28	only calcite

* Also includes Fe   ** by X-ray analysis

to be rhombohedral and are often associated
with microcrystalline ooze material.

RESULTS

Well-bedded limestones of significant
thickness containing chert, dolomite, montmorillonite, microcrystalline ooze and
coccoliths with high boron distribution
suggest that the carbonates are of marine
origin.   The broken ostracods and fragmentary remains of micro-vertebrates were probably transported before burial.   The
absence of laminae and the orientation deposition of larger ostracod shells in variously inclined positions seem to indicate
repeated reworking of the sediments.   It
is possible that plankton could live under
such conditions and become abundant in es-
tuaries as a result of tidal movements.
Presumably, the coccoliths underwent pro-

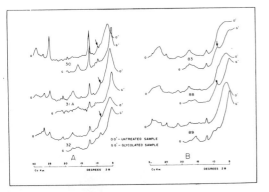

Fig.3--(A) Smoothed X-ray diffractometer
traces of red clay showing montmorillonite,
kaolinite and degraded illite (arrow).
(B) Smoothed X-ray diffractometer traces
of clay insoluble residues showing degraded illite (arrow), montmorillonite,
expandable mixed layers and kaolinite.
Specimen numbers are given between curves.

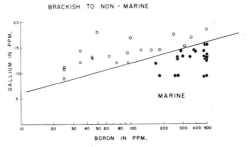

Fig.4--Scatter plots showing distribution
of boron and gallium in selected samples
of clays and insoluble (clay) residues of
limestones (size: -2µm).

137

gressive lithification as a function of the depth of burial (Scholle 1977) and the physio-chemical conditions of the environment. The absence of morphological details of the coccoliths could be explained by the facts that the forms were in very early stages of development, the features were destroyed by diagenesis, and the forms were poorly preserved. The well-defined bedding of the limestones probably reflects occurrence of coccoliths in distinct layers.

Clay mineral data presented in Fig. 3 indicate that terrigenous material came into the environment from the N and NW whereas the limestones in the S and SE have a more marine character. Abundant mixed layers in the clay bed above the limestone indicate that the sea was shallow. Insoluble residues of the limestones containing degraded illite and abundant expandable mixed layers (Fig. 3B) indicate that the clays were derived from land undergoing extensive weathering and alteration. SEM photographs indicate that a greater part of the montmorillonite may have been neoformed in a relatively alkaline environment leading to the dissolution of carbonates (Eberl 1978).

Dolomite and dolomitic limestones are dominant in the central, southern and southeastern part of the area in the direction of increasing marine influence. The originally precipitated magnesian calcites formed in shallow marine waters were later removed to the sea bottom where they changed to dolomite under an increased pressure that encouraged structural ordering and cation-exchange of magnesium from the brine. Thus we may infer from the total absence of chlorite and vermiculite in insoluble residues of limestones that wave activity and induced water circulation mechanically dehydrated a greater part of magnesium from brucite sheet and thus facilitated lattice ordering to form dolomites.

PALEOGEOGRAPHY

Jurassic marine transgressions over par of Gondwana were attributed by Ahmed (196 to widening of the Indo-Australian Gulf. This is supported by coccolith stratigrap (Bukry 1974). It is postulated that the limestones of the Kota Fm were deposited when the sea penetrated far inside into t subcontinent along a river estuary. Bar *et al.* (1978) concluded that the sea did exist before the Jurassic and described a close fit between India, Australia and Antarctica based on occurrences of Mesozo continental and marine Jurassic deposits northwest Australia. Their revised fit showed India and Australia as a continuous continental mass. Vuggy and Kankary carbonates were presumably formed during regressive intervals when more terrigenous muds accumulated.

SEDIMENTATION MODEL

Limestones of the Kota Fm are characteri by their lithographic character, extensive fracturing, and content of carbonate pelle dolomite, chert, broken ostracod shells an coccoliths. Burrowing structures are not uncommon, and gypsum has been reported. Associated clay mineral suite in the carbonates and accompanying clay beds containing high boron and low gallium concentr tions indicate intermixing of continental and marine environments, preferably within the marine-brackish water realm. Accordingly, an inter-tidal environment with a r estuary is proposed here to explain most of the anomalies. I believe carbonate sediments were deposited during transgressive cycles followed by temporary restrictions a the mouth of the estuary. Carbonate sedimentation became less active in regressive phases by progradation of temporary restric tions at the head of the estuary, resulting

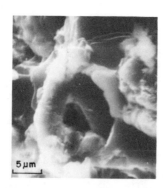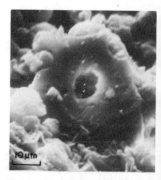

Fig. 5--SEM photographs of representative coccoliths.

n enclosed marine waters and gave rise to
deposition of clays and marls with gypsum
and/or dolomite under semi-evaporitic conditions.

CONCLUSIONS

Limstones of the Kota Fm are extremely fine-
grained, with good bedding and gentle dips;
they have few crystal forms and many carbonate pellets, chert and dolomite. They are
porous, fractured, and have high boron, broken
ostracod shells and coccoliths.

Important clay minerals are montmorillonite,
expandable mixed layers and kaolinite, occurring uniformly either in the associated clay
beds or in insoluble residues of carbonates.
The carbonates were deposited in partially
restricted marine transgressive intertidal
estuarine flats and, therefore, have both
marine and non-marine characteristics.

Marine transgression was possible from the
Indo-Australian Gulf in Jurassic times.
Australia was possibly in the area now
occupied by the Bay of Bengal. The Kota
Fm is given a special status as a marine
unit in the Upper Gondwana sequence of peninsular India.

ACKNOWLEDGMENTS

The problem was suggested by Mr S.N.
Talukdar. Samples were obtained from the
Oil & Natural Gas Commission. Drs K.
Satyanarayana and S.K. Sharma helped and
supplied data for chemical analysis of lime-
stones and clays. Dr P.G. Kale and Mr A.
Mukherjea extended supports for the study of
coccoliths. Drs William W. Hay, K. Perch
Nielson and Charles C. Smith examined the
SEM photographs and offered valuable com-
ments. The results were discussed with
officers of the Geological Survey of India.
The manuscript was critically read by Drs
C.G. Rao, S.K. Biswas and C.J. Burgess, and
is published by permission of the Director,
Institute of Petroleum Exploration, ONGC,
Dehra Dun.

REFERENCES

AHMED, F. 1961. Paleogeography of the Gondwana period in Gondwanaland, with special
reference to India and Australia and its
bearing on the theory of continental drift.
Mem. Geol. Surv. India 90: 142 pp.
BARRON, E.J.; C. G. A. HARRISON; W. W. HAY
1978. A revised reconstruction of the
southern continents. Trans. Amer. Geophys.
Union 59: 436-449.
BUKRY, D. 1974. Coccolith stratigraphy,
offshore Western Australia, DSDP Leg 27.
In Initial Reports of the DSDP 27: 623-630.

DATTA, P. M.; P. YADGIRI; B. R. JAGANNATHA
RAO 1978. Discovery of early Jurassic
micromammals from Upper Gondwana sequence. of
Pranhita-Godavari valley, India. Geol. Soc.
India 19: 64-68.
EBERL, D. 1978. The reaction of montmoril-
lonite to mixed-layer clay--the effect of
interlayer alkali and alkaline earth cations.
Geochim. Cosmochim. Acta 42: 1-27.
GOVINDAN, A. 1975. Jurassic freshwater os-
tracods from the Kota Limestone of India.
Palaeontol. 18: 207-216.
JAIN, S. L. 1973. New specimens of Lower
Jurassic holostem fishes from India.
Palaeontol. 16: 149-177.
KASHETTIWAR, B. Y.; M. K. BALAKRISHNAN 1972.
Report on the geological work in parts of
Godavari valley in Adilabad district,
Andhra Pradesh. Unpubl. Rep., ONGC, 1971-
72, Dehra Dun, 40 pp.
---; P. R. SURYANARAYAN; A. GOVINDAN 1973.
Report on the geological work in parts of
Godavari valley in Adilabad district,
Andhra Pradesh. Unpub. Rep. ONGC, 1972-73,
Dehra Dun, 41 pp.
---;---;--- 1974. Report on the geological
work in parts of Godavari graben in Adila-
bad district of Andhra Pradesh and Chandra-
pur district of Maharashtra State. Unpub.
Rep. ONGC, 1973-74, Dehra Dun, 32 pp.
KRISHNAN, M. S. 1960. Geology of India and
Burma. Madras: Higginbothams (Pvt.) Ltd.
604 pp.
PASCOE, E. H. 1959. A Manual of Geology of
India and Burma, vol. III. Delhi: Govern-
ment of India. 1343 pp.
SCHOLLE, P. A. 1977. Chalk diagenesis and
its relation to petroleum exploration - oil
from chalks, a modern miracle. Bull. Am.
Assoc. Petrol. Geol. 61: 982-1009.
YADGIRI, P.; K. N. PRASAD 1977. On the
discovery of new Pholidophorus fishes from
the Kota Formation, Adilabad district,
Andhra Pradesh. Geol. Soc. India 18: 436-
444.

Triassic fluvial depositional systems in the Fremouw Formation, Cumulus Hills, Antarctica

JAMES W. COLLINSON, K.O. STANLEY & CHARLES L. VAVRA
Ohio State University, Columbus, USA

The Early to Middle Triassic Fremouw Formation was deposited by the Fremouw river system, which drained highlands on the Gondwanian orogen. Lower and middle Fremouw rocks represent low sinuosity streams on well-vegetated floodplains and upper Fremouw represents a system of sand-dominant braided streams. Multiple large scour surfaces and large-scale trough cross-bedding are characteristic of channel-form sandstones and probably represent scour at high flow and fill by migrating dunes at lower discharge. Floodplain deposits include fining-upward greenish-grey, fine-grained argillaceous sandstone, siltstone, and silty mudstone with burrows, mudcracks, and root casts and red, sandy siltstone and mudstone, and tabular sheets of low-angle planar cross-bedded and horizontally bedded, medium-grained sandstone. Both fining-upward sequences represent flood deposits modified by soil processes and bioturbation. The sheets represent deposition during very high discharge when streams expanded to occupy part of the floodplain.

INTRODUCTION

The Fremouw Formation in the Cumulus Hills represents part of a major river system that drained the Gondwanian orogen and Antarctic craton during the Triassic (Fig. 1). A thick wedge of alluvial sediments accumulated in the Nilsen-Mackay basin (Elliot 1975), which was a foreland on the craton and adjacent to the orogen, which extended for at least 1800 km along the trend of the present Transantarctic Mountains. Triassic rocks occur in the central Transantarctic Mountains from Nilsen Plateau to Nimrod Glacier and in the Victoria Land sector of the Transantarctic Mountains from Mackay Glacier to upper Rennick Glacier. Fremouw drainage flowed away from the orogen and then longitudinally along the basin toward north Victoria Land (Elliot 1975; Collinson and Elliot, in press). Detrital rock fragments, quartz, feldspar, and heavy mineral grains in Fremouw sandstone indicate a source terrane dominated by flow and pyroclastic volcanic rocks with subordinate low-grade metamorphic, plutonic, and sedimentary rocks (Barrett 1969; Vavra et al. this vol.).

Depositional patterns in the central Transantarctic Mountains area changed from Permian to Triassic as the Nilsen-Mackay basin was formed by uplift along the Gondwanian orogen. During the Permian drainage flowed from a divide at about 78°S toward the Weddell Sea along the trend of the Transantarctic Mountains, and just before the Triassic it reversed (Barrett and Kohn 1975). In the central Transantarctic Mountains Permian tillite is overlain by black shale thought to have been deposited in an inland sea that existed before the uplift of the Gondwanian orogen (Lindsay 1969; Barrett 1969; Barrett and Faure 1973). Coal-bearing rocks of Late Permian age, indicative of fluvial deposition on a low-lying alluvial plain, are overlain disconformably, but with little apparent relief, by Triassic rocks (Barrett 1969).

STRATIGRAPHY

The Fremouw Formation was proposed by Barrett (1969) for a sequence of alternating sandstone and mudstone at Fremouw Peak in the Beardmore Glacier area. It has been traced from the Disch Promontory to the Nilsen Plateau (Fig. 1), 475 km (Barrett 1969; Collinson and Elliot, in press). La Prade (1970) proposed the name Mt Kenyon Formation for the same stratigraphic unit in the Shackleton Glacier area. The Fremouw Formation is 600-700 m thick and is separated by disconformities from the underlying Buckley Formation of Permian age and from the overlying Falla Formation of Late Triassic age (Kyle and Fasola 1978) and detailed in Barrett (1969) and Collinson and Elliot (in press).

Three informal members, although variable in thickness and bounded gradationally, can be identified. The lower consists of thick ledge-forming sandstone units alternating with equally thick, but non-resistant fine-grained units of fine- to very fine-grained sandstone, siltstone, and mudstone. The middle member, relatively nonresistant and generally poorly exposed, is predominantly fine-grained sandstone, siltstone, and mudstone. The upper member has more sandstone, much of

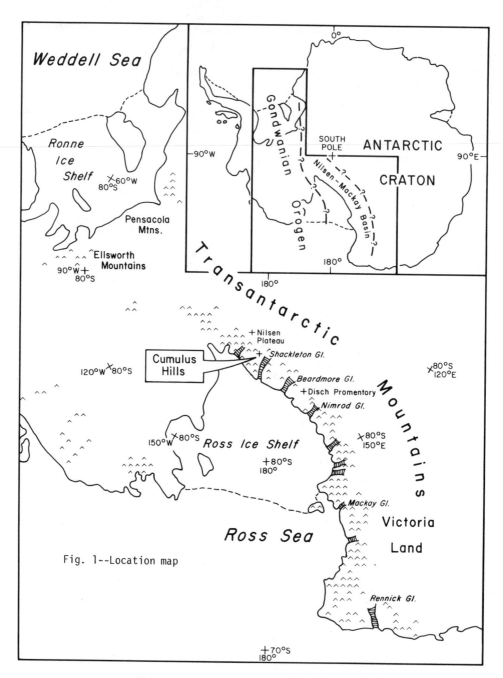

Fig. 1--Location map

which is slope-forming.

The most complete sequence of the Fremouw Formation in the Cumulus Hills is exposed along the NW face of Mt Kenyon where it is approximately 700 m thick (Collinson and Elliot, in press). This report is based primarily on data from five partial stratigraphic sections near Mt Kenyon, three of which are illustrated (Fig. 2).

The lower member consists of 40 to 50% coarse to medium-grained channel-form sandstone units as much as 17 m thick and several hundreds of metres wide. Sandstone units are thickest at the base and thin toward the top. Internally, sandstone bodies contain multiple large scours filled with large-scale trough cross-beds

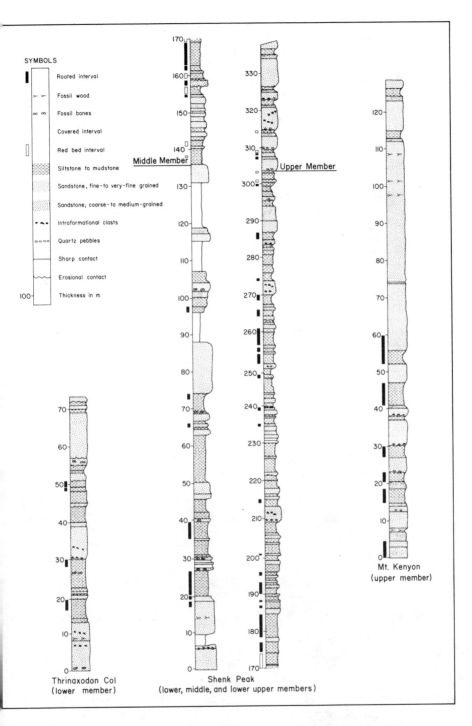

SYMBOLS

	Rooted interval
	Fossil wood
	Fossil bones
	Covered interval
	Red bed interval
	Siltstone to mudstone
	Sandstone, fine-to very-fine grained
	Sandstone, coarse-to medium-grained
	Intraformational clasts
	Quartz pebbles
	Sharp contact
	Erosional contact
100—	Thickness in m

Middle Member

Upper Member

Thrinaxodon Col
(lower member)

Shenk Peak
(lower, middle, and lower upper members)

Mt. Kenyon
(upper member)

Fig.2--Stratigraphic sections from the Cumulus Hills.

143

Fig. 3--Channel-form sandstone in the lower Fremouw member on Shenk Peak.

(Fig. 3). Trough heights average about 30 cm and are as great as 1 m. Basal scours and large scours within units are overlain by intra·formational conglomerate, quartz and phosphate pebbles, and rare vertebrate bones. Permineralized logs of *Antarcticoxylon* (identified by J. M. Schopf), averaging 50 cm diameter and 1 m long are abundant in sandstone near the base. Cross-bedding is commonly distorted by soft-sediment deformation. Some major scour truncation surfaces are covered by thin, silty mudstone drapes that are commonly burrowed and mudcracked (Fig. 4).

Fig. 4--Top of channel-form sandstone in lower Fremouw member on Thrinaxodon Col.

The transition from channel-form sandstone to overlying fine-grained rocks is generally abrupt. Tops of sandstone units fine upward from medium-grained sa⸱ stone into fine- and very fine-grained sandstone characterized by ripple lamina⸱ small-scale cross-bedding and (or) horizontal laminae with parting lineation. Fine-grained rocks are composed of finin⸱ upward sequences of laminated, massive, or brecciated greenish-grey fine-grained argillaceous sandstone, siltstone, and silty mudstone (Figs. 5, 6). Root cast⸱ and burrowing structures are abundant in some beds. Partially intact vertebrate skeletons occur here.

The middle member is dominated by finegrained rocks interbedded with relativel⸱ thin medium-grained sandstone units. Fine-grained rocks are composed of fining upward cycles, averaging 50 cm to 1 m in thickness, of greenish-grey and red fine- to very fine-grained sandstone, siltstone and mudstone (Figs. 5-8). Some beds di⸱ as much as $10°$ (Fig. 6) and pinch out laterally. Root casts and burrowing structures are most abundant in greenishgrey rocks; only the tops of red cycles c⸱ tain root casts.

Fig. 5--Greenish-grey fining-upward cycle (facies B1) overlain by sheet sandstone (facies C) in middle member on Shenk Peak

Fig. 6--Greenish-grey fining-upward cycles (B₁) with original dip suggestive of levee deposits in middle member on Shenk Peak.

Fig. 8--Red bed (B₂) fining-upward cycles in lower part of upper member in section at the head of LaPrade Valley.

The medium-grained sandstone units are channel-form and tabular-sheet. The former are similar to those in the lower member in external geometry, internal organization, and sedimentary structures, but are generally thinner and narrower. Tabular sheets of sandstone vary from 50 cm to 2 m thick and are at least 100 m wide. Bottom surfaces are sharp, but show little sign of scour; in some cases sandstone has filled mudcracks in the underlying mudstone. Irregularities on the bottoms of some suggest filling of local relief. Thicker tabular sheet sandstones contain low-angle planar cross-bedding at the base overlain by horizontal bedding and ripple laminae (Fig. 5); thinner units are only horizontally bedded (Fig. 6).

Fig. 7--Close-up view of B₁ in middle member on Shenk Peak.

Both fine upwards into siltstone and mudstone.

The upper member is marked by an increase in the proportion of sandstone. In the lower part, coarse- to medium-grained channel-form sandstone, containing coalified logs and quartz-pebble lenses, alternates with fining-upward sequences of greenish-grey or red, very fine-grained sandstone, siltstone, and mudstone. The sandstones are slope-forming because of a greater content of unstable volcanic rocks. Pumice fragments and light grey claystone beds that may represent altered vitric tuffs suggest an important pyroclastic contribution to this part of the sequence (Vavra *et al.* this vol.). A volcanic tuff was reported by Barrett (1969) in this part of the Fremouw at Wahl Glacier in the Beardmore Glacier area.

The uppermost part of the formation consists of multistorey sandstone units of coarse- to fine-grained sandstone up to 55 m thick. Sedimentary structures include abundant scours filled with low-angle trough cross-bedding and deformed cross-beds. Coalified wood, in some cases in growth position, and other plant material, including leaves of *Dicroidium*, occur locally.

DEPOSITIONAL SETTING

The depositional model (Fig. 9) proposed for the Fremouw Formation is similar to the hypothetical model developed by Moody-Stuart (1966) for low-sinuosity streams. The low variance of paleocurrent directions led Barrett (1970) to first recognize the low sinuosity of Fremouw streams. The channel form of Fremouw sandstone units

145

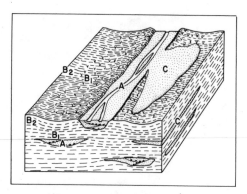

Fig.9--Depositional model of a low-sinuosity stream and floodplain for the Fremouw Fmn (see Fig.10).

and their lack of point-bar accretion bedding lend further support to this.

An idealized stratification sequence, derived from observations in the Cumulus Hills, is shown in Fig. 10 and defines (A) channel-form sandstone, (B) fine-grained beds, and (C) tabular sheet sandstone. Facies B can be subdivided into greenish-grey (B_1) and red (B_2) beds. Facies A and B together constitute a major fining-upward cycle; facies B contains minor fining-upward cycles. Facies C is intercalated in facies B.

Stratification sequences in the Fremouw Formation resemble parts of Devonian Old Red Sandstone sequences in Spitzbergen (Moody-Stuart 1966), southern Scotland (Leeder 1973), and the Battery Point Formation in Quebec (Cant and Walker 1976; Cant 1978). The Lower Triassic Middle Beaufort Group of South Africa, which may be similar to the Fremouw Formation, contains similar cycles which Hobday (1978) has interpreted as high sinuosity stream deposits. The Fremouw stratification sequence seems different from Miall's braided-stream vertical profile types (1977, 1978) but may represent a variation of the profile type modeled after the South Saskatchewan River in Canada (Cant 1978; Cant and Walker 1978). The difference is that Cant's facies model includes many features representing low flow modification of bedforms which are not preserved in the Fremouw Formation. In the classification of braided streams proposed by Rust (1978), the Fremouw Formation corresponds to the sand-dominant distal (S_{II}) and proximal (S_I) braided stream facies. The transition from distal to proximal braided facies is recorded in the upper part of the formation where floodplain deposits become uncommon.

The predominance of sandstone over con-glomerate indicates deposition in a par of the valley far away from the highland During deposition, the braided stream facies prograded basinward over low-sinusity stream facies, probably because of increase in sediment production related tectonism, volcanism, and (or) climatic change along the Gondwanian orogen.

Deposition and erosion in two straightened segments of the Rio Grande (Harms and Fahnestock 1965) provide a likely modern analogue for the channel-form sandstones Scours and trough cross-bedding characteristic of the channel-form sandstone facies were formed by erosion and downstream migration of dunes during floods. Large flows probably subsided abruptly, in some cases leaving scour surfaces and dunes preserved (Figs. 3, 4). Little evidence of low discharge modification of bedding structures is preserved. Mudst drapes across some stratification and truncation surfaces resulted from the deposition of suspended clay in waters stayi in low parts of the channel after floods Mudcracks and extensive burrowing along these surfaces indicate that parts of t stream bed were exposed between floods. As the stream bed was slowly aggrading, became more susceptible to abandonment.

The fine-grained and sheet sandstone facies represent floodplain deposits.

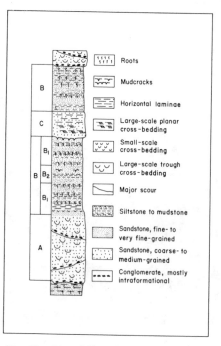

	Roots
	Mudcracks
	Horizontal laminae
	Large-scale planar cross-bedding
	Small-scale cross-bedding
	Large-scale trough cross-bedding
	Major scour
	Siltstone to mudstone
	Sandstone, fine- to very fine-grained
	Sandstone, coarse- to medium-grained
	Conglomerate, mostly intraformational

Fig.10--Stratification sequence.

Each fining-upward sequence represents a flooding episode and overbank deposition. The original dip of some beds (Fig. 6) suggests deposition on levees, and the abundant root structures suggest that vegetation stabilized banks. Mildly leached clayey sandstone to silty claystone units, either with abundant root structures and extensive burrow mottling (Fig. 7) or structureless, represent paleosols in the floodplain deposits. Polygonal fractures and fills indicate episodic exposure and desiccation.

Sheet sandstones probably represent extraordinarily large floods. Sand was apparently carried over levees onto the floodplain by events similar to the catastrophic flooding along Bijou Creek in Colorado reported by McKee *et al.* (1967). Agreement of paleocurrent directions, indicated by parting lineations on the upper surface of some sheet sandstones in the Fremouw with directions from channel deposits, indicates that water was moving directly down the floodplain during the floods.

PALEOGEOGRAPHY

As in the case of the Ganges-type model (Potter 1978), the Fremouw River valley was an elongate basin outside of and subparallel to an orogenic belt. Tributaries drained the Gondwanian orogen and joined the main river which flowed along the axis of the basin toward north Victoria Land. The known outcrop of the Triassic alluvial rocks indicates that the valley was at least 1800 km long and that the Fremouw river must have been one of the largest drainage systems in Gondwana. The Karroo basin in South Africa, the Tasmanian basin that probably bordered north Victoria Land, and the Nilsen-Mackay basin were probably parts of the same foreland structure, but it seems improbable that they were parts of the same valley.

Paleontologic evidence shows that the Pacific margin of Antarctica had a warm temperate climate during the Triassic despite its high-latitude position (Smith and Briden 1977). Thick and well-defined growth rings in fossil wood from the Fremouw Formation suggest a seasonal climate (Chaloner and Creber 1973). Vertebrate fossils from the Fremouw Formation (Colbert 1973, 1974; Colbert and Cosgriff 1974; Colbert and Kitching 1975, 1977; Cosgriff *et al.* 1978; Cosgriff and Hammer 1979) are part of the *Lystrosaurus* Zone, the diverse fauna of which was widely distributed in Gondwana, presumably because there were no climatic and physical barriers.

The abundant terrestrial vertebrate fossils in the lower part of the Fremouw are mostly herbivores, showing that the valley was well vegetated.

ACKNOWLEDGMENTS

We thank David H. Elliot, Richard J. Ojakangas, Norman D. Smith, and Charles H. Summerson for reviewing the manuscript. Research was supported by NSF grant DPP 76-23431.

REFERENCES

BARRETT, P. J. 1969. Stratigraphy and petrology of the mainly fluviatile Permian and Triassic Beacon rocks, Beardmore Glacier area, Antarctica. Institute of Polar Studies Report 34. Ohio State University, Columbus, Ohio. 132 pp.
--- 1970. Paleocurrent analysis of the mainly fluviatile Permian and Triassic Beacon rocks, Beardmore Glacier area, Antarctica. J. Sediment. Petrol. 40: 395-411.
---; G. FAURE 1973. Strontium isotope compositions of non-marine carbonate rocks from the Beacon Supergroup of the Transantarctic Mountains. J. Sediment. Petrol. 43: 447-457.
---; B. P. KOHN 1975. Changing sediment transport directions from Devonian to Triassic in the Beacon Super-Group of south Victoria Land, Antarctica. In K. S. W. Campbell (Ed.), Gondwana Geology. Canberra: ANU Press. Pp. 15-35.
CANT, D. J. 1978. Development of a facies model for sandy braided river sedimentation: comparison of the South Saskatchewan River and the Battery Point Formation. In A. D. Miall (Ed.), Fluvial Sedimentology, Memoir 5. Alberta: Canadian Society of Petroleum Geologists. Pp. 627-639.
---; R. G. WALKER 1976. Development of a braided fluvial facies model for the Devonian Battery Point Sandstone, Quebec. Can. J. Earth Sci. 13: 102-119.
---;--- 1978. Fluvial processes and facies sequences in the sandy braided South Saskatchewan River, Canada. Sedimentol. 25: 625-648.
CHALONER, W. G.; G. T. CREBER 1973. Growth rings in fossil woods as evidence of past climates. In D. H. Tarling and S. K. Runcorn (Eds.), Implications of Continental Drift to the Earth Sciences 1. London and New York: Academic Press. Pp. 423-436.
COLBERT, E. H. 1973. Continental drift and the distributions of fossil reptiles. In D. H. Tarling and S. K. Runcorn (Eds.), Implications of Continental Drift to the Earth Sciences 1. London and New York: Academic Press. Pp. 395-412.

147

--- 1974. Lystrosaurus from Antarctica. Am. Mus. Novit. 2535: 1-44.

---; J. W. COSGRIFF 1974. Labyrinthodont amphibian from Antarctica. Am. Mus. Novit. 2552: 1-30.

---; J. W. KITCHING 1975. The Triassic reptile Procolophon in Antarctica. Am. Mus. Novit. 2566: 1-23.

--- 1977. Triassic Cynodont reptiles from Antarctica. Am. Mus. Novit. 2611: 1-30.

COSGRIFF, J. W.; W. R. HAMMER 1979. New species of the Dicynodontia from the Fremouw Formation. Antar. J. U.S. 14:

---;---; J. M. ZAWISKIE; N. R. KEMP 1978. New Triassic vertebrates from the Fremouw Formation of the Queen Maud Mountains. Antar. J. U.S. 13: 23-24.

COLLINSON, J. W.; D. H. ELLIOT. In press. Triassic stratigraphy of the Shackleton Glacier region, Transantarctic Mountains. In M.D. Turner and J. F. Splettstoesser (Eds.), Geology of the Central Transantarctic Mountains, American Geophysical Union. Antarctic Research Series.

ELLIOT, D. H. 1975. Gondwana basins of Antarctica. In K. S. W. Campbell (Ed.), Gondwana Geology. Canberra: ANU Press. Pp. 493-536.

HARMS, J. C.; R. K. FAHNESTOCK 1965. Stratification bed forms and flow phenomena (with an example from the Rio Grande). In G. V. Middleton (Ed.), Primary Sedimentary Structures and Their Hydrodynamic Interpretation, Special Publication 12. Pp. 84-115.

HOBDAY, D. K. 1978. Fluvial deposits of the Ecca and Beaufort Groups in the eastern Karroo basin, southern Africa. In A.D. Miall (Ed.), Fluvial Sedimentology, Memoir 5. Alberta: Canadian Society of Petroleum Geologists. Pp. 413-429.

KYLE, R. A.; A. FASOLA 1978. Triassic palynology of the Beardmore Glacier area of Antarctica. Palinologia, num. extraord 1: 313-319.

LA PRADE, K. E. 1970. Permian-Triassic Beacon Group of the Shackleton Glacier, Queen Maud Range, Transantarctic Mountains, Antarctica. Geol. Soc. Am. Bull. 81: 1403-1410.

LEEDER, M. 1973. Sedimentology and paleogeography of the Upper Old Red Sandstone in the Scottish Border basin. Scott. J. Geol. 9: 117-144.

LINDSAY, J. F. 1969. Stratigraphy and sedimentation of lower Beacon Rocks in the central Transantarctic Mountains, Antarctica. Institute of Polar Studies Report 33. Ohio State University, Columbus, Ohio 58 pp.

McKEE, E. D.; E. M. CROSBY; H. L. BERRYHILL Jr. 1967. Flood deposits, Bijou Creek, Colorado, June 1965. J. Sediment. Petrol. 37: 829-851.

MIALL, A. D. 1977. A review of the braided-river depositional environment. Earth Sci. Rev. 13: 1-62.

--- 1978. Lithofacies types and vertical profile models in braided river deposits: summary. In A. D. Miall (Ed.). Fluvial Sedimentology, Memoir 5. Alberta: Canadian Society of Petroleum Geologists. Pp. 597-604.

MOODY-STUART, M. 1966. High- and low-sinuosity stream deposits, with examples from the Devonian of Spitsbergen. J. Sediment. Petrol. 36: 1102-1117.

POTTER, P. E. 1978. Significance and origin of big rivers. J. Geol. 86: 13-33.

RUST, B. R. 1978. Depositional models for braided alluvium. In A. D. Miall, (Ed.) Fluvial Sedimentology, Memoir 5. Alberta: Canadian Society of Petroleum Geologists. Pp. 605-625.

SMITH, A. G.; J. C. BRIDEN 1977. Mesozoic and Cenozoic paleocontinental maps. Cambridge: Cambridge University Press. 63 pp.

VAVRA, C.L.; K. O. STANLEY; J. W. COLLINSON. Provenance and alteration of Triassic Fremouw Formation, Central Transantarctic Mountains. This volume.

Institute of Polar Studies Contribution #

Provenance and alteration of Triassic Fremouw Formation, Central Transantarctic Mountains

CHARLES L. VAVRA, K. O. STANLEY & JAMES W. COLLINSON
Ohio State University, Columbus, USA

Detrital constituents of Fremouw sandstones indicate two distinct source areas; one dominated by sedimentary rocks and the other dominated by calcalkaline volcanic rocks with a minor contribution from low-grade metamorphic and plutonic rocks. The bulk of Fremouw sandstone was derived from a volcanic source area, probably a contemporaneous volcanic complex in the Gondwanian orogen. Paleocurrent data and distribution of volcanic and sedimentary detritus in Fremouw sandstones suggest that streams flowed from the volcanic complex and craton into and down the axis of a foreland basin.

Post-depositional alterations of Fremouw sandstone include grain dissolution, grain destruction and replacement, and formation of phyllosilicate, silica, and laumontite cements. These alterations resulted from (1) early diagenesis at near-surface temperature and pressure followed by (2) zeolite facies metamorphism and contact metamorphism at shallow depth with heat required for sandstone alteration supplied by Jurassic dolerite sills and dikes. Zeolite facies authigenic minerals, including laumontite pore-filling cements, were probably controlled largely by pore-fluid chemistry, in addition to temperature and pressure.

INTRODUCTION

The Fremouw Fm in the central Transantarctic Mountains is part of a Triassic clastic wedge of fluvial rocks that accumulated in the Nilsen-Mackay basin formed along the margin of the Antarctic craton, on the Lower Paleozoic Ross orogen, and adjacent to the Gondwanian Orogen (Fig. 1). The Nilsen-Mackay basin, Gondwanian Orogen, and Antarctic craton existed before the initial phase of rifting of Antarctica from other southern continents (Elliot 1975a, 1976). The Fremouw Fm records an influx of sediment related to major tectonic events and denudation of the craton and orogen, infilling of the Nilsen-Mackay basin, and post-depositional burial and thermal alteration of Fremouw rocks in the basin.

The history of Triassic tectonics and sedimentation in the central Transantarctic Mountains is preserved in the Fremouw and Falla Fms and Triassic-Jurassic Prebble Fm. These accumulated in the Nilsen-Mackay basin before eruption of basalt and emplacement of dolerite of the Ferrar Gp (Elliot 1975b). The Fremouw Fm rests disconformably on the Permian Buckley Fm, which pre-dates development of the Nilsen-Mackay basin, and is overlain disconformably by the Falla Fm (Barrett 1969; Collinson & Elliot, in press). In the Shackleton-Beardmore glacier region, the Fremouw Fm was deposited by low sinuosity, sand-dominant, braided streams (Collinson *et al.*, this vol).

DETRITAL MINERALOGY OF FREMOUW SANDSTONE

Fremouw sandstones vary from quartz arenite to volcanic litharenite and display variable amounts of grain alteration and cementation. Detrital constituents include quartz, feldspar, lithic fragments, biotite, muscovite, and minor amounts of magnetite, leucoxene, sphene, garnet, zircon, and tourmaline. Mono-crystalline

Fig.1--Present-day Antarctica showing the Shackleton Glacier-Beardmore Glacier study area with superimposed major Triassic tectonic elements (after Elliot 1975; 1976).

Table 1. Means and standard deviation of modal analyses of Fremouw sandstone

Member	N	Quartz	Feldspar	Lithic Fragments	Mica	P/F	V/L	Qp/Qt	Qr/Qt
Cumulus Hills Areas									
lower	45	41.4 ± 10	22.0 ± 7	34.6 ± 12	2.0 ± 2	.88	.80	.07	.06
middle	32	35.1 ± 5	28.6 ± 6	30.8 ± 11	5.5 ± 5	.94	.65	.08	.02
upper	17	28.2 ± 9	21.7 ± 6	45.1 ± 16	5.0 ± 4	.94	.80	.08	.03
Central Area									
lower	8	65.3 ± 18	18.1 ± 10	15.8 ± 7	0.8 ± 1	.92	.53	.06	.06
middle	17	38.0 ± 11	29.4 ± 7	30.6 ± 10	2.0 ± 1	.89	.87	.08	.03
upper	5	29.2 ± 6	27.1 ± 7	39.3 ± 9	3.4 ± 1	.90	.88	.07	.01
Beardmore Area									
lower	11	79.7 ± 18	13.0 ± 9	7.3 ± 8	0	.56	.52	.06	.31
middle	14	37.9 ± 11	29.8 ± 11	30.7 ± 9	1.6 ± 1	.88	.86	.07	.01
upper	15	28.7 ± 7	34.5 ± 8	35.3 ± 8	1.5 ± 1	.83	.84	.08	.01

N is number of thin sections counted. Values for quartz, feldspar, lithic fragments and mica are volumetric percentages of the total framework. Values for P/F, Qp/Qt, Qr/Qt and V/L are decimal fractions. Qt is total quartz, Qp is polycrystalline quartz, Qr is recycled quartz, L is lithic fragments, and V is volcanic rock fragments. Beardmore Area data from Barrett (1969).

quartz grains dominate the detrital quartz populations (Table 1) and include two distinctive forms: (1) well-rounded to rounded, unstrained grains, some with multiple syntaxial overgrowths, and (2) clear, unstrained grains, some with corrosion embayments and bipyramidal terminations. The former quartz grains were recycled from pre-existing sandstone, the latter were of volcanic origin. The greater the quartz content of sandstone, the greater is the ratio of recycled quartz to total quartz. Quartz-rich sandstone occurs in the lower member of the Fremouw in the Beardmore Glacier area but is sparse elsewhere (Table 1, Fig.2). Feldspars include plagioclase (oligoclase to andesine) and small amounts of orthoclase and microcline. Quartzose sandstone exhibits a low plagioclase to total feldspar ratio, whereas volcanic lithic sandstone exhibits a high plagioclase to total feldspar ratio.

There are many volcanic and sedimentary fragments and a few plutonic and low-grade metamorphic fragments. Sedimentary fragments are commonly mudstone, phosphatic mudstone, and sandstone. Most fragments are volcanic detritus from rhyolitic to andesitic flows and pyroclastic deposits. Pyroclastic material includes shards, pumice fragments, and rock fragments with relict vitroclastic texture, suggesting a possible ash flow origin. Volcanic detritus increases in abundance upwards in the sequence and laterally from the Beardmore to Shackleton glacier areas.

PROVENANCE

Sediment dispersal and sandstone composition data (Fig. 2) suggest that the bulk of the detritus was derived from the Gondwanian orogen and transported down the basin axis by a river system that flowed for nearly 1800 km subparallel to the orogen. Detrital modes indicate two distinct types of source areas. Sedimentary terranes included mudstone, phosphatic mudstone, quartz arenite, and probably some feldspathic sandstone or crystalline rocks. Well-rounded quartz grains were probably recycled from quartz arenite in the Taylor Gp, whereas the other detritus could have been derived from the Victoria Gp, Taylor Gp, rocks of the Ross orogen, and (or) crystalline rocks on the Antarctic craton. Such detritus characterizes the lower Fremouw, particularly in the Beardmore Glacier area, and probably came from local highs in the Nilsen-Mackay basin or from the adjacent orogen and craton.

Volcanic detritus from the Gondwanian orogen characterizes most of the Fremouw, and particularly the middle and upper part. Rhyolitic to andesitic lithic fragments of flow and pyroclastic origin suggest that the source area was a calc-alkaline volcanic complex. Small amounts of plutonic and low-grade metamorphic lithic fragments in volcanic litharenite indicate that crystalline rocks probably were exposed in the eroded volcanic complex. Vitric tuff and tuffaceous claystone associated with volcanic litharenite suggest contemporaneous volcanism. The increase in volcanic detritus upward could reflect increased volcanism, further uplift of the source area, increased runoff and sediment yield in upland areas caused by increased precipitation, and (or) shifts in the courses of streams that drained various parts of the orogen and local cratonic highs.

POST-DEPOSITIONAL MODIFICATIONS

Sandstone in the Fremouw Fm has undergone complex post-depositional modifications: (1) porosity reduction by phyllosilicate, calcite, and silica cementation and compaction of weak grains and minor porosity enhancement by dissolution of chemically unstable grains; (2) porosity reduction by laumontite pore filling and formation of prehnite concretions; (3) replacement of detrital grains by albite, laumontite, phyllosilicate, epidote, sphene, and (or) prehnite; and (4) laumontite and heulandite fracture fillings. These alteration fabrics and constituents can be related to early diagenesis, at near-surface temperature and pressure, and to zeolite facies metamorphism and superimposed contact metamorphism associated with emplacement of dolerite dikes and sills.

Early Diagenesis. The earliest phase of diagenesis observed is the formation of phyllosilicate cement coating grains and forming a lining for intergranular pores. Locally, these rim cements are followed by calcite, quartz, chert, and (or) phyllosilicate pore-filling cement. These cements appear to pre-date compaction of shale, volcanic, and phyllite grains, but compaction features commonly occur where cementation is absent or minimal. Most sandstone exhibits little evidence for mechanical alteration, and commonly, its fabric is characterized by tangential and long grain-to-grain contacts, reflecting depositional packing and fabrics. Discontinuous interstitial phyllosilicate paste formed by deformation of weak grains, and bent and wrinkled mica flakes reflect some compaction of weak grains. All this reflects reactions of interstitial fluids, probably ground waters, and sediment at near surface temperatures and pressures. Compaction of mechanically weak grains and the formation of quartz cements may reflect progressive, relatively shallow, sediment burial.

Zeolite Facies Metamorphism and Superimposed Contact Metamorphism. Pore-filling laumontite cement and prehnite concretions formed after early diagenesis of sandstone.

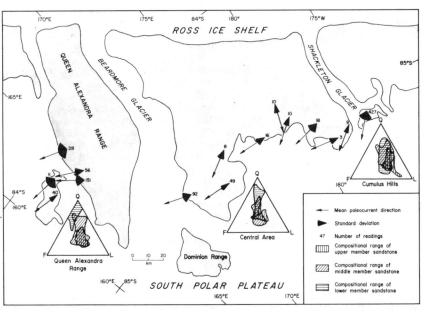

Fig.2--Shackleton Glacier-Beardmore Glacier area showing sediment dispersal directions for Fremouw Fm and the quartz (Q), feldspar (F), lithic fragment (L) modal abundance for the lower, middle and upper members. Dispersal directions were determined from 1025 sets of trough cross-stratification in fluvial channel-form deposits from the lower, middle and upper Fremouw, and no statistically significant differences were recognized. Q-F-L plots are based on data of Table 1.

Intergranular pore-space which remained after formation of authigenic quartz is filled with laumontite. This fabric selective cement is pervasive throughout the Fremouw in the Shackleton-Beardmore glacier region. Prehnite concretions also are common and replace laumontite-cemented volcanic lith-arenite.

In the central Transantarctic Mountains, Ferrar Dolerite sills and dikes were emplaced in Paleozoic and Mesozoic rocks 175 ± 5 m.y. ago (Fleck *et al.* 1977). These Jurassic intrusions make up 10 - 30% of Fremouw outcrop belts. They are commonly tens of metres in thickness; however, few are more than 100 m thick.

In much of the Fremouw, and particularly in the vicinity of Jurassic dolerite dikes and sills, laumontite-cemented sandstone was modified by destruction and replacement of grains and earlier cements by albite, laumontite, prehnite, epidote, chert, sphene, and sericite. In some cases alteration minerals and patterns form zones adjacent to dolerite that reflect contact metamorphic effects of the intrusion. Authigenic mineral assemblages in the Falla and Fremouw Fms correspond to metamorphic zones 1 and 2 described by Seki *et al.* (1968) from the Tanzawa Mountains, Japan, where emplacement of quartz diorite plutons resulted in metamorphism ranging from zeolite facies to amphibolite facies. In addition, adjacent to some dolerite sills and dikes, sandstone alteration may correspond to zone 3 of Seki *et al.*

Zeolite minerals in the Falla and Fremouw Fms are also similar to zeolite facies minerals of Hokonui sandstone reported by Coombs (1954) from the Southland Syncline, New Zealand.

Unlike Hokonui rocks, however, the Fremouw Fm was probably buried between 600 and 1000 m during the initial phase of igneous activity, judging from the stratigraphic thicknesses of overlying Triassic Falla and Triassic-Jurassic Prebble Fms. In the Jurassic, the intrusion of the Ferrar dolerite and the extrusion of the comagmatic Kirkpatrick basalt increased the depth of burial to about 2 km, which is probably a reasonable maximum depth of burial for the formation (Barrett 1969).

The illitization, albitization, sercitization, and chloritization of Fremouw sandstones adjacent to dolerite intrusions demonstrates the heating effects of the dikes and sills. The hydrochemical reactions were not isochemical, at least on a local scale, and K^+ and Ca^{++} at least were mobile. Alteration of Fremouw sandstones and mudstones in the presence of pore fluids remaining after early diagenesis could explain the authigenic minerals in the contact metamorphic aureoles adjacent to dolerite intrusions. It is not clear whether dolerite contributed constituents to the hydrochemical system, because mineral assemblages in the contact aureoles can be explained by reactions involving sediment and pore fluid without appealing to the dolerite for additional constituents.

Contact metamorphic aureoles characterized by albitization of plagioclase have been developed in sandstone in the Fremouw Fm and in the underlying Buckley Fm where thin dolerite intrusions are more common. Fremouw sandstones between contact metamorphic aureoles contain pervasive laumontite that occurs as a pore-filling cement with no obvious precursor in the host rock. The volumetrically significant dolerite intrusions in the Fremouw and underlying formations have altered sufficiently large volumes of sandstone in contact aureoles to account for all of the constituents needed to produce laumontite pore-filling cements and in particular, albitization of plagioclase could account for the Ca necessary for laumontite cements. Contact aureoles in the Fremouw are the most likely source of constituents, but similar constituents and fluids also could have been derived from metamorphic aureoles in Permian rocks beneath the Fremouw. Where dolerite intrusions and contact aureoles are sparse or absent as in the Falla and Prebble Fms, laumontite pore-filling cements are absent. Consequently, we believe that fluid chemistry was controlled by sediment-pore fluid reactions in the aureoles where dolerite supplied the heat required for albitization, seritization, illitization and chloritization of sandstone. High temperatures probably drove fluids into sandstone where laumontite cement was precipitated. The absence of such cements in Falla and Prebble rocks and the association of laumontite with sandstone containing siginificant dolerite intrusions underscores the importance of fluid chemistry in addition to pressure and temperature, as a control on the cement's distribution. Formation of authigenic minerals, including zeolite facies minerals in the Fremouw was probably controlled largely by pore fluid chemistry and high temperature gradients adjacent to intrusions.

CONCLUSIONS

Detrital constituents of Fremouw sandstone reflect debris weathered from upland terranes and transported into the Nilsen-Mackay basin by a major Triassic river system in Gondwana. The dominant source for river sand was probably a contemporaneous calcalkaline volcanic complex in the Gondwanian orogen. Subordinate source terranes were dominated by sedimentary rocks with minor amounts of crystalline rocks. Sedimentary detritus was derived from the Upper Paleozoic Victoria and Middle Paleozoic Taylor Gps, which are now exposed in the Transantarctic Mountains. Sedimentary and crystalline material also could have been derived from rocks of the Lower Paleozoic Ross orogen or Precambrian-Lower Paleozoic Antarctic craton. The sedimentary and crystalline rocks could have been exposed in local highs on the craton, in the Nilsen-Mackay basin, or on the margin of the Gondwanian orogen.

Alteration of the river sands record Triassic-Jurassic diagenesis during sediment accumulation in the Nilsen-Mackay foreland basin and later low-grade metamorphism associated with emplacement of dolerite during the initial phase of rifting of Gondwana. Early diagenesis of Fremouw sandstone records reactions of sediment and pore fluid, which probably was ground water, at near surface temperature and pressure. Early diagenesis was followed by zeolite facies metamorphism and superimposed contact metamorphism. Zeolite facies alterations of sandstone were controlled by pore fluid chemistry as well as temperature and pressure. Higher temperatures near dolerite intrusions influenced the stability of authigenic mineral phases, resulting in albitization, laumontization, sercitization, and (or) chloritization of sandstone. Dolerite supplied the heat required for the sediment-pore fluid reactions in the contact metamorphic aureoles and probably drove fluids produced by these reactions into adjacent sandstone, where laumontite pore-filling cement precipitated. Laumontite can be accounted for by albitization of plagioclase without addition of constituents from dolerite or other sources.

ACKNOWLEDGMENTS

We thank David H. Elliot, Richard J. Ojakangas and James R. Boles for reviewing the manuscript. Research was supported by National Science Foundation grant DPP76-23431.

REFERENCES

BARRETT, P. J. 1969. Stratigraphy and petrology of the mainly fluviatile Permian and Triassic Beacon Rocks, Beardmore Glacier Area, Antarctica. Inst. Polar Stud. Rep. 34: The Ohio State University: Columbus, Ohio. 132 pp.

COLLINSON, J. W.; D. H. ELLIOT. In press. Triassic stratigraphy of the Shackleton Glacier region, Transantarctic Mountains. In M. D. Turner and J. F. Splettstoesser (Eds.). Geology of the Central Transantarctic Mountains. American Geophysical Union Antarctic Research Series.

---; K. O. STANLEY; C. L. VAVRA. Triassic fluvial depositional systems in the Fremouw Formation, Cumulus Hills, Antarctica. This volume.

COOMBS, D. S. 1954. The nature and alteration of some Triassic sediments from Southland, New Zealand. Trans. R. Soc. N.Z. 82: 65-109.

ELLIOT, D. 1975a. Gondwana Basins of Antarctica. In K.S.W. Campbell (Ed.), Gondwana Geology. Canberra: ANU Press. Pp. 493-536.

--- 1975b. The tectonic setting of the Jurassic Ferrar Group, Antarctica. In O. Gonzales-F. (Ed.), Andean and Antarctic Volcanology Problems. Santiago: IAVCI. Pp. 357-372.

FLECK, R. J.; J. R. SUTTER; D. H. ELLIOT 1977. Interpretation of discordant $^{40}Ar/^{39}Ar$ age-spectra of Mesozoic tholeiites from Antarctica. Geochim. Cosmochim. Acta 41: 15-32.

SEKI, Y.; Y. OKI; T. MATSUDA; K. MIKAMI; K. OKUMURA 1969. Metamorphism in the Tanzawa Mountains, central Japan. J. Japan. Assoc. Mineral., Petrol., Econ. Geol. 61: 1-75.

Institute of Polar Studies Contribution #392

Jurassic turbidites in central western Argentina (Neuquén Basin)

ULRICH ROSENFELD
Geologisch-Pälaontologisches Institut,
Universität Münster, Germany

WOLFGANG VOLKHEIMER
Museo Argentino de Ciencias Naturales,
Buenos Aires, Argentina

Turbidites and related sediments have been studied in Sinemurian, Pliensbachian, Toarcian, Aalenian and Bajocian strata of the Argentine Neuquén Basin. Mud flows, low energy and high energy turbidites and olistostromes can be recognized. Most of the sediments are clearly related to the margin of the basin and indicate its rapid subsidence. Tuffites in the sequence indicate volcanic activities in the hinterland, for example in the Pliensbachian of Sierra de Chacai-Có and the Aalenian of the the upper Rio Atuel area.

The development of the Argentine Neuquén Basin in the early Jurassic, shown by gravitational mass movements described and other data, is discussed.

INTRODUCTION

In the Lower and Middle Jurassic of the Neuquén Basin (Argentina) "conglomerates" and "pebbly horizons" have repeatedly been mentioned or described which seemed to be turbidites or other types of submarine mass movements(e.g. Digregorio 1972; 465). Up to now detailed or systematic investigations of these phenomena do not exist.

During several field expeditions we both became aware of a number of such phenomena as well as of rhythmic sediments of an unknown genesis. That was the reason for this investigation in which gravitational mass movements in the Lower and Middle Jurassic of the Neuquén Basin are comprehensively described. A great number of Liassic and Dogger profiles and so far unknown profiles in the provinces of Mendoza and Neuquén were studied (see Fig. 2).

There is a preference for Liassic strata in the investigation. This is due to the assumption that if the subsidence of the Neuquén Basin began in the earliest Jurassic an accumulation of gravitational mass movements during certain stages could give hints as to the character of the subsidence. The Middle Jurassic represents a completely developed basin in which partial regressions were occurring. At the first sight it seemed more difficult to interpret the Middle Jurassic turbidites.

GRAVITATIONAL MASS MOVEMENTS

In the studied area three types of gravitational mass movements are to be found: mud flows, turbidites and olisto-stromes and other slump deposits. The mud flows are continental; the turbidites and olistostromes marine.

Mud Flows. In the Arroyo Poti Malal the Jurassic transgression begins with a basal conglomerate of several sequences: Well-rounded to subrounded quartz pebbles occur in the lowest sequence; well-subrounded volcanic pebbles in the middle; and non-rounded volcanic pebbles in the highest sequences. According to their components, habit and fabrics these sediments correspond to the descriptions given by Blissenbach (1954) or Bull (1960); layers with oblong clasts in an upright position are conspicious. These sediments are obviously bound to specific conditions, namely a coastal plain adjacent to highlands (Shepard & Dill 1966). They are likely to have been transported by sheet floods (Fairbridge & Bourgeois 1978; Hook 1967; Reineck & Singh 1975).

In the Arroyo Poti Malal it can be demonstrated that the first sheet flood came with high energy from the distant hinterland, a fact for which the occurrence of well-rounded to subrounded quartz may be a proof. The subsequent floods were in each case of lower energy. The coastal plain was slowly sinking, and near-shore volcanic areas delivered more material than the distant quartz source. Progressively increasing aquatic influences can be recognized. The rocks become more even-grained, better sorted and better bedded, and the mode of deposition is replaced by redeposition. Finally the first lumachelle indicates the permanent existence of marine conditions.

The outcrops in the Arroyo Poti Malal and at the Rio de Los Patos Sur, described by Volkheimer (1977), represent an extremely flat beach upon which the Lower Jurassic sea transgressed. The upward change provenance indicates either nearby active volcanism or rapid uplift of nearby pre-existing volcanic rocks. The second idea is supported by the assumption that the subsidence of the basin took place partially at synsedimentary faults.

In the Rio Atuel area the transgressive Jurassic begins with thick fluvial sediments of the El Freno Fmn (Volkheimer 1978), which also shows indications of mud flows. Under these high-energy beach conditions, mud flows from the hinterland are obviously possible, too. The sedimentary facies indicate rapid uplift of the hinterland coinciding with the subsidence of the Neuquén Basin.

Turbidites. The studied turbidites are mainly sandstone or tuffitic turbidites; calcareous turbidites are rare. Mostly they are thin, rather fine-grained and only in few cases regularly developed with all intervals of the Bouma sequence. Often they must be considered a distal facies or to have developed with rather low energy, because a coarse base or often other intervals are lacking. Thus, at the southern border of the Rio Atuel valley the sequences have with respect to the Bouma sequence the form $T_1 \rightarrow T_2 \rightarrow T_3 \rightarrow T_5$, while at the northern wall they generally begin with a more or less coarse parallel laminated unit, so that the sequence corresponds to the form $T_2 \rightarrow T_{?3} \rightarrow T_4 \rightarrow T_5$. Frequently the interval with ripple lamination or convolute bedding is not developed, although the appropriate grain-size distribution is present (see, however, Einsele 1963). An exception is the turbidites in the Arroyo Lapa in which convolute bedding is very common. On the whole, there are not only "high-velocity turbidites" in the sense of Moore (1969), but also very often "low-velocity turbidites" bound to shelf areas (Stanley 1969).

In all exposures the directional structures indicate that the flow and the material originated at a nearby basin margin. These are in most cases NE, E and S margins, but in some profiles (e.g. Chacai Melehue) an uplifted block situated within the basin may also be the origin. Exposures near Loncopue suggest a previously unknown western margin of the basin.

In no case could lateral changes in the character of a turbidite be observed. All the studied turbidite sequences change, however, in a vertical direction: In each case they begin with thicker and coarser, more completely developed turbidites, but upwards loose energy and become thinner, more fine-grained and less complete. In the upper parts of the sequences almost always a second high-energy phase begins in which slumping phenomena, olistolites and olistostromes occur. The latter are often considerably higher in energy than the underlying turbidites, i.e. they are thicker and coarser. Frequently they erode strongly into the underlying turbidite. It is evident that first the energy causing the mass movements fluidized a finer grained portion of sediments and then grew stronger without fluidizing completely further sediment material. Of great importance in this case are sediments rich in tuffs or sediments becoming richer in tuffs by continuous ash fall; these were unstable masses, from which the turbidites originated.

Olistostromes and Related Phenomena. Further submarine gravity flows are olistostromes, fluxoturbidites and phenomena of an unknown genesis. These masses are not single pebbles or olistolithes; they are paraconglomerates or breccias with clasts in a predominantly poorly sorted matrix. However, all gradations exist from completely disordered to sorted and/or graded sediments which may show particles more or less well oriented in the flow direction (Stanley 1969).

A genetic classification of these forms is not possible in each case. In our opinion all those masses which occur in close connection with turbidites can be identified as olistostromes. They are relatively small products of gravitative displacement in a not fully fluidized state, pebbly mudstones which may also contain reworked older turbidites and are not necessarily restricted to submarine canyons (Hsü 1974 or Fairbridge & Bourgeois 1978; keyword "gravity flows").

On the other hand, we call those slump masses which either occur alone without underlying turbidites or have a distinct lenticular shape (normal to the transport direction) canyon sediments or fluxoturbidites (Einsele 1963; Kuenen 1964; Bouma 1965). The reasons for this identification are, beside the lateral wedging out, the better degree of purity, the thicker bedding, the occurrence of faunal remainders and the retreat of sliding phenomena in the generally coarse-grained sediments.

It is remarkable that olistostromes overlying turbidite contain intraformational clasts of turbidite. Obviously the lithification of such turbiditic sequences rich in tuffs was taking place very quickly.

PALEOGEOGRAPHICAL REMARKS

Phenomena fall into two groups of different age. The older group, in which the turbidites become younger from N to S, ranges from Sinemurian to Toarcian; the younger one from Toarcian-Aalenian to Callovian (Fig. 1).

The phenomena differ in their paleogeographical positions. As already mentioned, the continental mud flows are related to a coastal plain and pass into littoral sediments. The older turbidites and olistostromes are always to be found in sediments which develop from continental or fluvial sediments to littoral and neritic sediments. They are assoc-

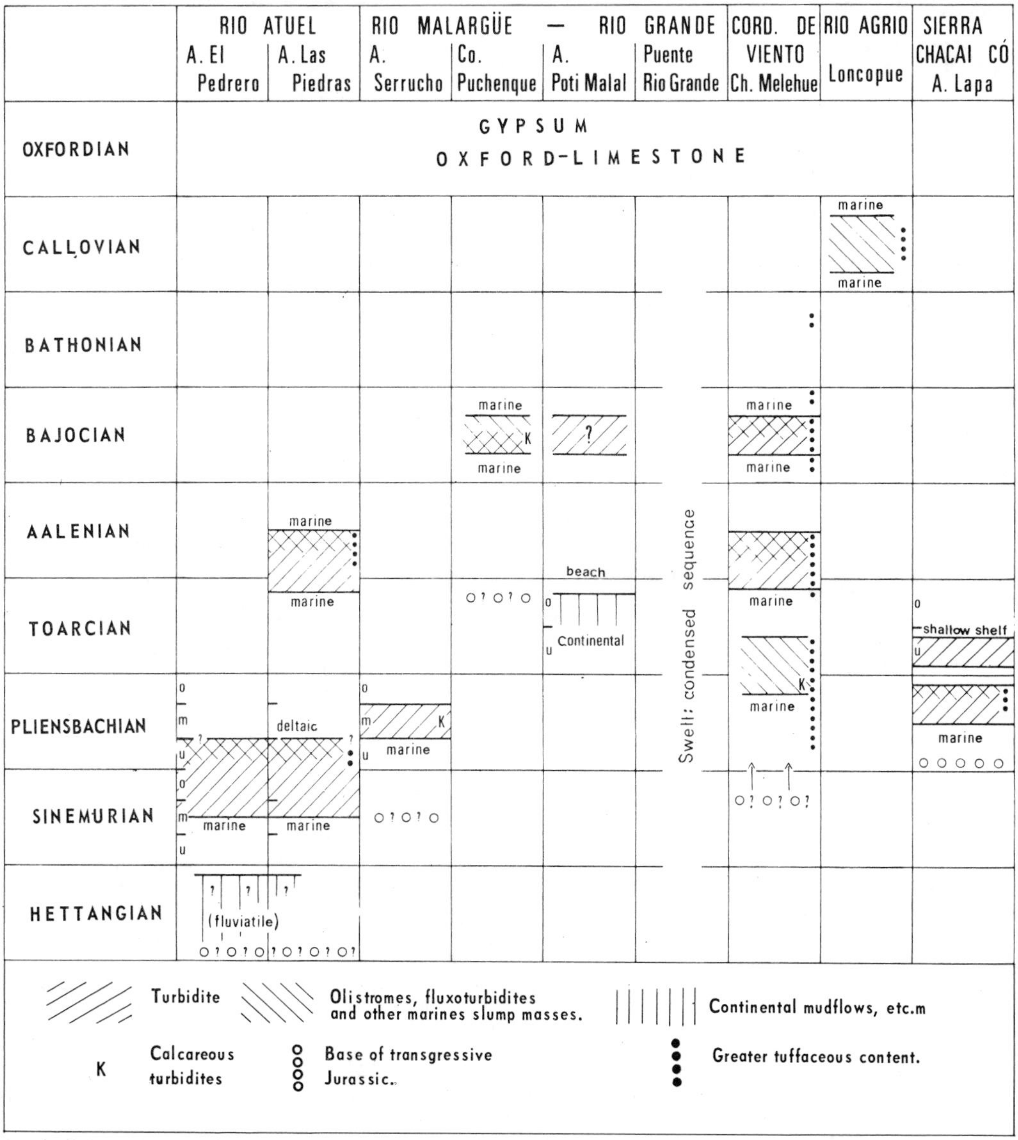

| | RIO ATUEL | | RIO MALARGÜE — RIO GRANDE | | | | CORD. DE VIENTO | RIO AGRIO | SIERRA CHACAI CÓ |
	A. El Pedrero	A. Las Piedras	A. Serrucho	Co. Puchenque	A. Poti Malal	Puente Rio Grande	Ch. Melehue	Loncopue	A. Lapa
OXFORDIAN	GYPSUM OXFORD-LIMESTONE								
CALLOVIAN								marine / marine	
BATHONIAN								(tuffaceous dots)	
BAJOCIAN				marine / K / marine	?		marine / marine	(dots)	
AALENIAN		marine / marine					marine		
TOARCIAN					o?o?o o / beach / Continental u		marine	Swell: condensed sequence	o / shallow shelf / u
PLIENSBACHIAN	o / m / ? / u	deltaic	o / m / ? K / u marine				K / marine		marine / o o o o o
SINEMURIAN	o / m marine / u	marine	o?o?o				o?o?o?		
HETTANGIAN	(fluviatile) o?o?o?o?o?o?								

Legend:
- (hatch) Turbidite
- (cross-hatch) Olistromes, fluxoturbidites and other marines slump masses.
- (vertical bars) Continental mudflows, etc.m
- K Calcareous turbidites
- o / o-o-o Base of transgressive Jurassic.
- (dots) Greater tuffaceous content.

Fig. 1--Sediment gravity flows in the Lower and Middle Jurassic of the Neuquén Basin/Argentina.

iated with the first transgression phase of the Jurassic sea. These turbidites are mainly low-velocity turbidites and characterize the first shelf or shelf margin of the transgressive Jurassic (Moore 1969). The observations at the Rio Atuel and in Chacai Melehue show that the shelf margin was partially developed as a synsedimentary fault. The older group of turbidites and olistostromes is furthermore characterized by the fact that the turbidite series are always overlain by deltaic or beach and near-shore shelf sediments. Obviously the basin, on the whole still shallow, had been partially filled by the described mass flow deposits. The first transgression was followed by a phase of slight subsidence (van Straaten 1970). In this connection the low-velocity turbidites can also be interpreted as local phenomena in the front of approaching deltas.

As the tuffite content of the sediments indicates, there is no doubt that all these events were influenced by volcanic activity in the hinterland.

The younger group of turbidites and olistostromes is, on the contrary, found in sediments which are "marine" in general. (Even the fluxoturbidites in the profile near Loncopue contain fragments of ammonites which could not have originated from littoral areas.) They are always associated with highly tuffaceous sediments and themselves contain a high proportion of tuffaceous material. These turbidites are mainly due to volcanic activity in the hinterland. Turbidites may have been generated on submarine slopes by earthquakes or other factors. Ashfall deposits may have been unstable and formed thick olistostromes. According to various references the tuffs in the Neuquén Basin are predominantly Pliensbachian, earliest Dogger, Bajocian and Callovian in age. The correspondence between volcanism and the occurence of turbidites is obvious.

The configuration of the Neuquén Basin and its characteristic features is well known. In Fig. 2 a contour line after Volkheimer (1970) shows the approximate Upper Liassic to Lower Dogger shore line. The Jurassic transgression began north of the Rio Atuel area in the Hettangian and advanced south to reach the regions between Sierra Chacai-Co and Rio Limay

in the Pliensbachian. In the southernmost part of the basin there may be no fully marine Liassic sediments but only paralic ones, as in the exposures near Alicura.

The Remoredo formation (Digregorio 1972: 456), and comparable sediments underlying the Jurassic are typical continental sediments deposited on widespread large lowlands and formed an almost-horizontal surface prior to the Jurassic transgression. Nevertheless, the area was deformed and showed relief differences which seem to have been caused by seismic activity that began before the transgression. The relief is indicated by thick fluvial sediments of the El Freno Fmn in the Rio Atuel valley and a condensed sequence in the Rio Grande area.

Further evidence of the shape of the basin is given by the turbidites and other mass movement deposits. They define the basin margins, including the previously unrecognized western margin of the basin in Loncopue and islands within the basin, at the Arroyo Poti Malal and Chacai Melehue, where basement massifs still exist.

The basin subsidence took place partly at synsedimentary faults. This conclusion can be drawn from the thickness and facies of the sediments in the Rio Atuel area (Volkheimer 1978) and from the profiles at the border of the Cordillere del Viento (Zöllner & Amos 1973). The strike of the synsedimentary faults agrees with the strike of large lineaments which have been mapped in the Neuquén Basin and which originated at different times (Ramos 1978). Near Puento Rio Grande a line of recent tectonic uplift coincides with one of early Jurassic uplift. It is likely that the major structures of the Neuquén Basin (Ramos 1978, Fig. 2) were all active during the early Jurassic and contributed to the division of the basin into uplifted and downfaulted blocks.

For all these reasons we believe that the Neuquén Basin is a fault trough in the Lower Jurassic, which is bordered and traversed by synsedimentary faults; it may be compared with the southern Penninic trough of the Alps during the Liassic (Gwinner 1978). Into this trough the Jurassic at first transgressed in a

Fig. 2--Paleogeography and development of the Neuquén Basin/Argentina in the Lower and Mi Jurassic. -- 1 = margin of the basin (contour) line); 2 = direction and time of transgre sion; 3 = near-shore sediments in general and above the older group of turbidites; 4 = transport direction of turbidites and continental mud flows; 5 = high tuffite content of sediments in the Pliensbachian (+), Aalenian (x), Upper Dogger (o); 6 = carbonate content in turbidites.

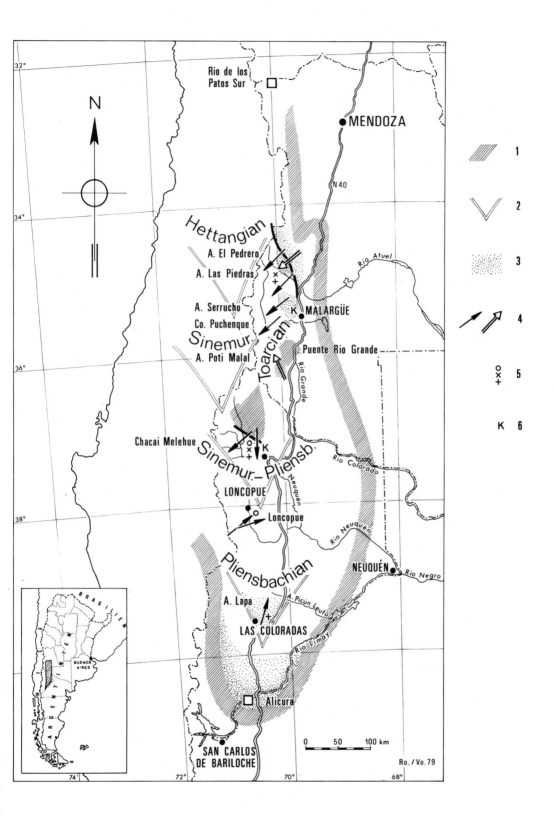

narrow linguoid form rapidly towards the south. Thereafter the transgression advanced eastward as shown by the out- crops at the Arroyo Poti Malal.

Basin subsidence and transgression were not only accompanied by more or less severe earthquakes, but also by strong volcanic activity in the marginal highlands. These hinterlands, however, must be characterized differently, as the southernmost part of the basin demon- strates: The sediments at the Rio Limay do not contain tuffites because no volcanic activity occurred in the Patagon- ian Massif. To the contrary, the rocks at the Arroyo Lapa, also derived from the south, have a considerable tuffite content. Therefore it is impossible that the latter had the same source area as the former. It follows that the unknown western borderland of the basin or a now- covered part of the eastern borderland were affected by volcanic activity.

Synsedimentary faults and volcanism are the origin for the older generation of gravitational mass movements. After the first transgressive phase a certain stagnation of the basin depression can be observed during which the basin was temp- orarily partially filled up to sea level. A younger generation of turbidites and related phenomena occur in the Aalenian to the Callovian. They are associated with a younger phase of intensified volcanic activity in the borderland of the basin.

ACKNOWLEDGMENTS

Both authors wish to thank the Museo Argentino de Ciencias Naturales "B. Rivadavia" for the generous disposal of a cross-country vehicle. U. Rosenfeld is grateful to the Deutsche Forschungs- gemeinschaft for financial support.

REFERENCES

BOUMA, A.H. 1965. Sedimentary character- istic of samples collected from some sub- marine canyons. Marine Geol. 3: 291-320.
DESSANTI, R.N. 1973. Descripción Geológica de la Hoja 29b, Bardas Blancas, Provincia de Mendoza. Ct. Geol. Econ. Rep. Arg., 1:200.000 Serv. Nac. Min. Geol., Bol. 139.
--- 1978. Descripción Geológica de la Hoja 28b, Malargue, Provincia de Mendoza. Ct. Geol. Econ. Rep. Arg., 1:200.000. Serv. Geol. Nac.
DIGREGORIo, J.H. 1972. Neuquén. In LEANZA, A.F.(Ed.), Geología Regional Argentina: 439-505. Córdoba (Acad. Nac. Cienc.).

FAIRBRIDGE, R.W.; J. BOURGEOIS (Ed.) 1978. The Encyclopedia of Sedimentology. Encyclopedia of Earth Sci. Ser. VI. Stroudsburg/Penn: Dowden, Hutchinson & Ross.
GROEBER, P., et al. 1953. Mesozoico. In: Geografía de la República Argentina, Soc. Arg. Est. Geogr. 2 (1): 1-541.
HSÜ, K.J. 1974. Melanges and their distinction from olistostromes. Soc. Econ. Pal. Min. Spec. Publ. 19: 321-333.
KUENEN, Ph. H. 1964. Deep-sea sands and ancient turbidites. Dev. Sedimentol. 3: 3-33.
RAMOS, V.A. 1978. Estructura. Relatorio Geología y Recursos Naturales del Neuquen: 99-118.
REINECK, H.-E; I.B. SINGH 1975. Depositional Sedimentary Environments. 2nd ed. Berlin/Heidelberg/New York: Springer-Verlag.
STRAATEN, L.M. van 1970. Holocene and late-Pleistocene sedimentation in the Adriatic Sea. Geol. Rundsch. 60: 106-131.
VOLKHEIMER, W. 1978. Descripción Geológica de la Hoja 27b, Cerro Sosneado, Provincia de Mendoza. Ct. Geol. Econ. Rep. Arg., 1:200,000. Serv. Geol. Nac., Bol. 151.
---; M. MANCENIDO; S. DAMBORENEA 1977. La Formación Los Patos (nov. form.), Jurásico inferior de la Alta Cordillera de la Provincia de San Juan (República Argentina), en su localidad tipo (Río de los Patos Sur). Rev. Asoc. Geol. Argent. 32 (4): 300-311.
ZAMBRANO, J.J.; C.M. URIEN 1970. Geological Outline of the Basins in Southern Argen- tina and their Continuation off the Atlan- tic Shore. J. Geophys. Res. 75 (8): 1363- 1396.
ZÖLLNER, W.; A.J. AMOS 1973. Descripción Geológica de la Hoja 32b, Chos Malal, Provincia de Neuquén. Ct. Geol. Econ. Rep. Arg., 1:200,000. Serv. Nac. Min. Geol., Bol. I43.

All further literature mentioned in the text is cited by H.-E. REINECK & J.B. SINGH 1975.

Petrography of the youngest known Murihiku Supergroup, New Zealand:
Latest Jurassic arc volcanism on the southern margin of Gondwana

PETER F. BALLANCE, ROBERT F. HEMING & TERUHIKO SAMESHIMA
University of Auckland, New Zealand

The Huriwai Group of SW Auckland, North Island, New Zealand, contains the youngest known rocks of the voluminous volcanogenic fore-arc sequence (in part the Murihiku Supergroup) which accumulated on the southern margin of Gondwana between the early Permian and latest Jurassic. The group comprises a prograding delta whose source presumably lay generally to present west, although known paleocurrent flow was towards present north and northwest. The sandstones are composed almost entirely of intermediate to acid ferromagnesian-free and plagioclase-phyric lava fragments (dominant), and plagioclase grains of volcanic (zoned) and possible plutonic (unzoned) derivation (minor to sub-dominant). Other very minor consituents include ferromagnesian grains, plutonic fragments, fine quartz sandstone, fine sutured quartzite, and mica schist (one grain). Very few single quartz grains were identified. All volcanic glass is devitrified, and most lava fragments are altered. Plagioclase ranges from fresh to extensively altered. Most sandstones are cemented by fine quartz/zeolite, and a few by calcite/chlorite. Secondary minerals include halloysite, metahalloysite, analcime, heulandite, chlorite, celadonite and epidote.

Pebbles include volcanic sandstones petrographically identical with Huriwai sandstones, but sometimes veined (>50%); lavas of intermediate (low silica) to intermediate (high silica) to dacitic composition, all ferromagnesian-free and plagioclase-phyric and altered (>30%); tuffs, also highly altered (>10%); and minor granite.

A source from both recycled and fresh volcanic products is inferred. The active volcanism was a latest Jurassic continuation of the arc which had been active, perhaps continuously, on the southern margin of Gondwana since the early Permian.

An original overburden is suggested by the secondary mineralisation, indicating that the Rangitata Orogeny and cessation of subduction may have taken place somewhat later than the latest Jurassic.

INTRODUCTION

The Huriwai Gp of latest Jurassic age outcrops only in SW Auckland (Fig. 1). The well-known Huriwai plant beds were discovered in 1859 by Hochstetter (1864). Maps and field descriptions of the group were given by Henderson & Grange (1926), Purser (1961), Kear (1966) and Waterhouse (1978).

The Huriwai Gp contains a lower sandy and conglomeratic formation (Mangatara Measures), and an upper finer grained formation (Matira Siltstone) (Kear 1966). Only the lower unit appears to be present in the well-exposed northern outcrops; the petrographic descriptions in this paper are all taken from it. The group comprises a prograding delta (Ballance 1977, in prep.).

PALEOGEOGRAPHY

The inset on Figure 1 shows what is normally inferred to be the Mesozoic regional paleogeography (e.g. Fleming 1970) The Murihiku Supergroup lies between Gondwana to present west, and the equival-

ent aged marine Torlesse rocks (Spörli 1978) to present east. A sediment source on Gondwana is inferred, encompassing the present Lord Howe Rise, and comprising part of the southern margin of the super-continent (e.g. Zeigler *et al.* 1979). However, paleocurrent readings in the Huriwai Gp indicate river flow towards present north and NW (Fig. 1; Ballance in prep.), but since the available data is restricted to the two small coastal exposures it may not represent the regional picture accurately. The paleolatitude is inferred to have been high (Stevens & Speden 1978).

DETRITAL PETROGRAPHY OF HURIWAI SANDSTONES

Auckland University petrographic collection numbers 23419-23433, 23468-23483, and 24662-24695.

The sandstones are overwhelmingly volcanistic. No lava flows or primary tuffs have been found, although altered tuff layers occur in the underlying marine mudstone formation.

Volcanic Components. The devitrification

Table 1--Huriwai Group Petrography.

	23425	24688	24682	24693	23470	PEBBLES (24677) 24	27	29	30	33
POINT-COUNT ANALYSES (%)										
Plagioclase	3	20	6	23					18	26
Porphyritic volc. r.f.	30	4	9	1					4	1
Non-porph. volc. r.f.	51	45	51	31					8	19
Ferromag. mins.	-	-	2	trace						2
Quartz/Qtz sdst.	-	-	-	1					2	8
Cement/matrix	16	31	32	44					67	44
XRD ANALYSES (approx. %)										
Quartz	20	20	>20	30	25	#	**	#	**	**
Plagioclase	40	30	**	40	20	**	**	**	*	*
K-Feldspar		-	-		5	**	**	#		
Illite				-					#	
Kaolinite					10					
Halloysite		10								
Metahalloysite	20			15(?)						
Celadonite					2					
Analcime		5			20					
Heulandite			#							
Chlorite	5	20		15	10				#	#
Calcite	2	8		-	-					

NOTES: 23425: Coarse stst., calcite/chlorite (?celadonite) cemented. 24688: Fine sdst., calcite/chlorite (?celadonite) cemented. 24682: Coarse pebbly sdst., quartz/zeolite cemented. Unusual concen. ferro. 24693: Fine sdst., quartz/zeolite cemented. 23470: Mudstone.
Pebbles 24677: 24: Porphyritic trachytic lava, v. few ferromags, extensively altered, ?andesite. 27: Tuff, no ferromags. Silicified. 29: Lava, ?andesite. Altered. Few Ferromags, ?originally augite. 30: V. fine sdst./coarse zst., veined. 33: V. fine sdst.
r.f. = rock fragments ** = dominant * = intermediate # = minor

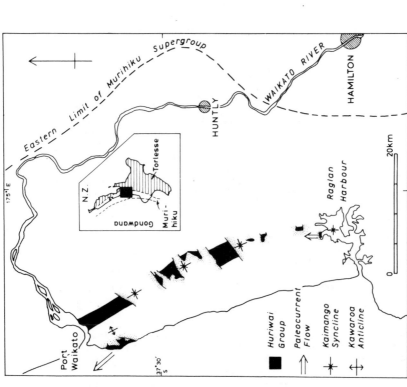

Fig. 1--Outcrop of the Huriwai Gp and paleocurrent directions. Inset shows location of study area in relation to regional paleogeography.

Fig. 2--Coarse, calcite/chlorite cemented, volcarenite. Specimen 23425 (Table 1). Note the predominance, and great variety, of lava fragments. Scale 0.5 mm. Crossed nicols.

Fig. 3--Fine, quartz-zeolite cemented, plagioclase volcarenite. Large amount of cement/matrix. Specimen 24693 (Table 1). Scale 0.1 mm. Crossed nicols.

and alteration of the rocks makes identifications imprecise. Most grains are of lava (Table 1); fine-sand-size grains are chiefly non-porphyritic, but larger grains contain a higher proportion of porphyritic types. Phenocrysts are chiefly plagioclase; ferromagnesian phenocrysts are uncommon, and where identifiable are clinopyroxene. The groundmass is devitrified glass containing plagioclase miocrolites. A very wide variety of textures is present, including spherulitic (Figs. 2,3). Clasts are mostly well rounded. Lava types appear to be andesite and dacite. The paucity of ferromagnesian minerals suggests high-silica andesites and dacites typical of island-arc volcanism.

Plagioclase crystals are common (Table 1), especially in fine sandstones and siltstones. The degree of alteration varies widely. Twinned grains often show zoning.

Some coarser sandstones contain clasts of fine volcanic sandstone, implying reworking of older Murihiku rocks, and pumice. Some of the latter are flattened, suggesting derivation from welded ignimbrite flows.

Ferromagnesian crystals are rare, and are normally altered to 'chlorite'. 24682 is exceptionally rich, with 2% ferromagnesian grains (Table 1).

Non-volcanic Components. These normally comprise less than 1% of the total rock. Commonest are fine quartz sandstones and sutured quartzites. Very few single grains of quartz were identified. A few possible plutonic igneous grains of intergrown quartz and feldspar were noted. One grain of muscovite schist, and a few grains of biotite and muscovite, were noted.

On the classification of Folk *et al.* (1970), the sandstones are immature and poorly sorted, calcite/chlorite--or quartz/zeolite-cemented, plagioclase volcarenites.

PETROGRAPHY OF PEBBLES IN THE HURIWAI GROUP

Conglomerate is abundant in the lower formation of the group. Pebbles are difficult to identify in the field. 33 pebbles collected at random from one exposure contained 18 sandstones (54%) 11 lavas (34%) and 4 tuffs (12%).

The lavas and tuffs are extensively altered. They include a porphyritic andesite with prominent glomeroporphyritic clots of plagioclase and augite; a basaltic or low-silica andesite with phenocrysts of plagioclase and augite, some in glomeroporphyritic clots, and a little ?olivine; a possible dacite; and two porphyritic andesites (Fig. 4).

The sandstones are mostly fine-grained and petrographically identical to those

163

of the Huriwai Gp (Fig. 5). Some are cut by veins of chlorite or zeolite.

CEMENTS AND SECONDARY MINERALS

Most Huriwai sandstones have a cement of fine grained, murky, quartz/zeolite. Comparison of the point-count and XRD data in Table 1 leaves an unanswered problem. The two quartz/zeolite cemented sandstones (24682 and 24693) each contain about 10% more quartz than the two calcite/chlorite cemented sandstones (23425 and 24688), suggesting that perhaps 10% of the cement is quartz. The remaining 20 to 30% of the cement, however, cannot all be zeolite according to the XRD data. Some of it is presumably fine volcanic dust. Some chlorite from the breakdown of the sparse ferromagnesian grains may also contribute.

Calcite-cemented sandstones are fewer. The calcite is usually associated with a brown, microcrystalline, birefringent mineral (the illite of the XRD data, possibly celadonite). There is an unexplained discrepancy between the volume percent of calcite/chlorite cement measured by the point-count and XRD data, particularly for 23425. Calcite can be seen to replace plagioclase.

Other secondary minerals observed include epidote, and zeolite in small radiating fibrous aggregates. One small lava pebble contains a cavity lined by zeolite and filled by fibrous chlorite/epidote (?).

The halloysite and metahalloysite are perhaps associated with the large amount of devitrified glass or are possibly weathering products.

DISCUSSION

The Huriwai Gp sandstones and conglomerates clearly indicate abundant volcaniclastic material, presumably derived from Gondawana to the present-day west. In view of the extensive alteration shown by all the lava fragments it is not possible to say what proportion of them was derived from contemporary activity. The pebbles and sand-sized grains of volcaniclastic sandstone, and the wide range of alteration shown by the plagioclases (from zero to extreme), clearly indicate reworking combined with input of fresh material. Thus the active volcanism which contributed voluminously to Permian, Triassic and Jurassic sedimentation on the southern margin of Gondawana (e.g. Boles 1974, South Island of N.Z.; Martin 1967, North Island of N.Z.; Campbell 1979, New Caledonia) was apparently still in progress in the latest Jurassic. The lava types--ferromagnesian-poor plagio-

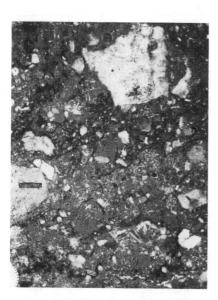

Fig. 4--Very fine, quartz-zeolite cemented plagioclase volcarenite, from a pebble. Grain boundaries indistinct. Specimen 24677 pebble 30 (Table 1). Scale 0.1 mm. Nicols partially crossed.

Fig. 5--Lava from a pebble. Plagioclase phenocrysts show varying states of alteration. No ferromagnesian phenocrysts. Specimen 24677 pebble 29 (Table 1). Scale 0.1 mm. Crossed nicols.

clase-phyric andesites and perhaps dacites--are typical of island arc activity. The 20% of quartz indicated by XRD analysis and not seen in the rocks is presumably present in the devitrified glass phase (Table 1).

Other very minor rock types contributing sediment were quartz sandstones and quartzites, and mica schist. A very minor plutonic contribution may be consanguineous with the arc volcanism, or may be much older. The pre-Mesozoic basement underlying the continental shelf to the west is known only in offshore drillhole Moa 1b, where it consists of fine-grained schists of inferred Lower Paleozoic age (Wodzicki 1974).

The question of how much overburden may once have been present on the Huriwai Gp is germane to the timing of the ensuing Rangitata Orogeny and cessation of subduction. The extensive alteration and cementation place the Huriwai Gp in the heulandite-analcime stage of zeolite facies metamorphism (Boles & Coombs 1977). In relation to possible depth of burial, analcime in the Niigata oilfield of Japan occurs only at depths greater than 2900 m (Iijima & Utada 1971), but in the Waitemata Gp of Auckland (Miocene volcaniclastic flysch) occurs in beds which may never have been buried deeper than 1000 m, but which on the other hand may have been subjected to contemporaneous hot spring activity (Sameshima 1978). Without further knowledge of the conditions of temperature and pressure within the Huriwai Gp, it is not yet possible to draw firm conclusions on the depth of burial of the group, but it is clear that there was an original overburden and that the end of volcanism and the ensuing Rangitata Orogeny may have occurred during the Cretaceous Period.

ACKNOWLEDGMENTS

Phillipa Black assisted with some XRD analyses. Field work was financed by a University of Auckland Research Grant.

REFERENCES

BALLANCE, P.F. 1977. The Huriwai Formation of Port Waikato - a Hokonui delta. Abstract, Geol. Soc. of N.Z. Queenstown Conference.

---; In prep. A volcanic-fed high-latitude prograding delta on the southern margin of Gondwanaland: Huriwai Group, Latest Jurassic, New Zealand.

BOLES, J.R. 1974. Structure, stratigraphy, and petrology of mainly Triassic rocks, Hokonui Hills, Southland, New Zealand. N.Z. J. Geol. Geophys. 17: 337-374.

---; D.S. COOMBS 1977. Zeolite facies alteration of sandstones in the Southland Syncline, New Zealand. Am. J. Sci. 277 982-1012.

CAMPBELL, H. 1979. Geology of Permian-Jurassic rocks of the Moindou-Téremba area, New Caledonia. Unpublished M.Sc. thesis, Univ. Auckland, N.Z. pp.273.

FLEMING, C.A. 1970. The Mesozoic of New Zealand: Chapters in the history of the Circum-Pacific mobile belt. Q. J. Geol. Soc. London 125: 125-170.

FOLK, R.L.; P.B. ANDREWS; D.W. LEWIS 1970. Detrital sedimentary rock classification and nomenclature for use in New Zealand. N.Z. J. Geol. Geophys. 13: 937-968.

HENDERSON, J.; L.I. GRANGE 1926. The Geology of the Huntly-Kawhia Subdivision, Pirongia and Hauraki Divisions. N.Z. Geol. Surv. Bull. 28.

HOCHSTETTER, F. von 1864. Geology of New Zealand. Translated C.A. Fleming, 1959, Wellington: Government Printer 320 pp.

IIJMA, A.; M. UTADA 1971. Present-day zeolite diagenesis of the Neogene geosynclinal deposits in the Niigata oil field, Japan. Adv. Chem. Ser. 101, Molecular Sieves Zeolites: 342-349.

KEAR, D. 1966. Sheet N55, Te Akau (1st Ed.), Geological Map of New Zealand 1:63360. Wellington: Government Printer.

MARTIN, K.R. 1967. The Mesozoic sequence at South-West Kawhia, New Zealand. Unpublished M.Sc. thesis, Univ. Auckland, N.Z. Pp.118

PURSER, B.H. 1961. Geology of the Port Waikato region. N.Z. Geol.Surv. Bull. 69.

SAMESHIMA, T. 1978. Zeolites in tuff beds of the Miocene Waitemata Group, Auckland Province, New Zealand. In L.B. Sand and Mumpton (Eds.), Natural Zeolites: Occurence, properties, use. Oxford: Pergamon: 309-317.

SPÖRLI, K.B. 1978. Mesozoic tectonics, North Island, New Zealand. Geol. Soc. Am. Bull. 89: 415-425.

STEVENS, G.R.; I.G. SPEDEN 1978. New Zealand. In M. Moullade and A.E.M. Nairn (Eds.), The Mesozoic, A. The Phanerozoic Geology of the World II. Amsterdam: Elsevier Scientific Publishing Co. pp. 251-328.

WATERHOUSE, B.C. 1978. Sheet N51, Onewhero. (1st Ed.) Geological Map of New Zealand 1: 63360 Wellington: Government Printer.

WODZICKI, A. 1974. Geology of the pre-
Cenozoic basement of the Taranaki-Cook
Strait-Westland area, New Zealand, based
on recent drill hole data. N.Z. J.
Geol. Geophys. 17: 747-757.
ZEIGLER, A.M.; C.R. SCOTESE; W.S. McKERROW;
M.E. JOHNSON: R.K. BAMBACH 1979. Paleozoic
Paleogeography. Ann. Rev. Earth Planet.
Sci. 7: 473-502.

2. Structure and paleogeology

Fold tectonics in Gondwana formations of India

RAJ RANJAN PRASAD VERMA & VIRENDRA KUMAR SINGH
Central Mine Planning & Design Institute, Dhanbad, India

Gondwana formations occur as outliers on Precambrian rocks along eight prominent river valleys of India. Large-scale fold tectonics are postulated for the coal belts of Mahanadi, Wardha-Godavari, Pench-Kanhar, Ganga (Rajmahal), and Brahmaputra (Eastern Himalaya), in addition to those earlier postulated for Damodar and Koel-Son Valley fields. It is also demonstrated that major faults, often forming present-day boundary faults, were formed by block-tectonics during the Tertiary period and are not only post-depositional but post-intrusive.

INTRODUCTION

Gondwana rocks, containing 95% of Indian coal reserves, are exposed along and named after 8 major river valleys: Damodar, Koel (Palamau), Son, Mahanadi, Wardha-Godavari, Pench-Kanhan (Satpura), Ganga (Rajmahal), and Brahmaputra (Eastern Himalaya).

These alignments have been interpreted as pre-depositional zones of weaknesses with continuing syntectonic activity (see references cited in Verma & Singh 1979) or as results of purely post-depositional faults. Each of these coal belts comprises numerous coalfields occurring as isolated outliers in the Precambrian areas. Their separation has been attributed by the one school to sub-basins within the broader basin and by another to post-depositional tectonism.

Since this could not explain the nature of tectonism responsible for isolation of the outliers, the authors, for the first time, came forward with the concept of large-scale fold-tectonics in coal belts (Verma & Singh 1978, 1979), which, in the past, had either been underplayed or completely ignored.

FOLD-TECTONICS

The present investigations indicate that the Gondwana formations have been affected by three sets of folds:

B_1 Folds. There is almost unanimity in considering the Narmada-Son-Brahmaputra lineament a pre-Gondwana tectonic feature with marine occupation to which the NW-flowing Mahanadi and Wardha-Godavari rivers discharged their waters. In the Damodar-Koel area, too, the eastward drainage previously envisaged was successfully demonstrated as going northwestward (Casshyap 1973).

This implies that deposition of Gondwana formations in Damodar, Koel, Mahanadi, and Wardha-Godavari took place in alluvial valleys trending SE-NW, whereas those in Son and Pench-Kanhan were deposited in a deltaic environment. Since the B_1 folds trending NW-SE developed as a result of syndepositional diastrophism parallel with the lineament of the depositional valley, they are found only in the Damodar, Koel, Mahanadi, and Wardha-Godavari coalfields and are conspicuous by their absence in the paleo-deltaic region of Brahmaputra, Son, and Pench-Kanhan valleys.

Damodar and Koel: Although E-W-trending post-depositional B_2 folds have obliterated the B_1 folds, axes of NW-SE-trending B_1 folds can still be seen forming zones of culminations and depressions across doubly-plunging E-W-trending B_2 folds (Figs. 1,2).

Gondwana formations have been mostly removed from the culminations which expose metamorphics as saddles between divergently-plunging coalfields, but depressional zones are present within the heart of the individual coalfields. The NW projections of Talchirs in Raniganj, Jharia, and N. Karanpura coalfields and the SE apex of the triangular Ramgarh fields are also likely manifestations of the B_1 folds in Damodar Valley coalfields.

Mahanadi: B_1 folds with two anticlines and two synclines along NW-SE axes are best developed in and around Raigarh, S. Raigarh, Mand River, and IB river coalfields (Fig. 3).

Wardha: The beds trend NW-SE with low easterly and westerly dips broadly indicating two anticlines and a syncline (Pande 1971). These folds must necessarily be related to the other B_1 folds discussed above.

Godavari: The Gondwana rocks all trend NW-SE with NE dips along the western margins, but in the bed of the Godavari River and some other places near the eastern margins WSW dips up to 15° (Rao 1971) are common. This clearly indicates development of a syncline caused by B_1 folding along NW-SE axes (Fig. 4).

Rajmahal: The general strike of the beds is NNW to NW and dips are gentle (Rao 1971). Since most of the field is covered by traps,

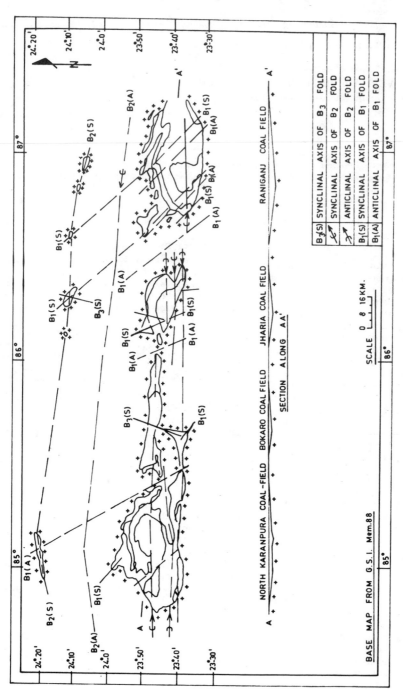

Fig. 1--Fold tectonic map of the Damodar Valley Coal Fields.

170

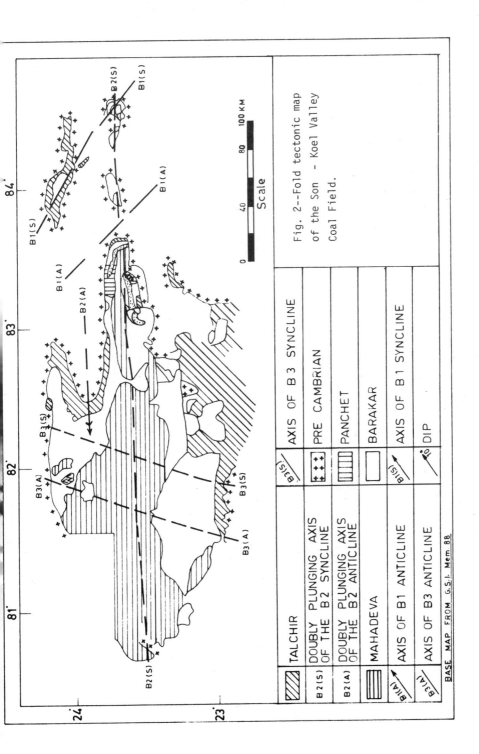

▨ TALCHIR	◢ AXIS OF B 3 SYNCLINE
B2(S) DOUBLY PLUNGING AXIS OF THE B2 SYNCLINE	⊞ PRE CAMBRIAN
B2(A) DOUBLY PLUNGING AXIS OF THE B2 ANTICLINE	▥ PANCHET
▥ MAHADEVA	□ BARAKAR
↗ B1(A) AXIS OF B1 ANTICLINE	↗ B1(S) AXIS OF B1 SYNCLINE
↗ B3(A) AXIS OF B3 ANTICLINE	↗ 10 DIP

BASE MAP FROM G.S.I. Mem 88

Scale

0   40   80   100 KM

Fig. 2.–Fold tectonic map of the Son – Koel Valley Coal Field.

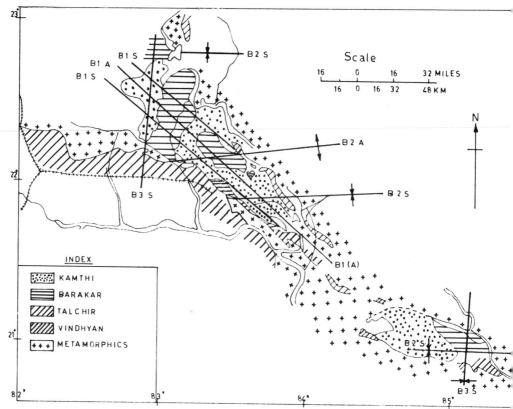

Fig. 3--Fold tectonic map of the Mahanadi Valley Coal Fields. B_1S - synclinal axis of B_1 folds; B_1A - anticlinal axis of B_1 folds; B_2S - synclinal axis of B_2 folds; B_2A - anticlinal axis of B_2 folds; B_3S - synclinal axis of B_3 folds. Base map from G.S.I. Mem. 88.

it is not possible to decipher the structure, but the fact that the Barakars flank on either side of the NW-SE-trending younger Dubrajpur formations is enough to allow recognition of a syncline presumably related to the B_1 folding.

B_2 *Folds.* This series of anticlinal and synclinal E-W-trending doubly plunging folds in the Damodar, Koel, and Son fields was treated earlier (Verma & Singh 1978, 1979). These folds have also been found in the coalfields of the Tawa-Pench-Kanhan, Wardha-Godavari, Mahanadi, and Brahmaputra valleys.

Damodar-Koel-Son: Doubly-plunging E-W-trending B_2 folds, from N to S, are: Deoghar-Giridih-Itkhori-Rajhara syncline; Singrauli anticline; Raniganj-Jharia-Bokaro-N. Karanpura-Auranga-Hutar-Ramkola-Tatapani-Korar syncline; Amlabad (Jharia CF)-Sohagpur-Johilla anticline; Tasra (Jharia CF)-Ramgarh-S. Karanpura-Sonhat-Jhilimili syncline; Bisrampur Block I (Son Vy CF) anticline; Kotma-Kurasia-Bisrampur Block II syncline.

Smaller B_2 folds are also present within individual coalfields. Although the plunge directions of the B_2 folds are largely governed by B_1 fold culminations and depressions, local variations due to B_3 folding are not uncommon.

Tawa: A doubly-plunging syncline flanked on each side by anticlines represents the B_2 folds (Fig. 5).

Pench-Kanhan: These are the extension of the southern line of the above (Fig. 5).

Mahanadi: These folds (Fig. 3) are more conspicuous in Talchir, Hingir, Mand River, Korba, Hasdo-Arand and Chirimiri (TISCO Block I) fields, where formations strike E-W with N or S dips forming synclines and anticlines related to B_2 folds.

Godavari: There seems to be an E-W anticline in Beddadanol ($17°14'$:$81°14'$), as evidenced by a closure towards the east and exposure of the upper Gondwanas towards the south, indicating southerly dips, but this is the only evidence so far for B_2 folds here.

Eastern Himalaya: The Damudas, mainly represented by sandstones, contain coal seams dipping steeply ($40-90°$) towards NW,

172

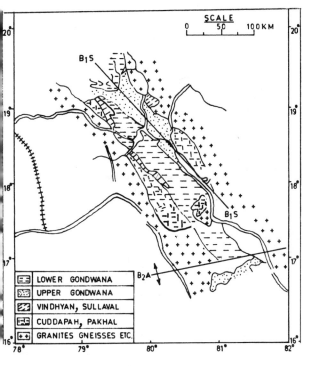

SCALE
0 50 100 KM

B₁S

LOWER GONDWANA
UPPER GONDWANA
VINDHYAN, SULLAVAL
CUDDAPAH, PAKHAL
GRANITES GNEISSES ETC.

B₁S

B₂A

Fig. 4--Fold tectonic map of the
Wardha Godawari Valley Coal Field.
B_2A - anticlinal axis of B_2 fold.
B_1S - synclinal axis of B_1 fold.
Base map from G.S.I. Mem. 88.

NNW, and NNE (Dutt 1971). The seams are
generally lenticular in habit and very
often contain intercalations of shale and
sandstone, most likely as a consequent to
severe overturned folding along ENE-WSW
axes related to B_2 folds.

Rajmahal: Large-scale extrusion of
trap, as well as scarcity of data, has so
far restricted recognition of B_2 folds in
the area.

B_3 *Folds*. These folds developed along
NNE-SSW axes by secondary tectonogenesis
after the B_2 folds.

Damodar, Koel, and Son: B_3 folds were
described by the authors (1978, 1979). In
Damodar Valley fields they are clearly noticed in Giridih and S. Karanpura coalfields. In Ramgarh, the northerly apex
probably represents synclinal closure of
a B_3 fold. In Jharia, easterly plunge of
the Baghmara-Muraidih syncline and Baghma-
ra-Barora anticline results from the synclinal trough near Tetulia ($86°16':23°47'$)
of the B_3 fold (Fig. 1).

In Son Valley there seem to be no B_1
folds, so reversal of plunge directions
of B_2 folds must be due to B_3 folds. These
B_3 fold axes have been postulated in
Fig. 2.

Koel: All three sets of folds are present, so it is difficult to determine the
exact position of B_3 axes. However, a B_3
synclinal axis clearly passes through Au-

ranga coalfield. More B_3 fold axes also
seem to cross other areas of the valley
(Fig. 2).

Tawa-Pench-Kanhan: At least one zone of
depression along a NNE-SSW axis is present as shown in Fig. 3. This has reversed
the plunge of B_2 folds and must be related
to B_3 folds (Fig. 5).

Mahanadi: B_3 folds are indicated by
the plunging B_2 folds described already
and by exposure of younger Kamtis between
the Mand River and Korba coalfields (Fig.3).

Godavari & Rajmahal: B_3 folds have so
far not been detected.

BLOCK-TECTONICS

After repeated folding the Gondwana formations stiffened and remained no longer
capable of refolding. During the Tertiary
major faults developed, and sections of
crustal blocks were raised or lowered. We
call these movements 'block-tectonics' or
'fault-tectonics', and they give rise to
so-called boundary faults.

A great controversy exists regarding the
age of the boundary faults, and has been
critically examined by the authors (1979).
Ahmad and Ahmad (1979) further agreed with
earlier theories that major faulting was
entirely post-depositional and that (in
the Damodar Valley) it took place during
the Paleocene, just before the intrusion
of ultrabasics. Likewise, Auden (1954) had

173

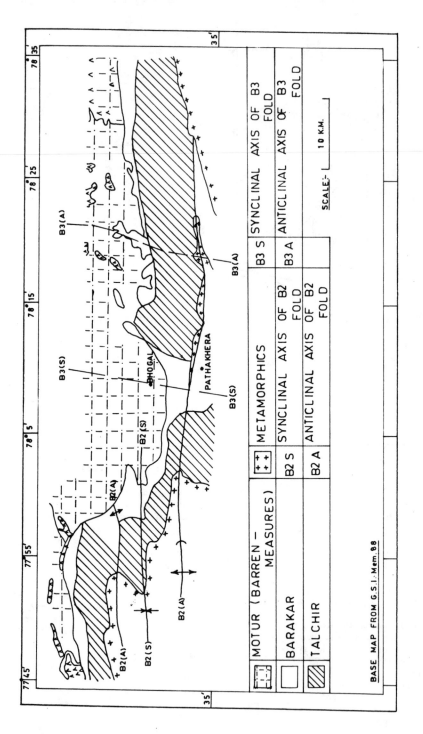

Fig. 5—Fold tectonics in the Pench – Kanhan – Tawa Valley Coal Fields.

174

strongly advocated a Tertiary age. It is noteworthy that major faults, including the boundary faults in different coalfields, are free from intrusives. Fox (1934) had maintained that at least the Satpura faults were Tertiary, and we are inclined to extend this age to the block-tectonics in all the Gondwana coalfields, although a few older pre-intrusive faults contemporary with the fold-tectonics also exist.

Basic intrusives are also emplaced along fractures and nowhere along faults. Our recent field observations in Jharia Coalfield indicate that faults with southerly throw displaced dextrally the steeply E-dipping Madhuband and Maheshpur dolerite dikes, and that the dextral shifts are not just kinks (GSI Sheet #2).

INTRUSIVES

There are two main periods of igneous activity envisaged in the Gondwana formations: (a) Paleocene, including the Rajmahal traps of the Rajmahal fields, and ultrabasic (mica-peridotites) sills and dikes of Raniganj and Jharia (Ahmad & Ahmad 1979). Obviously this took place during the last phase of the B_1 folding.

(b) Eocene, including Deccan traps and dolerite dikes crossing almost all the Gondwana coal belts, particularly near the Narmada-Son lineament. These intrusives appear to have been emplaced along the fractures generated by just-preceding and partly-overlapping B_2 and B_3 folds.

CONCLUSIONS

It therefore follows that the tectonic evolution we earlier proposed for the Damodar Valley fields holds good for the entire Gondwana coalfields in India. The chronology was broadly as follows:

1. Deposition of Gondwana sediments in fluvioglacial valleys (Upper Carboniferous - Upper Triassic).

2. B_1 folding along NW-SE axes and contemporaneous faulting (Permian - Lower Cretaceous), followed closely by intrusion of ultrabasics.

3. B_2 and B_3 folding along E-W and NNE-SSW axes respectively; fracturing, contemporaneous faulting and intrusion of basic intrusives/extrusives (Paleocene (Upper Cretaceous - Lower Eocene)).

4. Block-tectonic activity giving rise to major faults, including boundary faults (Tertiary).

5. Erosion and preservation of coalfields along folded downwarps and faulted down-thrown blocks.

REFERENCES

AHMAD, F.; Z. AHMAD 1979. Tectonic framework of the Gondwana basins of peninsular India. 4th Gondwana Symp., Calcutta, §8.

CASSHYAP, S. M. 1973. Paleocurrents and paleogeographic reconstruction in the Barakar (Lower Gondwana) sandstones of peninsular India. Sediment. Geol. 9: 283-303.

DUTT, G. N. 1971. Coal in the Eastern Himalayas. Mem. Geol. Surv. India 88: 100-104.

FOX, C. S. 1934. Lower Gondwana coalfields of India. Mem. Geol. Surv. India 59.

LASKAR, B. 1977. Evolution of Gondwana coal basins. 4th Gondwana Symp.,Calcutta.

PANDE, B. C. 1971. Wardha Valley coalfields. Mem. Geol. Surv. India 88: 324-34.

RAO, P. V. 1971. Godavari Valley coalfield. Mem. Geol. Surv. India 88: 335-348.

VERMA, R. P.; V. K. SINGH 1978. Tectonic evolution of Sone-Koel Valley coalfields. Minetech 3(2): 39-42.

---; --- 1979. A chronology of tectonic and igneous activity in Damodar Valley coalfields. 4th Gondwana Symp., Calcutta.

Structure and tectonics of Gondwana basins of peninsular India

T. N. BASU & B. B. P. SHRIVASTAVA
Central Mine Planning & Design Institute Ltd., Ranchi, India

The intracratonic Gondwana basins of peninsular India occur in long, narrow, well-defined belts on the Precambrian platform, corresponding to river valleys of Damodar-Son, Mahanadi and Godavari. Westward extension of the Son Valley marks the Satpura area and its further western extension coincides with the Narmada rift. The basins are half grabens or grabens, with margin faults near maximum sediment thickness. These faults, indicative of regional tension, played a dominant role in the basins' evolution. The terrestrial clastics and coal beds in the Damodar-Son and Satpura basins are cut by ultrabasic and basic dykes and sills; volcanic activity during the Upper Jurassic is also indicated near Rajmahal. Igneous intrusives are conspicuously absent in the Godavari and Mahanadi Valleys.

INTRODUCTION

The Peninsular Gondwanas are characteristically developed along well defined, long, narrow belts on the Precambrian platform, following prominent basement lineaments and coinciding with the present day river valleys from which they derive their names (Fig. 1), but, curiously, do not generally occur outside the main basin belts.

The most prominent is cradled by the Satpura folded belt of the Archeans, the dividing line between the Aravalli-Vindhyan block in the north and the Dharwarian block in the south. In this belt lies the E-W-trending Damodar-Son Valley, the western extension of which marks the Satpura area and whose westward extension coincides with the Narmada Rift. The Godavari and Mahanadi valleys trend NW-SE and seemingly merge with the Damodar-Son-Satpura alignment and lie in the folded belt of the Archeans. Lying parallel with and north of Damodar Valley is a chain of smaller basins, the Deoghar-Hazaribagh group; they are also regarded as part of it.

NE of the Damodar-Son-alignment, the Rajmahal Gondwanas are aligned in a N-S direction. Existence of the Gondwana formations towards north and south along this alignment below the Gangetic alluvium has been proved in the trough between Burdwan and Debagram. Gravity data indicate that 1.5 to 3 km of Gondwana strata may be present (Verma & Mukhopadhyay 1977).

The east-coast Gondwanas occur in detached patches and are exposed in the Atgarh, Krishna-Godavari, Palar and Cauvery basins, closely following the strike of the Eastern Ghat and Dharwarian metamorphosed folded belt.

STRATIGRAPHY

The Gondwana formations collectively represent about 6000-7000 m though a complete sequence is not exposed at any one place.

The column is divided into the Upper and Lower formations, the latter accounting for nearly 98% of India's total coal resources. The strata can be subdivided into three assemblages: glacial deposits (Permo-Carboniferous) a prolonged sequence of continental fluviatile deposits ranging from Permian through Late Jurassic, and fluvio-deltaic to paralic deposits (Late-Jurassic to Lower-Cretaceous) (Casshyap 1979). The stratigraphic succession is given in Sastri *et al.* (1977).

STRUCTURE OF GONDWANA BASINS

The basins occupy either isolated depressions or large tracts (Godavari, Son and Mahanadi Valleys). The isolated depressions are irregular, elongate and oval, with variable dimensions. Some of the larger basins contain several sub-basins. The basins do show mild folding, in general, exhibiting a strike parallel to the linear belts: EW to ENE-WSW (Damodar-Son-Satpura basin belt); NW-SE (Godavari and Mahanadi Valleys); NS to NNW-SSE (Rajmahal).

The basins' disposition within linear belts marked by prominent high angle normal faults is very striking. LANDSAT pictures show that these zones have suffered maximum fracturing compared with the rest of the peninsula (Fig. 1). Faulting is the most dominant structural element of the Gondwana basins, either one or both sides is marked by a fault with, generally, a 60° dip. Therefore, the basins are 'half-grabens' or 'grabens' which are asym-

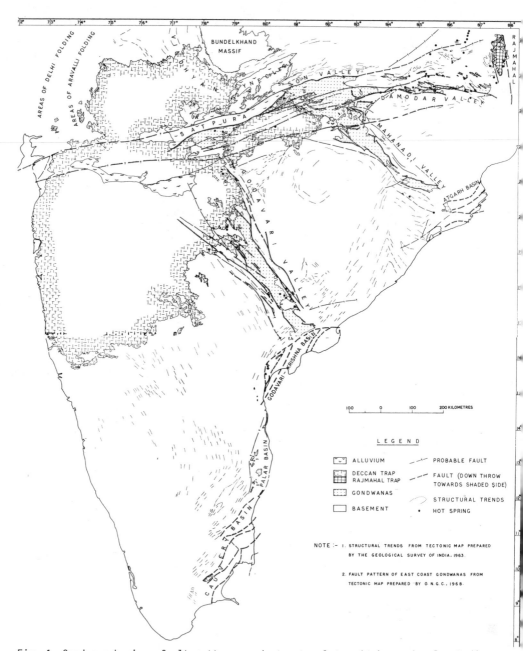

Fig. 1--Gondwana basins, fault patterns and structural trends in peninsular India.

metric in cross section (Fig. 2), but as the basement is tilted towards the basin-margin faults, these don't seem to be simple grabens.

Continued subsidence along the basin margin faults increased the tilt so strata generally dip towards the boundary fault. Thus the Damodar Valley generally dips to the south, while the Son and Satpura areas dip to the north. The Godavari and Mahanadi Valleys tilt NE and SW respectively. In these cases, greater amounts of sediment accumulated along the lines of subsidence.

Buried basement highs and sub-basins within larger basins can also be explained by differential subsidence along the marginal faults. Basement highs are reported in many basins and are aligned parallel to the

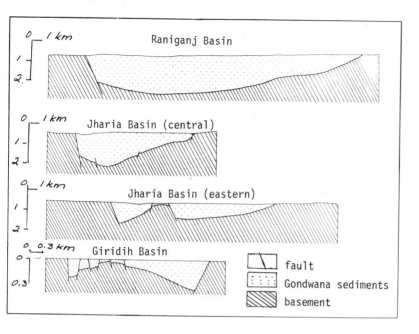

Fig. 2--Cross-section of some of the Gondwana basins.

Raniganj Basin

Jharia Basin (central)

Jharia Basin (eastern)

Giridih Basin

fault
Gondwana sediments
basement

half-graben and graben axes. However, synclinal axes are present on either side of the highs as in the Godavari Valley (Chanda-Wardha and Kamptee basins), Mahanadi Valley (Ib-basin), in Son Valley (Singrauli basin) and in some of the basins of the Damodar Valley e.g. North-Karanpura, West-Bokaro, Jharia, Giridih, etc. Similar features are reported from the Rhine Rift (Neugebauer & Braner 1977).

Boundary faults can be traced for long distances, not only marking the basin boundaries but traversing the intervening basement (Fig. 1). However, some of the shorter faults are arranged en echelon (e.g. southern margin of the Damodar Valley). Their orientation indicates that these are all closely related parts of a single system and tectonic feature. Some of the boundary faults enter the adjoining basin and assume intra-basinal characters (e.g. the southern basin-margin-fault bounding the basins of Ramgarh, South Karanpura, Auranga, Hutar (Damodar Valley) and Tatapani-Ramkola basins). The surface measurements indicate that these have dip angles of about 60°, but whether these develop listric surfaces at depth is not known.

Besides the characteristic boundary faults, the sediments are affected by an intricate network of intra-basinal normal faults of varying magnitude and with a discernable

pattern. In the Damodar Valley two prominent sets are developed, one trending NW-SE and the other NE-SW. In the Son Valley, these are aligned WNW-ESE (Dutta *et al.* 1979); in the Satpura area, while the major intra-basinal faults are aligned ENE-WSW, a second set is NW-SE or NE-SW. Some are arcuate, with the dip of the fault decreasing with depth (Savanur 1966); i.e. they are high-angle faults which develop listric-surfaces at depth. Detailed study of their behaviour, in some of the Damodar Valley basins, indicates that the change in dip may be controlled by the change in the lithology of the strata these faults traverse (Savanur 1971).

Several hot springs in these belts indirectly provide an idea about the geothermal field. The most prominent are in the Damodar and Godavari Valleys and the Rajmahal area; the hottest springs in the eastern belt are in Surajkund (24°09', 85°41') and Tatapani (23°41', 83°42') (surface temperature of 88°C). The hot springs invariably coincide with the major faults (Fig.1) and, therefore, do not seem to be an isolated phenomenon.

There is noticeable parallelism between the major structural trend of the basement and the direction of the boundary and some of the major intra-basinal faults (Fig. 1): the NE-SE Dharwarian and Mahanadi, the NW-SW and NS eastern Ghat and the EW to ENE-WSW Satpura structural trends parallel the boundary faults in the Mahanadi and most of the Godavari Valleys, but the southern part of Godavari Valley, however, cuts at right angles to the Eastern Ghat trend. The trend of faults in the Damodar-Son-Satpura belt is parallel to the trend of the Satpura strike. The faults in the east-coast Gondwana basins parallel the Eastern Ghat strike. Sastri *et al.* (1977) mention that "the fracture pattern within the Cauvery basin resembles that of the major trends in the adjoining shield area". Similar parallelism in other east-coast basins has been noted.

This indicates that the basement structure controlled the orientation of the faults and that a genetic relationship existed.

The strata in the Damodar-Son-Satpura belt are intruded by peridotite and lamprophyre sills, mostly in the lower Gondwana Group, and basic dykes and sills some of which are the same age as the Decan Traps. The Godavari and Mahanadi Valleys are free of intrusions except for a few dolerite dykes in the Hasdo-Arand Basin which is at the junction of the Mahanadi and Son Valleys. The dykes tend to parallel the Satpura trend and their emplacement may have been controlled by basement fracture zones. The Rajmahal traps are plateau basalts erupted from fissures aligned along the Rajmahal Valley and represent the only volcanism that occurred during the Gondwana eriod.

RIFT VALLEY STRUCTURE

The linearity, fault-bound nature and preferred orientation of the basin belts and the structural elements of the individual basins all suggest a rift-valley structure.

The physical dimensions of the various basin belts, their extent and linearity, and the striking similarity in their widths can be gauged from the following:

BELT	LENGTH (km)	WIDTH RANGE (km)
Damodar-Son	580	30-90
Satpura Area	220	30-60
Mahanadi	340	30-50
Godavari	580	30-65

This table does not include the area west of Satpura, which coincides with the Narmada Rift, as the existence of Gondwana formations is not known.

Including the Narmada Rift the total length of the Narmada-Satpura-Son-Damodar belt is 1280 km, with minimum and maximum width of 30 to 90 km. The Damodar Valley appears to have bifurcated into parallel sets of basins, of which the southern one is more prominently developed. Holmes (1975)

mentions that the width of a rift valley : of the same order as the thickness of the continental crust, and Hari Narain (1974) has indicated that the average crustal thi ness of the peninsular shield is 35 to 40 km. This would nearly correspond to the average width of these belts.

The parallelism discussed above is associated with some of the rift valleys of th world. Logatchev and Florensov (1978) me tion that rift faulting on the whole is dependent on the basement structure. The further mention that "the brittle fracturing of the lithosphere during the course of rifting must have been controlled by th crustal inhomogeneity which in the end det mined the high conformity of rift faulting and the basement structures", and this may account for the preferred orientation and longitudinal development of the Gondwana basin belts and the overall parallelism between the faults and basement trends. Gravity surveys of the Godavari Valley (Qureshy *et al.* 1968), Chanda-Wardha basin a part of Godavari Valley (Chakraborty *et al.* 1976) and the basins of Raniganj, Jharia and Bokaro in the Damodar Valley belt (Verma & Ghosh 1977) reveal that the basins are associated with gravity lows due to infilling by sediments. The Bouguer anomaly map of India published by the National Geophysical Research Institute (NGRI) also shows that other Gon dwana basins, not covered in the above studies, are also associated with gravity 'lows'.

Qureshy (1968) mentions that gravity studies are useful in elucidating rift-lik structure. Drawing an analogy from the East African and Rhine rifts associated wi (-)50 milligal lows, he infers a rift-like structure for the Godavari valley, which also is bounded by (-)50 milligal gravity contours, flanked by small highs. These are significant in that a general uplift o the rift shoulders is observed in other rift valleys. The NGRI map shows the gravity lows associated with Gondwana basi as separated closed minima, a feature noticed in other rift valleys. Most of the basins are bounded by (-)50 milligal contours excepting in the Damodar Valley area where it is about (-)30 milligal.

TECTONICS

The Gondwana basin belts are aligned along prominent basement lineaments (or weak zones) of Narmada-Son-Damodar, Mahanadi and Godavari, which may have been caused by different basement development rates in the Archaean times. Because of their inherent weakness these zones were the most suitable sites for the concentra-

TABLE 1
STRATIGRAPHIC SUCCESSION AND CORRELATION OF PENINSULAR GONDWANAS
(FROM GEOLOGICAL SURVEY OF INDIA, MISC. PUB., NO. 36, 1977)

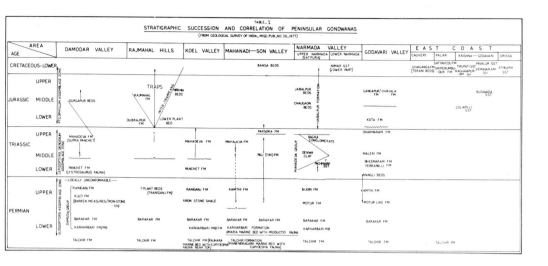

AGE	DAMODAR VALLEY	RAJMAHAL HILLS	KOEL VALLEY	MAHANADI—SON VALLEY		NARMADA VALLEY UPPER NARMADA (SATPURA)	LOWER NARMADA	GODAVARI VALLEY	EAST COAST CAUVERY	PALAR	KRISHNA—GODAVARI	ORISSA
CRETACEOUS-LOWER				BANSA BEDS			NIMAR SST (LOWER PART)		SATYAVEDU FM SIVAGANGA FM (TERANI BEDS)	SRIPERUMBU-DUR FM	TIRUPATI SST RAGHAVAPUR-AM SH	PAVALUR SST VEMAVARAM SH	ATHGARH SST
JURASSIC UPPER		TRAPS RAJMAHAL FM NIPANIA BEDS				JABALPUR BEDS		GANGAPUR / CHIKIALA FM FM				
JURASSIC MIDDLE	DURGAPUR BEDS					CHAUGAON BEDS				BUDAVADA SST	GOLAPILLI SST	
JURASSIC LOWER		DUBRAJPUR FM LOWER PLANT BED						KOTA FM				
TRIASSIC UPPER	MAHADEVA FM (SUPRA PANCHET)		MAHADEVA FM	MAHADEVA FM	PARSORA FM	BAGRA CONGLOMERATE		DHARMARAM FM				
TRIASSIC MIDDLE					PALI (TIKI) FM	DENWA CLAY	PACHMARI SST	MALERI FM				
TRIASSIC LOWER	PANCHET FM (LYSTROSAURUS FAUNA)		PANCHET FM					BHEEMARAM FM YERRAPALLI FM MANGLI BEDS				
PERMIAN UPPER	RANIGANJ FM KULTI FM (BARREN MEASURES/IRON-STONE SH)	?PLANT BEDS (?RANIGANJ FM)	RANIGANJ FM IRON STONE SHALE	KAMTHI FM		BIJORI FM	MOTUR FM	KAMTHI FM MOTUR LIKE FM				
PERMIAN LOWER	BARAKAR FM KARHARBARI FM/MB	BARAKAR FM	BARAKAR FM KARHARBARI MB/FM	BARAKAR FM KARHARBARI FORMATION UMARIA MARINE BED WITH PRODUCTIO	BARAKAR FM	BARAKAR FM KARHARBARI MB		BARAKAR FM				
	TALCHIR FM (RAJHARA MARINE BED WITH EURYDESMA FAUNA NEAR TOP)	TALCHIR FM	TALCHIR FORMATION (MANENDRAGARH MARINE BED WITH EURYDESMA FAUNA)			TALCHIR FM		TALCHIR FM				TALCHIR FM

tion of the rifting forces.

The first phase of rifting was when, in the post-Dharwar period, the Cuddapahs were laid down in a trough faulted basin-rift valley (Dar & Viswanathan 1964).

Towards the Late Carboniferous, when Gondwana sedimentation started, great orogenic movements (the Hercynian revolution) were taking place in Asia and Europe (Krishnan 1953). These and the regional uplift after the removal of ice sheets, appear to have reactivated ancient faults along the Narmada-Son-Damodar, Mahanadi and Godavari--the areas where the Gondwana sediments were laid down. Thus, the tectonic framework was largely influenced by trends on the basement we have discussed. After the initial phase the Gondwana sedimentation began with the deposition of the glacial tillites, varves and sand-shale alternations --the Talchir Fm.

The second phase was possibly at the dawn of the Karharbari sedimentation, recorded in the dominantly coarse clastics. Initially the general uplift rate increased, but was arrested for restricted periods resulting in swamps and development of thin coal seams. Thereafter and through the rest of the Permian movement along the marginal faults continued pari-passu with sedimentation and was characterised by weak and moderate tectonic movement accompanied by slow plastic deformation of the basement. This was when the buried basement highs parallel to the graben axis came into existence. The rift zone at this time was marked by vast basins surrounded by subdued uplands. Lake and river sediments were deposited in slowly subsiding basins. This general scenario seems to have persisted throughout the Permian in the Damodar Valley, Godavari Valley and Satpura area. However, after the deposition of the Barakar Fm, the movement along the marginal faults seems to have been arrested in the Mahanadi and Son Valley and Rajmahal area in response to tectonic adjustments following continued subsidence in other areas. This led to widespread erosion of the Barakar Fm in the Mahanadi and Son Valley and Rajmahal area.

At the end of the Permian the marginal faults were reactivated, as shown by coarse clastics of the Triassic strata. During this phase all the areas were involved and the bulk of the intrabasinal faulting took place. This is corroborated by the fact that the Permian strata in general are more severely faulted than the overlying Triassic strata. By this time, the limits of the Gondwana basins were perfectly demarcated.

The sedimentation in the Rajmahal area ceased after the deposition of the Barakar Fm and recommenced only in the Lower Jurassic, coinciding with the development of the east coast Gondwana basins which preserve the upper Gondwana sediments. This phase coincides with the first sign of rifting between Antarctica and India and fragmentation-rifting of Sri Lanka from SE India (Katz 1978) and was also when basaltic lava flows occurred in the Rajmahal area. The volcanic activity in the Rajmahal was preceded by the intrusive phase in the eastern basins of Damodar Valley. The intrusives in the western Son Valley and Satpura area are later and coincide with the Deccan Traps.

The above is a general framework of the evolution of the Gondwana basins, although some regional variation can be expected.

The preponderance of normal faults and the absence of structural elements related to compressive forces indicates an essentially non-orogenic tensional field, resulting in crustal stretching.

CONCLUSION

The disposition of the Gondwana basins within linear belts, with boundaries marked by prominent high angle normal faults, is very striking and has on the whole a rift configuration. The Gondwana basins are not simple grabens or half-grabens as the basement is tilted towards the boundary faults and the basin floor is asymmetric. The trend of the boundary faults is dominantly controlled by the basement trends. Subsidence along the marginal faults caused accumulation of greater thickness of sediments and also gave rise to buried basement highs. However, on either side of these synclinal axes are present. The movement along the marginal faults seems to have been spasmodic. Hot springs along the major faults provide indirect evidence about the geothermal province.

ACKNOWLEDGMENT

We thank Mr Naeem Ahmad, Geologist, CMPDI, for his all-round assistance in the preparation of this paper and Mr T.K. Mukherjee, Sr. Geologist, CMPDI, for his valuable assistance in the interpretation of LANDSAT images.

We are grateful to Mr A.N. Banerjee, Chairman-Cum-Managing Director, CMPDI, for his kind permission to present the paper at the Fifth Gondwana Symposium. The views expressed in this paper are authors' own and not necessarily those of CMPDI Ltd.

REFERENCES

CHAKRABORTY, R. N.; M. V. JOGARAO; A. K. SEN; R. K. SARKAR; D. GHOSH 1976. Report on gravity, magnetic (V.F.) and deep electrical resistivity (DES) Surveys for subsurface structural features and associated coal in Chanda-Wardha Valley. Geological Survey of India (unpubl.).

CASSHYAP, S. M. 1979. Pattern of sedimentation in Gondwana Basins. 4th Gondwana Symp., India.

DAR, K. K.; S. VISWANATHAN 1964. On the possibility of the existence of a Cuddapah rift valley in the Penganga Godavari basin. Geol. Soc. India Bull. 1(2): 1-6.

DUTTA, N. R.; A. K. DE; S. K. CHAKRABORTI 1979. Environmental interpretation of Gondwana coal measures in Peninsular India. 4th Gondwana Symp., India.

HARI NARAIN 1974. Crustal structure of the Indian subcontinent. The structure of the Earth's crust. Elsevier: Amsterdam 249-258.

HOLMES, A. 1975. Principles of Physical Geology. Thomas Nelson and Sons: London

KATZ, M. B. 1978. Sri Lanka in Gondwanaland and the evolution of the Indian Ocean Geol. Mag. 115 (4): 237-316.

KRISHNAN, M. S. 1953. The structural and tectonic history of India. Mem. Geol. Surv. India 81.

LOGATCHEV, N.A.; N. A. FLORENSOV 1978. The Baikal system of Rift Valleys. Tectonophysics 45 (1): 1-11.

NEUGEBAUER, H. J.; B. BRANER 1977. Crustal doming and the mechanism of rifting Part II. Tectonophysics 46: 1-20.

QURESHY, M. N.; N. KRISHNA BRAHMAN; S. C. GARDE; B. K. MATHUR 1968. Gravity anomalies and Godavari rift, India. Geol. Soc. Am. Bull. 79: 1221-1230.

SASTRI, M. V. A.; S. K. ACHARYA; S. C. SHA; P. P. SATSANGI; S. C. GHOSH; P. K. RAHA; GOPAL SINGH; R. N. GHOSH 1977. Stratigraphic lexicon of Gondwana formations of India. G.S.I. Misc. Publ. 36.

SASTRI, V. V.; A. T. R. RAJU; R. N. SINHA B. S. VENKATACHALA; R. K. BANERJI 1977. Biostratigraphy and evolution of the Cauvo basin. J. Geol. Soc. India 18(8): 355-3

SAVANUR, R. V. 1966. A study of the fau pattern in some of the blocks of South Karanpura and Ramgarh Coalfields. J. Mi Metal. Fuel 14 (9): 280.

--- 1971. The low dipping faults in Saunda and Bhurkunda Collieries, South Ka pura Coalfield. J. Mines Metal. Fuel.: 39.

VERMA, R. K.; M. MUKHOPADHYAY 1977. An analysis of the gravity field in NE India Tectonophysics 42: 290-291.

---; D. GHOSH 1977. Gravity field, Structure and tectonics of some Gondwana basin of Damodar Valley, India. 1: 97-112.

Elastic properties of Lower Gondwana rocks of eastern India (Abstract)

C. RAMACHANDRAN & M. RAMACHANDRAN NAIR
Geological Survey of India, Calcutta

Compressional (P) and shear wave (S) velocities have been measured in the laboratory in about 150 rock samples representative of the different geological formations of the Lower Gondwana rocks of eastern India. The P-wave velocity was measured along the bedding plane and perpendicular to it. The anisotropic factor $\parallel V_P / \perp V_P$ was estimated. The absorption coefficient, α, expressed as Db/cm was also determined. The effect of fluid saturation, with water and oil, on velocity, absorption and anisotropy was evaluated. The elastic constants were calculated, and the entire data were statistically analysed to obtain results relating the elastic parameters to the fluid saturation, absorption and anisotropy.

The basic data on the elastic properties furnish useful information for seismic prospecting in the coal bearing Gondwana rocks. Fluid saturation, in general, increased the P-wave velocity. The absorption coefficient decreases with increasing compressional wave velocity. Absorption increased in medium and fine-grained sandstones with saturation with water, while it decreased in coarse-grained sandstones and shales. $\parallel V_P$ is noted invariably greater than $\perp V_P$. In general, fluid saturation decreased the anisotropic factors in sandstones and shales alike.

An attempt has been made to compare the elastic properties of the Gondwana sandstones with those of the older and younger sandstones in India.

Development of Permian intracratonic basins in Australia

HELMUT WOPFNER
University of Cologne, Germany

The Permian in the intracratonic basins consists of basal glacial to periglacial deposits overlain by marine and finally freshwater deposits.   Ice movement was NW and S
in the western basins and probably NE in the eastern basins and was influenced by
several local factors.   Postglacial paleogradient directions were similar and suggest
a ridge separating the eastern and western basins.   During the Sakmarian the sea
transgressed from the west and from the east, without covering either the ridge, or the
northern basins, then retreated again.   Post-Sakmarian deposits are entirely non-
marine.

Basin shape was controlled mainly by NW- and NE-trending structures that first appeared
in the late Carboniferous or early Permian and were intermittently active until the end
of the Permian.   The structures resulted from compression generated mainly by plate
convergence on the eastern craton margin (Tasman Fold Belt) and to a lesser extent by
taphrogenesis on the western side of Australia.

INTRODUCTION   (Fig. 1)

Between two meridional Permian belts on
the eastern and western margins of Australia
are several intracratonic Permian basins.
These basins might be expected to have been
landlocked during Permian times, but some
contain marine sediments.   They are com-
plexly structured, and tectonism controlled
paleogeography.   Owing to post-Permian erosion basin outlines of today (Fig. 3 below)
seldom correspond with original margins.
The Permian deposits are wholly or largely
covered by younger sedimentary rocks but
are well known from drill-holes and geophysical exploration.

STRATIGRAPHY

The simplified Permian stratigraphy is
summarised in Fig. 2.

Invariably, Permian sedimentation was
initiated by glacial deposits (David &
Howchin 1923; Ludbrook 1967; Thornton 1974,
1978; Wopfner 1965, 1970, 1978; Youngs
1975).   Surface exposures of tillite are
shown in Fig. 3 and include an isolated
outcrop of possibly Permian till reported by
Coats (1962) from the northern Adelaide
Fold Belt.

Glaciated pavements in the Troubridge
Basin, together with intercalated fluviatile
sands and conglomerates elsewhere, show that
most of the foreland of the terminal glacier-
tongues was land.   Graded diamictitic sequences in the east of the Arckaringa Basin
indicate that there, glaciers issued into an
aqueous, probably marine environment
(Wopfner & Allchurch 1967).   Marine micro-
plankton occur also in glacigene strata of

the Denman and Renmark Basins (Ludbrook
1967; Thornton 1974).

In all basins, palynology indicates an
early Sakmarian age for the upper part of
the glacigene sequence.   The base is invariably unfossiliferous and may be late
Stephanian (Fig. 2).

Early Sakmarian conglomerates and sandstones, some with gypsiferous and kaolinitic
matrix mark the glacial retreat in the
Arckaringa and Pedirka Basins (Wopfner 1970;
Youngs 1975).

In the Denman, Arckaringa, Troubridge and
Renmark Basins, glacial and periglacial
sediments are blanketed by marine, Sakmarian
mudstones.   Near the base, dropstones indicate a periglacial position.   Fossils include mostly agglutinated foraminifera
(*Hyperammina*, *Ammodiscus* and *Hemodiscus*)
and, in the Arckaringa Basin, Conularia and
gastropods (Ludbrook 1967).   The low
diversity of the fauna has been cited as
evidence for cold water, but Harris and
McGowran (1973) attributed it to restricted
marine conditions.   No marine sediments
are known in the Pedirka Basin, and a single
Fenestella from the Merrimelia Fm suggests
only a brief incursion of the sea into the
Cooper Basin (Fig. 2).

In the Troubridge, Denman and Renmark
Basins, erosion has removed any Permian
strata younger than the Sakmarian marine
deposits.   Artinskian sediments are represented in the Arckaringa and Pedirka
Basins, and an almost continuous sequence up
to lower Triassic is preserved in the
Cooper Basin (Fig. 2).   The Artinskian and
younger parts of the Permian succession are
entirely freshwater, comprising sandstones,

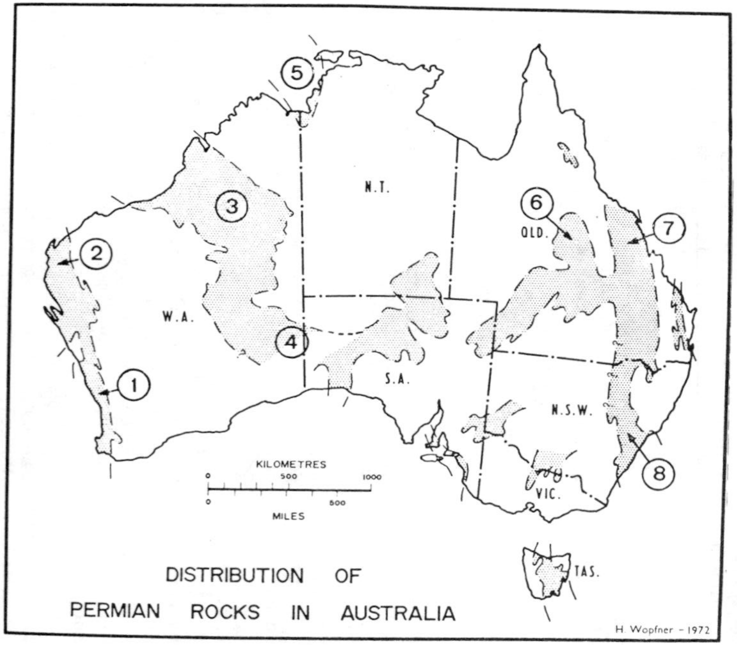

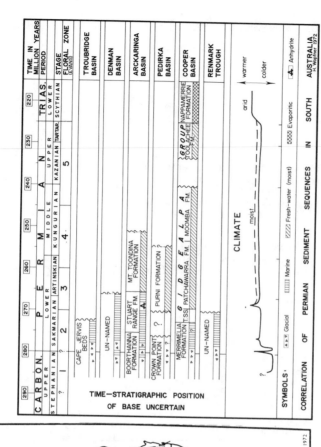

Fig.2. Stratigraphy, depositional environment and paleo-climate of intracratonic Permian basins.

Fig.1. Generalised map of Australian Permian basins: 1 - Perth; 2 - Carnarvon; 3 - Canning; 4 - Permian basin overlying Officer; 5 - Bonaparte Gulf; 6 - Galilee; 7 - Bowen; 8 - Sydney. The intracratonic basins in South Australia and adjacent parts of the Northern Territory, Queensland and New South Wales are in Figure 3.

186

siltstones, shales and thick coal seams. In the Cooper Basin they include both source and reservoir rocks for large hydrocarbon deposits (Wopfner 1965; Martin 1967; Kapel 1972; Thornton 1978).

PALEOGEOGRAPHY

From glacial pavement features and sediment transport directions ice-flow was N to NW in the Troubridge Basin and W to NW in the Arckaringa and Denman Basins, but erratics in the Pedirka Basin clearly indicate S to SE direction. The tongues of the glaciers must either have joined near the northern Arckaringa Basin and then flowed west or have terminated before meeting. The evidence of an aqueous environment in the eastern Arckaringa Basin suggests that they terminated.

Paleogradients suggest NE transport direction in the Cooper Basin, whilst erratics in drill cores indicate a N direction in the Renmark Basin.

The data indicate two major ice-bodies divided roughly along the axis of the Adelaide Fold Belt. Most of the ice certainly came from Antarctica, then adjacent to Australia, but considerable contributions of local, plateau ice are indicated by glaciated valley topography and composition of erratics. Independent centres of glaciation must have existed N and NW of the Pedirka Basin. The suggestion by Barrett et al. (1972) that the Permian glaciation of Antarctica resulted from a temperate not polar ice sheet is consistent.

The direction of marine transgression following ice retreat and eustatic rise of sea-level, was roughly opposite to that of ice-flow. A slight increase in the diversity of faunas in the Arckaringa and Denman Basins compared with the Troubridge Basin, suggests a connection to open sea to the west. Open marine conditions did exist in the northern Perth Basin but not in the southern Perth Basin (Johnstone et al. 1973).

The northern Perth Basin was connected to the Carnarvon and Canning Basins (Playford et al. 1975). From the Canning Basin open marine conditions decreased to the SE and a long embayment probably reached from West Australia to the Denman, Arckaringa and Troubridge Basins. The Pedirka Basin was a coastal plain with meandering streams issuing into the embayment (cf. Fig. 1). The paleogradients continued unchanged after the sea retreated and freshwater conditions became established (Fig. 3).

Most reconstructions place the Wilkes Coast of Antarctica against the Great Australian Bight with the Ross Orogen (Craddock 1972) continued by the Adelaide Fold Belt

(Johnstone et al. 1973). A SW ice-flow in the Transantarctic Mountain area (Barrett et al. 1972) locates the glaciation centre in Victoria Land. Rocks of the Ross and adjacent Borchgrevink orogenes would have provided the elevated platform for the formation of the ice cap. Ice issued from there towards South Australia where an elevated block formed a median ridge that divided it and provided loci for local ice accumulations. The dividing ridge remained during the later marine and fluviatile episodes of the Permian.

TECTONIC CONTROL

The intracratonic Permian basins of Australia have often been regarded as erotional features formed by glacial scour, but differences in shape and dimensions preclude that explanation (Wopfner 1970). Seismic investigations and drilling have shown that the basins are structural not erosional features.

All the basins rest on consolidated crust of varying composition and age, ranging from Precambrian crystalline rocks to folded Paleozoic sediments (Wopfner 1969). The relation of the Permian basins to pre-Permian structures is shown in Fig. 3. The youngest pre-Permian rocks are Devonian to early Carboniferous (Pedirka Basin). Granites beneath the Cooper Basin are clearly of pre-Permian age (probably late Precambrian; Wopfner 1969; cf. Scheibner 1978) and do not intrude the Permian as suggested by Harrington (1974).

Rocks folded during the late Devonian-early Carboniferous Alice Springs Orogeny not only became the floor of both Cooper and Pedirka Basins but also formed large tracts of the elevated area to the NW of the Pedirka Basin, showing that substantial tectonic movements must have been involved both in the formation of the basins, and of the adjacent highlands. The movements began before glaciation and continued intermittently during the Permian.

Basin shape was controlled primarily by marked linear elements (Fig. 4). Troughing, along NE and NW trends, resulted from the initial phase of movements. Deepening of the troughs, anticlinal growth and additional faulting, occurred during the Artinskian and again during the latest Permian (Kapel 1972; Paten 1969; Wopfner 1965, 1970), leading to stronger basin differentiation (Thornton 1978) and development of second-order structures.

En-echelon folds, structural growth-summation and abrupt trend-changes indicate transcurrent movements at depth (Harding 1976), NE and NW shear-directions being influenced by major pre-Permian lineaments

187

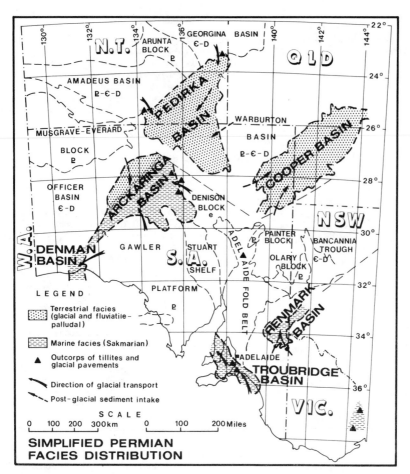

Fig. 3. Position of Permian intracratonic units in relation to pre-Permian geotectonic units. Ice-flow directions are indicated by bold arrows; postglacial paleogradients by dashed arrows. Facies boundary in NW Arckaringa Basin is conjectural.

(Figs. 3, 4).

Times of tectonism coincided with compressive deformation resulting from plate convergence in the Tasman Orogene (Packham & Leitch 1974; Scheibner & Glen 1972), as well as with taphrogenesis along the western margin of Australia (cf. Johnstone *et al.* 1973; Playford *et al.* 1975). Compressive forces, emanating from these regions are held responsible for the tectonism which created the Permian intracratonic basins and adjacent highlands.

REFERENCES

BARRETT, P. J.; G. W. GRINDLEY; P. N. WEBB 1972. The Beacon Supergroup of East Antarctica. In R. J. Adie (Ed.), Antarctic Geology and Geophysics. Oslo: Scandinavian University Books. Pp 319-332.

COATS, R. P. 1962. Permian deposits, possible occurrence near Blinman. Geol. Surv. S. Australia, Quar. Geol. Notes 1: 449-455.

DAVID, T. W. E.; W. HOWCHIN 1923. Glacial Research Committee. Rep., 16th AAAS, Wellington, New Zealand. Pp. 74-94.

HARDING, T. P. 1976. Predicting productive trends related to wrench faults. World Oil 182: 64-69.

HARRINGTON, H. J. 1974. The Tasman Geosyncline in Australia. In The Tasman Geosyncline - a symposium. Brisbane: University of Queensland. Pp 383-407.

HARRIS, W. K.; B. McGOWRAN 1973. Cootanoorina No.1 well - Upper Palaeozoic and Lower Cretaceous micropalaeontology. Geol. Surv. S. Aust., Rep. Investig. 40: 59-89.

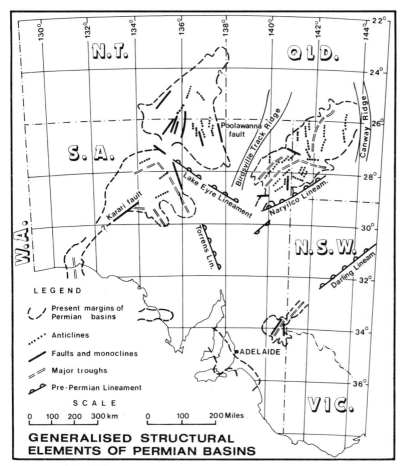

**GENERALISED STRUCTURAL
ELEMENTS OF PERMIAN BASINS**

Fig. 4. Major structural elements of Permian basins and some
of the more important pre-Permian lineaments. Birdsville
Track Ridge and Canaway Ridge are subsurface features. The
uneven density of structural elements shown within each basin
is dependent on exploration density.

JOHNSTONE, M. H.; D. C. LOWRY; P. M. QUILTY
1973. The geology of south-western
Australia - A review. R. Soc. W. Aust.
J. 56: 5-15.
KAPEL, A. J. 1972. The geology of the
Patchawarra area, Cooper Basin. Aust.
Petrol. Expl. Assoc. J. 12: 53-57.
LUDBROOK, N. H. 1967. Permian deposits
of South Australia and their fauna. Trans.
R. Soc. S. Aust. 91: 65-87.
MARTIN, C. A. 1967. A descriptive summary
of the Moomba gasfield. Australas. Oil Gas
J. 13: 23-26.
PACKHAM, G. H.; E. C. LEITCH 1974. The
role of the plate tectonic theory in the
interpretation of the Tasman orogenic zone.

In The Tasman Geosyncline - a symposium.
Brisbane: University of Queensland. Pp
129-155.
PATEN, R. J. 1969. Palynologic contribu-
tions to petroleum exploration in the
Permian formations of the Cooper Basin,
Australia. Aust. Petrol. Expl. Assoc. J.
9: 79-87.
PLAYFORD, P. E.; R. N. COPE; A. E. COCKBAIN;
G. H. LOW; D. C. LOWRY 1975. Phanerozoic.
In Geology of Western Australia. Geol.
Surv. Australia. Pp 223-433.
SCHEIBNER, E. 1978. Tasman fold belt
system in New South Wales - General Descrip-
tion. Tectonophysics 48: 207-216.
---; R. A. GLEN 1972. The Peel Thrust and

its tectonic history. Geol. Surv. NSW,
Quar. Geol. Notes 8: 2-14.
THORNTON, R. C. N. 1974. Hydrocarbon potential of western Murray Basin and infra-
basins. Geol. Surv. S. Aust., Rep. Investig. 41.
--- 1978. Regional lithofacies and palaeogeography of the Gidgealpa Group. Aust.
Petrol. Expl. Assoc. J. 18: 52-63.
WOPFNER, H. 1965. Case history of the
Gidgealpa Gas Field, South Australia.
In U.N. Econ. Comm. Asia Far East, 3rd
Petrol. Symp.. Tokyo: Document I & NR/PR
3/110. 19 pp.
--- 1969. Depositional history and tectonics of South Australian sedimentary
Basins. In U.N. Econ. Comm. Asia Far East,
4th Petrol. Symp. Canberra: Document I &
NR/PR. 4/57.
--- 1970. Permian palaeogeography and depositional environment of the Arckaringa Basin,
South Australia. In 2nd Gondwana Symp.
Proc. Pap. Johannesburg: Pp32-50.
--- 1978. Die Klimaentwicklung des Perms
in Süd- und Zentralaustralien. In
Schwarzbachfestschrift. Köln: Sonder-
veröff. Geol. Inst. Univ. 33: Pp. 259-279.
---; P. D. ALLCHURCH 1967. Devonian sediments enhance petroleum potential of Arckaringa Sub - Basin. Austral. Oil Gas J.
14: 18-32.
YOUNGS, B. C. 1975. The geology and hydrocarbon potential of the Pedirka Basin.
Geol. Surv. S. Aust., Rep. Investig. 44:
47 pp.

Geochronological correlations of Precambrian and Paleozoic orogens in New Zealand, Marie Byrd Land (West Antarctica), Northern Victoria Land (East Antarctica) and Tasmania

CHRISTOPHER JOHN ADAMS

Institute of Nuclear Sciences, DSIR, Lower Hutt, New Zealand

In Tasmania and New Zealand, late Precambrian low-grade metasedimentary sequences and higher-grade paragneisses yield Rb-Sr and K-Ar ages in the range 550-800 m.y. Similar gneisses may occur in North Victoria Land but appear to be absent from Marie Byrd Land.

In Western Tasmania and Northern Victoria Land, metamorphic age differences in Cambrian rocks delimit intra- and post-Cambrian orogenic phases. Late Precambrian and early Cambrian metasediments and volcanics all yield cooling ages in the range 450-505 m.y. (mid-Ordovician-late Cambrian) as a result of mid-late Cambrian (510-520 m.y.) metamorphism ($\equiv$ Ross Orogeny in Antarctica). Mid-Cambrian-Ordovician metasediments in New Zealand, North Victoria Land and Marie Byrd Land give younger ages, 400-450 m.y. (Silurian-Ordovician). In Antarctica these are related to metamorphism in the Borchgrevink Orogeny (Ordovician-Silurian?). In New Zealand, this event post-dates Greenland Group sedimentation and hence metamorphism probably occurred in late Ordovician-early Silurian times, i.e., an early phase of the Tuhua Orogeny.

Late Devonian-early Carboniferous plutonism was widespread and the regional thermal effect has resulted in the superimposition of a fourth age pattern upon older metamorphic rocks. Associated with this is Devonian regional metamorphism of the Tuhua Orogeny in New Zealand and Tabberabberan Orogeny in Australia.

INTRODUCTION

In any reconstruction of Gondwana, the reassembly of Australia, New Zealand/ Campbell Plateau, and West and East Antarctica is particularly complex. Geological checks of morphological fit are difficult since large areas intervene that are either submerged and mantled by younger sediments, e.g., Campbell Plateau, or extensively ice covered, e.g., Marie Byrd Land (Fig. 1). Correlations rely on generalised geological and geophysical characteristics of orogenic belts or sedimentary basins and in the former case, radiometric age data, particularly of metamorphic rocks, provide a powerful tool for correlation. Geological and geophysical data pertinent to such reconstructions have been discussed by Grindley and Davey (in press), Cooper (1975), Craddock (1975), Laird *et al.* (in press), and Wade and Long (in press). Metamorphic age (mainly K-Ar) data relevant to the problem are presented here to define the main episodes of tectonism and low-grade regional metamorphism associated with late Precambrian and Paleozoic orogens in four areas crucial to the reconstruction of the eastern margin of Gondwana - 1) South Island, New Zealand, 2) Marie Byrd Land, W. Antarctica, 3) North Victoria Land, E. Antarctica, 4) Tasmania, Australia. The data are used to trace the patterns of Paleozoic orogenies and combined to test the continental reassembly.

Ages used here are mainly K-Ar total-rock ages of slates and phyllites which are specifically chosen as low-grade metamorphic rock types with relatively potassic assemblages, quartz ± muscovite ± biotite ± chlorite. They are uniform, finely crystalline rocks that do not contain minerals of high argon retentivity, e.g., hornblende, that might retain pre-metamorphic ages, or low retentivity, e.g., plagioclase or clays. Their slaty cleavage is important since it may be attributed to a tectonic-metamorphic event in an orogeny. Since slates are low-grade metamorphic rocks of the prehnite-pumpellyite or pumpellyite-actinolite facies, they are rarely deeply buried and thus cool and begin to retain radiogenic argon soon after their formation. Their K-Ar ages are thus close to the time of metamorphism. In contrast, coarsely crystalline and higher grade metamorphic (and plutonic) rocks form at greater depths and yield mineral ages significantly younger than the time of recrystallisation, because cooling to the temperature of argon retention is much more protracted.

AGE DATA FOR LATE PRECAMBRIAN AND PALEO-ZOIC OROGENS AT THE EASTERN MARGIN OF GONDWANA

Several authors mentioned above have given detailed correlations of Precambrian-Paleozoic rocks between New Zealand, Australia and Antarctica, and the geological summaries below outline only the major sedimentary, tectonic and metamorphic cycles relevant to the metamorphic age data.

New Zealand. Late Precambrian-Paleozoic rocks occur principally in the NW part of the South Island, New Zealand, where they form three N-S sedimentary belts (Grindley 1978; Cooper 1979).

A mature quartzo-feldspathic sedimentary suite (early-late Ordovician) occurs in the northern part of the western belt and to the south, is related with widespread Cambro-Ordovician quartzose greywackes of the Greenland Gp (Laird 1972; Adams *et al.* 1975; Cooper 1975b). Greenland Gp slates yield K-Ar total-rock ages, 407-438 m.y. (Adams *et al.* 1975), which relate to an early Silurian or latest Ordovician metamorphism associated with the Tuhua Orogeny (Silurian-Devonian).[*] On the west coast, Precambrian basement (Charleston Metamorphic Gp) is dated by the Rb-Sr whole-rock method at 680±21 m.y. (Adams, 1975).

The central and eastern belts contain more varied sediments (early Cambrian-early Ordovician and early Ordovician-early Silurian, respectively). In the central belt early Cambrian sediments and mid-Cambrian volcanics are unconformably overlain by Cambrian-Ordovician conglomerate, flysch and some limestones. Major fault block tectonism, the 'Haupiri Disturbance' (Grindley 1978) occurred within the Cambrian. The eastern belt is dominated by Ordovician graptolitic shale and limestones, overlying late Cambrian clastics and underlying Silurian quartzites. Grindley (1978) considered the central belt allochthonous and thrust northwards to its present position during an early Tuhuan tectonic phase. Alternatively, Cooper (1979) considered that the central belt is a basal section of the eastern belt, thrust westward.

In the western and eastern belts, late Ordovician and Silurian sediments respectively, are overlain by early Devonian sediments which were folded under low T/P

---

[*]Radiometric ages referred to the Phanerozoic time scale of Armstrong (1978), with modifications to the Cambrian time scale as suggested by Cowie & Cribb (1978).

conditions during later (Devonian) phases of the Tuhua Orogeny. Thus, radiometric dating of the earliest Tuhuan events is complicated by younger thermal overprints. In addition, there is a thermal metamorphic overprint associated with late Tuhuan granitic plutonism, 280-370 m.y. (Aronson 1968). A similar age range occurs in Paleozoic slates of the central and eastern belts (Aronson 1968; author, unpubl.) and occasionally in Greenland Gp slates of the western belt (Adams *et al.* 1975). No 400-440 m.y. ages, typical of the Greenland Gp, occur in Paleozoic slates to the east.

Possible correlatives of the Greenland Gp occur south and east of New Zealand on Campbell Island and near the Bounty Islands. Quartzose metagreywacke and phyllite occur in both areas and on Campbell Island yield a maximum K-Ar age of 443 m.y. (Adams *et al.* 1979).

Marie Byrd Land, West Antarctica. Seafloor spreading data for the SW Pacific Ocean (Weissel *et al.* 1977) allow accurate pre-Cretaceous reconstructions which indicate important late Paleozoic and Mesozoic correlations between southern New Zealand, Campbell Plateau, and Marie Byrd Land, West Antarctica (Grindley & Davey, in press). In the Ford Ranges, Marie Byrd Land, the oldest rocks are late Precambrian, low-grade metagreywackes and slates (Swanson Gp) (Wade 1969) similar to the Greenland Gp in New Zealand. Possible Late Precambrian microfossils have been reported from these rocks but without precise details of location (Iltchenko 1972). Ages of Swanson Group slates, 410-450 m.y. (Krylov *et al.* 1970, author, unpubl.), indicate a minimum, late Ordovician age for low-grade metamorphism. Younger ages, 350-380 m.y., and 110-145 m.y., are related to the widespread thermal overprint of younger granites. Gneisses originally considered Precambrian basement (Fosdick Metamorphic Gp) are probably migmatised Swanson Gp injected by Devonian-Carboniferous (350-380 m.y.), and Cretaceous (110-145 m.y.) granites (Halpern 1968, 1972; Wade 1969; Wilbanks 1972).

In the Kohler Range, western Marie Byrd Land (Wade 1968), Swanson Gp correlatives may occur with high grade paragneisses. Here, and on Thurston Island further east, the oldest K-Ar ages are in the range 420-450 m.y. (Craddock 1969).

Devonian sediments (not *in situ*) occur on tne Ruppert Coast, west of Marie Byrd Land (Grindley *et al.*, this volume).

North Victoria Land, East Antarctica.
Late Precambrian and Paleozoic sedimentary, metamorphic and plutonic rocks form

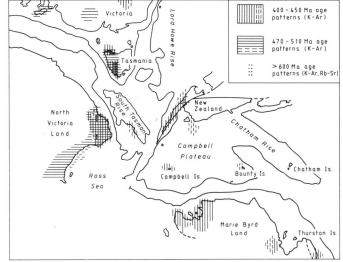

Fig. 1--Sketch reconstruction of E margin of Gondwana after Weissel *et al.* (1977) and Grindley & Davey (in press). Areas of late Precambrian, Cambrian-Ordovician, and Ordovician-Silurian metamorphic age (K/Ar & Rb/Sr) patterns are shown, reflecting the extent of late Precambrian, intra-Cambrian and Ordovician-Silurian orogens respectively.

basement along the Transantarctic Mountains to North Victoria Land. Detailed studies have been made in the Nimrod Glacier area in the central Transantarctic Mountains and the Bowers Mountains, North Victoria Land, a critical region in Australia-New Zealand-East and West Antarctic correlation (Laird *et al.*, Grindley and Davey, Bradshaw *et al.*, Adams *et al.*, all in press). The oldest rocks are Precambrian (?) high-grade para-gneisses (Wilson Gp) which possibly predate lower-grade, late Precambrian (Vendian?) quartzose metasediments (Robertson Bay Gp), and slightly younger Vendian (?)-early Cambrian rocks (Sledgers Gp). Throughout the Transantarctic Mountains and Victoria Land widespread granitic-granodioritic plutons (Granite Harbour Intrusives) in-trude Precambrian-early Cambrian rocks.

An apparent Precambrian Rb-Sr rock age, 770±20 m.y. (Faure & Gair 1970), has a much larger age error when recalculated with a reasonable range of initial $^{87}Sr/^{86}Sr$ ratio (Adams *et al.* in press). A Pre-cambrian age is thus not really proven and all other age data for the Wilson, Robert-son Bay and Sledgers Gps are much younger with a maximum age group at 470-510 m.y. (Cambro-Ordovician).

Fossil evidence suggests a stratigraphic hiatus between early Cambrian, Sledgers Gp and mid-late Cambrian, Mariner Gp. Whilst rocks older than the Mariner Gp yield main-ly 470-510 m.y. ages, Mariner Gp and young-er rocks all form a 380-420 m.y. group (Silurian-Devonian((Adams *et al.* in press). The mineral age range for the Granite Harbour Intrusives, 450-500 m.y.,is essentially similar to that of the pre-Mariner Gp rocks (Hulston & McCabe 1972; Adams *et al.* in press). Adams *et al.* (in press) consider that these age data thus reflect two main orogenic events--1) the Ross Orogeny, which could be intra-Cambrian, rather than Cambro-Ordovician, affecting Wilson, Robertson Bay and Sledgers Gp rocks and culminating in Granite Harbour pluton-ism, 2) the Borchgrevink Orogeny (probable late Ordovician-Silurian) principally affecting Mariner and Leap Year Gps (and to a lesser extent older rocks) and cul-minating in Devonian plutonism of the Ad-miralty Intrusives, which yield cooling ages in the range 310-380 Ma (Devonian-Carboniferous) (Gair *et al.* 1969; Faure & Gair 1970; Hulston & McCabe 1972).

Tasmania. In many ways, the geology and geochronology of the late Precambrian and early Paleozoic in Tasmania provide important data for correlation between New Zealand and Antarctica.

The oldest rocks occur in the Tyennan complex of central and SW Tasmania (Spry 1962) and yield a wide range of Rb-Sr whole-rock and mineral ages, 350-800 m.y., (Raheim & Compston 1977), which reflects the complicated tectonic-metamorphic his-tory of the region. The ages have been subdivided by Raheim and Compston into subgroups at ∿800 m.y., 550-630 m.y., 480-520 m.y., and 350-380 m.y., the first two reflecting Precambrian, regional metamor-phism in the Tyennan complex. K-Ar total-rock ages of Tyennan schists and slates also fall in the 540-630 m.y. group (Adams & Corbett, in prep.). Slightly older late Precambrian ages, 600-660 m.y., are obtained from lower-grade Vendian slates of the Oonah Fm, western Tasmania, and Rocky Cape Gp and Burnie Fm on the north coast (Adams & Corbett, in prep.). These ages are related to the late Precambrian Penguin Orogeny, whereas the older Tyennan Rb-Sr dates, close to 800 m.y., are more properly referred to the Frenchman Orogeny (Spry 1962).

Cambrian volcanics (Mt Read Volcanics) and sediments (Dundas Gp) overlie these rocks and are in turn unconformably overlain by late Cambrian conglomerates (Owen Conglomerate) which signal minor late Cambrian-Ordovician tectonism (Corbett *et al.* 1977). K-Ar mineral and Rb-Sr whole-rock ages of minor plutonic rocks, probably associated with the Mt Read Volcanism range from 460-520 m.y. (McDougall & Leggo 1965; Jago *et al.* 1977). K-Ar ages of slates in the lower Mt Read Volcanics, group similarly at 450-470 m.y. (Adams & Corbett, in prep.) and all these ages must reflect uplift and cooling of the Cambrian-Precambrian sequences in latest Cambrian and early Ordovician times after a phase of tectonism and low-grade metamorphism at least 490 m.y. ago (late Cambrian). Late Cambrian (Dundas Gp) slates yield younger K-Ar ages, 440 m.y., similar to the Greenland Gp slates in New Zealand. Some ages of west Tasmanian granites also fall in the same age range, 410-450 m.y. (McDougall & Leggo 1965).

The Owen Conglomerate is succeeded by a continuous Ordovician-Silurian-early Devonian sedimentary succession which was affected, with most older rocks, by widespread mid-Devonian deformation, low-grade regional metamorphism and plutonism of the Tabberabberan Orogeny.

Numerous K-Ar mineral ages of Tabberabberan granites and metamorphic rocks fall in the range 330-390 m.y. (McDougall & Leggo 1965; Adams & Corbett, in prep.) and reflect prolonged uplift and cooling of the orogen in late Devonian and Carboniferous times.

METAMORPHIC AGE PATTERNS AND PRECAMBRIAN-EARLY PALEOZOIC OROGENS

The age data presented above place constraints in space and time on Precambrian-early Paleozoic orogens in New Zealand, West and East Antarctica and Tasmania. No single area records a complete orogenic history, but the combination of all four areas resolves some ambiguities in specific cases.

The metamorphic ages are arranged into four groups, each reflecting an orogenic event(s).

Late Precambrian 'Tyennan' Type. The Tyennan gneisses and lower grade late-Precambrian metasediments of Tasmania provide the best evidence for late Pre-cambrian at the eastern margin of Gondwana. The oldest ages for Tyennan gneisses, 800 m.y., and King Island granites of Bass Strait, 715-835 m.y. (McDougall & Leggo 1965), indicate an early (Frenchman), orogenic event at least 800 m.y. ago. A younger, late Precambrian (Penguin) orogeny is reflected in Rb-Sr ages, 530-660 m.y. (Figs. 1 and 2) in the Tyennan and K-Ar slate ages, 600-660 m.y., in Vendian (?) lower-grade metamorphic rocks in NW Tasmania. The Penguin orogeny thus intervenes between the Oonah, Rocky Cape and Burnie sedimentation (Vendian?) and that of the Dundas Trough (Vendian?-Cambrian).

Elsewhere, it is only in New Zealand that Precambrian ages are also definitely proven. The Charleston Metamorphic Gp, South Island, yield only Precambrian Rb-Sr whole rock ages, 680 m.y.

The Wilson Gp of North Victoria Land is *possibly* Pre-cambrian (Fig. 1), although the evidence remains inconclusive. Similarly, in Marie Byrd Land, the original

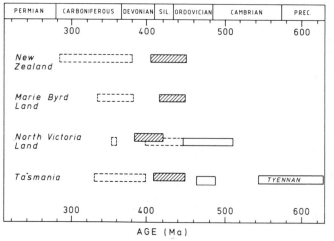

Fig. 2--K/Ar age patterns, principally of low-grade metamorphic rocks (slates) from N.Z., Marie Byrd Land, N. Victoria Land, & Tasmania grouped into late Precambrian-early Cambrian (open boxes) and mid-Cambrian-Ordovician(?)(shaded boxes). Extent of Devonian orogenic overprint on older rocks shown as dashed boxes.

suggestion of Precambrian basement below the Swanson Fm is unlikely. The Rb-Sr ages and stratigraphic evidence now indicate a Paleozoic age.

Mid-Cambrian 'Ross' Type. The late Precambrian-Cambrian in North Victoria Land, provides the most tightly constrained age evidence for regional metamorphism, tectonism and plutonism of the Ross Orogeny. The numerous mineral ages 450-510 m.y., with an abrupt maximum at 505-510 m.y., reflect uplift and cooling after the Ross Orogeny which occurred at least 505-510 m.y. age, i.e., mid-Cambrian (Figs. 1, 2).

A similar age pattern, 450-510 m.y., occurs in Precambrian-Cambrian metasediments and Cambrian granites in Tasmania (Fig. 1) and reflects a minor phase of tectonism, low-grade metamorphism and plutonism in the late Cambrian (Figs. 1, 2).

In New Zealand and Marie Byrd Land, no 'Ross' age pattern is yet seen.

Late Ordovician-Silurian, 'Early Tuhuan' Type. The Tuhua Orogeny in New Zealand is perhaps rather loosely defined and includes several phases from late Ordovician to Carboniferous. It is becoming clear that a major orogenic event, involving regional deformation, thrusting, nappe formation and metamorphism is pre-Devonian (Grindley 1978).

Greenland Gp slate ages, 407-438 m.y., imply that these low-grade metamorphic rocks were uplifted and cooling in latest Ordovician or Silurian times, and since they are indirectly related to a continuous Ordovician succession nearby then they must have been metamorphosed in the latest Ordovician.

A similar pattern occurs in the Swanson Gp slate ages, 410-450 m.y., in Marie Byrd Land and the Mariner/Leap Year Gps slate ages, 380-420 m.y., in North Victoria Land (Figs. 1 2). This seems a good and consistent point of correlation between New Zealand and West and East Antarctica.

In Tasmania there are very few ages in the range 410-450 m.y. from early Paleozoic slates and granites and the presence of an early Tuhuan group is doubtful. There is no geological evidence for metamorphic events in the Ordovician or Silurian and it is possible that the few ages are partial overprints of early Cambrian events by the Tabberabberan (Devonian) orogeny.

Mid-Devonian-Carboniferous, 'Tabberabberan' Type. Mid-Devonian-Carboniferous, 300-380 m.y., mineral ages occur in all areas but mainly record the widespread thermal effect of Devonian and Carboniferous plutons--Tuhua and Karamea Granites in

New Zealand, Ford Granodiorites in Marie Byrd Land, Admiralty Intrusives in North Victoria Land, Tabberabberan granites in Tasmania. Comprehensive Rb-Sr age data on Tasmanian examples indicate late Devonian emplacement (McDougall & Leggo 1965).

In Tasmania there is also age evidence for Tabberabberan low-grade regional metamorphism. Slates from early Devonian seuences (Bell Slate) yield 390-400 m.y. ages (mid-Devonian) (Adams & Corbett, in prep.).

In New Zealand and Antarctica a combination of regional metamorphism and local contact metamorphism has produced 'Tabberabberan' type age patterns.

CONCLUSIONS

In the late Precambrian-Cambrian of New Zealand, Tasmania and Antarctica at least three orogenic cycles are recognised in cooling age patterns of metamorphic rocks (Fig. 1).
1). Late Precambrian orogens, between 800 and 700 m.y. ago, in Tasmania and New Zealand. possibly absent in North Victoria Land and probably absent in Marie Byrd Land. Where they have survived the superimposition of younger events, mineral ages form a 'Tyennan' type, 550-700 m.y. (Fig. 2).
2) A mid-late Cambrian orogen, the Ross Orogen, North Victoria Land. Present in Tasmania and possibly also in New Zealand (Fig. 1). Mineral ages form a 'Ross' type, 450-510 m.y. (Fig. 2).
3) A late Ordovician-early Silurian orogen possibly related to the earliest phase of the Tuhua Orogeny in New Zealand and producing mineral ages in the range 400-450 m.y. (Fig. 2). This is well seen in New Zealand, Marie Byrd Land and North Victoria Land, but is doubtfully present in Tasmania.

In all areas upper Paleozoic (Devonian-Carboniferous) plutonism and low-grade regional metamorphism have superimposed a fourth age pattern, 300-400 m.y. (Fig. 2).

Although the age data are incomplete in some areas they may suggest limits to the extent of these orogens (Fig. 1):
1) the Precambrian orogens are, at present, proven only in the northern areas, Tasmania and New Zealand;
2) the mid-Cambrian orogen is proven only in a meridional belt in the Transantarctic Mountains, North Victoria Land and Tasmania;
3) the late-Ordovician-early Silurian orogen might occur in all four areas and could be correlated with the Benambran orogen (late Ordovician-Silurian) in southeast Australia;
4) a major Devonian orogen occurs in all four areas and continues northwards into eastern Australia.

ACKNOWLEDGMENTS

G.W. Grindley and R. Grapes are thanked for suggesting several important improvements to the manuscript.

REFERENCES

ADAMS, C.J. 1975. Discovery of Precambrian rocks in New Zealand: Age relations of the Greenland Group and Constant Gneiss, West Coast, South Island. Earth Planet.Sci. Lett. 28(1): 98-104.

---; C.T. HARPER, M.G. LAIRD 1975. K-Ar ages of low grade metasediments of the Greenland and Waiuta Groups in Westland and Buller, New Zealand. N.Z. J. Geol. Geophys. 18(1): 39-48.

---; P. A. MORRIS; J. M. BEGGS 1979. Age and correlation of volcanic rocks of Campbell Island and metamorphic basement of the Campbell Plateau, South-West Pacific N.Z. J. Geol. Geophys. (22):

---; J.E. GABITES; A. WODZICKI; J. D. BRADSHAW in press. Potassium-argon geochronology of the Precambrian-Cambrian Wilson and Robertson Bay Groups and Bowers Supergroup, North Victoria Land, Antarctica. Proc. Third SCAR Symp. Antar. Geol. Geophys., Madison, Wisconsin.

ARMSTRONG, R. L. 1978. Pre-Cenozoic Phanerozoic Time-Scale: Computer file of critical dates and consequences of new and in progress decay constant revisions. In: G. Cohee (Ed.), Contributions to the Geologic Time Scale Studies in Geology No.6, AAPG. Pp 73-91.

ARONSON, J. L. 1968. Regional geochronology of New Zealand. Geochim. Cosmochim. Acta 32: 669-97.

BRADSHAW, J.D.; M. G. LAIRD, A. WODZICKI in press. Structural style and tectonic history in North Victoria Land. Proc. Third SCAR Symp. Antar. Geol. Geophys., Madison, Wisconsin.

COOPER, R. A. 1974. Age of the Greenland and Waiuta Groups, South Island, New Zealand (Note). N.Z. J. Geol. Geophys. 17: 955-62.

--- 1975. New Zealand and south-east Australia in the early Palaeozoic. N.Z. J. Geol. Geophys. 18(1): 1-20.

CORBETT, K. D.; G. R. GREEN; P. R. WILLIAMS 1977. Geology of central western Tasmania. In, Landscape and Man. Roy. S. Tasmania Symp. Pp 7-27.

COWIE, J. W. & CRIBB, S. J. 1978. The Cambrian System. In G. Cohee (Ed.) Contributions to the Geologic Time-Scale. Studies in Geology 6, AAPG Pp 355-62.

CRADDOCK, C. 1969. Radiometric age map o Antarctica: Plate 19. In V.C. Bushnell, C. Craddock (Eds.) Antarctic Map Folio Series. Folio 12: Am. Geog. Soc.

CRADDOCK, C. 1975. Tectonic evolution of the Pacific margin of Gondwanaland. In K.S.W. Campbell (Ed.), Gondwana Geology, ANU Pp 609-18.

DRYLOV, A. Y.; B. G. LOPATIN; T. I. MAZINA 1970. Age of rocks in the Ford Ranges and on Ruppert Coast (western part of Marie Byrd Land). Soviet Antar. Exped. Info. Bull. 80: 64-6.

FAURE, G., H.S. GAIR 1970. Age determina tions of rocks from Northern Victoria Land Antarctica. N.Z. J. Geol. Geophys. 13(4) 1024-6.

GAIR, H.S.; A. STURM: S. J. CARRYER: G. W. GRINDLEY 1969. The geology of Northern Victoria Land, Plate 12. In V.C. Bushnel C. Craddock (Eds.). Antarctic Map Folio Series, Folio 12: Am. Geog. Soc.

GRINDLEY, G.W. 1978. In R. P. Suggate et al. (Eds.), The Geology of New Zealand. Government Printer, Wellington.

---; F. J. DAVEY in press. Some aspects of the reconstruction of New Zealand, Australia and Antarctica (review). Proc. Third SCAR Symp. Antar. Geol. Geophys., Madison, Wisconsin.

HALPERN, M. 1968. Ages of Antarctic and Argentine rocks bearing on continental drift. Earth Planet. Sci. Lett. 5(3): 159-67.

--- 1972. Rb-Sr total-rock and mineral ages from the Marguerite Bay area, Kohler Range and Fosdick Mountains. In R.J. Adie (Ed.) Antarctic Geology and Geophysics. Universitetsforlaget, Oslo. Pp197-204.

HULSTON, J. R.; W. J. McCABE 1972. New Zealand potassium-argon age list-1. N.Z. J. Geol. Geophys. 15(3): 406-32.

ILTCHENKO, L. N. 1972. Late Precambrian acritarchs of Antarctica. In R.J. Adie (Ed.) Antarctic Geology and Geophysics. Universitetsforlaget, Oslo. 599-602.

JAGO, J. B.; J. A. COOPER; K. D. CORBETT 1977. First evidence for Ordovician igneous activity in the Dial Range Trough, Tasmania. J. Geol. Sci. Australia 24(2): 81-6.

LAIRD, M. G. 1972. Sedimentology of the Greenland Group in the Paparoa Range, West Coast, South Island. N.Z. J. Geol. Geophys. 15: 372-93.

---; R. A. COOPER; J. B. JAGO in press. New data on the lower Palaeozoic sequence of Northern Victoria Land, Antarctica, and its significance for Australian-Antarctic relations in the Palaeozoic. Nature 265: 107-110.

McDOUGALL, I.; P.J. LEGGO 1965. Isotopic age determination on granitic rocks from

Tasmania.   J. Geol. Soc. Australia 12(2): 295-332.

RAHEIM, A.; W. COMPSTON 1977.   Correlations between metamorphic events and Rb-Sr ages in metasediments and eclogite from Western Tasmania.   Lithos 10: 271-89.

SPRY, A. H. 1962.   The Precambrian Rocks. In A.H. Spry, M. R. Banks (Eds.)   The Geology of Tasmania.   J. Geol. Soc. Australia 9(2): 107-126.

WADE, F. A. 1968.   Geology of the Hobbs and Bakutis Coasts sectors of Marie Byrd Land.   Antar. J. U.S.: 89-90.

--- 1969.   Geology of Marie Byrd Land. Plate 18.   In: V.C. Bushnell, C. Craddock. (Eds.)   Antarctic Map Folio Series, Folio 12: Am. Geog. Soc.

---; D. R. LONG in press.   The Swanson Formation, Ford Ranges, Marie Byrd Land - Evidence for direct relationship with Robertson Bay Group, Northern Victoria Land.   Proc. Third SCAR Symp. Antar. Geol. Geophys., Madison, Wisconsin.

WEISSEL, J. K.; D. E. HAYES; E. M. HERRON 1977.   Plate tectonics synthesis: The displacements between Australia, New Zealand and Antarctica since the late Cretaceous Mar. Geol. 25: 231-277.

WILBANKS, J. R. 1972.   Geology of the Fosdick Mountains, Marie Byrd Land.   In R.J. Adie (Ed.)   Antarctic Geology and Geophysics.   Universitetsforlaget, Oslo. 277-84.

Late Precambrian – Early Paleozoic geological relationships between Tasmania and northern Victoria Land

J. B. JAGO
South Australian Institute of Technology, Adelaide

The Bowers Trough of northern Victoria Land has been linked with the Dundas Trough of Tasmania on the evidence that, in both basins, there is a Late Precambrian-?Early Cambrian sequence disconformably overlain by a fossiliferous Middle to Late Cambrian sequence which in turn is unconformably overlain by dominantly continental coarse quartzose sandstones and conglomerates of Late Cambrian or Early Ordovician age. Both basins were rapidly subsiding trough-like structural depressions bounded by Precambrian blocks; the Cambrian rocks of both areas are intruded by Devonian-Carboniferous granites.

The absence in northern Victoria Land of equivalents to the acid to intermediate ?Late Precambrian-?Late Cambrian Mt Read Volcanics and the Cambrian ultramafic complexes of Tasmania suggests that there may not have been a direct alignment as previously proposed. It is suggested that the sediments of the Bowers and Dundas Troughs were deposited in different, but parallel troughs, which developed by rifting of the Precambrian crust in the Late Proterozoic and that the Bowers Trough lay to the west of the Dundas Trough.

INTRODUCTION

Before separation of Australia and Antarctica the two continents had been joined together since at least the Late Precambrian (McElhinny *et al.* 1974). Recent continental reconstructions of Antarctica and Australia place Tasmania close to northern Victoria Land, but the exact relationship between these two areas has been disputed.

The reconstruction by Laird *et al.* (1977) links the Bowers Trough of northern Victoria Land with the Dundas Trough of Tasmania, but significant differences between the Late Precambrian and Early Paleozoic sequences suggests that the two troughs were not directly aligned.

LATE PRECAMBRIAN-PALEOZOIC ROCKS OF NORTHERN VICTORIA LAND

The Bowers Supergroup contains all known Early Paleozoic sedimentary rocks of northern Victoria Land (Laird *et al.* 1977). It outcrops in the 300 km long and 20-25 km wide NW-SE-striking Bowers Trough (Laird *et al.* 1977). It extends from the Mariner Glacier region (about 72°40'S, 166°00'E) in the SE to the Edlin Névé region (about 71°21'S, 163°50'E) in the NW. To the SW the rocks are probably faulted against the Precambrian Wilson Group which consists of metamorphic rocks of amphibolite facies (Laird *et al.* 1976), which have undergone at least three periods of folding. To the NE the rocks of the Bowers Supergroup are in inferred faulted contact with those of the Robertson Bay Group. The Robertson Bay Group has undergone two periods of

folding and comprises at least 5000 m of greywacke, argillite and some volcanics, which show metamorphic grades from pumpellyite-actinolite to greenschist facies (Laird *et al.* 1976). Both Riphean (Iltchenko 1972) and Vendian (Cooper *et al.* in press) acritarchs have been reported from the Robertson Bay Group.

The Bowers Supergroup contains the Sledgers Group, Mariner Group and Leap Year Group, separated by unconformities (Laird *et al.* in press). The Sledgers Group consists of more than 3000 m thick basic Glasgow Volcanics to the NW and the interfingering clastic marine sediments of the Molar Formation to the SE. Vendian acritarchs and probable inarticulate brachiopods indicate a late Precambrian to early Cambrian or younger age (Cooper *et al.* in press). There appears to be a major break, representing most of the Early and Middle Cambrian, between the Sledgers and Mariner Groups.

The Mariner Group comprises over 3000 m of sandstone, mudstone and limestone forming a transgressive-regressive sequence dominated by shallow and marginal-marine facies. Parts of the Mariner Group are richly fossiliferous, with trilobites, brachiopods, molluscs, sponges and trace fossils which range in age from late Middle Cambrian to mid Late Cambrian (Cooper *et al.* in press).

The mainly fluvial Leap Year Group is exposed in two separate strips of contrasting lithologies (Laird *et al.* 1976, in press). The eastern strip (Camp Ridge Quartzite) comprises over 3 km of quartz sandstone and minor conglomerate, with marine intercalations near the base containing *Scolithus*, ?*Monocraterion* and other trace fossils probably of Late Cambrian or Early Ordovician age. The western strip lies in the NW of the Bouwers Mountains, where the locally deeply-channeled Mariner Group is overlain by up to 800 m of polymict conglomerate (Carryer Conglomerate) and minor sandstone (Laird *et al*. in press). The conglomerate is overlain by quartzose conglomerate and sandstone. Laird *et al*. (1977) concluded that paleocurrent data and facies relationships within the Bowers Supergroup suggest that the Sledgers and Leap Year Groups and perhaps also the Mariner Group were deposited in an elongated NW-SE-trending basin (the Bowers Trough).

The change to continental deposition represented by the Leap Year Group reflects the onset of the Ross Orogeny, a major Late Cambrian to Ordovician tectonic event which resulted in the uplift, folding and in places the metamorphism of most of the Precambrian and Cambrian rocks of the Transantarctic Mountains. The Ross Orogeny was accompanied and succeeded by the intrusion of the very widespread Granite Harbour Intrusives (granite, granodiorite, tonalite, etc.). Most of the granite ages fall into 520-450 m.y. range, as do the ages on metamorphic events. This suggests that the Ross Orogeny culminated in the Ordovician (Craddock 1972).

In northern Victoria Land the Granite Harbour Intrusives do not intrude either the rocks of the Robertson Bay Group or the Bowers Supergroup. Two K-Ar whole rock dates (Gair *et al*. 1969) from phyllites of the Robertson Bay Group near the Zykov Glacier suggest a Silurian metamorphic event and possibly that the main folding of the Robertson Bay Group and of the Bowers Supergroup took place within the Silurian. Craddock (1972) called this event the Borchgrevink Orogeny. Gair *et al*. (1969) have suggested that the Admiralty Intrusives (granite, granodiorite) of Devonian-Late Carboniferous age (385-300 m.y.), which intrude the Robertson Bay Group are related to the Borchgrevink Orogeny. The rhyolitic volcanics of the Gallipoli Porphyries (375 m.y.) are probable extrusive equivalents of the Admiralty Intrusives. This mid-Paleozoic orogenic period could be considered as either a separate orogeny, i.e. the Borchgrevink Orogeny, or as a continuation of the Ross Orogeny, with the orogenic axis shifted 200 km to the east. The Borchgrevink Orogeny may be related to some of the mid-Paleozoic tectonic events of Eastern Australia.

LATE PRECAMBRIAN-EARLY PALEOZOIC GEOLOGY OF TASMANIA

The main area of Cambrian sedimentation in Tasmania, the Dundas and Fossey Mountain Troughs, lay between the Precambrian Rocky Cape and Tyennan crustal blocks (Fig. 1) and the Forth Nucleus, forming an arcuate belt around the northern and western margins of the Tyennan Geanticline (Corbett *et al*. 1972; Williams 1978). The Tyennan Geanticline and the Forth Nucleus consist mainly of multiply-folded and metamorphosed sequences of interbedded siltstone and orthoquartzite (Williams 1978). Regional metamorphism reached the upper greenschist facies during the Proterozoic Frenchman Orogeny. The Rocky Cape Geanticline consists of two areas of unmetamorphosed sediments and minor volcanics separated by a narrow NE-SW-trending belt of metamorphic rocks of up to middle greenschist facies. The rocks of the Rocky Cape Geanticline were folded in the 700

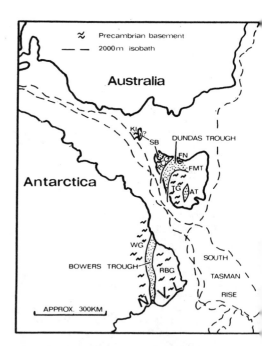

Fig. 1--SE Australia-Antarctica reassembly after Laird et al. 1977. NVL, Northern Victoria Land. RBG, Robertson Bay Group. WG, Wilson Group. AT, Adamsfield Trough. FMT, Fossey Mountain Trough. TG, Tyennan Geanticline. FN, Forth Nucleus. RCG, Rocky Cape Geanticline. SB, Smithton Basin. KI, King Island.

m.y. Penguin Orogeny, which is considered to be younger than the Frenchman Orogeny (Williams 1978).

The axis of the Dundas trough contains two main marine sedimentary sequences separated by an inferred unconformity (Williams 1978). The older sequence (Crimson Creek Formation, Success Creek Group and correlatives) consists of sandstones, mudstones and occasional lithic wacke and basic volcanic horizons. Vendian acritarchs have been reported from this sequence (Cooper and Grindley 1978). A number of ultramafic and mafic complexes occur within the unfossiliferous mudstone of the Crimson Creek Formation and at the disconformable boundary between the Crimson Creek Formation and the younger Dundas Group and correlatives. The Dundas Group and correlatives consist of turbidite lithic wacke, mudstone and conglomerate with some acid volcanic layers. It contains fossils which range in age from middle Middle Cambrian to middle Late Cambrian (Jago 1979). Between the N and W margins of the Tyennan Geanticline and the Dundas Trough accumulated thick acid to intermediate volcanics, constituting the Mt Read Volcanic Belt (Williams 1978). The volcanic rocks are mainly altered rhyolite and dacite, with minor andesite and basalt. Several small associated granite bodies include the Murchison Granite which has a minimum radiometric age of 515^{+} 15 m.y. (McDougall and Leggo 1965). The maximum age of the Mt Read Volcanics is unknown, but may be Late Proterozoic. In the Queenstown area most of the Mt Read Volcanics are older than the late Middle Cambrian (Corbett *et al.* 1974), although the volcanism continued until very late in the Cambrian. Both the Dundas Group and the Mt Read Volcanics are overlain either conformably or unconformably by terrestrial fans of siliceous conglomerate and shallow-marine conglomerate and quartz sandstone. The succeeding Ordovician marine sediments are transgressive on to the Tyennan Geanticline.

Other areas of Cambrian sedimentation in Tasmania include the Smithton Basin and the Adamsfield Trough (Fig. 1). In the Smithton Basin to the NW of the Rocky Cape Geanticline the late Proterozoic Smithton Dolomite unconformably overlies the rocks of the Rocky Cape Geanticline. The dolomite is unconformably overlain by a dolomite breccia which passes up into a sequence of tholeiitic lavas, mudstone, and greywacke with minor breccia and tuffs (Griffin and Preiss 1976). A siltstone containing late Middle Cambrian fossils (Jago 1979) probably belongs in this sequence. No post-Cambrian Paleozoic sediments are known from the Smithton Basin.

In the Adamsfield Trough, east of the Tyennan Geanticline, there appears to be a Late Proterozoic/Early Cambrian sedimentary sequence unconformably overlain by a fossiliferous Middle Cambrian sedimentary sequence which at least in one place is intruded by an ultramafic body. Other ultramafic bodies occur in the Adamsfield Trough. Both the Middle Cambrian sediments and the ultramafic body are overlain unconformably by rocks of the Denison Group which contains middle Late Cambrian fossils at the base and comprises a proximal quartz-wacke flysch at the base overlain by shallow marine and non-marine sediments, which pass up into marine Early Ordovician sediments (Corbett 1975). Small amounts of acid to basic volcanics are associated with the lower part of the sedimentary sequence.

On King Island, granites which have been dated from 715 to 835 m.y. intrude sediments of the Rocky Cape Geanticline. Near the northern margin of the Tyennan Geanticline three small granitic plutons of probable Late Cambrian age (McDougall and Leggo 1965) intrude the Precambrian rocks; they are probably related to the Mt Read Volcanics. The Late-Cambrian Early Ordovician siliceous sediments noted above resulted from uplift of the Tyennan and Rocky Cape Geanticline in the Late Cambrian. In the mid-Devonian, the Tabberabberan Orogeny affected the Proterozoic and Early Paleozoic rocks, with two phases of folding occurring in Western Tasmania. Following the Tabberabberan Orogeny granitic rocks, ranging in age from 375 - 335 m.y. intruded the older rocks (Williams 1978).

NORTHERN VICTORIA LAND - TASMANIA COMPARISON

In the following discussion it is assumed that the present position of Tasmania with respect to mainland Australia is similar to its Proterozoic-Early Paleozoic position (Webby 1978).

In both the Dundas and Bowers Troughs there is a Late Proterozoic-?Early Cambrian clastic sequence unconformably overlain by a Middle to middle Late Cambrian fossiliferous sequence which in turn is overlain unconformably by coarse quartzose sandstones and conglomerates of probable latest Cambrian to earliest Ordovician age. In northern Victoria Land the coarse clastics are dominantly of continental origin with only a minor marine influence, but in Tasmania there is a greater marine influence. Both Cambrian successions were apparently deposited in rapidly subsiding troughs flanked by Precambrian blocks. Both the northern Victoria Land and Tasmanian areas were similarly affected by Late Cambrian-Early Ordovician tectonic activity which led to the deposition of coarse quartzose deposits derived from nearby uplifted blocks of Precambrian rocks. The Precambrian-Early Paleozoic rocks of both areas are intruded by Devonian to Carboniferous granites. On the basis of the above similarities Laird *et al.* (1977) suggested that in the Early Paleozoic the Dundas and Bowers Trough were in direct alignment.

One of the major differences between the sequences in the two areas is the presence of the acid to intermediate volcanics of the Mt Read Volcanic Belt in western Tasmania; there is no equivalent in the Bowers Trough. A second important difference is the presence of quite a number of mafic to ultramafic complexes within the Late Proterozoic-?Early Cambrian Crimson Creek Formation of Western Tasmania. There are no equivalents in the Sledgers Group of northern Victoria Land. It also should be noted that while the Sledgers Group contains a considerable thickness of basic volcanics (Glasgow Volcanics) the Crimson Creek Formation and equivalents contain only minor basic volcanics, except for the Motton Spilite of the Dial Range Trough, a northern extension of the Dundas Trough. Another difference between northern Victoria Land and Tasmania is with respect to the timing of the major mid-Paleozoic folding event. The Proterozoic-Early Paleozoic rocks of northern Victoria Land were folded in the Late Silurian-Early Devonion Borchgrevink Orogeny which has been dated at 420-380 m.y. (Adams *et al.* 1977), whereas the major mid-Paleozoic folding in Tasmania was caused by the middle Devonian Tabberabberan Orogeny (Williams 1978). It is worth noting that a weak metamorphic event in the SE of South Australia took place in the Early or Late Silurian (Webb 1976; Milnes *et al.* 1977) and could be related to either the Borchgrevink Orogeny or the Benambran or Bowning Orogenies of eastern Australia.

In the Laird *et al.* (1977) reconstruction there is a gap of about 200 km between the southern end of the Dundas Trough and the northern end of the Bowers Trough, and alternatives to a Dundas-Bowers Trough alignment should be considered.

One possibility is alignment of the Bowers Trough with the Adamsfield Trough of Tasmania. The Adamsfield Trough contains only small quantities of Mt Read type volcanics, but contains several ultramafic bodies which have no equivalent in the Bowers Trough. It was folded in the Tabberabberan Orogeny.

The Smithton Basin, like the Bowers Trough, contains basic rather than acid volcanics but is closed to the south and cannot be a direct continuation of the Bowers Trough. It also was folded in the Tabberabberan Orogeny. Thus there are problems in directly aligning the Bowers Trough with any of the Cambrian basins of Tasmania. Other possibilities include that the Bowers Trough may have been linked with possible Cambrian rocks to the east of the Adamsfield Trough or to the west of the present known Cambrian rocks in Tasmania. The former possibility does not seem likely on the grounds of the differences in timing of the mid Paleozoic folding. It would also have the effect of aligning the Cambro-Ordovician Granite Harbour Intrusives with the post-Tabberabberan granites of Tasmania, rather than with the Cambro-Ordovician Delamerian granites of south-eastern South Australia and western Victoria.

It is possible that the alignment of the Bowers Trough was to the west of Tasmania (Fig. 2). In this position the Granite Harbour Intrusives are in alignment with the Delamerian granites of South Australia. The more basic Late Proterozoic-Early Cambrian volcanics of the Smithton Basin compared with the majority of Cambrian volcanics in Tasmania also suggest association with the Bowers Trough. A further point of interest is that in the Dundas Trough, the fossiliferous Cambrian sediments immediately beneath the contact with the Late Cambrian-Early Ordovician siliceous coarse clastics progressively decrease in age over a distance of 150 km from very early Late Cambrian in the north to middle Late Cambrian in the south (Jago 1973). In the Bowers Trough there is a similar decrease in age from north to south over a distance of about 200 km (Cooper *et al.* 1976). This similarity seems to be better explained in two essentially parallel troughs rather than in a single trough.

It is suggested that the sediments of the Bowers and Dundas Troughs were deposited in adjacent, but non-aligned troughs, which developed by rifting of the Precambrian crust in the Late Proterozoic, and that the Bowers Trough was to the west of the Dundas Trough. The troughs were probably rift valleys of the type proposed by Campana and King (1963); Corbett *et al.* (1972) and Williams (1978) for the Dundas Trough. It is possible that the Bowers Trough started to develop in the same way as the develop-

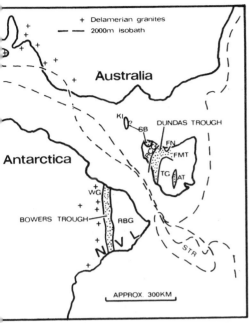

Fig.2--SE Australia-Antarctica reassembly after Crook and Belbin (1978). STR, South Tasman Rise. NVL, Northern Victoria Land. RBG, Robertson Bay Group. WG, Wilson Group. AT, Adamsfield Trough. FMT, Fossey Mountain Trough. TG, Tyennan Geanticline. FN, Forth Nucleus. RCG, Rocky Cape Geanticline. SB, Smithton Basin. KI, King Island.

ment of the Dundas Trough Williams (1978, Fig. 3), but did not proceed past the extension stage, whereas the Dundas Trough extension stage was followed by slight compression, accompanied by the extrusion of the Mt Read Volcanics. The configuration shown in Fig. 2 implies that in the northern Victoria Land-Tasmanian part of Gondwana tectonism proceeded from west to east with the Cambro-Ordovician Ross Orogeny folding the rocks to the west of the Bowers Trough; the Granite Harbour Intrusives are associated with this orogeny. The Late Silurian-Early Devonian Borchgrevink Orogeny affected the rocks of the Bowers Trough and the Robertson Bay Group in northern Victoria Land. The mid-Devonian Tabberabberan Orogeny affected the rocks of Tasmania.

REFERENCES

ADAMS, C.J.; A. WODZICKI; J.D. BRADSHAW; M.G. LAIRD (in press). K-Ar Ages of the Wilson, Robertson Bay, Sledgers and Mariner Groups, Northern Victoria Land, Antarctica. In C. Craddock (Ed.), Proc. 3rd Symp. Antarct. Geol. Geophys., Madison.

CAMPANA, B.; D. KING, 1963. Paleozoic Tectonism, Sedimentation and Mineralization in West Tasmania. J. Geol. Soc. Aust. 10: 1-53.
COOPER, R.A.; G. GRINDLEY 1978. South West Pacific basement correlation: Project 7. Pp. 36-38. In M.G. Bassett (Ed.), International Geological Correlation Programme (IGCP) Scientific achievements 1973-1977. Geological Correlation (special issue) 1-120. UNESCO: Paris.
---; J.B. JAGO; D.I. MacKINNON; J.H. SHERGOLD; G. VIDAL (in press). Late Precambrian and Cambrian fossils from Northern Victoria Land and their implications. In C. Craddock (Ed.), Proc. 3rd Symp. Antar. Geophys., Madison.
---; ---; ---; J. F. SIMES; P.E. BRADDOCK 1976. Cambrian fossils from the Bowers Group, Northern Victoria Land. Antarctica (preliminary note). N.Z. J. Geol. Geophys. 19: 283-288.
CORBETT, K.D. 1975. The Late Cambrian to Early Ordovician sequence on the Denison Range, southwest Tasmania. Pap. Proc. R. Soc. Tasm. 109: 111-120.
---; M.R. BANKS; J.B. JAGO 1972. Plate tectonics and the Lower Paleozoic of Tasmania. Nature (London) Phys. Sci.240:9-11.
---; K. O. REID; E. B. CORBETT; G.R. GREEN; K. WELLS; N.W. SHEPPARD 1974. The Mount Read Volcanics and Cambro-Ordovician Relationships at Queenstown, Tasmania. J. Geol. Soc. Aust. 21: 173-186.
CRADDOCK, C. 1972. Antarctic Tectonics. In R.J. Adie (Ed.). Antarctic Geology and Geophysics. Universitetsforlaget, Oslo. Pp. 449-455.
CROOK, K.A.W.; L. BELBIN 1978. The Southwest Pacific Area during the last 90 million years. J. Geol. Soc. Aust. 25 : 23-40.
GAIR, H.S.; S.J. CARRYER; G.W. GRINDLEY 1969. The Geology of Northern Victoria Land. Plate XII, Folio 12 - Geology, Antarctic Map Folio Series, American Geographical Society.
GRIFFIN, B.J.; W.V. PREISS 1976. The significance and provenance of stromatolitic clasts in a probable Late Cambrian diamictite in northwestern Tasmania. Pap. Proc. R. Soc. Tasm. 110: 111-127.
ILTCHENKO, L.N. 1972. Late Precambrian acritarchs of Antarctica. In R.J. Adie (Ed.). Antarctic Geology and Geophysics. Universitetsforlaget, Oslo. Pp.599-602.
JAGO, J.B. 1973. Paraconformable contacts between Cambrian and Junee Group sediments in Tasmania. J. Geol. Soc. Aust. 20: 373-277.
--- 1979. Tasmanian Cambrian biostratigraphy--a preliminary report. J. Geol. Soc. Aust. 26: 223-230.
LAIRD M.G.; J.D. BRADSHAW; A. WODZICKI 1976. Re-examination of the Bowers Group (Camb-

rian), Northern Victoria Land, Antarctica.
N.Z. J. Geol. Geophys. 19: 275-282.

---; J.D. BRADSHAW; A. WODZICKI (in press)
Stratigraphy of the late Precambrian and
early Paleozoic Bowers Supergroup, North-
ern Victoria Land. In C. Craddock (Ed.),
Proc. 3rd Symp. Antarct. Geol. Geophys.,
Madison.

---; R.A. COOPER; J.B. JAGO 1977. New data
on the lower Palaeozoic sequence of northern Victoria Land, Antarctica, and its
significance for Australian-Antarctic relations in the Palaeozoic. Nature 265:
107-110.

McDOUGALL, I.; P.J. LEGGO 1965. Isotopic
age determinations on granitic rocks from
Tasmania. J. Geol. Soc. Aust. 295-322.

McELHINNY, M.W.; J.W. GIDDINGS; B.J.J.
EMBLETON 1974. Palaeomagnetic results and
late Precambrian glaciations. Nature 248:
557-561.

MILNES, A.R.; W. COMPSTON; B. DAILY 1977.
Pre to syn-tectonic emplacement of Early
Palaeozoic granites in southeastern South
Australia. J. Geol. Soc. Aust. 24: 87-106.

WEBB, A.W. 1976. The use of the potassium argon method to date a suite of granitic rocks from south-eastern South Australia. Amdel. Bull. 21: 25-35.

WEBBY, B.D. 1978. History of the Ordovician Continental platform shelf margin
of Australia. J. Geol. Soc. Aust. 25:
41-63.

WILLIAMS, E. 1978. Tasman Fold Belt System in Tasmania. Tectonophysics 48: 159-
205.

Observations on the Ross Orogen, Antarctica

EDMUND STUMP

Arizona State University, Tempe, USA

An orogenic sequence of sedimentary, volcanic, metamorphic and plutonic rocks, truncated by a major middle to upper Paleozoic erosion surface occurs throughout the Transantarctic Mountains. Along its length the belt shows considerable variations in both tectonic style and timing. Relationships discussed in the paper suggest a migrating tectonic cycle from the African to the Australian sector of Gondwana from the late Precambrian through the middle Paleozoic.

An orogenic sequence of sedimentary, volcanic, metamorphic and plutonic rocks (the Ross Orogen) of upper Precambrian to lower or middle Paleozoic age occurs throughout the Transantarctic Mountains. These rocks accumulated along the Pacific margin of the East Antarctic craton, which crops out adjacent to the Trans-antarctic Mountains in the Wilson Hills and the Miller Range. The orogenic sequence is truncated by a major middle to upper Paleozoic erosion surface over-lain by gently tilted sedimentary units of the Devonian to Triassic Beacon Super-group and basalt flows and diabase sills of the Jurassic Ferrar Gp. In a recon-structed Gondwana the Ross Orogen of Antarctica connects to sequences in southern Africa and southeastern South America in the zone where the subsequent continental split occurred, and to sequences in eastern Australia.

The initial phase of sedimentation in the Ross Orogen (Robertson Bay Gp, Beard-more Gp, Patuxent Fmn) is characterized by accumulations of greywacke and shale, which at many places show features of turbidites. These rocks are preserved in discontinuous areas throughout the range, but their widespread distribution suggests a throughgoing basin of deposit-ion, probably a series of interconnected deep-sea fans (Stump 1976). However, relatively shallow-water conditions existed at times in portions of the basin as demonstrated by arenaceous and calcareous sedimentary rocks in the sequence (Katz and Waterhouse 1970; Laird *et al.* 1971). Rare mafic and felsic volcanic or hypabyssal rocks are also intercalated (Schmidt *et al.* 1965).

A Rb-Sr whole-rock isochron date of 1210 ± 76 m.y. on felsic rocks in the Patuxent Fmn of the Pensacola Mountains sets a minimum age for the onset of

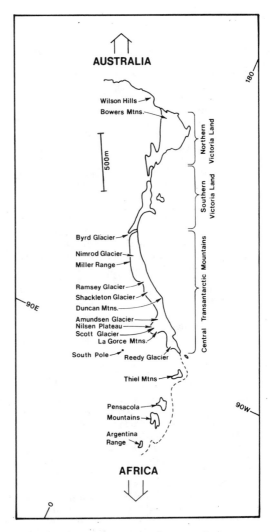

FIG. 1 -- Location map, Antarctica.

deposition (Eastin and Faure 1972).

Tectonism, called the Beardmore Orogeny, began late in the Precambrian from the Pensacola Mountains to the central Transantarctic Mountains (Grindley and McDougall 1969). The only bit of evidence for activity in southern Victoria Land at this time is a concordia plot of 610 m.y. on zircons from the Olympus granite-gneiss (Deutch and Grögler 1966). Southern Victoria Land was intensely affected by the subsequent Ross Orogeny, so the extent or existence of Beardmore tectonism in that area is not now known.

In northern Victoria Land evidence of any deformational activity associated with this period is sparse and equivocal. In the Bowers Mountains, the Robertson Bay Gp contains a schistosity, which predates cleavage associated with folding assigned to the Borchgrevink Orogeny (Bradshaw and Laird, in press). This may or may not be interpreted as having originated during the Beardmore Orogeny. Sedimentation and basaltic volcanism in the trough filled by the Bowers Supergroup had begun in Vendian time and may represent subsidence and uplift as a local manifestation of the Beardmore Orogeny (Laird *et al.*, in press).

In the Pensacola Mountains and Nimrod Glacier area, the Beardmore Orogeny caused compressional deformation which produced upright isoclinal folding (Schmidt *et al.* 1965; Laird *et al.* 1971). Thrust faulting also occurred in the Miller Range at this time (Grindley 1972).

Late Precambrian (670-600 m.y.) magmatism between Amundsen Glacier and the Thiel Mountains has also been related to the Beardmore Orogeny. This includes portions of the Wisconsin Range Batholith (Murtaugh 1969), extrusive porphyries in the Thiel Mountains (Ford 1964), and a variety of silicic porphyries collectively called the Wyatt Fmn in the upper Amundsen and Scott Glaciers area (Minshew 1967).

The Wyatt Fmn is a massive, silicic porphyry where it is found at the head of Reedy Glacier, the west side of upper Scott Glacier and the west side of Nilsen Plateau. At the Nilsen Plateau locality it discordantly intrudes Beardmore Gp (Stump 1976). At Mt Wyatt (the type locality, Minshew 1967) it appears to be a shallow intrusion, but in outcrops farther to the north it seems to be extrusive, in part confirming Minshew's (1967) original interpretation (Stump *et al.* 1979). Whether the Reedy Glacier occurrences are intrusive or extrusive has not been determined (Murtaugh 1969).

In the LaGorce Mountains the Wyatt Fmn and Beardmore Gp are in conformable contact. Unlike the aforementioned localities, here the Wyatt Fmn contains a variety of volcanic, volcaniclastic and clastic units. Unfortunately for this part of the story, Katz and Waterhouse (1970) and the author (Stump 1976) disagree on the stratigraphic order of the two units in the LaGorce Mountains.

One anomalous area in the central Transantarctic Mountains is the Duncan Mountains, where the Beardmore Gp apparently was not affected by the Beardmore Orogeny (Stump, in press).

The folded metasedimentary rocks in the Pensacola Mountains and Nimrod Glacier area were eroded before the Cambrian when a new phase of sedimentation and volcanism commenced. Between these areas no contacts of the late Precambrian and Cambrian, other than faults, are known.

During the Early and Middle Cambrian the region of the Transantarctic Mountains from Byrd Glacier to the Pensacola Mountains was a site of active volcanism and carbonate deposition (Byrd Gp, Liv Gp, Neptune Gp). From Shackleton to Reedy Glacier are extensive felsic volcanic rocks, erupted subaerially, but in association with carbonate and subaqueous clastic and epiclastic rocks (Stump 1976). Minor mafic volcanic rock are also associated. In the Byrd-Nimrod Glaciers area thick deposits of the Shackleton Limestone began accumulating in the Early Cambrian (Laird 1963; Skinner 1965). Early Cambrian fossils have been collected in the Argentina Range (Palmer and Gatehouse 1972), but marine transgression did not occur in the Pensacola Mountains until Middle Cambrian (Schmidt *et al.* 1965).

Elsewhere I have proposed the model of a linear chain of volcanic islands bounding a back-arc basin with primarily carbonate sedimentation (Stump 1976). However, there appears to have been very little influx of volcanic or clastic material from the volcanic region, so perhaps the basin of the Shackleton Limestone (Byrd-Nimrod Glaciers area) was separated from it by a ridge of eroded Beardmore Gp, outcropping today between Nimrod and Ramsey Glaciers.

An association of volcanic and clastic rocks continues into southern Victoria Land (Blank *et al.* 1963; Findlay 1978), but no fossils have been recovered to assign an age, and isotopic dates record the succeeding orogenic event.

In northern Victoria Land, deposition of the Bowers Supergroup began with coarse, clastic sedimentation and basaltic volcanism of the Sledgers Gp (Laird et al., in press). A period of non-deposition followed for much of the Early and Middle Cambrian. But two conformable episodes of sedimentation (Mariner and Leap Year Gps), separated by a disconformity, occurred from late Middle Cambrian through Late Cambrian or even into the Ordovician. No volcanism accompanied the deposition of these shallow marine and alluvial sediments.

The Late Cambrian-Early Ordovician Ross Orogeny was felt throughout the Transantarctic Mountains. A single, major phase of deformation occurred from south of Byrd Glacier to the Pensacola Mountains, apparently parallel to trends established during the Beardmore Orogeny. An exception to this is cross-folding in the LaGorce Mountains (Katz and Waterhouse 1970). In the Duncan Mountains a minimum of 3000 m of high-angle reverse faulting occurred between the Duncan and Fairweather Formations (Stump, in press). Metamorphism accompanied deformation, and an extensive complex of batholiths was emplaced, in part syn-tectonically, but largely post-tectonically. Isotopic systems closed or homogenized by about 500-450 m.y. (Adams et al., in press; Grindley and McDougall 1969).

In northern Victoria Land a limited number of plutons was emplaced in Late Cambrian or Early Ordovician, but compressive deformation apparently did not occur and sedimentation was active in the Bowers Trough. The main deformational event in northern Victoria Land was the Borchgrevink Orogeny of Siluro (?)-Devonian age. Both Bowers Supergroup and Robertson Bay Group were folded, and post-tectonic plutons were emplaced. Accompanying the event were limited extrusions of rhyolite (Dow and Neall 1972; Sturm and Carryer 1970). It is of note that structural trends in northern Victoria Land are oriented 15° counterclockwise from trends of the Ross Orogeny elsewhere in the range.

Southern Victoria Land links the disparate terrains of northern Victoria Land and the central Transantarctic Mountains. There, the Ross Orogeny produced multiple folding episodes, high-rade metamorphism, and extensive plutonism (Gunn and Warren 1962; Findlay 1978, Smithson et al. 1970). Two or three phases of folding are recorded at different localities, indicating that effects of the Ross Orogeny were more complex in southern Victoria Land than in the regions on either side.

The pattern that emerges from the above discussion is one of a tectonic cycle migrating from the African end of the Transantarctic Mountains to the Australian end. The Beardmore Orogeny significantly affected the Pensacola Mountains and central Transantarctic Mountains, but apparently had little or no effect in southern and northern Victoria Land. Sedimentation apparently had ceased in the central Transantarctic Mountains before Late Cambrian, but continued through Late Cambrian, perhaps into the Ordovician, in northern Victoria Land. Felsic volcanism preceded the Ross Orogeny in the Pensacola and central Transantarctic Mountains, and preceded or accompanied the Borchgrevink Orogeny in northern Victoria Land. Most importantly the Cambro-Ordovician Ross Orogeny was the terminal event from southern Victoria Land to the Pensacola Mountains, whereas the Siluro (?)-Devonian Borchgrevink Orogeny was the terminal event in northern Victoria Land.

The polyphase deformation in southern Victoria Land may represent anticipation of the tectonic realignment that caused structures in northern Victoria Land to be oriented 15° counterclockwise from trends elsewhere in the Transantarctic Mountains. But the polyphase effects may also be somehow related to the major discordance between the crystalline rocks north of Byrd Glacier, and the deformed and only slightly metamorphosed limestones south of it.

Erosion of the resultant mountain chain also apparently proceeded from the African to the Australian end of the Transantarctic Mountains, as reflected in the oldest sedimentary sequences which unconformably overlie the Ross Orogen. Devonian age sedimentary rocks accumulated in isolated basins from the Pensacola Mountains to southern Victoria Land, with widespread deposition following the Permo-Carboniferous glaciation, whereas sedimentation did not begin until Permian in northern Victoria Land and was not appreciable until Triassic (Barrett et al. 1972).

Support from NSF Grant, DPP76-82040

REFERENCES

ADAMS, C.J.D.; J. GABITES; G.W. GRINDLEY In press. Orogenic history of the Central Transantarctic Mountains: New K-Ar age data on the Precambrian-Early Paleozoic Basement. In C. Craddock (Ed.), Antarctic Geosciences. Madison: University of Wisconsin Press.

BARRETT, P.J.; G.W. GRINDLEY; P.N. WEBB 1972. The Beacon supergroup of East Antarctica. In R.J. Adie (Ed.), Antarctic Geology and Geophysics. Oslo: Universitetsforlaget.

BLANK, H.R.; R.A. COOPER; R.H. WHEELER; I.A.G. WILLIS 1963. Geology of the Koettlitz-Blue Glacier region, southern Victoria Land, Antarctica. Trans. R. Soc. N.Z., Geol. 2: 79-100.

BRADSHAW, J.D.; M.G. LAIRD In press. Structural style and tectonic history in northern Victoria Land, Antarctica. In C. Craddock (Ed.), Antarctic Geosciences. Madison: University of Wisconsin Press.

DEUTCH, S.; N. GRÖGLER 1966. Isotopic age of Olympus Granite-gneiss, Victoria Land, Antarctica. Earth Planet, Sci. Lett. 1: 82-84.

DOW, J.A.S.; V.E. NEALL 1972. A summary of the geology of the lower Rennick Glacier, Northern Victoria Land, Antarctica. In R.J. Adie (Ed.), Antarctic Geology and Geophysics. Oslo: Universitetsforlaget. 339-344.

EASTIN, R.; G. FAURE 1972. Geochronology of the basement rocks of the Pensacola Mountains, Antarctica. Geological Society of America, Abstracts with programs 4: 496.

FINDLAY, R.H. 1978. Provisional report on the geology of the region between the Renegar and Blue Glaciers, Antarctica. N.Z. Antarct. Rec. 1: 39-44.

FORD, A.B. 1964. Cordierite-bearing, hypersthene-quartz-monzonite porphyry in the Thiel Mountains and its regional importance. In R.J. Adie (Ed.), Antarctic Geology. Amsterdam: North Holland Publishing Company. 429-441.

GRINDLEY, G.W. 1972. Polyphase deformation of the Precambrian Nimrod Group, Central Transantarctic Mountains. In R.J. Adie (Ed.), Antarctic Geology and Geophysics. Oslo: Universitetsforlaget.

---; I. McDOWELL 1969. Age and correlation of the Nimrod Group and other Precambrian Rock Units in the Central Transantarctic Mountains, Antarctica. N.Z. J. Geol. Geophys. 12: 391-411.

GUNN, B.M.; G. WARREN 1962. Geology of Victoria Land between Mawson and Mullock Glaciers, Antarctica. N.Z. Geol. Surv. Bull. 71: 157.

KATZ, H.R.; B.C. WATERHOUSE 1970. A geological reconnaissance of the Scott Glacier area, south-eastern Queen Maud Range, Antarctica. N.Z. J. Geol. Geophys. 13: 1030-1037.

LAIRD, M.G.; 1963. Geomorphology and stratigraphy of the Nimrod Glacier-Beaumont Bay region, Southern Victoria Land, Antarctica. N.Z. J. Geol. Geophys. 6: 465-84.

---; J.D. BRADSHAW; A. WODZICKI In Press Stratigraphy of the late Precambrian and early Paleozoic Bowers Supergroup, northern Victoria Land, Antarctica. In C. Craddock (Ed.), Antarctic Geosciences Madison: University of Wisconsin Press.

---; G.D. MANSERGH; J.M.A. CHAPPELL 1971 Geology of the central Nimrod Glacier area, Antarctica. N.Z. J. Geol. Geophys 14:427-468.

MINSHEW, V.H. 1967. Geology of the Scott Glacier and Wisconsin Range areas, centr Transantarctic Mountains, Antarctica. Unpublished Ph.D. dissertation, Ohio Sta University, Columbus, Ohio. 268 pp.

MURTAUGH, J.G. 1969. Geology of the Wisconsin Range batholith, Transantarcti Mountains. N.Z. J. Geol. Geophys. 12: 526-550.

PALMER, A.R.; C.G. GATEHOUSE 1972. Earl and Middle Cambrian Trilobites from Antarctica. U.S. Geological Survey Professional Paper 456-D: 37 pp.

SCHMIDT, D.L.; J.H. DOVER; A.B. FORD; R.N BROWN 1964. Geology of the Patuxent Moun ains. In R.J. Adie (Ed.), Antarctic Geolo gy. Amsterdam: North-Holland. Pp. 276-28

---; P.L. WILLIAMS; W.H. NELSON; J.R. EGN 1965. Upper Precambrian and Paleozoic str tigraphy and structure of the Neptune Ran Antarctica. U.S. Geol. Surv. Prof. Pap. 525-D: D112-D119.

SKINNER, D.N.B. 1965. Petrographic crite-ria of the rock units between the Byrd an Starshot Glaciers, South Victoria Land, A arctica. N.Z. J. Geol. Geophys. 8: 292-30

SMITHSON, S.B.; P.R. FIKKEN; D.J. TOOGOOD 1970. Early geologic events in the icefree valleys, Antarctica. Geol. Soc. Am. Bull. 81: 207-210.

STUMP, E. 1976. On the late Precambrian-early Paleozoic metavolcanic and meta-sedimentary rocks of the Queen Maud Mountains, Antarctica, and a comparison with rocks of similar age from southern Africa. Ohio State Inst. Polar Stud. Rep. 62: 212.

---; In press. Structural relationships in the Duncan Mountains, central Trans-antarctic Mountains, Antarctica. N.Z. J. Geol. Geophys.

---; M.F. SHERIDAN; S.G. BORG; P.N. LOWRY P.V. COLBERT 1979. Geological investig-ations in the Scott and Byrd Glacier areas. Antarct. J. U.S. 14:

STURM, A.G.; S. CARRYER 1970. Geology of the region between Matusevich and Tucker Glaciers, northern Victoria Land, Antarctica. N.Z. J. Geol. Geophys. 12: 408-435.

Regional metamorphism in Antarctica

E. N. KAMENEV & V. S. SEMENOV
SEVMORGEO, Leningrad, USSR

The main metamorphic features of the Archean crystalline basement rocks of Antarctica are: (1) two-pyroxene (granulite) facies metamorphism (accompanied by widespread development of charnockites in Katarchean and lower Archean rocks; (2) upper amphibolite facies metamorphism (accompanied by intense granitization) in upper Archean rocks.

Proterozoic and Phanerozoic fold belts are characterized by early-stage prehnite-pumpellyite and greenschist facies metamorphism and high-pressure eclogite-glaucophane facies series metamorphism. Later stages are characterized by metamorphism in the inter-mediate pressure greenschist facies that has been partially overprinted by zonal metamorphism of kyanite-sillimanite (lower Proterozoic) and andalusite-sillimanite (upper Proterozoic-Phanerozoic) facies series metamorphism. The final stages of the metamorphic evolution were characterized by high-pressure diaphthoresis and dynamic metamorphism. In general, the grade of metamorphism in lower Proterozoic fold belts was higher than in the upper Proterozoic and Phanerozoic belts.

INTRODUCTION

From regional and local mapping (Craddock 1972; Ravich & Grikurov 1976) a 1:5 000 000 map of metamorphic facies in Antarctica has been produced (Kamenev & Ravich 1979).

This map demonstrates that within each geotectonic province metamorphism took place in a definite succession with each stage characterized by distinct thermodynamic regimes.

DISTRIBUTION OF METAMORPHIC ROCKS IN ANTARCTICA: EAST ANTARCTIC CRATON

The basement of the craton consists of crystalline rocks of Archean age that are best exposed in Enderby Land, the northern Prince Charles Mountains, and in Queen Maud Land. Supracrustal sequences form either Proterozoic - early Paleozoic fold systems in aulacogens (southern Prince Charles Mountains, Shackleton Range, region of Mount Amundsen and Mount Sandow, Jutulstraumen) or are flat lying and practically unmetamorphosed cover rocks (Ritscher Upland, Beaver Lake area, &c.).

Crystalline basement. Katarchean and early Archean basement forms the Enderby Land shield extending from the Porthos Range in the east to the Molodezhnaya Station area in the west and probably outcrops in western Princess Elizabeth Land (Amery Ice Shelf area) and on the Pravda Coast (Bunger Hills). The rocks are poorly exposed and have not been well dated. However, relics of protocrustal Katarchean rocks (termed the Napier Complex) have been re-cognized and tentatively assigned to the early Archean Rayner Complex. A Pb-isochron date from these rocks gives an age of 4000 ± 200 m.y. The Napier Complex comprises enderbites, mesoperthitic charnockites, pyroxene-plagioclase schists, and mesoperthitic garnet granite-gneisses, i.e. plutonic rocks metamorphosed in the two-pyroxene facies (T = $900°$ - $1000°$C; P = 3 - 4 kbar, P_{H_2O} = 0.2 of P_{total} with low alkalinity) The presence of osumulite in rocks of this facies (Grew, written comm.) supports their high-temperature origin.

Isotope dates indicate that the Rayner Complex, which laterally replaces the rocks of the Napier Complex, may have formed in the early Archean about 3000-3500 m.y. ago. The rocks comprise enderbites, charnockites, garnet gneisses, pyroxene-plagioclase schists, carbonate and high alumina rocks, and migmatite with charnockitic neosome assemblages. Temperature-pressure conditions were those of the two-pyroxene facies (T = $700°$ - $900°$C; P = 8 - 13 kbar; P_{H_2O} = 0.3 - 0.4 of P_{total}).

Mineral assemblages of the highest pressure two-pyroxene facies (with high Na alkalinity) replacing those of intermediate pressure two-pyroxene facies (with high K alkalinity) rocks are restricted to narrow zones along the margins of the Rayner and Napier complexes (i.e., Wilma and Robert Glaciers area, southern Sakellary Peninsula, Vernadsky Peninsula, Scott Mountains, Beaver Glacier area).

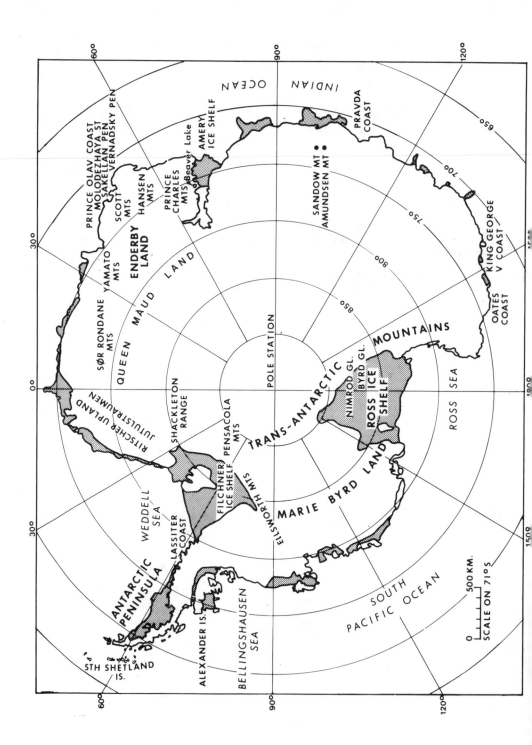

Rocks of the two-pyroxene facies are laterally replaced by mixed polyfacial zones of two-pyroxene facies and intermediate pressure amphibolite facies rocks (e.g., the Humboldt Complex in Queen Maud Land).

Complexes of the Humboldt type are in turn replaced by those of the Insel type or have tectonic boundaries with them (type locality in Queen Maud Land). Rocks of the Insel type have been metamorphosed with the intermediate pressure amphibolite facies (T = 650° - 800°C, P = 7 - 9 kbar, P_{H_2O} = 0.8 of P_{total} with high potassium alkalinity) and are intensely granitized and migmatized. Isotope ages range from 3000 m.y. to 2500 m.y.

Insel type rocks occur in all basement exposures in the central and western areas of the East Antarctic craton (central Prince Charles Mountains, Prince Olav Coast, Sør Rondane Mountains, central Queen Maud Land, some massifs in the Shackleton Range). The most abundant rock varieties are biotite-, garnet-biotite-, and biotite-amphibole gneisses, amphibolites with minor quartzites, carbonate and high-alumina rocks.

Bulk analyses of rocks from the Insel, Humboldt, Rayner, and Napier type complexes are similar and have the average composition of a granodiorite.

Metamorphism of basement rocks subsequent to the formation of the Insel Complex was of the high pressure greenschist and epidote-amphibolite facies localized along fault zones both inside the above crystalline complexes and especially in the regions bordering Proterozoic-early Paleozoic fold systems. Mineral assemblages of such rocks indicate that they are of high pressure kyanite-sillimanite facies series.

Kyanite-bearing rocks of the high pressure epidote-amphibolite and amphibolite facies outcrop in: (a) the western Fyfe Hills, where they replace two-pyroxene facies rocks of the Rayner and Napier Complexes along faults separating these complexes; (b) the Shackleton Range, where they replace polymetamorphic rocks of the Humboldt Complex along the boundary with Proterozoic folded formations; (c) the Jutulstraumen area in Queen Maud Land, where they replace amphibolite facies rocks of the Insel Complex at the boundary with supracrustal Proterozoic folded formations; (d) the Bunger Hills, where they replace two-pyroxene facies rocks within fault zones; (e) the Polkanov Hills, where they replace amphibolite facies rocks as products of the post-migmatite stage of acidic leaching.

Non-kyanite-bearing rocks of high pressure greenschist and epidote-amphibolite facies are known in: (a) the Sør Rondane

Mountains, where they replace rocks of the Insel Complex; (b) the Knuckey Peaks in Enderby Land, where they replace rocks of the Rayner Complex; (c) the Vestfold Hills, where they replace rocks of the two-pyroxene facies; (d) the Grearson Hills on the Budd Coast, where they replace polymetamorphic rocks. These rocks are typically associated with zones of cataclasis and mylonitization of basement rocks.

Proterozoic-early Paleozoic fold belts. In the southern Prince Charles Mountains, the best exposures of supracrustal metasedimentary and metaigneous (Ravich *et al.* 1978) rocks lie on the eroded surface of late Archean granite gneisses. Two complexes with the total thickness reaching 6-7 km are recognized. The *lower Ruker Complex* directly overlying the Archean basement is characterized by regional metamorphism in the greenschist facies and by the presence of basal jaspilites. Mica-chlorite-, mica-chlorite-carbonate-quartz-, quartz-carbonate-, and garnet-biotite-quartz schists with subordinate interbeds of epidote-biotite- and mica-microcline-quartz schists, metasandstones, metaconglomerates, and metabasites are characteristic rock types.

Basal jaspilites have been formed under high-pressure conditions. The upper part of the complex consists of mineral assemblages of the intermediate pressure greenschist facies, and there is a lateral zonation into rocks of the high pressure epidote-amphibolite facies (of the kyanite-sillimanite series), indicating P-T changes from 4-8 kbar, 550°-630°C. Migmatites are present in the highest temperature zone, where there are small intrusive bodies of granites and crystalline basement inliers of the craton. Late stage metamorphism was characterized by greenschist facies diaphthoresis and the development of zones of cataclasis and mylonitization. The upper age limit of the second stage of metamorphism is determined by isotope dates of dikes and interbedded bodies of amphibolite ranging from 1000 to 900 m.y.

The *Ross Complex* overlies the Ruker Complex, probably with an unconformity, and contains metaconglomerates with pebbles of Ruker Complex rocks and granitoids dated at 800 m.y. Low-grade metamorphics of quartz- and quartz-chlorite-sericite-, sericite-carbonate-quartz-, and epidote-stilpnomelane-quartz schists with albite and chlorite, quartzose metasandstones, and marmorized limestones dominate the lithology. Metamorphism was transitional between the epidote-chlorite subfacies of the greenschist facies and the pumpellyite-stilpnomelane subfacies of the prehnite-pumpellyite facies. Rb-Sr dates on phyllites suggest that the age of metamorphism and folding was 490 m.y.

In the Shackleton Range, supracrustal rocks of the Skidmore Complex are represented by kyanite-staurolite-, and kyanite-garnet-biotite schists with orthoclase, garnet-mica schists , and gneisses with interbedded amphibole schists, quartzites, and marbles. Another sequence consists of garnet-clinopyroxene amphibolites, schists, plagiogneisses, and amphibole schists.

In general highest-grade rocks are adjacent to the surrounding crystalline basement or developed near small granite bodies and reach the migmatite grade of the amphibolite facies (700°-740°C and 9-10 kbar with high Na alkalinity and the formation of second-generation andalusite.

Lowest-grade zones (520°-600°C and 4-5 kbar) are represented by the epidote-amphibolite facies. Relics of such assemblages are preserved in the higher-grade rocks. Diaphthoresis followed by cataclasis and mylonitization of greenschist facies grade took place along the fault zones and near the area of the intracratonic branch of the Ross System in the Shackleton Range.

The intracratonic branch of the Ross System in the southern Shackleton Range consists of a thick, essentially terrigenous sequence (phyllites, chlorite schists, quartzites, metasandstones, metaconglomerates, weakly marmorized dolomites and limestones) metamorphosed to the same grade as the Ross Complex in the Prince Charles Mountains. The sequence is dated by Riphean stromatolites in the lower part and Cambrian fossils in the upper part.

Metamorphism of other Proterozoic-early Paleozoic complexes of the craton is poorly known. However, in the Prince Charles Mountains and parts of the Shackleton Range, high-pressure amphibolite facies rocks (kyanite-sillimanite-facies series) and greenschists outcrop in the Jutulstraumen area. High pressure (glaucophanic facies) and intermediate pressure greenschist facies rocks are known in the southern Sør Rondane Mountains, while low-grade greenschist facies rocks outcrop at Mount Sandow and Mount Amundsen.

THE PACIFIC MOBILE BELT

The Pacific mobile belt of Antarctica borders the East Antarctic craton in the west and is subdivided into a number of fold systems belonging successively in the Ross, Hercynian, and Mesozoic epochs of tectogenesis.

The oldest fold system comprises the Transantarctic Mountains and was metamorphosed during the Ross epoch. The folded

rocks in the Ellsworth Mountains were pro bably consolidated during the Hercynian orogeny. Hercynian tectogenesis also affected the Pensacola Mountains fold syste which is probably a branch of the Transan arctic Mountains Ross system. The younge regions of the Antarctic Pacific belt are fold systems in Marie Byrd Land, Ellsworth Land, and the Antarctic Peninsula with adjacent islands, where cores of the Ross and Hercynian metamorphic rocks are exposed.

Because of poor exposure of the Hercyni systems, only the Ross and Mesozoic syste are discussed here.

The Ross Fold System of the Transantarc tic Mountains. The present stage of exploration of the system enables recognition of: (a) pre-Ross metamorphic complexes comparable with the early Precambrian crystalline basement of the craton, but probably remetamorphosed in the Ross orogeny; (b) Ross folded complexes proper separated by a locally marked unconformit into the early Ross geosynclinal and late Ross orogenic formations.

Early Precambrian metamorphics of the p Ross basement are exposed in the regions the Nimrod and Byrd Glaciers, Dry Valleys Priestly Glacier, and on the Oates Coast. The association of intermediate pressure amphibolite facies of granite-gneisses an schists is best known in the Dry Valleys gion. The presence of some migmatization however, suggests somewhat lower pressure conditions in places.

The Nimrod block consists essentially o marble, amphibolite, and calc-silicate. Folding was accompanied by migmatization and potassium metasomatism. Abundant bla toclasites (with new metamorphic minerals and eclogite-like bodies are associated w zones of subsequent tectonism. Isotope ages range from 3700 to 450 m.y. and sug gest repeated folding and metamorphism. The last Ross event is marked by blastoclasis and mylonitization.

On the Oates Coast, interbedded high-alumina schists, biotite-amphibole plagi gneisses, amphibolites and marbles are e posed and are commonly diaphthorized to epidote-amphibolite and greenschist faci High pressure conditions under which the diaphthorites formed is suggested by a z of high pressure greenschist facies on th George V Coast containing lawsonite and glaucophane. This zone is probably cont led by deep faults along the boundary sep rating the Transantarctic Mountains from craton and is tentatively regarded as par the Ross metamorphism. Isotope dates of rocks in the region range from 480 to 45 m.y.

MINERAL SYMBOLS

Ab - albite
Act - actinolite
Amph - amphibole
And - andalusite
Bi - biotite
Ca - calcite
Carb - carbonate
Chl - chlorite
Cord - cordierite
Cpx - clinopyroxene
(monoclinic pyroxene)
Di - diopside
Ep - epidote
Gr - garnet
Graph - graphite
Gros - grossular
Hb - hornblende
Hmt - hematite
KFsp - potassium
feldspar
Ky - kyanite

Mi - microcline
Mt - magnetite
Mu - muscovite
Olig - oligoclase
Parg - pargasite
Pl - plagioclase
Pr - prehnite
Pump - pumpellyite
Q - quartz
Rieb - riebeckite
Scap - scapolite
Ser - sericite
Sid - siderite
Sill - sillimanite
Sph - sphene
Stlp - stilpnomelane
Tourm - tourmaline
Trem - tremolite
Woll - wollastonite
Zo - Zoisite

The early Precambrian complexes of the pre-Ross basement in the Transantarctic Mountains are therefore similar to metamorphic rocks of the Insel Complex type in the East Antarctic craton.

Ross folded geosynclinal rocks are widespread in northern Victoria Land and in the near-pole part of the Transantarctic Mountains. They consist chiefly of weakly metamorphosed slate-greywacke flysch and associated volcanic and carbonate rocks.

In the Dry Valleys, earliest metamorphism of Ross folded terrigenous-carbonate rocks is represented by medium-grade intermediate pressure greenschist facies rocks with assemblages of Mu+Ep+Trem+Carb, Graph+Act+Ab+ Q+Chl, Ser+Q+Act (+Carb), Bi+Q+Sph+Hb. In the central Dry Valleys regions where numerous pre-Ross basement inliers and various intrusions of granites occur, the regionally metamorphosed greenschist sequence is overprinted by the effects of contact metamorphism. Mineral assemblages allow recognition of zones of pyroxene-hornfels facies (Gros+Di+Pl+KFsp+Woll+Ca, Woll+Di+ Pl+Q, and others), amphibole-hornfels facies (Di+Pl+Hb+Bi+Sph (+Q) (+KFsp), Gr+Bi+ Cord+Q+Pl+KFsp, Q+Pl+Bi (+Mi) (+Tourm) (+Mu)), and muscovite-hornfels facies (Q+Act+KFsp+Ep+Ca+Sph, Q+Act+Bi+KFsp+Sph).

Within fault zones, the hornfelses, rocks of the pre-Ross basement and granites are transformed into typical blastomylonites represented by fine-grained biotite-quartz-feldspar, actinolite-quartz-feldspar schists with lenses of granulite-like bodies. Mineral assemblages of blastomylonites are Q+Pl+KFsp+Gr (with 20% glossular molecule), Q+Pl+KFsp+Bi+Cpx+Gr, Q+Pl+Act+Bi+Ca, and

Q+Pl+KFsp+Act+Ca+Sph, i.e., of high pressure amphibolite facies.

In the Schmidt and Williams Hills of the Pensacola Mountains outcrops of intensely schistose rocks of the eugeosynclinal spilite-diabase formation occur with the mineral assemblage Ab+Chl+Act+Pr+Zo+Stlp+ Pump, indicating metamorphic conditions transitional between the high pressure prehnite-pumpellyite and greenschist facies. Zonal contact metamorphism of the low pressure amphibolite facies occurs near small granite intrusions.

The early Paleozoic orogenic complex of the Ross system is widely developed in the southern Transantarctic Mountains, in the Bowers Mountains, northern Victoria Land, and in the Neptune Range, where it is separated from the geosynclinal complex of the Ross system by an unconformity. The presence of zeolite, chlorite, prehnite, epidote, and albite suggests pressure-temperature conditions of the laumontite-prehnite-pumpellyite facies.

Mesozoic Fold System of the Southern Branch of the Scotia Arc. The Mesozoic fold system consists of two zones: (1) a zone of early Mesozoic folding; (2) a zone of late Mesozoic folding.

The zone of early Mesozoic folding includes the South Orkney and South Shetland Islands, Antarctic Peninsula (except for its SE and S margins) and Alexander Island. In these areas there are extensive outcrops of crystalline rocks which are interpreted as intensely reworked inliers of the pre-Mesozoic, probably early Precambrian, infrastructure. The age of the folded geosynclinal complex is ambiguous but is likely to be Triassic.

Early Precambrian complexes of the pre-Mesozoic basement are represented by amphibolite facies gneisses and schists which outcrop over several hundred square kilometres in the axial zone of the Antarctic Peninsula. The rocks are associated with Cretaceous intrusions and have therefore suffered contact metamorphism. Direct contacts of metamorphic rocks with other formations have not been observed. Intermediate pressure amphibolite facies biotite- and biotite-amphibole gneisses, granite-gneisses, biotite-amphibole-feldspar schists and amphibolites are widely distributed, and migmatization is locally found in basic and dioritic rocks. The metamorphic complex is believed to have been formed from the early Precambrian crystalline basement similar to the pre-Ross basement of the Transantarctic Mountains fold system but somewhat reworked due to geosynclinal development of the Mesozoic system. The epoch of most intense reworking

213

is recorded in isotope dates ranging from 200 to 100 m.y., but maximum age values of the metamorphic infrastructure do not exceed 380 m.y.

Early Mesozoic (Triassic) geosynclinal folded complexes are best exposed in the northern Antarctic Peninsula and on Alexander Island (Dalziel 1974; De Wit 1977; Rivano & Cortes 1976; Edwards, in press). Early stages in the metamorphic evolution are indicated by conversion of the prehnite-pumpellyite facies rocks into high pressure greenschist facies (glaucophane facies). Glaucophane-bearing schists occur intermittently in a narrow zone stretching from the South Orkney Islands through the South Shetland Islands (Clarence, Elephant, Gibbs Islands) to eastern Alexander Island, where they are associated with serpentinized dunites and peridotites together with phyllonites, quartz-micaceous and epidote-chlorite-actinolite schists, metacherts, marbles and garnet amphibolites. The time of primary metamorphism is determined by isotope dates ranging from 180 - 240 m.y.

The main part of the geosynclinal complex of the early Mesozoic system consisting of slate-greywacke and chert-volcanic formations is weakly metamorphosed to conditions transitional from the prehnite-pumpellyite facies to the greenschist facies. Stratigraphic sections suggest that the orogenic complex is practically unmetamorphosed. Zonal contact metamorphism of the andalusite-sillimanite series is associated with emplacement of Jurassic granites.

The zone of late Mesozoic folding includes the eastern zone of the Antarctic Peninsula south of 72°S (Black Coast, Lassiter Coast) representing so far the only zone in Antarctica where the age of the fold structure is younger than the Triassic-Jurassic boundary. Metamorphism of this zone is well studied on the Lassiter Coast.

Relics of early Precambrian complex rocks on the Lassiter Coast probably include a small outcrop of intermediate pressure amphibolite facies rocks in the region north of Mount Grimminger. The rocks are gneissic with the mineral assemblages: Pl+Mu+Bi+Sill+Q, Pl+Mu+Bi+Cord+Sill+Gr+KFsp+Q, and Pl+Hb+Cord+Gr+Q. Migmatization has also occurred.

The late Mesozoic geosynclinal complex is represented by metavolcanics and metasedimentary rocks. The metavolcanics consist mostly of bluish-green amphibole, stilpnomelane, biotite, pumpellyite, ilmenite, albite, in some cases chlorite, which replace relics of magmatic minerals, suggesting metamorphic conditions of the high pressure prehnite-pumpellyite and greenschist facies.

Metasedimentary rocks corresponding mainly to slate-greywacke assocations are represented by phyllites, slates, metasandstone metamorphosed to the low-grade greenschist facies. Metasedimentary sequences are well exposed at Rivera Peaks and in the Werner Mountains. Common mineral assemblages are Ab+Ca+Chl, Ab+Ser+Chl+Q, Ser+Bi+Ep+Q, and Ab+Act+Bi+Chl+Q+Ca, indicating temperature of 370° to 470°C and pressures from 3 to 7 kbar, indicating intermediate pressure greenschist facies.

Contact metamorphism in aureoles around late Cretaceous gabbro-granite intrusions overprint the lower-grade rocks. The width of these contact aureoles varies from several hundred metres to a few kilometres. Metamorphic rocks belong to the muscovite and hornblende hornfels facies with typical assemblages of: Pl+Cpx+Amf+Bi+Q, Pl+Amf+Ser+Chl+KFsp+Q, Ca+Ser+Chl+Q, Olig+Bi+Mu+Q, Q+Mu+And+Bi+Chl+Pl, Pl+Cpx+Amf+Bi+And, Pl+Amf+Bi+Gr (+Chl), Pl+Cord+Ser+Q, Cord+And+Sill+Amf+Mu+KFsp+Q.

The last metamorphic event on the Lassiter Coast was dynamic metamorphism along fault zones.

CONCLUSIONS

The crystalline basement of the East Antarctic craton has undergone polymetamorphism over a very long period (over 3000 m.y.).

Two distinct periods of regional metamorphism are recognized: (1) Katarchean-early Archean period of metamorphism in the two-pyroxene facies accompanied by intense charnockitization; (2) late Archean-early Proterozoic period of metamorphism in the amphibolite facies accompanied by granitization. Each period embraced vast areas but terminated in narrow linear zones under conditions of high pressure, increased Na alkalinity and relatively low temperatures. Early stages of each period are characterized by a uniform regime of metamorphism, while final stages are characterized by differentiation of pressure regime. Complexes of amphibolite facies rocks appeared first in zones where, during the first period, transformations took place in the high pressure two-pyroxene facies (i.e., in Queen Maud Land).

The earliest metamorphism of Proterozoic and Phanerozoic fold belts was burial. Rocks of the early Proterozoic mobile belt were initially subjected to burial metamorphism under greenschist facies conditions, while those in the late Proterozoic-Phanerozoic belts underwent low and intermediate pressure laumontite-prehnite-pumpellyite facies metamorphism. During this period, high-pressure prehnite-pumpellyite (or greenschist facies) metamorphism developed in the deepest parts of eugeosynclinal troughs. The high-pressure zones are characterized by intensive folding.

With later inversion of folding the rocks were subjected to metamorphism of the intermediate pressure greenschist facies through the entire mobile belt.

Horizontal zonation developed during the subsequent metamorphic stage. The zonation was paragenetically (and genetically in young belts) associated with granite intrusion. In the early Proterozoic system zonal metamorphism took place in the regime of the kyanite-andalusite series, while in the late Proterozoic system andalusite-sillimanite series, metamorphism was dominant.

Low temperature and somewhat high pressure (near-fault) metamorphism occurred at a late stage in the fold mobile belts. In the Proterozic systems metamorphism took place in the greenschist and even epidote-amphibolite facies, and in the Phanerozoic systems metamorphism was essentially dynamic.

Crystalline basements of all platforms, including Gondwana ones, consist of rocks of the two-pyroxene and amphibolite facies, the former being older. Cyclic metamorphic evolution appears to be an important characteristic of fold mobile belts, as exemplified in Antarctica.

REFERENCES

CRADDOCK, C. 1972. Geological map of Antarctica 1:5 000 000. New York: American Geographical Society.

DALZIEL, I.W.D. 1974. Margins of the Scotia Sea. In C. A. Burke, C. L. Drake (Eds.), The Geology of Continental Margins. New York: Springer-Verlag. Pp. 567-580.

DE WIT, M. J. 1977. The evolution of the Scotia Arc as a key to the reconstruction of southwestern Gondwanaland. Tectonophysics 37: 53-81.

EDWARDS, C. W. in press. Further paleontological evidence of Triassic sedimentation in Western Antarctica. In C. Craddock (Ed.), Antarctic Geosciences. Madison: University of Wisconsin Press.

KAMENEV, E. N.; M. G. RAVICH 1979. Map of metamorphic facies of Antarctica 1 : 5 000 000. Leningrad: Ministry of Geology USSR.

RAVICH, M. G.; G. E. GRIKUROV 1976. Geologic map of Antarctica 1 : 5 000 000. Leningrad: Ministry of Geology USSR.

RIVANO, S; R. CORTES 1976. Note on the presence of the lawsonite-sodic amphibole association on Smith Island, South Shetland Islands, Antarctica. Earth Planet. Sci. Lett. 29: 34-36.

ZWART, H. J.; J. CORVALAN; H. E. JAMES; A. MIYASHIRO; E. P. SAGGERSON; V. S. SOBOLEV; A. P. SUBRAMANIAN; T. G. VALLANCE 1967. A scheme of metamorphic facies for the cartographic representation of regional metamorphic belts. Geol. Newsl. 2: 57-72.

Carboniferous to Cretaceous on the Pacific margin of Gondwana: The Rangitata Phase of New Zealand

J. D. BRADSHAW
University of Canterbury, Christchurch,
New Zealand

P. B. ANDREWS
N. Z. Geological Survey, Lower Hutt, New
Zealand

C. J. ADAMS
Institute of Nuclear Sciences, DSIR, Lower
Hutt, New Zealand

The Carboniferous to Lower Cretaceous rocks comprise sub-parallel terranes which abut the Lower Paleozoic margin of Gondwana. The latter constitutes a foreland and is cut by late Mesozoic granitoids.

The terranes contain rocks of overlapping ages but dissimilar sedimentary facies, petrographic affinity, and structural history. From the foreland eastwards they represent a) island arc; b) fore-arc basins, c) ultramafic basin floor slices, d) trench slope basins and e) allochthonous continent-derived quartzo-feldspathic clastic wedge.

It is generally accepted that a-d are terranes developed on a convergent margin, with a history that culminated in the Rangitata Orogeny in the early Cretaceous. It is suggested that the clastic wedge should be divided into at least four sub-terranes with an important orogenic episode (Rangitata I) at the end of the Triassic which marks the accretion of the older part of the wedge to the trench slope suite. The younger parts of the clastic wedge were deformed by the early Cretaceous main phase of the Rangitata Orogeny.

SUMMARY

The Tuhua Orogeny (Cooper 1979) saw the consolidation of New Zealand's Lower Paleozoic rocks as part of the Pacific edge of the Australian-Antarctic segment of Gondwana. These rocks form the SW quarter of New Zealand, and the remainder is underlain by the Rangitata Orogen developed between the late Carboniferous and early Cretaceous.

When the results of post-Rangitata strain (related to the current Indian-Pacific plate boundary) are removed (Walcott 1978) the Rangitata Orogen can be seen as a number of terranes subparallel to the Paleozoic Gondwana margin (Fig. 1).

The Western Province or foreland comprises folded Paleozoic sediments and granites cut by late Mesozoic granitoids with associated high temperature metamorphism. The terranes to the east lie in the zone of high pressure metamorphism and can be divided into two groups. Those (1-5) that lie west of the Livingstone-Waipa Fault (Fig. 1) are characterised by simple, upright, gently plunging folds, and metamorphism related to stratigraphy. Those to the east (5-8) show polyphase deformation with metamorphic isograds that post-date one or more phases of folding.

From west to east the terranes are:
1) Brook Street terrane; Early Permian calc-alkaline volcanic rocks and related plutons - a Permian volcanic arc;
2) Productus Creek terrane; Late Permian carbonate and detrital rocks on a subsiding arc margin;
3) Murihiku terrane; thick (15 km) Triassic and Jurassic volcaniclastic sediments of a fore-arc basin;
4) Maitai terrane; Late Permian to Early Triassic carbonate and volcaniclastic sediments over 4 km thick. They are partly contemporaneous with Productus Creek (3) and early Murihiku (4) rocks and appear to be part of an outer fore-arc basin.
5) Dun Mountain terrane; mafic volcanic and ultramafic ophiolite. The original floor of the Maitai basin.
6) Caples terrane; volcaniclastic sediments and metavolcanics. They are mainly Permian, equivalent in age to Maitai rocks and probably of trench-slope basin origin. The Morrinsville sub-terrane (6a) lies in an analagous position in the North Island but consists of Jurassic rocks.
7) Haast Schist terrane; metamorphic tectonites of Caples and Torlesse (8) parentage, in a post-metamorphic arch. Contacts with Caples and Torlesse rocks are gradational.
8) Torlesse terrane; mainly quartzo-feldspathic sandstone and mudstone with minor metavolcanics, chert and conglomerate.

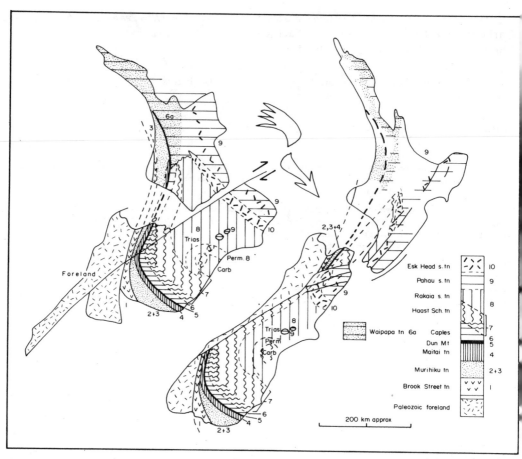

Fig.1--Simplified sketch of the Carboniferous to Cretaceous geology of New Zealand shown as a series of terranes on the Gondwana margin. Left, possible original arrangement.   Right, current configuration due to dextral faulting and strain related to the Indian-Pacific plate boundary.

This unit is more extensive than terranes 1-7 combined.

Descriptions of terranes 1-7 can be found in Coombs *et al.* 1976 and the Torlesse terrane in Andrews *et al.* 1976 and Speden 1976.   A conventional description of the Mesozoic rocks may be found in Stevens and Speden 1978.   All have extended references.

The broad interpretation of the Late Paleozoic-Mesozoic geology of the foreland and terranes 1-6 is straightforward. These terranes are characterised by volcanics and volcaniclastics of calc-alkaline affinities and can be related to a convergent plate boundary in Permian to Lower Cretaceous times.   The dominant structural grain was established by the Rangitata Orogeny in the early Cretaceous but with lesser orogenic pulses in the Late Permian and Late Triassic.

The main geotectonic problem in New Zealand is the presence of a great volume of predominantly continent-derived quartzofeldspathic material on the Pacific side of the convergent margin suite. This problem has typically been solved by arguing that the Torlesse rocks are exotic and have been rafted towards the convergent margin to collide with the convergent margin suite during the Rangitata Orogeny (e.g. Blake *et al.* 1974; Coombs *et al.* 1976) although some disagree (Spörli & Bell 1976; Wood 1978).

The nature of the source for the Torlesse rocks has provided a problem which has been compounded by recent geochronological work on schist cooling ages (e.g. Adams & Robinson 1977) which suggests a more complex history for the Caples, Haast Schist and Torlesse terranes.   This history is

218

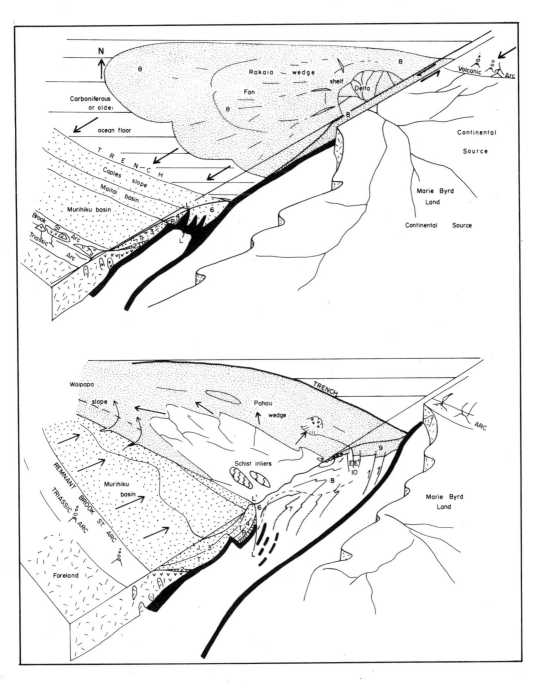

Fig.2--Block diagrams of Gondwana margin. Top, Late Triassic showing the convergent margin suite, trench and encroaching quartzo-feldspathic sediments derived from Marie Byrd Land.   Below, after deformation of older Rakaia rocks in Rangitata I, Pahau sediments accumulate on Pacific facing slope to the SE of a new trench.   The fore-arc basin continues to fill in the SW.

more readily understood if the Torlesse terrane is divided into a minimum of four sub-terranes (Fig.1).

8a) Rakaia sub-terrane; late Carboniferous to late Triassic mainly quartzo-feldspathic sandstone and mudstone of marginal marine to sub-marine fan origin. The rocks typically show polyphase deformation with early isoclinal folding. Mélange is present locally. They grade by increased metamorphism into Haast Schist with K-Ar cooling ages up to 200 m.y.

8b) Pahau sub-terrane; mainly late Jurassic and early Cretaceous sediments with rare early Jurassic. The sediments resemble Rakaia sub-terrane rocks but include recycled Rakaia material in addition to quartzo-feldspathic and some volcanic detritus. Pahau rocks are locally unconformable on the Rakaia sub-terrane, but generally the contact is tectonic. The Pahau sub-terrane was strongly deformed in the early Cretaceous.

8c) Esk Head sub-terrane; the Esk Head Mélange and adjacent belts of broken formation and tectonic slices. The sub-terrane comprises material derived from both the Rakaia and Pahau sub-terranes and forms a complex tectonic contact between them. Although it contains no truly exotic blocks, the Esk Head Mélange is richer in basalt, chert, limestone and conglomerate than typical Torlesse rocks.

8d) Hunua sub-terrane; mainly Late Jurassic sediments but including allochthonous Permian masses. Although the Hunua and Pahau sub-terranes are contemporaneous, their relationship is uncertain. Spörli (1978) grouped these rocks with the Morrinsville sub-terrane as a discrete Waipapa terrane.

The Rangitata Orogeny has traditionally been placed in the early Cretaceous, an interpretation which accords well with the evidence from the foreland and the western domain (i.e. west of the Livingstone Fault). However geochronology strongly suggests that east of the Livingstone Fault, Caples terrane and Rakaia sub-terrane rocks were folded and metamorphosed to form the Haast Schist by the end of the Triassic. This early orogenic episode is labelled Rangitata I. It is suggested that this early episode is the collision event and marks a major addition of continent-derived sediment to the accretionary prism on the convergent margin (Fig.2). Subsequently the trench appears to have re-established itself farther to the east and the Pahau sub-terrane represents sediments deposited on an east (Pacific) facing slope (Fig.2).

The axial zone of the fore-arc basin was not folded by Rangitata I, but the fore-

land and the whole mobile belt were affected by Rangitata II towards the end of the early Cretaceous. In the foreland there was a widespread phase of granitic magmatism followed by uplift. The terranes of the mobile belt were folded (and refolded) although towards the eastern margin the break between Pahau sub-terrane sedimentation and post-Rangitata clastics appears surprisingly brief (Laird, this volume).

Consideration of the geological history of adjacent regions in the SW Pacific suggests that the quartzo-feldspathic Rakaia sub-terrane was derived from what is now Marie Byrd Land across a transcurrent section of the Gondwana-Pacific margin. It was deposited on the outer (Pacific) side of the calc-alkaline suite and brought into contact by convergence. There is no evidence that the Rakaia sub-terrane extended beyond the central North Island and the fundamentally different arrangement of terranes in the north reflects the absence of Rakaia rocks and the early collision event.

ACKNOWLEDGMENTS

We would like to thank the many colleagues whose contributions to Permian-Mesozoic geology are not individually referenced. We hope to enlarge this summary and provide full reference lists in the near future.

REFERENCES

ADAMS, C. J.; P. ROBINSON 1977. Potassium-argon ages of schists from Chatham Island, New Zealand Plateau, Southwest Pacific. N.Z. J. Geol. Geophys. 20: 287-302.

ANDREWS, P. B.; I. G. SPEDEN; J. D. BRADSHAW 1976. Lithological and paleontlogical content of Carboniferous to Jurassic Canterbury Suite, South Island, New Zealand. N.Z. J. Geol. Geophys. 19: 791-819.

BLAKE, M. C.; D. L. JONES; C. A. LANDIS 1974 Active continental margins: contrasts between California and New Zealand. In Burk, C.A., C.L. Drake (Eds.), The Geology of Continental Margins. Springer-Verlag, New York. Pp. 853-872.

COOMBS, D. S.; C. A. LANDIS; R. J. NORRIS; J. M. SINTON; D. J. BORNS; D. CRAW 1976. The Dun Mountain Ophiolite belt, New Zealand, its tectonic setting, constitution and origin, with special references to the southern portion. Am. J. Sci. 276: 561-603.

COOPER, R. A. 1979. Lower Paleozoic rocks of New Zealand. J. R. Soc. N.Z. 9: 29-84.

LAIRD, M. G. 1980. The Late Mesozoic fragmentation of the New Zealand segment of Gondwana. This volume.

SPEDEN, I. G. 1976. Fossil localities in Torlesse rocks of the North Island, New Zealand. J.R. Soc. N.Z. 6: 73-91.

SPÖRLI, K. B. 1978. Mesozoic tectonics, North Island, New Zealand. Geol. Soc. Am. Bull. 89: 412-25.

---; A.B. BELL 1976. Torlesse mélange and coherent sequences, eastern Ruahine Range, North Island, New Zealand. N.Z. J. Geol. Geophys. 19: 427-47.

STEVENS, G. R.; I. G. SPEDEN 1978. New Zealand. In Moullade, M.; A.E.M. Nairn (Eds), The Mesozoic, A. The Phanerozoic Geology of the World II. Elsevier: Amsterdam.

WALCOTT, R. I. 1978. Present tectonics and Late Cenozoic evolution of New Zealand. Geophys. J. R. Astron. Soc. 52: 137-64.

WOOD, B.L. 1978. The Otago Schist megaculmination: its possible origins and tectonic significance in the Rangitata Orogen of New Zealand. Tectonophysics 47: 339-68.

Significance of Tethyan fusulinid limestones of New Zealand

K. B. SPÖRLI & M. R. GREGORY
University of Auckland, New Zealand

The "Marble Bay" section is the principal North Auckland locality for Permian fusulinid limestones. New data demonstrate that the fossils date the enclosing volcanic rocks only and not the overlying "greywacke" type clastic sequence, which was deposited on the volcanics much later, possibly in the Jurassic, before being welded as a series of imbricate thrusts onto the margin of Gondwana in the early Cretaceous Rangitata Orogeny.

The fusulinid limestones were deposited in an equatorial belt of open ocean, distant from the New Zealand Gondwana margin, where climate contrasted markedly with the cold environment of continental New Zealand represented by the Parapara Peak sequence. They were part of a Tethyan area of deposition now widely dispersed in North America, New Zealand, and Southeast Asia by subsequent sea-floor spreading and accretion onto the continents bordering the Paleo-Pacific.

The other New Zealand fusulinid occurrences, in the South Island, can be similarly interpreted.

INTRODUCTION

Since their discovery in 1950 (Hornibrook 1951; Leed 1951; Hay 1960), considerable significance has been attached to the Permian fusulinid- and coral-bearing limestones of Northland, the oldest fossiliferous rocks known from the North Island. On their presence much of the Waipapa terrane (Spörli 1978) has been assumed to be Permian (Kear & Hay 1961; Thompson 1961; Speden 1975). The Waipapa terrane is one of several belts of Mesozoic and older, mainly "greywacke"-type rocks, finally metamorphosed and structurally juxtaposed onto the Gondwana margin in the early Cretaceous Rangitata Orogeny (Coombs *et al.* 1976; Spörli 1978) and subsequently forming the basement over much of New Zealand.

From our lithological and structural mapping of key exposures near Marble Bay (Fig. 1A), reconsideration of the faunal evidence, and re-assessment of the rock collection by Maehl (1970), and examination of new material, we propose that these Permian fusulinid limestones are exotic to the New Zealand margin of Gondwana. A dynamic-stratigraphic model is presented to explain their emplacement.

LITHOLOGIC TYPES

Near Marble Bay we recognise 5 lithologic types. Four of these, green spilites, cherts, green argillites, and terrigenous "greywackes", form stratigraphic sequences with basal spilite and "greywackes" at the top. The fifth type, red mélange, has been sheared over "greywacke" and is either in stratigraphic or tectonic contact with overlying green spilite (see Fig. 1). Limestones and marbles are restricted to the green spilites and red mélange associations.

Green Spilites. The green-grey spilites, pilotaxitic, trachytic and intergranular in microtexture, consist mainly of altered plagioclase, augitic pyroxene and opaques, with a few amygdules of calcite and chlorite (Maehl 1970). At Wherowhero Point (Fig. 1A), 10-30 cm irregular layers of spilite are intercalated with red and green argillites, volcaniclastic conglomerates with calcite matrix, red cherts, and limestone (or marble) lenses. In the promontory E of Orua Bay, a sequence of pillow breccias grades E into flows up to several metres thick. Well-preserved pillow lavas are exposed on the E side of Kairawaru Bay.

At Marble Bay, the green spilites stratigraphically underlie thin cherts, which in turn underlie terrigenous sandstones and argillites (Figs. 1A, 1B). Red mélange underlies the green spilites on both sides of Kairawaru Bay, the contact either being gradational or a shear at low angle to the layering. The nature of this contact is a major unsolved problem in this area.

Red Mélange. Whether this unit is sedimentary in origin and represents a breccia or conglomerate, or whether it is tectonic is uncertain. The term "mélange" emphasizes its disrupted fabric and accords with an at least partially tectonic origin, which we favour. Lensoid bodies of various lithologies ranging from centimetre to metre size lie in a red mudstone matrix. Mélange-type fabric is best developed at lower levels. Towards the top, bodies of hematitic, vesicular spilite become predominant, apparently forming a more coherent fabric. These spilites are differentiated from the green ones by colour and more abundant calcitic amygdules. Well-developed pillow lavas are found within the mélange exposed on the SW of Kairawaru Bay. The hematitic spilites are chemically similar to the green spilites, both having 46-51% SiO_2, 4-6% NaO_2, and high TiO_2 (1.2-2.5%). The obvious difference lies in the oxidation state of iron.

Blocks of chert, fossiliferous limestone and marble, graded and hematite-rich calcarenite, and volcaniclastic breccia are also found in the red mélange. The noteworthy absence of terrigenous detritus in this association we interpret to indicate

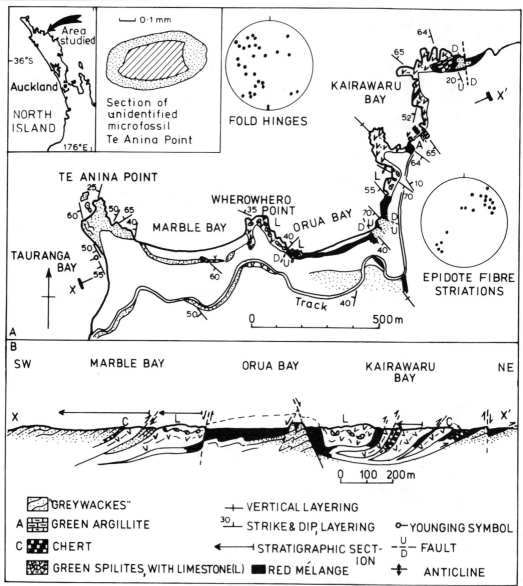

Fig. 1--A. Geological map of the Marble Bay area. Fold hinges and striations are shown on lower hemisphere equal area nets. Younging symbol: line points to top of beds. B. Schematic section, projected onto line XX' in A. Deeper parts of section are entirely speculative.

an oceanic depositional environment far from any terrigenous source. The high oxidation state, expressed by the striking red colour in many lithologies, may reflect prolonged sea-floor weathering and alteration (halmyrolysis).

A very sharp but irregular contact against terrigenous "greywackes" marks the base of the red mélange NE of Kairawaru Bay and within the "greywacke" horst E of Orua Bay (Fig. 1). Mélange and "greywackes" are not mixed, and the latter were apparently little disturbed by the emplacement of the disrupted material above. The red mélange association is overlain by green spilites.

Cherts. Well-bedded alternating siliceous argillites and radiolarian cherts, mostly 5-30 cm thick, with extensive manganese mineralisation (Maehl 1970; Sekula 1972) are prominent around Kairawaru Bay and at Marble Bay. The chert association overlies green spilites and is overlain by green argillites or "greywacke"-type clastic sequences. NE of Kairawaru Bay red cherts are intruded by a sill or dike of hematitic spilite. Maehl (1970) noted spherulitic cherts interbedded with spilites at Kairawaru Bay.

Green Argillites. A sequence of well- and thinly-bedded alternating green argillites and grey-green cherts overlies the red cherts at Kairawaru Bay. Maehl reported glass shards in these rocks, justifying interpretation of at least some as tuffaceous; however, some may be recrystallised deep-sea oozes. Radiolaria are not evident in the cherts. Isolated layers of green argillite also occur in lower parts of the "greywacke" association.

Terrigenous Clastics ("Greywackes"). Well-bedded fine-grained quartzo-feldspathic sandstones displaying indistinct grading and bioturbation, alternate with dark grey to black argillites. Individual beds range from 5-40 cm thick and have some characteristics of distal flysch. Petrographic evidence suggests a plutonic and metamorphic provenance, supplemented by acid and intermediate volcanic sources (Maehl 1970). Other than an unidentified microfossil found at Te Anina Point (see inset of Fig. 1A), no fossils have been found in these "greywackes".

Limestones and Marbles. The calcareous rocks vary from massive cream to grey, fossiliferous, veined, and stylolitic limestones, with increasing recrystallisation, to reddish rocks with anastomosing sutures and tightly interlocking wholly crystalline texture, which have been termed marbles. They occur as lensoid bodies up to several metres across and more laterally extensive sheets, generally conformably intercalated

with the green spilites.

Apparently unconformable relationships are due to subsequent folding and faulting. Some marbles associated with spilites are silicified along bedding and cross-cutting fractures, and contacts with enclosing volcanics are suture-like, highly irregular on a small scale. Limestone clasts are also found in the red mélange association. The very large, often richly fossiliferous, erratic blocks of limestone at Orua Bay are probably derived from the mélange association.

The limestones contain a varied fossil fauna, including fusulinid foraminifera (Hornibrook 1951) and reef-building corals (Leed 1956). Other taxa include small calcareous and agglutinated foraminifera, crinoids, bryozoa, ostracods, and a brachiopod (Hay 1960). Recrystallisation has totally or partly destroyed this skeletal detritus in many thin sections examined. Many freshly-broken limestone surfaces have faintly blotchy or mottled texture at a scale reminiscent of corals. In two an abrupt lateral textural transition could be distinguished from obvious corallites through a clearly mottled to a non-mottled lithology, suggesting reef-building corals were originally far more extensive than is now evident.

Limestones (and marbles) at the Whangaroa localities are restricted to green spilite and red mélange. No trace of carbonate has been found in the adjacent "greywackes". Conversely, the limestones examined contain no terrigenous detrital material except volcanic fragments identical to the surrounding spilites.

The fossils are late Permian (Hornibrook 1951; Leed 1956; Waterhouse 1970; Speden 1976). The fauna is typically Tethyan and indicative of a shallow, warm clear-water marine environment in which reef-building corals (Leed 1956) and fusulinids (Gobbett 1967) thrived. Ross (in Speden 1976) suggested that the assemblage at Kairawaru Bay belongs to a fore-reef facies, whilst that of Wairoa Bay, Bay of Islands, belongs possibly to a back-reef facies. Blocks of brecciated limestone at the E side of Orua Bay also suggest fore-reef facies.

STRUCTURE

Strata and volcanic layers trend mainly NW and dip mostly SW. Younging directions at Te Anina Point indicate that the sequence is not inverted. Two fold axis directions, almost at right angles, NE and NW, can be recognised (Fig. 1A). NE-trending folds mainly verge SE and include all folds in the chert sequences, whereas NW-trending folds all verge NE. Folds of both directions influence the shape of some of the limestone lenses.

Juxtaposition of red mélange onto "greywackes" NE of Kairawaru Bay (Figs.1A, 1B) is an early structure. The contact has been reactivated by a dextral shear and is cut by NE-trending fractures. Fabric within the red mélange indicates low-angle shearing on E-W planes, with transport of higher parts southwards, postdated by similarly E-W-striking low-angle extension faults downthrown to N. Some shears have well-developed breccia zones.

Red mélange is traversed by calcareous mylonite zones containing rotated fragments of hematitic spilite. Mylonite on the SW side of Kairawaru Bay has been folded dextrally in a second phase, and sheared and veined by sinistral movement in a third.

Epidote fibre striations occupy a NE-SW girdle pattern normal to the NW-trending folds, with which they appear to be genetically linked.

Steep faults are mostly normal and strike NW, N-S, and NE. Both NW-SE and E-W extension directions have been obtained from conjugate couples. N-S and NE-trending faults show evidence for an additional earlier(?) phase of dextral movement.

From cross-cutting relationships and general considerations the following sequence of events is tentatively proposed:

(1) Juxtaposition of red mélange on the "greywackes"; shearing of mélange with thrusting towards S; formation of calcareous mylonites; development of S-verging, NE-trending folds in all rock units.

(2) Extensional, low-angle faulting in red mélange, downthrow to N; development of dextral folds with NE-trending axes as in mylonite at Kairawaru Bay.

(3) NW-SE folding, verging to NE; formation of epidote veining and fibre striations indicating compression from NE; metamorphism (?); dextral movement on N-S and NW-SE-trending faults.

(4) Normal movement on steep faults; sinistral movement of mylonite at Kairawaru Bay; calcite veining of calcareous rocks indicates NW-SE extension, corresponding to that deduced from one set of conjugate faults.

No direct evidence relates these events to an absolute time scale. If metamorphism is placed correctly in the sequence, phases 1-3 probably predate or are contemporaneous with the Rangitata Orogeny, whilst phase 4 is later.

DISCUSSION

Of the lithological types discussed, green spilites, cherts, green argillites, and terrigenous "greywackes" occur in stratigraphic sequences with spilite at the base and "greywacke" at the top. Red mélange has been sheared over "greywacke" and is in stratigraphic or tectonic contact with over lying green spilite (Fig. 2). This sequence has been repeated by thrusting and imbrication. The earlier phase of deformation, with NE-SW fold hinges and S vergence, may indicate the direction of imbrication. Such a pattern would be similar to that recognised elsewhere in New Zealand, with evidence for plate convergence oblique to the Gondwana margin (Spörli 1978; Ward & Spörli 1979).

Alternatively, the basal contact of the red mélange near Marble Bay could be interpreted as a normal stratigraphic feature. This would call for extremely abrupt vertical and lateral facies changes from terrigenous to non-terrigenous sedimentation and for similar spasmodic interruptions to the supply of terrigenous detritus if volcanics were intercalated elsewhere in the Waipapa terrane. This seems to us paleogeographically untenable and contrary to structural and sedimentological evidence.

Repetition of a standard stratigraphic sequence fits more readily into the plate tectonic model (Spörli 1975, 1978). The volcanics and their accompanying limestones represent the very top of an oceanic crust, formed either at mid-ocean ridges or by intraplate volcanism. The oceanic material is transported towards a convergent margin, is first covered by deep-sea sediments represented by cherts and green argillites, and is finally scraped off and imbricated in the subduction zone along the margin. Volcanics would eventually come to lie with thrust contacts on "greywacke" below but could still be in stratigraphic sequence (via chert and green argillite) with the "greywacke" above.

The limestones (or marbles) were deposited in shallow water far removed from any terrigenous influence, accumulated as thin and irregular carbonate banks or reefs capping remote submarine highs of volcanic origin. Fossiliferous limestone blocks may have been incorporated into the red mélange in part by slumping and sliding from the crest down the flanks of these seamounts or guyots (Fig. 2).

According to our model, the volcanics can be stratigraphically intercalated with sediments deposited in the open ocean, especially green and red mudstones, tuffs and cherts, as on the nearby Cavalli Is, which therefore fits our proposed model despite statements to the contrary (Moore & Ramsay, in press). Intercalation with terrigenous clastics, however, will only be possible in those special cases where a spreading ridge

226

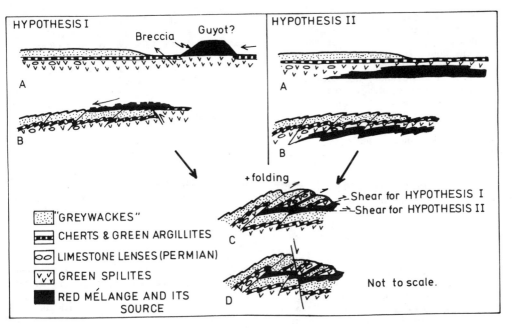

HYPOTHESIS I

Breccia Guyot?

A

B

+folding

C

→ Shear for HYPOTHESIS I
→ Shear for HYPOTHESIS II

D

Not to scale.

HYPOTHESIS II

A

B

"GREYWACKES"

CHERTS & GREEN ARGILLITES

LIMESTONE LENSES (PERMIAN)

GREEN SPILITES

RED MÉLANGE AND ITS
    SOURCE

Fig. 2--Hypothetical development of structure near Marble Bay during under-
thrusting and imbrication of uppermost oceanic crust in a subduction zone along
the Gondwana margin. I: Both top and bottom contacts of red mélange are shear
zones. II: Shear along base of red mélange, top contact stratigraphic. For
stages A and B in both hypotheses, the join between converging plates (trench?)
can be visualized to the left of the diagram; in subsequent stages the material
has been transferred to the Gondwana side. Note that direction of folding and
imbrication in stage B may be at right angles to that in stage C.

or other source of ocean-floor volcanism
lies near a continental margin, or an island
arc, or in a basin formed by back-arc
spreading, or where a large submarine fan
reaches far out into the open ocean. Thus
it is acceptable that an igneous dike or
sill cuts the chert NE of Kairawaru Bay,
but such intrusions should not occur and
have in fact not been seen within the
terrigenous clastics of the Waipapa terrane (Spörli 1978).

As a consequence of our model, the age of
the Permian fusulinid limestones cannot *a
priori* be transferred to the other lithological associations and especially not to
the terrigenous clastics. The clastics may
be much younger than Permian; for instance
Jurassic, like the clastics of the Waipapa
terrane of the Auckland area (Spörli 1978).
An analogous age difference is known from
near Hawarden in the South Island, where
limestones associated with volcanics are
Triassic and overlying terrigenous sediments are late Jurassic (Bradshaw 1972).

Due to the conveyor-belt mechanics of
plate movement, lithological units which
now overlie each other in a stratigraphic
column may have been deposited originally

hundreds of kilometres apart. The fusulinids of the Permian limestone at Marble
Bay, typically Tethyan (Hornibrook 1951;
Gobbett 1967), and the reef-building corals
(Leeds 1956) suggest deposition near the
Permian equatorial belt. On the other hand,
paleomagnetism and plate reconstruction
(Crook & Belbin 1978) show that New Zealand
was then at a relatively high latitude, although the non-fusulinid Permian marine
faunal succession has been taken to indicate
alternating cool and warm dominance consistent with glacial and interglacial climates
on nearby Gondwana (Waterhouse 1970, 1975;
Waterhouse & Bonham-Carter 1975; Stevens &
Speden 1978).

Thus, in our interpretation, the fusulinid
limestones at Marble Bay, clearly exotic to
the New Zealand Gondwana margin, are best
interpreted as remnants of Permian oceanic
crust, perhaps welded onto Gondwana as late
as the Rangitata Orogeny, having been moved
over large distances by sea-floor spreading
and after serving as "basement" for the deposition first of oceanic and then of terrigenous clastic sediments.

Isolated and sporadic exposures of fusulinid limestones have been traced some 65

227

km SE of Marble Bay (Hay 1960; Amos 1979; Moore & Ramsay, in press). The only other known New Zealand occurrences of fusulinids are in the South Island, at Benmore (Hornibrook & Khoon, in Campbell & Warren 1965), where Permian limestones, associated with siltstone, sandstone, conglomerate and volcanic rocks, lie faulted against Mesozoic plant-bearing sandstones, and in Rakaia Valley (N. de B. Hornibrook, pers. comm.). In all these instances, geological relationships are quite compatible with our model for the Marble Bay area. Possibly, too, the Upper Carboniferous marbles at Kakahu in the South Island (Hitching 1979) record even older ocean floor.

Exotic, Devonian to Permian fusulinid Tethyan limestones, comparable to those found in New Zealand, are widespread around the eastern border of the north Pacific Ocean (Thompson 1967; Danner 1977). These are considered allochthonous to the Western Cordillera, having been accreted onto the North American continent during the Mesozoic, having originated some distance away in the Paleozoic Pacific Ocean (Monger & Ross 1971; Danner 1977). Conceivably the Permian (and Carboniferous) Tethyan limestones exotic to New Zealand, together with similar rocks of NW North America, were deposited in the same general (equatorial?) region of the Paleozoic Pacific Ocean. Subsequent sea-floor spreading dispersed the once-contiguous sediments, until they became accreted onto opposite margins of the Pacific.

ACKNOWLEDGMENTS

Both authors have been supported by funds from the University of Auckland Research Committee.

REFERENCES

AMOS, L. M. 1979. Basement geology of the Whananaki-Whangaruru region, Northland. Unpublished thesis, University of Auckland.
BRADSHAW, J. D. 1972. Statigraphy and structure of the Torlesse Supergroup (Triassic-Jurassic) in the foothills of the Southern Alps near Hawarden (S60-61), Canterbury. N.Z. J. Geol. Geophys. 15: 71-87.
CAMPBELL, J. D.; G. WARREN 1965. Fossil localities of the Torlesse Group in the South Island. Trans. R. Soc. N.Z., Geol. 3: 99-137.
COOMBS, D. S.; C. A. LANDIS; R. J. NORRIS; J. M. SINTON; D. J. BORNS; D. CRAW 1976. The Dun Mountain ophiolite belt, N.Z., its tectonic setting, constitution and origin, with special reference to the southern portion. Am. J. Sci. 276: 561-603.
CROOK, K.A.W.; L. BELBIN 1978. The South west Pacific area during the last 90 m.y. J. Geol. Soc. Aust. 25: 23-40.
DANNER, W. R. 1977. Paleozoic rocks of northwest Washington and parts of British Columbia. In J. H. Stewart et al. (Eds.), Paleozoic Paleogeography of the Western United States. Soc. Econ. Paleontol. & Mineral.: Los Angeles.
GOBBETT, D. J. 1967. Palaeozoogeography of the Verbeekinidae (Permian Foraminifera In C. G. Adams, D. V. Ager (Eds.), Aspects of Tethyan Biogeography. System. Assoc. Publ. 7.
HAY, R. F. 1960. The geology of the Mangakahia subdivision. N.Z. Geol. Surv. Bull. (n.s.) 61.
HITCHING, K. D. 1979. Torlesse geology of Kakahu South Canterbury. N.Z. J. Geol. Geophys. 22: 191-197.
HORNIBROOK, N. de B. 1951. Permian fusulinid foraminifera from the North Auckland Peninsula, New Zealand. Trans. R. Soc. N.Z. 79: 319-321.
KEAR, D; R. F. HAY 1961. Sheet 1--North Cape. Geological Map of New Zealand 1 : 250 000. Government Printer: Wellington.
LEED, H. 1951. Permian reef-building corals from the North Auckland Peninsula, New Zealand. N.Z. J. Sci. Tech. B33: 126-128.
--- 1956. Paleozoic corals from New Zealand, Part 2. Permian reef-building corals from North Auckland Peninsula, New Zealand. N.Z. Geol. Surv. Paleontol. Bull. 25.
MAEHL, H.W.R. 1970. Geology of the Whangaroa-Marble Bay-Te Ngaere district, Northland. Unpublished thesis, University of Auckland.
MONGER, J.W.H.; C. A. ROSS 1971. Distribution of fusulinaceans in the western Canadian Cordillera. Canad. J. Earth. Sci. 8: 259-278.
MOORE, P. R.; W.R.H. RAMSAY, in press. Geology of the Cavalli Islands, northern New Zealand. Tane
SEKULA, J. 1972. The manganese deposits and associated rocks of Northland, New Zealand. Unpublished thesis, University of Auckland.
SPEDEN, I. G. 1976. Fossil localities in Torlesse rocks of the North Island, New Zealand. J. R. Soc. N.Z. 6: 73-91.
SPÖRLI, K. B. 1975. Waiheke and Manaia Hill Groups east Auckland. Comment. N.Z. J. Geol. Geophys. 18: 757-762.
--- 1978. Mesozoic tectonics, North Island, New Zealand. Bull. Geol. Soc. Am. 89: 415-415.
STEVENS, G. R.; I. G. SPEDEN 1978. New Zealand. In M. Moullade, A.E.M. Nairn (Eds.),

The Mesozoic: The Phanerozoic Geology of the
world II. Elsevier: Amsterdam.

THOMPSON, B. N. 1961. Sheet 2A--Whanga-
rei. Geological Map of New Zealand 1 :
250 000. Government Printer: Wellington.

THOMPSON, M. L. 1967. American fusulina-
cean faunas containing elements from other
continents. In C. Teichert; E. L. Yochel-
son (Eds.), Essays in Paleontology and
Stratigraphy. Univ. Kansas Spec. Pub. 2:
102-112.

WARD, C. M.; K. B. SPÖRLI 1978. Exceptionally large steeply plunging folds in the
Torlesse Terrane, New Zealand. J. Geol.
87: 187-193.

WATERHOUSE, J. B. 1970. The world significance of New Zealand Permian stages.
Trans. R. Soc. N.Z., Earth Sci. 7:97-109.

--- 1975. The Rangitata Orogen. Pacific
Geol. 9: 35-73.

---; G. F. BONHAM-CARTER 1975. Global
distribution and character of Permian
biomes based on brachiopod assemblages.
Canad. J. Earth Sci. 12: 1085-1146

Metallogenic provinces, mineral occurrences, and geotectonic settings in New Zealand

FRANCO PIRAJNO

Gold Mines of New Zealand Ltd., Nelson

In New Zealand, three main orogenies are recognised: early Paleozoic (Tuhua), Mesozoic (Rangitata) and Tertiary to present-day (Kaikoura). Each brought about a particular metallogenic episode. The Tuhua Orogeny introduced an epoch of predominantly Au mineralisation characterised by hydrothermal activity related to calcalkaline magmatism. The Rangitata Orogeny resulted in Mo, Ni, Cu, and Cr minerals being brought into the upper crust in events ranging from high-level granitic intrusions to ophiolite emplacement and regional metamorphism of thick sediments. During the Kaikoura Orogeny, Au, Ag, and base-metal mineralisation was generated by hydrothermal systems developed during the waning stages of felsic to intermediate volcanism above a west-dipping subduction zone. The metallogenic provinces thus formed are categorised according to type.

METALLOGENESIS IN NEW ZEALAND

Most of the present-day New Zealand micro-continent, as defined by the 2000 m isobath, is thought to be a fragment of Gondwana. The continental fragments that make up New Zealand broke away from Australia and Antarctica during a series of rifting and sea-floor spreading events between the Jurassic and early Tertiary periods. The episodes are conventionally subdivided into three main orogenies: Tuhua Orogeny (Lower Paleozoic), the Rangitata Orogeny (Permo-Carboniferous to Cretaceous), and the Kaikoura Orogeny, which continues to affect New Zealand today.

Each brought about a well-defined metallogenic epoch with characteristic mineralisation expressible as provinces and districts (Fig. 1).

THE TUHUA OROGENY

Lower Paleozoic terranes west of the Median Tectonic Line form the Western Province or Foreland recognised by Howell (1979) as the Tuhua Plate, a Gondwana segment, whose oldest basement rocks are Precambrian. West of the Alpine Fault in the Nelson-Westland district the Tuhua Plate contains thick intercalated argillite and greywacke (Greenland and Waiuta Gp); a belt of calcalkaline rocks (Devil River Volcanics) and mafic-ultramafic rocks (Cobb Intrusives) thought to represent floored magma chambers (Hunter 1977); sedimentary formations of basinal and reef facies flanking the volcanic belt. Metamorphosed equivalents of the above rock groups (metavolcanics, ultramafic schist, marble, paragneiss) occur east of the Alpine Fault in Fiordland.

A major granite batholith (Karamea batholith) extends some 250 km, N-S, if the Nelson-Westland and Fiordland districts are restored to their positions before transcurrent movement along the Alpine Fault. It includes low-soda, peraluminous potash granites (Eggers 1978) and is Devonian to Carboniferous (360-280 m.y. BP; Aronson 1968).

Mineralisation. Known metalliferous mineral deposits related to the Tuhua Orogeny are: Au, minor Sb, hydrothermal quartz lodes associated with granitic activity; Au, minor base metals, associated with carbonaceous sediments and possibly related to calcalkaline volcanism; Cu, Pb-Zn mineralisation of sedimentary association; Cu, Ni, and Cr as magmatic segregations in mafic and ultramafic rocks.

All these are found in the Nelson-Westland and Fiordland districts, but the most important economically is the gold mineralisation

The Tuhua gold (minor antimony) mineralisation, somewhat reminiscent of that in the Rhodesian greenstone belts, involves proximity to granite batholith, hydrothermal alteration of wall rocks, gold-bearing quartz veins emplaced generally in shear and crush zones near synclinal axes in the Waiuta and Greenland Gps. Minor pyrite, arsenopyrite and stibnite generally accompany the gold. Stibnite locally reaches ore concentrations.

Gold and minor base metals in carbonaceous argillites and graphitic schists close to the Devil River Volcanics may be related to island-arc volcanism and may have an exhalative-sedimentary origin.

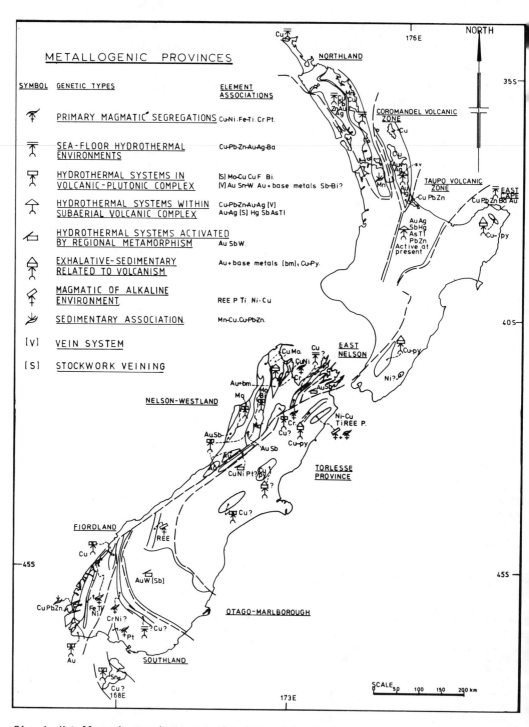

Fig. 1--Metallogenic provinces and districts of New Zealand. — — — — = boundaries of metallogenic province. ————— = boundaries of metallogenic district within a given province. (Genetic symbols are taken from the Metallogenic Map of Australia and Papua-New Guinea, R. G. Warren, BMR Canberra, 1972.)

Copper and lead/zinc are found in reef facies rocks in Fiordland ,and uneconomic amounts of copper, nickel, and chromium in the Cobb Intrusives in the Nelson-Westland area.

THE RANGITATA OROGENY

Following Howell (1979), the main Rangitata tectonic elements are, from W to E: the Hokonui Plate, which includes a calc-alkaline volcanic arc and the Dun Mountain ophiolite belt; the Caples Plate (an arc-trench system); and the Torlesse Plate, predominantly quartzo-feldspathic sediments.

The orogeny began towards the end of the Paleozoic and lasted for most of the Mesozoic era. The Stoke Magnetic Anomaly System (Hunt 1978; Christoffel 1978) shows that the calcalkaline arc and the Dun Mountain ophiolite belt are at least 1500 km long. Both have been displaced, 460 km and 330 km respectively, by the Alpine and Campbell right slip faults. The Hokonui Plate and the Caples Plate were formed by multi-stage subduction of oceanic lithosphere (Coombs *et al.* 1976). Partial melting connected with the subduction caused plutonic activity during Permian, Triassic, and Cretaceous times. A variety of plutons were emplaced, from granite batholiths and stocks, to mafic and ultramafic layered bodies. The Separation Point Granite and the Darran Complex were intruded along the boundary between the Hokonui Plate and the Tuhua Plate. Smaller stocks and plutons penetrated the Tuhua block.

A mafic-ultramafic, funnel-shaped, layered intrusion (Riwaka Complex) was emplaced in Lower Paleozoic, Tuhuan rocks. In Nelson and Westland it extends discontinuously for about 50 km, generally N-S. It continues east of the Alpine Fault in Fiordland as a belt of mafic and ultramafic rocks 80 - 30 km long between Mt Soaker in the north and Lake Hauroko in the south. Soda-rich, calc-alkaline and porphyritic granitic stocks intrude areas of the Devonian Karamea batholith and adjacent Greenland Gp sediments (Eggers 1978). These may be due to marginal re-activation of the Karamea batholith late in the Rangitata Orogeny.

The Torlesse Plate, a large prism of quartzo-feldspathic sediments with minor intercalated spilitic volcanics, collided with the eastern side of the Caples Plate as a result of subduction and plate convergence. The plate boundary may be represented by the Pounamu Ultramafic rocks now shredded and deformed into small lenses and pods by the Alpine transcurrent movement. The Torlesse Plate was penetrated during the late Cretaceous by alkaline volcano-plutonic complexes, forming flood-type basalt, and ring and layered complexes.

Mineralisation: Tuhua, Hokonui, and Caples Plates. In the Riwaka Complex, Nelson-Westland district, sulphides occur in the middle and late magmatic pulses (Bates 1978). The usual sulphide assemblage is pyrrhotite-chalcopyrite with minor pentlandite and pyrite. Vanadiferous magnetite and ilmenite-rich pods are found in the N sectors of the Mt Soaker and Mt George belt in Fiordland (Williams 1974).

Pods of chromite in the basal dunite and harzburgite of the Dun Mountain ophiolite belt possibly originated as magmatic segregations below a spreading ridge (Johnston 1977).

In D'Urville Island some copper is associated with pillow lavas. In the Northland peninsula and adjacent islands small pockets and lenses of manganese and copper in Permian and Jurassic spilitic rocks are thought to be sea-floor hydrothermal deposits (Stonaway *et al.* 1978).

During Cretaceous times molybdenum-bearing soda-rich, calcalkaline porphyry stocks (Eggers 1978) intruded the Karamea (Devonian) granites, mostly along their margins. In the Bald Hill Prospect the stocks are I-type granites (Chappell & White 1974), whereas the Karamea granites belong to the S-type. The same association of porphyry mineralisation and I-type granite is found in Eastern Australia (Chappell & White 1974; Eggers 1978). The porphyry in the granite belts of Nelson-Westland and Fiordland is mainly plutonic and hypabyssal. It has predominant molybdenum mineralisation with minor copper and bismuth and occurs in brecciated and hornfelsed Greenland Gp sediments near granite contacts or in granite. Sulphides are associated with quartz veins and stockwork type veinlets through the porphyry intrusives and shattered roof rocks. The main assemblage is molybdenite-pyrite-chalcopyrite and, locally, bismuth (Eggers & Adams 1979). Hydrothermal alteration zones are usually well developed.

Most of the known porphyry-molybdenum in the Nelson-Westland district is near the margins of polygonal features or at their intersections (Eggers 1978). These structures, which are visible on LANDSAT photographs, range in diameter from 8 - 30 km. They appear to be structurally controlled in a predominantly block-faulting environment. Evidence gathered from Reefton (Westland) suggests that they may represent magmatic columns. Following the subsidence of the magma body, stresses set up in the overlying rocks effected polygonal block faulting. Nearly contemporary high-level late differentiates were emplaced, taking advantage of weakness lines established by the collapsing superstructure (pers. obs.).

Torlesse Plate. Hydrothermal vein type
Au-Sb and Au-W with minor sulphides typically occur in the Haast Schist Gp. The mi-
neralised quartz lodes are thought to be derived from hydrothermal systems activated by
regional metamorphism (Henley *et al.* 1976;
Pirajno 1979a).

Massive and/or disseminated sulphides in
metavolcanic bands in the Haast Schist are
exhalative-volcanic on geological and geochemical evidence, slightly modified by later
metamorphism. Pyrrhotite, chalcopyrite, and
some sphalerite are the main sulphide
species (Wright 1966; Henley 1975). Similar
sulphide deposits in prehnite-pumpellyite
grade Torlesse spilites are represented by
the pyrite-chalcopyrite pair (Pirajno 1979b).

Close to and almost parallel with the Alpine Fault, the Pounamu Ultramafics may represent the remnants of a suture line between the Caples Plate in the west and the
Torlesse Plate in the east. They contain
sulphides, gold, and platinum. It is a matter of debate whether this mineralisation was
primary magmatic segregation later modified
by the intense dynamo-metamorphism along the
fault, or whether the metals were concentrated during metamorphism.

Cretaceous layered plutons and ring-type
complexes in Torlesse strata contain disseminated Cu-Ni sulphide mineralisation. In the
Blue Mountain ring complex mineralisation
consists of a primary assemblage made up of
pyrrhotite-chalcopyrite-pentlandite emplaced
with the magma as an immiscible phase, concentrated by gravity (McKenzie 1979; pers.
obs.).

THE KAIKOURA OROGENY

The Kaikoura Orogeny gave New Zealand its
present configuration.

The Northland peninsula is thought to represent a sialic fragment broken off Eastern
Australia during the opening of the Tasman
Sea (Brothers 1972). On this block are
tectonically emplaced allochthonous ophiolitic sequences and sedimentary formations.
Calcalkaline volcanic products also penetrated, in response to magma generation above
the west-dipping subduction of the Pacific
Plate under the Indo-Australian Plate. Ballance (1976) distinguished "six separate
arcs". He sees the progressive development
of northern New Zealand calcalkaline magmatism as due to a gradual steepening of the
Benioff zone. His six arcs can be grouped
into a Northland arc, Coromandel Volcanic
Zone, and Taupo Volcanic Zone. The last two
are thought to be related to areas of tensional stress with possible collapse of sedimentary blocks into a zone of partial melting (Ewart & Stipp 1968).

Mineralisation. In Northland and East C
sulphide mineralisation is found at a numb
of localities in allochthonous ophiolitic
type terranes (Tangihua and Matakaoa Volca
ics). It is always associated with argill
ceous sediments and radiolarian chert, int
calated with mafic volcanics of tholeiitic
affinity. The chief ore minerals are pyri
chalcopyrite, galena and sphalerite, at so
places contained in a barite gangue, and
some gold. Colloidal fabrics, including
sulphides with spherical forms, were seen
one locality. The ore minerals are consid
ered part of the sediment between submarin
lava flows and may have been formed by hyd
thermal activity on the sea floor (Pirajno
1979c).

One of the most interesting old mining
districts in New Zealand is that of the
Coromandel Volcanic Zone. Here are severa
mineral deposits whose mechanism of origin
is undisputedly connected with hydrotherma
circulation in subaerial volcanic complexe
of felsic to intermediate composition.

Mineralisation comprises hydrothermal
veins, either predominantly Au-Ag or with
minor Au and Ag; sinterous and quartz vein
stockwork Au-Ag deposits; or disseminated
sulphides (Cu, Mo) related to the subvolca
nic porphyry type.

These mineral localities mark veins clos
to the eastern edge of the Thames Graben
(Hauraki Fault). Most of these veins occu
in andesitic rocks of the Beesons Island G
practically all are associated with intens
alteration, and all appear to trend N or N
parallel with the regional structure. Pyrite is abundant, tellurides and selenides
localised. The source of sulphur is believed to be partly magma and partly sulpha
in the basement greywackes. Metals were
leached by the circulating brines, mostly
from basement rocks. The veins were formed
by open space filling and locally by replacement.

Studies of volatile and precious-metal
zoning in the Broadland Geothermal Field
(Taupo Zone) conducted by Ewers and Keays
(1977) indicate that hydrothermal solutions
in active geothermal systems at Taupo are i
the process of "making ores". Dilute hot
water gives precipitates high in Au, Ag, As
and Tl. The field is the present-day equivalent of the Coromandel Volcanic Zone.

REFERENCES

ARONSON, J. L. 1968. Regional geochrono-
logy of New Zealand. Geochim. Cosmochim.
Acta 32: 660-697.
BALLANCE, P. F. 1976. Evolution of the
Upper Cenozoic magmatic arc and plate bound

ary in northern New Zealand. Earth Planet. Sci. Lett. 28: 356-370.

BATES, T. E. 1978. Riwaka Complex. Interim report, G.M.N.Z. Ltd., Nelson.

BROTHERS, R. N. 1972. Kaikoura Orogeny in Northland, New Zealand. N.Z. J. Geol. Geophys. 17(1): 1-18.

CHAPPELL, B. W.; A.J.R. WHITE 1974. Two contrasting granite types. Pacific Geol. 8: 173-174.

CHRISTOFFEL, D. A. 1978. Interpretation of magnetic anomalies acorss the Campbell Plateau, south of New Zealand. Bull. Aust. Soc. Explor. Geophys. 9(3): 143-145.

COOMBS, D. S.; C. A. LANDIS; R. J. NORRIS; J. M. SINTON; D. J. BORNS; D. CRAW 1976. The Dun Mountain ophiolite belt, New Zealand, its tectonic setting, constitution and origin, with special reference to the southern portion. Am. J. Sci. 276: 561-603.

EGGERS, A. J. 1978. Geology and mineralisation of the Bald Hill molybdenum occurrence, Buller district, west Nelson, New Zealand. Unpublished thesis, Victoria University of Wellington.

---; C. J. ADAMS 1979. Potassium-argon ages of molybdenum mineralisation and associated granites at Bald Hill and correlation with other molybdenum occurrences in the South Island, New Zealand. Econ. Geol. 74: 628-637.

EWART, A.; J. J. STIPP 1968. Petrogenesis of the volcanic rocks of the Central North Island, New Zealand, as indicated by a study of $^{87}Sr/^{86}Sr$ ratios, and Sr, Rb, K, U, and Th abundances. Geochim. Cosmochim. Acta 32: 699-736,

EWERS, G. E.; R. R. KEAYS 1977. Volatile and precious metal zoning in the Broadlands Geothermal Field, New Zealand. Econ. Geol. 72: 1337-1355.

HENLEY, R. W. 1975. Metamorphism of the Moke Creek Lode, Otago, New Zealand. N.Z. J. Geol. Geophys. 18(2): 229-239.

---; R. J. NORRIS; C. J. PATERSON 1976. Multistage ore genesis in the New Zealand geosyncline. A history of post-metamorphic lode emplacement. Mineral. Dep. 11: 180-196.

HOWELL, D. G. 1979. Mesozoic microplates of New Zealand. Geol. Soc. N.Z. Newslett. 47: 24-26.

HUNT, T. 1978. Stoke magnetic anomaly system. N.Z. J. Geol. Geophys. 21(5): 595-606.

HUNTER, H. W. 1977. Geology of the Cobb Intrusives, Takaka Valley, NW Nelson, New Zealand. N.Z. J. Geol. Geophys. 20(3): 469-502.

JOHNSTON, M. R. 1977. Stratigraphy and mineralisation of the Upper Paleozoic rocks, east Nelson. Australas. Inst. Mining Metall., Ann. conference papers, Nelson.

McKENZIE, F. I. 1979. Sulphide petrology of the Blue Mountain complex. Unpublished honours project, Victoria University of Wellington.

PIRAJNO, F. 1979a. Geology, geochemistry, and mineralisation of the Endeavour Inlet antimony-gold prospect, Marlborough Sounds, New Zealand. N.Z. J. Geol. Geophys. 22(2): 227-237.

--- 1979b. Geology, geochemistry, and mineralisation of a spilite-keratophyre association in Cretaceous flysch, Tapuaeroa Valley and Te Kumi area, East Cape, New Zealand. N.Z. J. Geol. Geophys. 22(3): 307-328.

--- 1979c. Sub sea-floor mineralisation in rocks of the Matakaoa Volcanics around Lottin Point, East Cape, New Zealand. N.Z. J. Geol. Geophys.

STONAWAY, K. J.; H. W. KOBE; J. SEKULA 1978. Manganese deposits and the associated rocks of Northland and Auckland, New Zealand. N.Z. J. Geol. Geophys. 21(1): 21-33.

WILLIAMS, G. J. 1974. Economic geology of New Zealand. Australas. Inst. Mining Metall. Monog. Ser. 4. Parkville.

WRIGHT, J. B. 1966. Studies on the pyrrhotite and paragenesis of the Moke Creek sulphide lode, Wakatipu district. N.Z. J. Geol. Geophys. 9: 301-322.

235

Pre-Andean orogenies of southern South America in the context of Gondwana

HUBERT MILLER
Westfälische Wilhelms-Universität, Münster, Germany

In southern South America, Precambrian rocks are known with some certainty only in the early consolidated regions of Uruguay and the Sierra de Tandil, in the North Patagonian massif and in a few parts of the Sierras Pampeanas. Metamorphism and folding of the Sierras of northwestern Argentina occurred largely during the Middle Cambrian (Pampean cycle). The Caledonian orogenic period (Famatinian cycle) led to some consolidation in NW Argentina. In other regions it is weakly developed or has largely been overprinted by later movements. A strong Hercynian orogeny influenced part of the Andean region and the Chilean coastal chain. At the end of the Permian the Gondwanide orogeny affected large parts of the Andes.

Common features become now evident between the Argentine Lower Paleozoic orogenies, the Ross orogeny and the Australian Delamerian and Tyennan orogenies. The SW decrease in age of the South American pre-Andean orogenies suggests a revision of the relative positions of South America and West Antarctica within Gondwana.

PRE-ANDEAN OROGENIES OF SOUTHERN SOUTH AMERICA

Introduction. Until some ten years ago most metamorphic rocks of the southern Andes were taken for Precambrian, so few attempts were made to consider the relation of South American pre-Andean rocks to the Paleozoic mobile belts of Australia and Antarctica. Radiometric age determinations and fossils led to a revision of the ages of sedimentation, folding, metamorphism and magmatism of the "crystalline basement" of the southern Andes. The difficulty of interpretation of radiometric dates in geologically and petrologically poorly known regions and the scarcity of fossils cause serious problems for the reconstruction of ancient geosynclinal troughs and orogenic belts. Above all the lack of structural work on the internal fabrics of pre-Andean rocks prevents reliable reconstructions of orogenes. Nevertheless, efforts have been made to get a more detailed concept of the Paleozoic orogenic history of the southern Andes (Vicente 1975; Miller 1973, 1979b; Aceñolaza & Toselli 1976). Though the times of orogenic activity are rather well known, their regional distributions (Fig.1) will certainly be revised by future research.

Precambrian. The basement rocks of Uruguay and the Sierra de Tandil in Buenos Aires province are shown to be Precambrian by their radiometric ages and their relations to the Brazilian shield. In the Northern Patagonian Massif most granites have been shown to be Upper Paleozoic, but the various high-grade metamorphic rocks surrounding them are partly Precambrian (Miller 1976). In Central and West Argentina (Sierra Pampeanas) only a few radiometric ages (Sierra Chica de Córdoba) demonstrate Precambrian metamorphism. The "Upper Precambrian" dates of NW Argentina are insufficient to demonstrate the existence of Precambrian metamorphism and tectonism.

On the other side of the Cordillera, in Northern Chile, recently Precambrian metamorphism has been dated at 1,000 m.y. by Cordani *et al.* (pers. comm.) in schists of the Belén Fm near Arica. Possibly this complex belongs to the Arequipa massif of Perú and continues to the south as far as the Mejillones Peninsula.

The formerly so-called "crystalline basement" of most of Chile is without doubt mostly Paleozoic and in certain parts possibly Triassic. Thus the Brazilian shield including its youngest orogene (Brazilian Cycle) ends far from the modern Andes.

Pampean cycle. In NW Argentina Aceñolaza & Toselli (1976) defined the "Pampean cycle", with sedimentation of clastic and rarely carbonate rocks in the latest Precambrian and early Cambrian and ending in the Middle Cambrian with polyphase folding, metamorphism and some magmatic events. Lithological and structural comparisons (Aceñolaza *et al.* 1978) illustrate the importance of that Cambrian orogenic event. Structures trend generally about NE-SW, but detailed structural data are scarce (Miller 1979c). The effects of this orogeny are identifiable from the Bolivian border to Mendoza. The metamorphism varies from very low grade

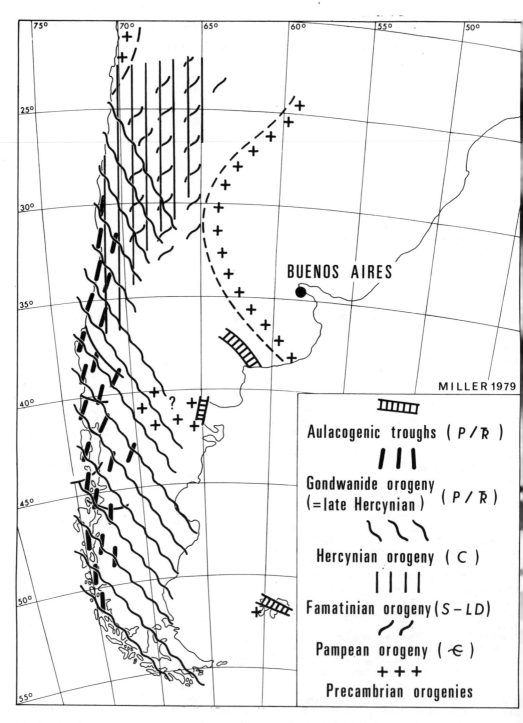

Fig.1--Overlapping and eastward shifting of pre-Andean orogenic activity in the southern Andes. The signs are adjusted as well as possible to the mean fold directions.

(Puncoviscana Fm) in the north to medium and high grade in most of the Sierras Pampeanas. The Rb/Sr age of basement cores east of the Falkland Islands (Beckinsale *et al.* 1976) fits well in the Pampean cycle.

Famatinian cycle. Upper Cambrian to Devonian sedimentary rocks are widespread in NW Argentina, disconformably overlying the Pampean orogene. Metamorphism is absent or weak. There are abundant igneous events, stratigraphic breaks and weak deformation during the Silurian, but sedimentation, partly in molasse facies, and deformation lasted until the Devonian. In Chile the effects of the Famatinian orogeny are generally overprinted by the strong Hercynian orogeny, but there are signs of pre-Devonian deformation and metamorphism at many places, probably due to the Famatinian orogeny (Miller 1976, 1979a). The folds of the Famatinian orogene strike generally N-S in NW Argentina and Chile. In the south of Chile E-W trends prevail (Fig.1). Effects of the Famatinian orogeny are also visible at the Atlantic side in the Sierra de la Ventana (Kilmurray 1975) and Sierra Grande regions (Braitsch 1965).

Hercynian cycle. In NW Argentina the pre-Andean orogenic history ends with the Silurian and Devonian folding of the Famatinian cycle. In West and South Argentina and in Chile from Antofagasta to the south, Devonian sediments of great thickness are widespread. They were folded and metamorphosed during the early Carboniferous. Hercynian magmatism is documented here and as far away as the Atlantic coast in Patagonia and the Sierra de Córdoba in Central Argentina. The Hercynian orogeny is further documented by typical flysch and molasse facies of the Sierra de Languiñeo in Central Patagonia (Miller 1976) and the Paganzo Group molasse in NW Argentina. Folding occurred predominantly along NW-SE striking axes. This typifies only the Carboniferous orogeny and not the older foldings or the more recent Gondwanide orogeny.

Gondwanide cycle. This term was introduced by du Toit (1927: 105), who said "...at the close of the Permian a lengthy arc of compression (the Gondwanides) came into being, extending east-west through the south of the Cape, the Sierra de la Ventana,...". To avoid errors only orogenic events of approximately Late Permian to Early Triassic age should be named "Gondwanide". Events of this age are widespread in Chile and southern Argentina (Hervé 1976; Thiele *et al.* 1976; Hervé *et al.* 1976; Miller 1976, 1979a). I named

them hitherto "tardivariscic", but because of their autonomous development they deserve a proper name: "Gondwanide" in the meaning of du Toit. These youngest pre-Andean folds strike about NNE-SSW.

Paleozoic orogenic and aulacogenic belts. Figure 1 shows clearly that the four pre-Andean phanerozoic orogenic cycles are not only widely overlapping but also become younger towards the SW, i.e., from the Brazilian shield towards the Pacific, and progressive consolidation from the shield area to the Pacific Ocean is evident.

Only the Atlantic side (Sierra de la Ventana, Sierra Grande) does not fit. There, post-Ordovician rocks are not metamorphic, in spite of the thickness of the sedimentary pile. The sediments are mostly well-sorted quartzites; greywackes and impure pelites are not as common as in the west. Folding is strong, but the trends of the fold axes do not match those of the Pacific side although the times of folding coincided with those of the Pacific side. Harrington (1970) was the first to consider the Sierra de la Ventana as an "aulacogenic" chain, and I agree with him (Miller 1979b).

GONDWANA RECONSTRUCTION

NW Africa is oriented to Laurasia and will not be considered here. The only Paleozoic fold belt of Gondwana-Africa is the Cape belt, a long persistent Paleozoic sedimentary trough folded at the end of the Permian (du Toit 1927). The most often-mentioned link to South America is the continuation of the Cape Mountains into the Sierra de la Ventana.

Peninsular India has no Paleozoic orogenes. The reconstruction of the India--Sri Lanka--Antarctica fit is made here after Dietz *et al.* (1972).

In Eastern Australia during the Paleozoic a number of orogenies took place. They overlap frequently like those in southern South America and on the whole they are younger towards the Pacific (Plumb 1979).

In the Transantarctic Mountains of Antarctica three overlapping orogenes are known: the Late Precambrian Beardmore orogene, the late Cambrian-early Ordovician Ross orogene and the Silurian-Lower Devonian Borchgrevink orogene. The relations of Antarctica to Australia are well known and generally interpreted as linking Eastern Australia with the Ross and Borchgrevink orogene (Craddock 1972). In the Weddell Sea area the Ellsworth Mountains are considered to have been folded mainly at the end of the Permian, with

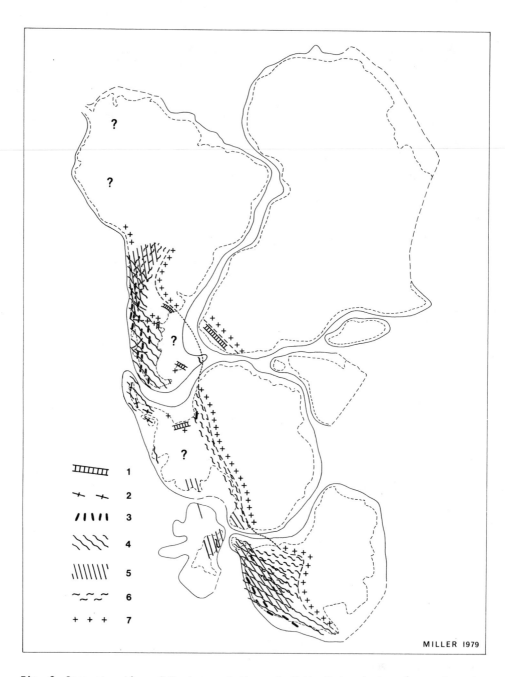

Fig. 2--Reconstruction of Gondwana at the end of the Triassic based on Paleozoic orogenic belts of southern South America, Antarctica and Australia. 1: Aulacogenic troughs. Last folding at Permian/Triassic boundary. 2: Late Triassic to Lower Jurassic orogenies. 3: Late Permian to Lower Triassic ("Gondwanide") orogenies. 4: Carboniferous ("Hercynian") orogenies. 5: Ordovician to Devonian ("Caledonian") orogenies. 6: Cambrian orogenies. 7: Precambrian rocks.

folding continuing possibly until the end of the Triassic. A corresponding deformation ("Weddell orogeny") took place in the Pensacola Mountains, overprinting older structures (Ford 1972a). In the Antarctic Peninsula assumed Precambrian and Lower Paleozoic rocks and orogenies are very poorly documented. The Upper Paleozoic(?) Trinity Peninsula Series continues up to the Triassic (Thomson 1977). Its metamorphism and folding at least in parts occurred between latest Triassic and earliest Jurassic, accompanied by magmatic events (Gledhill *et al.*1977).

There are three ways to reconstruct pre-Andean Gondwana relations of Antarctica, Africa and South America: many authors place the Antarctic Peninsula east of the southern tip of South America to get a good fit between the "Gondwanide chains" of the Sierra de la Ventana, the Cape Mountains and the Ellsworth Mountains (A). In this model the Ross orogene ends abruptly at the east coast of Africa. A more or less continuous transition of southernmost South America to the Antarctic Peninsula is preferred by many authors working in the Andean orogene (B). A position of the Peninsula west of Tierra del Fuego is preferred by many paleomagnetic workers (C).

The remarkable analogy of the Paleozoic orogenic history of southern South America to those of Australia and the Transantarctic Mountains makes the last-mentioned model (C; Fig.2) appear to be the most satisfactory (see also Ford 1972b). Above all, the Late Triassic age of the last pre-Andean orogeny in the Antarctic Peninsula fits very well with the general decrease of the ages of pre-Andean orogenies from the Brazilian shield towards the SW, assuming this position of the Peninsula until the end of the Triassic. Certainly geosynclines and orogenic belts are not necessarily obliged to go around the earth without any interruption and local differences. But if we admit that the shield-bounding Paleozoic orogenes of Gondwana have a unified relation to the Pacific-Gondwana border, the abrupt termination of the Australia-Transantarctic orogene complex (first-named model A) is difficult to accept. It is true that in the model preferred here (C), the Ellsworth Mountains cannot be linked directly with the famous Sierra de la Ventana-Cape Mountains chain, but they can be reasonably well correlated (Fig. 2). All these short fold belts show the same peculiarities: very thick sedimentary sequences, little magmatism, simple, but strong folding, culminating at the end of the Permian. All are probably surrounded by Precambrian rocks.

Therefrom I conclude that these "aulacogenic" chains had had a separate history in the development of Gondwana, different from the "orogenic" belts of the Pacific coast. Interpreted as aulacogenes they constitute subparallel striking structures related to the later South Atlantic-Weddell Sea region. In Mesozoic and Cenozoic times such aulacogenes are widespread in the Argentine Atlantic border (Salado, Colorado and San Jorge basins).

In the late Jurassic and early Cretaceous the Magallanes and South Georgia "marginal basins" or a "Proto-Weddell-Sea" evolved. Suárez & Pettigrew (1976) reconstructed a continuous island arc system from the southernmost Andes to the Antarctic Peninsula. But the Antarctic Peninsula is geologically not the southern continuation of the Patagonian Cordillera. It was formerly the western borderland of the Early Cretaceous ocean floor of SW Patagonia, shifted away from Patagonia during the Mid Cretaceous towards the SE along a basin-parallel transform fault. So the apparent differences between the southernmost Andes and the Antarctic Peninsula (Miller 1977), above all the lack of a flysch sequence east of the Peninsula and of a "tectonic arc" west of the Magallanes Andes are easily understood: a former complete orogene has been split up along an orogene-parallel sinistral transform fault.

ACKNOWLEDGMENTS

I am indebted to my colleagues in South America, above all to F. Hervé (Santiago) and F.G. Aceñolaza and A. Toselli (Tucumán) for many discussions and for oral communications of unpublished data. The recent work in Argentina was conducted within the convention of Lower Paleozoic studies operated by the universities of Tucumán and Münster. The work was supported by the Deutsche Forschungsgemeinschaft (DFG), the Deutscher Akademischer Austauschdienst (DAAD), the univerities of Santiago, Valdivia and Tucumán, and the Geological Surveys of Chile and Argentina.

REFERENCES

ACEÑOLAZA, F.G.; H. MILLER; A. TOSELLI 1978. Das Altpaläozoikum der Anden Nordwest-Argentiniens und benachbarter Gebiete. Muenstersche Forsch. Geol. Palaeontol. 44/45: 189-204.
ACEÑOLAZA, F.G., A. TOSELLI 1976. Consideraciones estratigráficas y tectónicas sobre el Paleozoico inferior del Noroeste Argentino. Mem. Seg. Congr. Latinoamericano Geol., Caracas 1973, 2: 755-764.

BARKER, P.F.; D.H. GRIFFITHS 1977. Towards a more certain reconstruction of Gondwanaland. Phil. Trans. R. Soc. London B 279: 143-159.

BECKINSALE, R.D.; J. TARNEY; D.P.F. DARBYSHIRE; M.J. HUMM 1976. Rb/Sr and K-Ar age determinations on samples of the Falkland Plateau basement at site 330, DSDP. In P.F. Barker *et al.* (Eds.), Initial reports DSDP 36: 923-927.

BRAITSCH, O. 1965. Das Paläozoikum von Sierra Grande (Prov. Río Negro, Argentinien) und die altkaledonische Faltung im östlichen Andenvorland. Geol. Runds. 54: 698-714.

DALZIEL, I.W.D.; R.H. DOTT jr.; R.D. WINN; R.L. BRUHN 1975. Tectonic relations of South Georgia Island to the southernmost Andes. Geol. Soc. Am. Bull. 86: 1034-1040.

DIETZ, R.S.; J.C. HOLDEN; W.P. SPROLL 1972. Antarctica and continental drift. In R. Adie (Ed.), Antarctic Geology and Geophysics. Oslo: Universitetsforlaget. Pp. 837-842.

DU TOIT, A.L. 1927. A geological comparison of South America with South Africa. Publ. Carnegie Inst. 381: 158 pp., 16 tables.

FORD, A.B. 1972a. Weddell orogeny -- Latest Permian to Early Mesozoic deformation at the Weddell Sea margin of the Transantarctic Mountains. In R. Adie (Ed.), Antarctic Geology and Geophysics. Oslo: Universitetsforlaget. Pp. 419-425.

---1972b. Fit of Gondwana continents -- drift reconstruction from the Antarctic continental viewpoint. 24 th Int. Geol. Cong., sect. 3: 113-121.

GLEDHILL, A.; D.C. REX; P.W.G. TANNER 1977. Rb-Sr and K-Ar geochronology of rocks from the Antarctic Peninsula between Anvers Island and Marguerite Bay. III. Symp. Antarct. Geol. Geophys. Madison, preprint, 22 pp.

HERVÉ, F. 1976. Superimposed folding and metamorphism in the Laraquete - Colcura area (metamorphic basement of Central Chile). Muenstersche Forsch. Geol. Palaeontol. 38/39: 99-110.

---; R. THIELE; M.A. PARADA 1976. Observaciones geológicas en el Triásico de Chile Central entre las latitudes 35°30' y 40°00' sur. I. Congr. Geol. Chileno 1: A 297-313.

KILMURRAY, J.O. 1975. Las Sierras Australes de la provincia de Buenos Aires. Las fases de deformación y nueva interpretación estratigráfica. Rev. Asoc. Geol. Argentina 30: 331-348.

MILLER, H. 1973. Características estructurales del basamento geológico chileno. Actas V. Congr. Geol. Argentino 4: 101-115.

--- 1976. El basamento de la provincia de Aysén (Chile) y sus correlaciones con las rocas premesozoicas de la Patagonia Argentina. Actas VI. Congr. Geol. Argentino 1: 125-141.

---1977. Geological comparison between the Antarctic Peninsula and southern Sou[] America. III. Symp. Antarct. Geol. Geophys. Madison, preprint, 18 pp.

---1979a. Das Grundgebirge der Anden i[] Chonos-Archipel, Region Aisén, Chile. Geol. Runds. 68: 428-456.

---1979b. Orógenos pre-Mesozoicos en e[] cono Sur de Sudamérica (Argentina, Chile[] Mem. III. Congr. Latinoamericano Geol. Trinidad. In press.

---1979c. Las características estructurales de la Fm. Puncoviscana (Cámbrico inferior, Noroeste Argentino). -- Un primer intento. Bol. Acad. Nac. Cien. Córdoba. In press.

PLUMB, K.A. 1979. The tectonic evoluti[] of Australia. Earth Sci. Rev. 14: 205-[]

SUÁREZ, M.; T.H. PETTIGREW 1976. Upper Mesozoic island-arc-back-arc system in t[] southern Andes and South Georgia. Geol. Mag. 113: 305-328.

THIELE, R.; F. HERVÉ; M.A. PARADA 1976. Bosquejo geológico de la Isla Huapi, Lag[] Ranco, provincia de Valdivia: Contribuci[] al conocimiento de la Formación Panguipu[] (Chile). Actas I. Congr. Geol. Chilen[] 1: A 115-136.

THOMSON, M.R.A. 1977. Mesozoic palaeoge[] graphy of Western Antarctica. III. Sym[] Antarct. Geol. Geophys. Madison, preprint 13 pp.

VICENTE, J.C. 1975. Essai d'organisati[] paléogéographique et structurale du Paléozoique des Andes Méridionales. Ge[] Runds. 64: 343-394.

Geology of the Ruppert Coast, Marie Byrd Land, Antarctica

K. B. SPÖRLI
University of Auckland, New Zealand

C. CRADDOCK
University of Wisconsin, USA

There are 5 groups of rocks in the Ruppert Coast area (listed by probable age): Layered metagabbros; foliated felsic plutonic rocks trending E to NE; calcalkaline volcanic rocks metamorphosed to greenschist facies, trending NW to NNW; an "epi-zone" plutonic rock suite of alkali syenite, alkali granite, quartz monzonite, and granite and minor garnet-two mica granite; and predominantly E-W-striking mafic and felsic dikes, associated with the plutonic rocks.

Except for abundant Cretaceous plutonic rocks, the Ruppert Coast differs markedly from the adjacent Ford Ranges and Rockefeller Mountains. The layered gabbros and mafic gneisses may have their equivalent in some of the basement rocks of Fiordland, New Zealand. The intrusive rocks are part of the circum-Pacific plutonic suite. Orientation of the dikes may indicate the direction of incipient rifting of Gondwana during separation of the Campbell Plateau and New Zealand from Marie Byrd Land.

INTRODUCTION

The Ruppert Coast area lies between 74°30'S and 76°30'S and 136°W and 143°W (Fig. 1). The purpose of this paper is to supplement information published on the area by Spörli and Craddock (1967, 1968), Wade (1969), Lopatin and Polkov (1976), and Metcalfe *et al*. (1978). The detailed maps presented are based on observations made during the 1966-67 U.S. helicopter-supported Marie Byrd Land survey.

OUTLINE OF GEOLOGY

Layered Gabbros. These are restricted to the NE part of the area (Fig. 1) and mainly consist of plagioclase, clinopyroxene, olivine, opaque minerals, and subophitic kaersutite and titaniferous biotite. Textures suggest that they are basic cumulates (Wager *et al*. 1960). The layering is due to variation in the modal proportions of mafic minerals and plagioclase. Locally developed augen textures may have resulted from shearing of partially crystallised cumulates. Strikes of the layering appear to swing from E-W at Mt Giles to NE-SW at Cape Burks, where a synformal structure is exposed (Fig. 1).

Foliated Felsic Plutonic Rocks. In the S part of the Mt Gray area and at Mt Steinfeld (Figs. 1, 4) hydrothermally altered (with alkali metasomatism) and foliated granodiorites and quartz diorites occur, but their relationship to the other groups is not yet clear. They may be either a separate gneissic terrane (Lopatin & Orlenko 1972), or they may represent a foliated phase of the Cretaceous plutonic intrusives discussed below. The foliation strikes E-W in the Mt Gray area.

A combined Rb/Sr whole-rock isochron of 154±35 m.y. was obtained from gabbro and quartz monzonite at Mt Giles and quartz diorite at Mt Gray. The significance of this apparent Jurassic age is not yet clear, but it may be a minimum age for the gabbros.

Metavolcanic Rocks. To the SW of the basement rocks lies a suite of calcalkaline metavolcanic and volcaniclastic rocks whose regional metamorphic grade, although largely masked by contact metamorphism from the intrusion of cross-cutting batholiths, may be as high as greenschist facies. Composition of the volcanic rocks ranges from andesitic to rhyolitic. Interbedded sedimentary rocks include some coarse conglomerate and agglomerate units with clasts of mafic and felsic volcanic rocks, quartz diorite, more felsic plutonic rocks, slates, and argillites. Some of the units are strongly sheared and display augen texture. Folds with axial trends between N and NW have been recognized at Pearson Peak, to the south of Mt McCoy, and at the summit of Mt Vance. Regional fold axis trends may swing continuously from NW in the eastern areas to NNW in the SW.

K/Ar whole-rock analyses on specimens from Mt McCoy and Mt Shirley give ages of 103±4 m.y. and 100±2 m.y. respectively. Because of the proximity to cross-cutting felsic plutonic rocks of Cretaceous age, it is very likely that the apparent ages in the metavolcanic rocks reflect this intru-

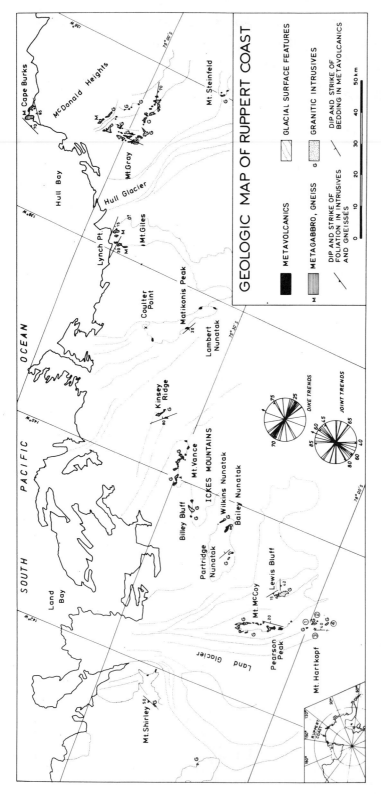

Fig. 1--Physiography and general geology of the Ruppert Coast. Note name change: Billey Bluff is Landry Peak of previous publications.

244

sive event. Rb/Sr whole-rock determinations were not successful because Rb/Sr ratios were too low, and the three samples used did not produce an interpretable isochron.

Distribution of the gabbros and the metavolcanic rocks defines two distinct areas, probably separated by a NW-trending boundary between Coulter Point and Mt Giles (Fig. 1).

Felsic Plutonic Rocks. A diverse suite of felsic plutonic rocks makes up the largest proportion of outcrops on the Ruppert Coast (Metcalfe *et al.* 1978). Where the plutonic rocks are not exposed, their secondary effects, such as contact metamorphism ranging from albite-apidote-hornfels facies to hornblende-hornfels facies and the K-feldspar-cordierite-hornfels facies (Winkler 1967), accompanied by quartz veining, epidotisation, and iron staining, are easily recognized.

The attitudes of the intrusive contacts range from vertical to horizontal. They are commonly sharp but locally assume complex patterns. Only at the S end of Mt McCoy at at Mt Vance (Fig. 3b) do large plutonic masses overlie their country rock. Veins and dike-like apophyses are partially oriented after structural features (Wilkins Nunatak, Fig. 3a). At the S end of Mt McCoy at Lewis Bluff (Fig. 2a) and possibly at Mt Hartkopf, the transition into the host rocks is more gradual through thick zones of inhomogeneous hybrid rocks.

A domed shape for one single intrusive may be indicated by the attitude of the contacts in the outcrops at Wilkins Nunatak, Bailey Nunatak, and in the exposures to the west. At the N end of Mt Gray the upper surface of the felsic acid pluton undulates with a relief of about 200 m.

Internal structures of the intrusives include linear and tabular mafic schlieren, autoliths, and xenoliths, and, at Billey Bluff (Fig. 1) proto-orbicules as described by Leveson (1966). Joint surfaces are commonly strongly chloritised and epidotised.

The weak development of lineations, the nearly-complete absence of foliation or flowage in the country rock parallel with the contacts, and the occurrence of granophyre and of miarolitic cavities indicate that the intrusives were emplaced at relatively shallow depths in the crust.

Alkali syenites and alkali granites are exposed at Mt Gray and Mt Vance, in the NE part of the area, and can be subdivided into zones of successively stronger contamination by country rock. Calcalkaline quartz monzonites and granites are restricted to the area SW of Coulter Point, approximately coinciding with the areal extent of the metavolcanic suite. Garnet-two mica granites were collected at Mt Shirley.

Alkaline granite from the Ickes Mountains

yielded a K/Ar age of 96 m.y. A calcalkaline granite gave a Rb/Sr age of 92 m.y. (Metcalfe *et al.* 1978).

Mafic and Felsic Dikes. Mafic and felsic dikes cut all the other rock units. The dominant trend of the dikes is E-W, and dips are uniformly steep (Fig. 1). At some localities the felsic dikes postdate, and at others predate, the mafic dikes. Similarly ambiguous age relationships exist between the dike suite, epidote veining, iron-rich alteration zones, and faulting, thus indicating that formation of all these features was more or less contemporaneous. Composite dikes are quite common.

Colour indices of dike rocks mainly range between 40 and 50, but extremes of 7 and 78 were also encountered. Textures in the felsic dikes are granular, porphyritic, granophyric, and spheroidal. The chemical composition of the felsic dikes appears to be transitional between alkaline and calcalkaline.

Mafic dikes have intergranular to porphyritic textures. Plagioclase composition ranges from An_{30} to An_{60}. K/Ar whole-rock determinations yielded an age of 98±1 m.y. on a rhyolite porphyry from Bailey Nunatak and an age of 113±1 m.y. for a porphyritic basalt from Billey Bluff (Ickes Mountains). This, together with chemical and mineralogical data, indicates that the dikes are closely related to the felsic plutonic suite.

A relatively smooth, subhorizontal erosion surface tops a number of nunataks (Lewis Bluff; Mt Gray, Nunataks 17 & 18); the height of the surface varies between groups of nunataks. Similar surfaces have been recognized elsewhere in Marie Byrd and Ellsworth Lands (Le Masurier 1972), and in the areas adjacent to the Ruppert Coast they are overlain by volcanic rocks which may be as old as Eocene. The height differences of the surfaces suggest block faulting of a once regionally extensive surface.

DISCUSSION

An outstanding problem in the Ruppert Coast area concerns the relation between the gabbros (and gneisses?) in the E part and the metavolcanic rocks in the W part. K/Ar and Rb/Sr age determinations have been inconclusive, and merely indicate that the age of the gabbros must be Upper Jurassic or older and that of the metavolcanics Lower Cretaceous or older. Correlation with other areas may give indications of possible ages for these rock units. The association of granitic rocks, metavolcanics, gneisses, and gabbros is very similar to that on Thurston Island, where a Carboniferous age has been reported for a gneiss terrane

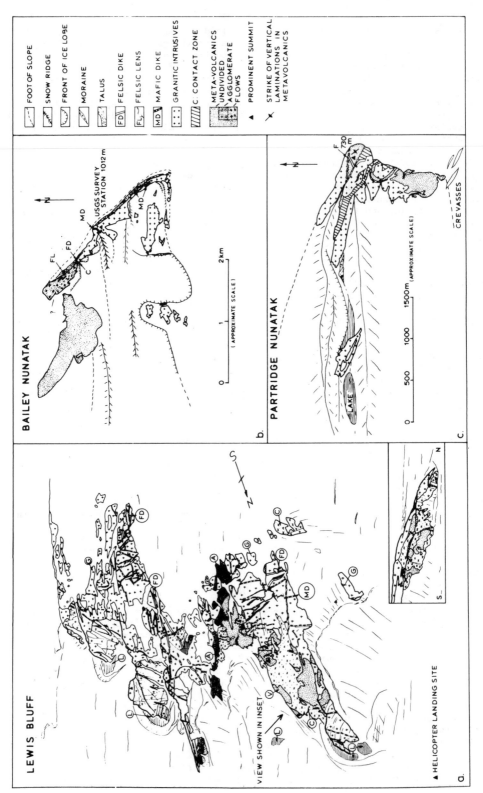

Fig. 2--Lewis Bluff (oblique view), Bailey and Partridge Nunataks (maps). A = alteration zone; C in Fig. a is a contact zone. V = metavolcanics. L = lake.

246

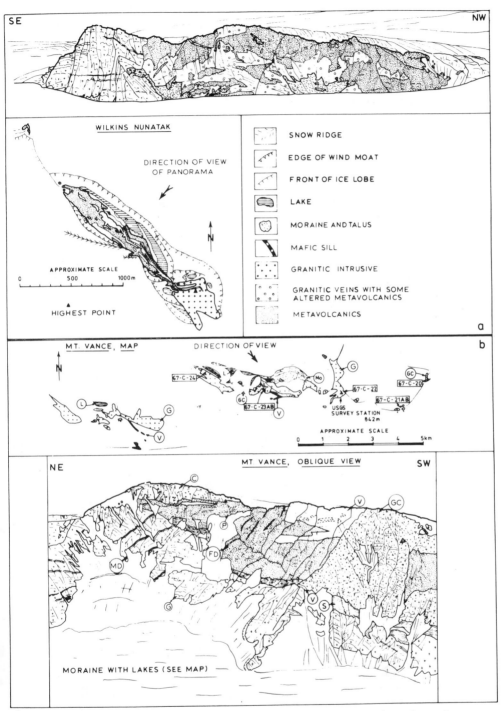

Fig. 3--Wilkins Nunatak and Mt Vance. V = metavolcanics. G = plutonic rocks.
C = foliated plutonic contact rocks. S = sheared plutonic rocks. P = green
porphyritic rocks. GC = contaminated plutonic rocks. MD = mafic dikes. FD =
felsic dikes. MO = moraine. L = lake. Boxed numbers indicate University of
Wisconsin specimen numbers.

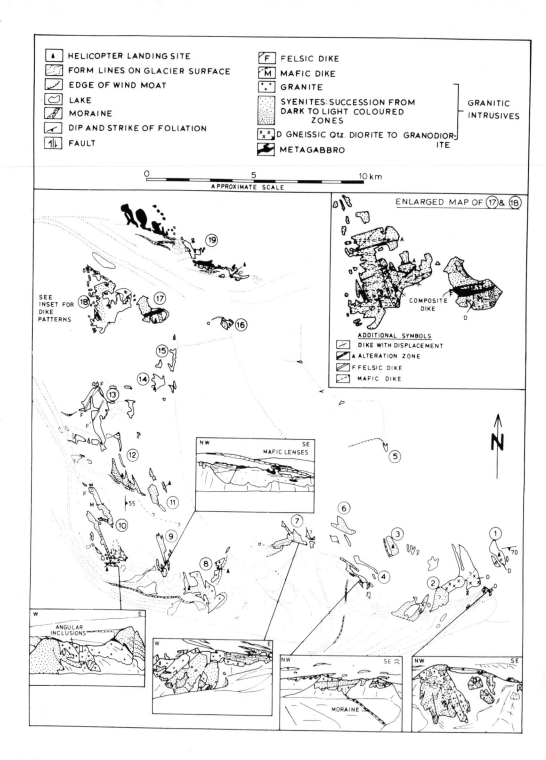

Fig. 4--Mt Gray area. Circled numbers identify individual nunataks.

(Wade & Wilbanks 1972). It is therefore possible that the gabbros on the Ruppert Coast are part of an Upper Paleozoic terrane, as has also been suggested by Lopatin and Orlenko (1972).

The calcalkaline metavolcanic rocks of the Ruppert Coast are very similar to the suite of Mesozoic volcanic rocks exposed in the Antarctic Peninsula (Metcalfe et al. 1978), although Wade and Wilbanks (1972) and Lopatin and Orlenko (1972) consider the entire metavolcanic complex of Marie Byrd Land and Eights Coast to be Paleozoic. Klimov (1967, transl.) reports that unconformably underlying the metavolcanics on the Ruppert Coast there is a schist/greywacke complex correlated with the thick and regular late Precambrian or Paleozoic sequence in the Ford Ranges.

The dominant basement rocks in the Ruppert Coast area are significantly different from those in the Ford Ranges. We agree with Lopatin and Orlenko (1972) that an important regional tectonic boundary must pass between the two regions. Similarities in the basement geology of the Ruppert Coast and the Hobbs Coast to the east are far greater.

The cross-cutting felsic plutons of the Ruppert Coast were emplaced in early Cretaceous time and are similar to batholiths to the east and west. Differences in chemistry of these intrusives may be due to differences in the country rocks traversed by the plutons. The batholiths form part of the "Andean Intrusive Suite" (Adie 1954; Halpern 1968).

In reconstructions of Gondwana, the Ruppert Coast lies along the join between Marie Byrd Land and the Campbell Plateau of New Zealand, well to the south of the Stokes and Campbell magnetic anomalies which separate the largely Paleozoic foreland province of New Zealand from the younger Pacific belt, including the collision suture of the Mesozoic Rangitata Orogeny (Grindley & Davey, in press).

The gabbroic rocks of the Ruppert Coast may have their equivalents in yet undated gabbroic and gneissic suites in eastern Fiordland, New Zealand (Grindley & Davey, in press).

If the metavolcanic rocks of the Ruppert Coast are assumed to be Mesozoic, they have no obvious equivalent within the foreland province of New Zealand. However, similar calcalkaline volcanic and volcaniclastic rocks form the Pacific rim of the foreland (Murihiku terrane, Coombs et al. 1975). These were deposited along the oceanic side of a Mesozoic island arc associated with a subduction system. In contrast to this, the fact that the Ruppert Coast metavolcan-

ics lie on the continental crust of the Gondwana continent (if a late Paleozoic age is accepted for the Ruppert Coast gabbros) may indicate the transition from an island arc which was intra-oceanic or marginal to the continent in the New Zealand sector to a calcalkaline volcanic chain positioned within the continent, as is the case in the Antarctic Peninsula and the Andes.

The felsic plutonic suite of the Ruppert Coast is part of a belt of batholiths which can be traced along the Cordilleran system of North and South America, through West Antarctica, and into New Zealand (Katz, in press) and was also associated with Mesozoic (and earlier) subduction. It is puzzling that in the reconstruction of Gondwana (Grindley & Davey, in press), the zone of Cretaceous plutons in Marie Byrd Land does not trend parallel with the margin of the continent, as it should if the intrusions are the result of subduction of the Pacific Plate under Gondwana. Rather, the trend of the plutons appears to swing 400 km into the interior of the continent along the Ford Ranges. This may either be due to nonrecognition of post-early Cretaceous oroclinal bending (in sympathy with possible anticlockwise rotation of the Ellsworth Mountains?) in Marie Byrd Land or, as suggested by Katz (in press), a very complex relation between plutonism, subduction, and continental break-up.

The main orientation of dikes on the Ruppert Coast is E-W, subparallel with the inferred boundary between the Campbell Plateau and Marie Byrd Land. These dikes may therefore represent the extensional regime prevailing during initiation of fragmentation of Gondwana.

ACKNOWLEDGMENTS

We are grateful for financial support of our field and laboratory work by the U.S. National Science Foundation, and for logistic support by the U.S. Navy and the U.S. Army. Special thanks are extended to Courtney Skinner, logistics coordinator at Camp I, Marie Byrd Land Survey, and to the Helicopter Detachment, U.S. Army, under the command of Major Hawkins. R. Harris grappled with the problem of reducing the figures to manageable size.

REFERENCES

ADIE, R. J. 1954. The petrology of Graham Land II, the Andean granite-gabbro suite. Falkland Is. Dep. Surv. Sci. Rep. 12: 39.

249

COOMBS, D. S.; C. A. LANDIS; R. J. NORRIS;
J. M. SINTON; D. J. BORNS; D. CRAW 1976.
The Dun Mountain ophiolite belt, N.Z., its
tectonic setting, constitution, and origin
with special reference to the southern por-
tion. Am. J. Sci. 276: 561-603.
GRINDLEY, G. W.; F. J. DAVEY, in press.
Some aspects of the reconstruction of N.Z.,
Australia, and Antarctica (review). 3rd
SCAR Symp. Antar. Geol. Geophys., Madison,
1977.
HALPERN, M. 1968. Ages of Antarctic and
Argentine rocks bearing on continental
drift. Earth Planet. Sci. Lett. 5(3):
159-167.
KATZ, H. R., in press. West Antarctica and
New Zealand. A geological test of the model
of continental split. 3rd SCAR Symp. Antar.
Geol. Geophys., Madison, 1977.
KLIMOV, L. V. 1967. Hektorye rezul'taty
geologicheskitch issledovaniye na Zemle
Meri Berd v. 1966-67 g.g. (Some results of
geological observations in Marie Byrd Land
1966-67.) Int. Byull. Sov. Antarkt. Exped.
66: 18-25 (English transl. 6(6): 555-559)
LE MASURIER, W. E. 1972. Volcanic record
of Cenozoic glacial history of Marie Byrd
Land. In R. J. Adie (Ed.), Antarctic Geo-
logy and Geophysics. Oslo: Universitets-
forlaget. Pp. 251-60.
LEVESON, D. J. 1966. Orbicular rocks: A
review. Bull. Geol. Soc. Am. 77(4): 409.
LOPATIN, B. G.; E. M. ORLENKO 1972. Out-
line of the geology of Marie Byrd Land and
the Eights Coast. In R. J. Adie (Ed.),
Antarctic Geology and Geophysics. Oslo:
Universitetsforlaget. Pp. 245-50.
---; M. M. POLKOV. 1976. Geology of Marie
Byrd Land and the Eights Coast (in Russian).
Moscow. Pp. 144-47; 42-50. Acad. Nauk.
METCALFE, A; K. B. SPÖRLI; C. CRADDOCK 1978.
Plutonic rocks from the Ruppert Coast, West
Antarctica. Antar. J. U.S. 13(4): 5-7.
SPÖRLI, B.; C. CRADDOCK 1967. Geology of
the Ruppert Coast. Antar. J. U.S. 2(4): 94.
---;--- 1968. Analysis of Ellsworth Moun-
tains and Ruppert Coast geologic data.
Antar. J. U.S. 3(5): 179.
WADE, F. A. 1969. Geology of Marie Byrd
Land (Sheet 18--Marie Byrd Land). In
V. C. Bushnell; C. Craddock (Eds.), Geologic
Maps of Antarctica. Antar. Map. Folio Ser.,
Fol. 12. Pl. XVII.
---; J. R. WILBANKS 1972. Geology of Marie
Byrd and Ellsworth Lands. In R. J. Adie
(Ed.), Antarctic Geology and Geophysics.
Oslo: Universitetsforlaget. Pp. 207-214.
WAGER, L. R.; C. M. BROWN; W. J. WADSWORTH
1960. Types of igneous cumulates. J.
Petrol. 1(1): 73-85.
WINKLER, H.G.T. 1967. Petrogenesis of
Metamorphic Rocks. Springer: N.Y. P. 237

Magmatic evolution of the northern Antarctic Peninsula

STEPHEN D. WEAVER
University of Canterbury, Christchurch,
New Zealand

JOHN TARNEY
University of Leicester, UK

ANDREW D. SAUNDERS
Bedford College, London, UK

Since the early Mesozoic, steady state subduction of oceanic lithosphere has operated along that section of the Gondwana continental margin which has become the Antarctic Peninsula.   In the Mesozoic calc-alkaline igneous rocks of the Antarctic Peninsula and the South Shetland Island, silica mode increases from west to east with basalt lavas predominating on the South Shetland Is. and highly silicic rocks on the east coast of the Peninsula.   Large ion lithophile elements (K, Rb, Ba, Th and LREE) exhibit K-h type variations whereas high field strength elements (Zr, Nb, Y and HREE) do not. Such patterns have important implications for the petrogenesis of calc-alkaline magmas.

Subduction ceased progressively from south to north along the Antarctic Peninsula as spreading sections of the post-Paleocene Aluk Ridge arrived at the continental margin. Spreading in Bransfield Strait, within the last 2 m.y., followed the cessation of subduction at the South Shetland trench.   Deception, Bridgeman and Penguin I. volcanoes have affinities with both calc-alkaline magmas and abyssal tholeiites, in accord with the view that Bransfield Strait represents the initial stages of back-arc basin development.   Together with recent alkali basalt activity around James Ross I. and at Seal Nunataks, the Bransfield Strait volcanics reflect a change from a 180 m.y. history of compressional tectonics to the current extensional regime.

INTRODUCTION

Throughout most of the Mesozoic, and probably from early Paleozoic times, the west coast of South America, the Antarctic Peninsula, Marie Byrd Land and the New Zealand continental block formed a continuous margin of the supercontinent Gondwana.   The Mesozoic-Cenozoic Andean orogenic belt may be considered to extend throughout the length of this margin from South America to New Zealand.   Subduction-related igneous rocks characterize this belt and testify to the consumption of great volumes of oceanic crust from at least early Mesozoic times.

TECTONIC EVOLUTION OF THE NORTHERN ANTARCTIC PENINSULA

The Antarctic Peninsula separated from East Antarctica in the mid-Jurassic (Jahn 1978) producing a narrow sliver of continental crust beneath which Pacific ocean floor was being subducted.   The asymmetric pattern of ages of Pacific ocean crust (Heezen & Fornari 1975) indicates that great volumes of Cretaceous and Jurassic lithosphere have been consumed beneath South America and the Antarctic Peninsula. In the SE Pacific, late Cretaceous ocean crust generated at the Pacific-Antarctic

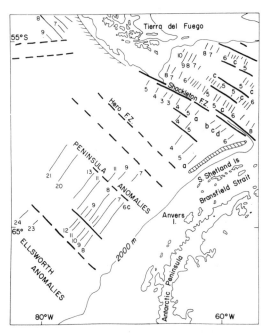

Fig.1--Magnetic anomalies in the SE Pacific and Drake Passage, after Herron & Tucholke (1976) and Barker & Burrell (1977).   Diagonal ruling = South Shetland trench.

Ridge is separated from the Antarctic Peninsula by Cenozoic ocean crust (Fig. 1) generated at the Aluk Ridge. The magmatic record of subduction beneath the west coast of the Antarctic Peninsula extends at least back to 185 m.y. BP. Jurassic and early Cretaceous ocean crust appears therefore to have been consumed along the Antarctic Peninsula before the initiation of the Aluk Ridge.

Since the Paleocene, there has been progressive consumption from south to north of the spreading sections of the Aluk Ridge with concomitant cessation of subduction and magmatism along the western margin of the Peninsula as the SE Pacific and Antarctic joined to form a single Antarctic Plate. Both the "Ellsworth" and "Peninsula" anomalies of Herron and Tucholke (1976) young towards the Antarctic Peninsula, indicating consumption of the ridge (Fig. 1). Rough estimates of the times the ridge arrived at the trench may be made: 50 m.y. off south Alexander I., 33 m.y. off north Alexander I., and 20-23 m.y. off south Anvers I. Subduction appears to have ceased once the ridge sections reached the trench, and the progressive consumption of the ridge is reflected in the northward younging of ages of calc-alkaline plutons along the Antarctic Peninsula (Saunders *et al.* in press).

The Antarctic Peninsula and South America maintained a narrow continental connection until the opening of Drake Passage about 30 m.y. ago (Barker & Burrell 1977). Fragments of this connection became dispersed into the extended loop of the Scotia Ridge, and the East Scotia Sea formed by secondary extension behind the intraoceanic South Sandwich arc during the last 8 m.y. (Barker & Griffiths 1972).

In west Drake Passage, between the Hero and Shackleton Fracture Zones (Fig. 1), both flanks of the recently active Drake Passage spreading centre have been identified. Oceanic crust generated in Drake Passage has been subducted beneath the South Shetland arc. Spreading ceased at about 4 m.y., and the South Shetland trench has become partially buried by sediments and is seismically quiet.

The 100 km wide Bransfield Strait, which separates the ensialic South Shetland Is. from the Antarctic Peninsula (Fig. 1), has an asymmetric graben structure. It contains a narrow trough along which runs a line of seamounts including the active and recently active volcanoes of Deception and Bridgeman Is. The Recent volcano Penguin I. lies off this axis on the NW fault-controlled margin of the trough. The seismic structure of the Strait resembles that of an oceanic axial zone (Barker 1976) and

magnetic lineations identified by P.J. Roach (pers. comm.) suggest opening within the last 2 m.y. Bransfield Strait therefore represents the earliest stages of the formation of a back-arc marginal basin and appears to be related to the waning of subduction beneath the South Shetland Is. arc. Recent volcanic activity at James Ross I. and Seal Nunataks may also reflect the change to extensional tectonics

MESOZOIC-CENOZOIC IGNEOUS GEOLOGY

The pre-Jurassic geological history of the Antarctic Peninsula has been summarized by Dalziel (in press).
Mesozoic Igneous Activity. The calc-alkaline basalt-andesite-rhyolite suite which occurs extensively throughout the Antarctic Peninsula is now referred to as the 'Antarctic Peninsula Volcanic Group' (Thomson, in press). Stratigraphic ages range from Middle Jurassic to Aptian and K/Ar dates fall in the range 186-86 m.y. (Rex 1976; Pankhurst *et al.*, in press). The Antarctic Peninsula volcanism may have occurred either in a number of discrete pulses or more or less continuously throughout the Jurassic and Cretaceous.

There is a marked change in composition of the lavas across the Antarctic Peninsula (Weaver *et al.* in press). On the South Shetland Is., closest to the Mesozoic trench, basalts predominate, whereas on the Danco Coast a much higher proportion of andesites and dacites occurs . On the east coast of the peninsula rhyolites are dominant (Fig. 2).

The geology of the Antarctic Peninsula is dominated by calc-alkaline plutonic rocks ranging from gabbro to granite or trondhjemite. Radiometric data on the plutonic rocks are reviewed by Saunders *et al.* (in press). K/Ar ages, mostly from Rex (1976), fall in four groups of ages (Fig. 3). Plutons of the 185-155 m.y. group are restricted to the east coast of the Peninsula, and those of the youngest 60-45 m.y. group occur only in the northern part of the west coast and in the South Shetland Is. (Fig. 4). Ages are concentrated in the range 110-90 m.y. which may represent a mid-Cretaceous peak of plutonic activity. The plutonic foci seem to have migrated from east to west. The compositional variations across the peninsula recognized in the volcanic rocks are paralleled by the plutonic rocks (Fig. 2). Silicic rocks on the east coast of the peninsula constitute a 'normal' calc-alkaline tonalite-granodiorite-granite series whereas those on the west coast are trondhjemitic granites and tonalites (Saunders *et al.*,

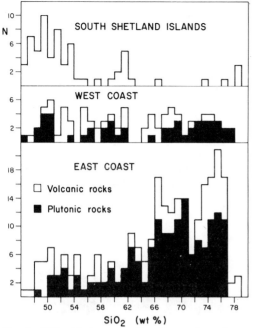

Fig.2--SiO$_2$ histograms for Mesozoic volcanic and plutonic rocks.

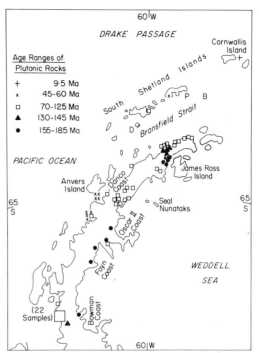

Fig.4--Locality map for distribution of the plutonic rocks of Fig. 3.

in press).

Tertiary Igneous Activity.   Tertiary plutons are restricted to the northern part of the west coast of the Antarctic Peninsula and the South Shetland Is.   The granodiorite of Cornwallis I. has a K/Ar radiometric age of 9.5 m.y. (Rex & Baker 1973).   Tertiary volcanic rocks are abundant on the South Shetland Is.   Grikurov *et al.* (1970), Watts (in press) and Pankhurst (pers. comm.) report ages in the range 20-60 m.y. for the basalts and andesites of Fildes Peninsula, King George I. Late Tertiary basalts, andesites and tephra on the south coast of King George I., South Shetland Is. (Weaver *et al.*, in press), represent remnants of volcanoes that have been largely down-faulted beneath Bransfield Strait.

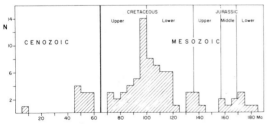

Fig.3--Radiometric dates determined on plutonic rocks from the northern Antarctic Peninsula and South Shetland Islands.

Quaternary Igneous Activity.   The active Deception I. and the recently active Bridgeman and Penguin Is. are associated with the opening of Bransfield Strait (Gonzalez-Ferran & Katsui 1970; Baker *et al.* 1975; Weaver *et al.* 1979). Pre-caldera pyroclastic deposits of Deception Island are mostly basaltic whereas post-caldera eruptions have been distinctly more silicic, reaching dacite in composition.   Bridgeman I. is a small remnant of an eroded volcano composed of basalts and basaltic andesites.   Penguin I., about 1 km south of King George I., is composed largely of olivine-basalt lavas.

Alkali olivine-basalts and hawaiites occur on and around James Ross I. and farther south along the east coast of the Antarctic Peninsula at Seal Nunataks. Rex (1976) reports K/Ar ages of 6-1 m.y. for the James Ross I. volcanics and <1 m.y. for those at Seal Nunataks.

GEOCHEMICAL PATTERNS WITHIN THE ANTARCTIC PENINSULA

About 500 igneous rocks from the South Shetland Is. and the Danco, Oscar II, Foyn and Bowman Coasts (Fig. 4), have been chemically analysed.   Preliminary evaluation of the data, details on analytical methods and further references are given by Saunders *et al.* (in press) and Weaver *et al.* (in press).

MESOZOIC-TERTIARY IGNEOUS GEOCHEMISTRY

Little time-dependent chemical change has been recognised in the 180 m.y. span of igneous activity. On the South Shetland Is. where the volcanic record is most complete, Mesozoic (Byers Peninsula) and Tertiary (Fildes Peninsula) lavas are chemically indistinguishable, yet distinct from lavas on the peninsula mainland, erupted farther from the trench. The geochemistry may be described in terms of transverse variations across the peninsula (Saunders *et al.* 1980).

The igneous rocks are typically calc-alkaline, displaying low Fe-enrichment on AFM diagrams, high levels of K, Rb, Ba, Sr and Th, low K/Rb ratios and light rare earth element (LREE) enriched patterns. They closely resemble the Andean rocks of South America. South Shetland I. lavas have lower K, Rb, Th, Ba and LREEs than Antarctic Peninsula rocks, and in these respects may be likened to the island arc tholeiite series (Jakeš & Gill 1970) which are typically located close to the trench. We consider that the South Shetland Is. volcanics represent an end member of a broadly calc-alkaline series of magmas rather than a separate series.

There are no significant chemical differences between volcanic and plutonic rocks from the same area (Figs. 2 and 5). The trondhjemitic character of west coast plutons compared with the tonalite-granodiorite-granite series of the east coast corresponds to a west to east decrease in Na/K ratio. The progressive west to east increase in SiO_2 is paralled by K_2O, Rb, Ba (Fig. 5), Th and LREEs. Moreover, at a given SiO_2 content there is a progressive increase in average K_2O (Fig. 6a), Rb etc. across the peninsula. This pattern corresponds to the K-h (K = potassium content, h = depth to Benioff zone) correlation recognized in many other arc systems (Dickinson 1975). In the Antarctic Peninsula, only the large ion lithophile elements (LILEs), K, Rb, Ba, Th, La and Ce follow this pattern of continuous enrichment with increasing SiO_2, both in volcanic and plutonic rocks. The small high field strength (HFS) elements, Zr, Nb, Y and heavy rare earth elements (HREEs) do not behave in an incompatible manner in these calc-alkaline magmas and do *not* exhibit K-h type variations (Saunders *et al.* in press). For example, Zr begins to decrease at intermediate SiO_2 levels (Fig. 6b).

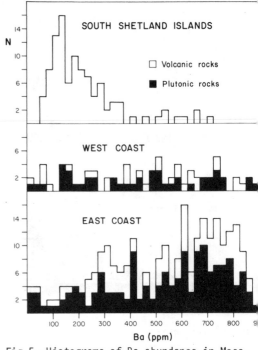

Fig.5--Histograms of Ba abundance in Mesozoic volcanic and plutonic rocks from the South Shetland Islands and the Antarctic Peninsula.

Ce_N/Yb_N ratios of plutonic rocks (Saunders *et al.* 1980) range from 2.5 to 6.1 and are typical of other calc-alkaline rocks such as those from the Andes. Initial $^{87}Sr/^{86}Sr$ ratios for Mesozoic volcanics from the South Shetland Is. are in the range 0.70397 - 0.70434. Some samples have Rb/Sr ratios too low to account for their Sr isotopic ratios, which may imply incorporation of radiogenic Sr derived from sea-water-altered, subducted, ocean crust. Ce_N/Yb_N ratios of these Mesozoic volcanics are in the range 1.8 to 3.0 and in terms of REE chemistry are intermediate between typical island arc tholeiites and normal calc-alkaline magmas (Jakeš & Gill 1970).

Tertiary lavas on the South Shetland Islands show the same geochemical features as the Mesozoic volcanics. Early Tertiary, Fildes Peninsula lavas are best regarded, like the Byers Peninsula lavas, as representing low-K calc-alkaline magmas. Late Tertiary volcanics around Bransfield Strait have higher K, Rb, Ba and REE levels approaching those of normal calc-alkaline rocks, but Quaternary lavas revert to markedly low-K compositions.

The pattern of geochemical variation

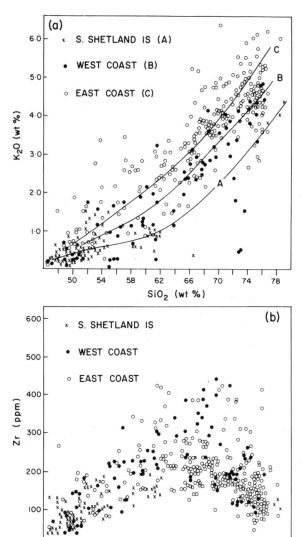

Fig. 6--(a) Graph of K_2O versus SiO_2 for Mesozoic igneous rocks of the Antarctic Peninsula and the South Shetland Islands. Although data are very scattered, curves A, B, and C represent the trend for each region. (b) Graph of Zr versus SiO_2 for the same samples shown in Fig. 6a.

recognized in the Antarctic Peninsula has important implications concerning the origin of calc-alkaline rocks. Detailed discussion is presented elsewhere (Saunders *et al.* 1980).

GEOCHEMISTRY OF THE LAVAS OF BRANSFIELD STRAIT, JAMES ROSS I., AND SEAL NUNATAKS

All the Bransfield Strait lavas are characterized by low K and extremely high Na/K ratios. The extreme Na/K ratios of Deception basalts are matched only by those of depleted ocean-floor basalts. The lavas of Deception I., ranging from olivine-tholeiite to dacite, appear to be related by the high level fractionation of phenocryst minerals. They show a low to moderate trend of Fe-enrichment and both LILEs and HFS elements increase progressively with fractionation and SiO_2 content (Weaver *et al.* 1979). In this latter respect Deception lavas differ from the typical calc-alkaline rocks of the Antarctic Peninsula.

Ce_N/Yb_N ratios in Deception and Bridgeman lavas range from 1.7 to 2.5, and $^{87}Sr/^{86}Sr$ ratios for both are in the range 0.70336-0.70358. Modelling suggests that both derived from a spinel-peridotite source, Bridgeman basaltic magmas being generated under more hydrous conditions and representing a higher degree of melting (10-20%) compared with Deception basaltic magmas (5%) (Weaver *et al*. 1979).

The mildly ne-normative basalts of Penguin I. are more primitive than Deception or Bridgeman basalts in having high levels of Cr (500 ppm) and Ni (170 ppm). They have K, Rb, Ba and Sr levels similar to Deception basalts, but Zr, Y and HREE concentrations are low and probably were retained by refractory garnet in the mantle. Also, the rare earth patterns of Penguin basalts are considerably more fractionated, with Ce_N/Yb_N of 4.0. Modelling suggests that Penguin magmas formed by about 3% melting of a garnet-peridotite source. The relatively high $^{87}Sr/^{86}Sr$ ratios of Penguin Island basalts (0.7039) may be explained as contamination by radiogenic Sr derived from subducted ocean crust.

Bransfield Strait lavas represent magmatism associated with the initial stages of back-arc spreading and in several respects their chemistry is unusual. The high K/Rb, Na/K, Zr/Nb and low Rb/Sr ratios are characteristic of abyssal tholeiites yet LREE enrichments (Ce_N/Yb_N>1.7) and $^{87}Sr/^{86}Sr$ ratios (>0.7034) are significantly higher than average ocean floor basalt. Low Fe-enrichment and high Ba and Sr suggest calc-alkaline affinities, although low K and Rb are similar to those of abyssal tholeiites.

The alkaline olivine-basalts of James Ross I. and Seal Nunataks have similar major element compositions to Penguin I. basalts but higher total alkalies and lower Na/K ratios. K, Rb, Zr, Nb and REE levels are considerably higher than at Penguin I., and the low Zr/Nb ratios (7.0) match those of alkaline volcanics from ocean islands and continental rifts (Weaver *et al*.1972), whereas Penguin basalts have high Zr/Nb ratios (80) typical of calc-alkaline and abyssal tholeiitic magmas (Erlank & Kable 1976). James Ross I. basalts have strongly fractionated REE patterns, with Ce_N/Yb_N in the range 7.2-8.2 (Pankhurst, in press). All this suggests small degrees of melting of a garnet peridotite source. The James Ross I./Seal Nunataks volcanics are of a type frequently associated with ensialic rifting which may be taking place as a secondary extensional feature related to active diapirism in Bransfield Strait.

CONCLUSIONS

The last few million years have seen the end of a 180 m.y. history of Pacific lithosphere subduction along the west coast of the Antarctic Peninsula. The long-lived nature of geochemical patterns of the Mesozoic-Cenozoic igneous rocks implies th operation of steady-state subduction. Such patterns identified in the relatively tectonically uncomplicated region of the northern Antarctic Peninsula may be useful in future studies for deciphering magmatic tectonic regimes in other sectors of the Gondwana continental margin, such as Marie Byrd Land and New Zealand.

Only broad generalizations have been elucidated here, but we have no doubt that detailed co-ordinated geological, geochemical and geochronological work will reveal much complexity not here identified.

REFERENCES

BAKER, P. E.; I. McREATH; M. R. HARVEY; M. J. ROOBOL; T. G. DAVIES 1975. The geology of the South Shetland Islands: V. Volcanic evolution of Deception Island. Brit. Antar. Surv. Sci. Rep. 78: 81 pp.
BARKER, P. F. 1976. The tectonic framework of Cenozoic volcanism in the Scotia Sea region: a review. In O. Gonzalez-Ferran (Ed.), Symposium on Andean and Antarctic volcanology problems. Rome: IAVCEI. Pp. 330-346.
---; J. BURRELL 1977. The opening of Drake Passage. Marine Geol. 25: 15-34.
---; D. H. GRIFFITHS 1972. The evolution of the Scotia Ridge and Scotia Sea. Phil Trans. R. Soc. London A271: 151-183.
DALZIEL, I. W. D. (in press). The pre-Jurassic history of the Scotia Arc: a review and progress report. In C. Craddock (Ed.), Antarctic Geoscience. Madison: University of Wisconsin Press.
DICKINSON, W. R. 1975. Potash-depth (K-h) relations in continental margin and intra-oceanic magmatic arcs. Geol. 3: 53-56.
ERLANK, A. J.; E. J. D. KABLE 1976. The significance of incompatible elements in Mid-Atlantic Ridge basalts from 45°N with particular reference to Zr/Nb. Contrib. Mineral. Petrol. 54: 281-291.
GONZALEZ-FERRAN, O.; Y. KATSUI 1970. Estudio integral del volcanismo cenozoico superior de L as Islas Shetland de Sur, Antarctica. Inst. Antar. Chileno Ser. Cien. 1: 123-174.
GRIKUROV, G. E.; A. Ya. KRYLOV; M. M. POLYAKOV; Ya.N. TSOVBUN 1970. The age of rocks in the northern part of the Antarctic Peninsula and the South Shetland Islands (by the K-Ar method). Soviet Antar. Exped. Info. Bull. 80: 30-34.
HEEZEN, B. C.; D. J. FORNARI 1975. Geological map of the Pacific Ocean. In J.E. Andrews et al., Init. Rep. Deep Sea Drill. Proj. 30. Washington: U.S. Government Printing Office.

HERRON, E. M.; B. E. TUCHOLKE 1976. Sea-floor magnetic patterns and basement structure in the southeastern Pacific. In C.D. Hollister et al., Init. Rep. Deep Sea Drill. Proj. 35. Washington: U.S. Government Printing Office. Pp. 263-278.

JAHN, R. 1978. A preliminary interpretation of Weddell Sea magnetic anomalies. Geophys. J.R. Astron. Soc. 53: abstract 4.21.

JAKEŠ, P.; J. GILL 1970. Rare earth elements and the island arc tholeiite series. Bull. Geol. Soc. Am. 83: 29-40.

PANKHURST, R.J., in press. Sr-isotope and trace element geochemistry of Cenozoic volcanics from the Scotia Arc and northern Antarctic Peninsula. In C. Craddock (Ed.), Antarctic Geoscience. Madison: University of Wisconsin Press.

---; S. D. WEAVER; M. BROOK; A. D. SAUNDERS, in press. K-Ar chronology of Byers Peninsula, Livingston Island, South Shetland Islands. Brit. Antarc. Surv. Bull.

REX, D. C. 1976. Geochronology in relation to the stratigraphy of the Antarctic Peninsula. Brit. Antar. Surv. Bull. 43: 49-58.

---; P. E. BAKER 1973. Age and petrology of the Cornwallis Island granodiorite. Brit. Antar. Surv. Bull. 32: 55-61.

SAUNDERS, A. D.; J. TARNEY; S. D. WEAVER. 1980. Transverse geochemical variations across the Antarctic Peninsula: implications for the genesis of calc-alkaline magmas. Earth Planet. Sci. Lett. 46: 344-60.

---; S. D. WEAVER; J. TARNEY in press. The pattern of Antarctic Peninsula plutonism. In C. Craddock (Ed.), Antarctic Geoscience. Madison: University of Wisconsin Press.

THOMSON, M. R. A., in press. Mesozoic palaeogeography of western Antarctica. In C. Craddock (Ed.), Antarctic Geoscience. Madison: University of Wisconsin Press.

WATTS, D. R., in press. Potassium-argon and palaeomagnetic results from King George Island, South Shetland Islands. In C. Craddock (Ed.), Antarctic Geoscience. Madison: University of Wisconsin Press.

WEAVER, S. D.; A. D. SAUNDERS; R. J. PANKHURST; J. TARNEY 1979. A geochemical study of magmatism associated with the initial stages of back-arc spreading: Quaternary volcanics of Bransfield Strait from South Shetland Islands. Contrib. Mineral. Petrol. 68: 151-169.

---;---; J. TARNEY in press. Mesozoic-Cenozoic volcanism in the South Shetland Islands and the Antarctic Peninsula: geochemical nature and plate tectonic significance. In C. Craddock (Ed.), Antarctic Geoscience. Madison: University of Wisconsin Press.

---; J. S. C. SCEAL; I. L. GIBSON 1972. Trace-element data relevant to the origin of trachytic and pantelleritic lavas in the East African Rift System. Contrib. Mineral. Petrol. 36: 181-194.

3. Break-up of Gondwana

Models for the fragmentation of Gondwana

DONALD H. TARLING
University of Newcastle upon Tyne, UK

While oceanic magnetic anomalies provide constraints on the Cretaceous-Tertiary formation of the Indian and Atlantic Oceans, such observations are sparse for the earliest phases of opening and thus for the actual shape of Gondwana in Jurassic and earlier times. There is general agreement, using paleomagnetic, paleoclimatic and geological matching, for the previous relationship of South America with Africa, and also for Australia, India, Antarctica and New Zealand, but there are major differences in proposed fits of these 'Western' and 'Eastern' Gondwanan fragments. Geological evidence is not consistent with a derivation of Madagascar-India from the East African area, which is the basis of most extant reconstructions. Paleomagnetic data appear equivocal but consistent with a reconstruction in which Madagascar has retained its present relationship with Africa. Such a reconstruction appears to be consistent with most available observations, but matching of geological features from northern India with those of the Middle East indicates that major revisions are required in this region. Provisional observations in southeastern Asia suggest that these blocks were close to Australia during much of the Paleozoic and early Mesozoic. The proposed reconstruction also has major impact on Paleozoic relationships between Gondwana, Siberia and Laurentia.

MALAGASY: THE KEY TO GONDWANA (FIG.1)

Malagasy (Madacascar) is almost the same length as the western Indian coastline and some 590,000 km^2 in area. The paleontological links between Malagasy and India led to the very early recognition of a connection expressed as the Lemurian continent (Schuchert 1928). Most reviewers have considered that Malagasy and India were once adjacent, with the Mascarene Ridge, the Chagos-Laccadive Ridge and the Seychelles between them, India and Malagasy cannot be jig-saw fitted uniquely as their opposed edges are nearly straight, but the fundamental problem is to determine the fit of Malagasy to Africa, as this defines the fit of 'Eastern' and 'Western' Gondwana. Three positions for this fit are generally considered: (1) against the East African coast, (2) against Mozambique, and (3) leaving Malagasy close to its present position.

The East African Fit. On the Tanzanian coast are 7 to 10 km of Karroo and Mesozoic sediments, with Jurassic sediments extending at least 500 km from the coast and Cretaceous sediments almost as far as the Seychelles. The ages have been confirmed by commercial drilling at and near the coast (Kent *et al.*1971; Kent 1972; Kent & Perry 1973) and by deep-sea drilling in water depths of over 4,000 m. Similar paleogeographical conditions persisted from at least the time of the major vertical movements of 7-10 km which took place before mid-Jurassic times (Kent & Perry

1973). There is no geological evidence for crustal thinning or other significant deformation since Jurassic times (Kent & Perry 1973). The crust of the Somali basin (Francis *et al.* 1966; Sowerbutts 1969) is neither truly oceanic nor continental, but is most similar to the quasi-continental Lau and Lord Howe rises in the SW Pacific. The basin has only average heat flow, and no magnetic anomalies have so far been identified (Francis *et al.* 1966; Bunce *et al.*1967; Fisher *et al.* 1971; Green 1972). It was clearly already in existence by at least Jurassic times and has since played only a quiescent role, with the possible exception of subduction along the small now extinct Amirantes trench which contains basalts some 82 m.y. old (Fisher *et al.* 1968). There is no possibility that Malagasy was rifted from the present Kenya-Tanzanian coast during the Mesozoic or Cenozoic.

The Mozambique Fit. Baker (1912) and Wegener (1915) both placed Malagasy against Mozambique. Flores (1970, 1973) summarised the stratigraphic similarities between southern Africa and Malagasy, pointing out that the first major difference arose when the Stormberg lavas were erupted over southern Africa, but not in Malagasy, although the mid Cretaceous Lupata volcanics in Mozambique are unique along the entire East African coast and possibly correspond with mid Cretaceous volcanics in Malagasy. Dinosaurs moved freely between Africa, Malagasy and India in Cretaceous times (Keast 1973). The Mozambique Channel is floored

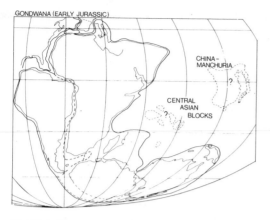

Fig. 1--A postulated Gondwana reconstruction. The overall relationships between the continents are shown above, and the proposed relationship between 'Eastern' and 'Western' Gondwana is shown below.

divides the Zambesi and Limpopo sediments into two basins. It is flanked by asymmetric magnetic anomalies with amplitudes of several hundred gammas (Green 1972; Scrutton 1978) and a large free-air gravity anomaly lies along one side of the ridge (Vening Meinesz 1948). Seismically, the area has been considered oceanic (Ludwig *et al.* 1968). The presence of many plateaux indicates block faulting (Simpson & Needham 1972). Darracott (1974) considered the crust to be intermediate rather than oceanic, and Scrutton (1973) suggested that the ridge is either a continental fragment or uptilted thickened oceanic crust. However, there is a definite change in the magnetic nature of the crust from the Mozambique Channel to the Mozambique Basin between 21°S and 27°S (H. Burgh, pers. comm.). Oceanic-type magnetic anomalies are present and indicate formation by ocean-crust accretion at a time when geomagnetic polarity changes were occurring. However, the Mozambique Ridge is dated by the age of older sediments trapped by it. The oldest rifting in this region was between 200 and 160 m.y. ago when N-S grabens formed and were subsequently covered by the Limpopo and Zambesi sediments (Dingle & Scrutton 1974). The age of the graben-filling non-marine sediments is uncertain but probably Jurassic (Flores 1973), and the oldest dated marine sediments are upper Lower Cretaceous. The ridge is thus probably of mid to late Jurassic age.

Along the southern tip of Africa, the continental shelf is narrow, with the exception of the Agulhas Bank and the Agulhas Plateau (Tucholke *et al.* 1978). The Agulhas Bank is separated from the mainland by a marginal ridge (Scrutton & du Plessis 1973; Scrutton & Dingle 1976) and comprises continental crust that includes parts of the Triassic-Lower Cretaceous Outeniqua sedimentary basin. The sedimentary basins along the Natal and southern coast of Africa (Dingle *et al.* 1978; Scrutton & Dingle 1976) were subaerial until the early Cretaceous, after which sea encroached on the outer margins. The basins appear to have been truncated by faulting between 140 and 120 m.y. ago (Dingle & Scrutton 1974).

Malagasy Fixed to Africa. There is no direct evidence that Malagasy moved relative to Africa. The paleontological relations between Malagasy and both East and Southern Africa could be explained by minor changes in sea-level providing occasional free faunal and floral migration (Förster 1975). The geological and geophysical evidence therefore only indicates that Malagasy could not have been close to the

by 14 km of sediments deposited since late Carboniferous times (Dixey 1960; Talwani 1962). DSDP hole 242 was still in Upper Eocene sediments when drilling stopped at nearly 700 m (Anon.1972). The whole area has only very low magnetic relief (~120 gammas), except near the volcanic Comores, Aldabra and Comores Is. Sedimentary inclusions in lavas of the Comores Is. indicate sedimentary rocks at depth (Flower & Strong 1969).

Farther south, up to 4 km of sediment have accumulated on the western side of the Mozambique Ridge (Dingle *et al.* 1978). The Mozambique Ridge appears to extend into Mozambique as a NNW basement high that present coastline of either East Africa or Mozambique.

The paleomagnetism of early Mesozoic rocks in Malagasy (Embleton & McElhinny 1975) was considered as showing that Malagasy lay adjacent to East Africa (Embleton & McElhinny 1975; McElhinny *et al.* 1976), but was shown to be inconclusive by Tarling & Kent (1976).

POLAR WANDERING (FIGS. 2, 3)

Accumulations of paleomagnetic data for the different continents has now allowed polar wandering paths to be estimated for most Gondwana continents. The extensive data for Africa and Australia vary in quality. Although most of the paleomagnetic data has been of a high standard, the age of the remanence is uncertain in many cases because of either lack of stratigraphic age control or the possibility of post-stratigraphic age remagnetisation. Observations have to be assessed by their consistency with other paleomagnetic observations and also with independent criteria such as paleoclimatic indicators.

When South America and Africa are united as 'Western' Gondwana their Paleozoic polar wander paths are similar and consistent with the paleoclimatic evidence for migration of a southern polar conditions from the Sahara, in the Ordovician, to just off southern Africa by the Early Permian (Fig. 2a). The agreement with the polar wander paths for the 'Eastern' Gondwana is poorer but reasonable considering the sparse data from India and Antarctica (Fig. 2b). However, the match of the 'Eastern' and 'Western' Gondwana polar wandering paths, while still excellent when compared with the polar wandering paths for the continents in their present positions, is inadequate for precise testing of the Smith & Hallam model (1970) and the 'Malagasy fixed' model (Fig. 2c; Table 1). Such a test may be diagnostic when there is a better data base, but at this stage it is useful only to indicate a broad agreement with some form of Gondwanan continent.

THE EXTENT OF GONDWANA

The similarities of the geology of Arabia, Afghanistan and Iran all seem to indicate that these have formed a single unit for most of the Phanerozoic. This seems to indicate that the Zagros Thrust is not a major tectonic boundary when considered on a Gondwana scale. Such an interpretation would indicate caution in any assessment of the gross significance of similar ophiolitic belts in northern India. There also appear to be strong similarities between NE India and parts of SE

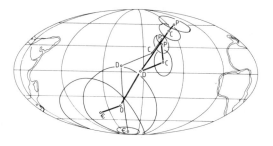

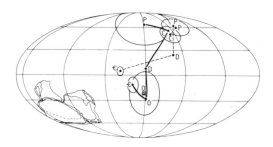

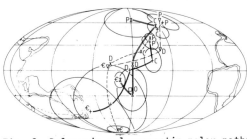

Fig. 2--Paleozoic paleomagnetic polar paths for Gondwana. A. The South American (single line) and African pole paths (double lines) are nearly identical, considering the uncertainty about the location of each polar path. B. These eastern paths are not in very good agreement but are reasonably consistent in that the Australian (solid) path and Indian (dashed) are broadly parallel and the Ordovician Antarctic pole (square) is very consistent with that of Australia. C. The combined paths again show reasonable agreement, while differences in relative positions are small. Є = Cambrian; O = Ordovician; S = Silurian; D = Devonian; C = Carconiferous; P = Permian.

Asia (Ridd 1971) and there is sparse paleomagnetic, but strong paleoclimatic and paleontological evidence that much of China and Manchuria, as well as southeastern Asia, formed integral parts of Gondwana during different periods of the Phanerozoic. However, these areas appear to

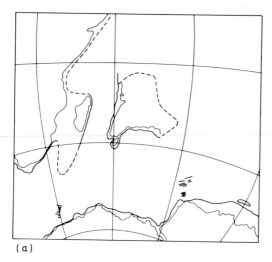

(a)

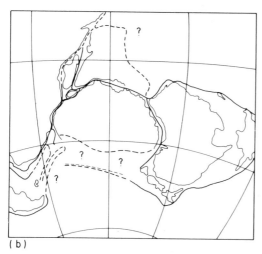

(b)

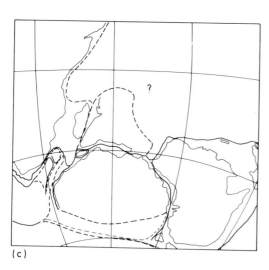

(c)

Table 1--Euler poles and rotations for Gondwana reconstruction relative to a fixe Africa-Malagasy continent

	N	E	Rotation
S. America	43.9	-30.1	56.1
Arabia	36.5	18.0	-6.1
India	15.2	31.1	-68.8
Ceylon	14.6	34.3	-75.3
Australia	13.3	98.5	-40.9
New Zealand	-34.8	-66.1	44.2
E.Antarctica	9.5	-41.3	39.5
W.Antarctica	8.2	155.5	-42.0

have changing affinities, indicating that the actual connections changed from direct continuous terrestrial routes at some time to complete marine barriers at others. Farther west, the paleomagnetic data, combined with geological and paleomagnetic observations, suggest that Gondwanan Iran extended towards Europe, including most areas south of the Zagros Mountains, and that the Menderes Block, in Turkey and Greece, formed part of this unity as well as the Adriatic Peninsula and the Pannonian Basin in SE Europe.

UPPER PALEOZOIC MOVEMENTS

Although the polar wandering paths are poorly defined, they are adequate for assessing the general motions of Gondwana during the late Paleozoic (Fig. 2), particularly when supported by paleoclimatic evidence. For example, both considerations are consistent with the location of the southern rotation pole in the Sahara in Ordovician times (Llanvirn -Llandeilo times) (Beuf *et al*. 1968), and again in Southern Africa for early Permian times (Anderson & Schwyzer 1977), with the implication that the southern rotational pole migrated from NW. to SE Africa during the Silurian and Devonian (Tarling 1978).

It was during the late Paleozoic that the Caledonian and Taconic orogenies resulted in the formation of the Laurentian continent (North America-Greenland and NW Europe), and this was in collision with NW Africa forming the Taconic orogeny. North America and Africa were fairly close

Fig. 3--Gondwana reconstruction based on oceanic magnetic anomalies. (a) The 75 m.y reconstruction, showing older anomalies of possible Jurassic age. This is based on an omalies and rotations defined by Sclater & McKenzie (1971) and must have evolved from reconstructions similar to that of Smith & Hallam (1970)(b) or that proposed here (c) for early Jurassic times.

to each other for much of the late Paleozoic, although possibly separated at times
by as much as 500 km of ocean (Tarling
1979). After the Taconic collision, both
supercontinents were rotating clockwise,
although the rate of rotation of Laurentia
was less than that of Gondwana, and both
continents had a predominantly northerly
motion in the 'Mediterranean' region until
their collision in Middle-Upper Devonian
times, resulting in the Acadian Orogeny.
This collision had little effect on the
motion of Gondwana, but the rotation of
Laurentia reversed to become anticlockwise.
Both continents collided against the Iber-
ian Peninsula in the Carboniferous, and
then with each other at the end of the
Carboniferous in the area of the present
Gulf of Mexico. This major impact reversed the sense of motion of Gondwana,
closing the 'Atlantic' and 'Mediterranean'
and resulting in the Hercynian Orogeny and
temporarily forming the single continent
of Pangea - although the Siberian block
was still separate. The Hercynian collision did not unite the two supercontinents
as a single tectonic entity, as both continued to rotate independently, causing
right lateral slip between them, until the
impact of the Adriatic Promontory against
Iberia. The Hercynian event was of major
importance, not merely because it initiated
the Gulf of Mexico, but also because the
drastic change in the sense of motions of
the two supercontinents resulted in the
fracturing, or renewing of appropriately
aligned old fractures, on both continents.
These fractures were to be loci for subsequent disruptions of the two supercontinents when convective motions began to
operate on the weaker areas of the continental lithosphere in Mesozoic times.

REFERENCES

ANDERSON, J.M.; R.V. SCHWYZER 1977. The
biostratigraphy of the Permian and Tri-
assic. Trans. Geol. S. Africa 80: 211-234.
ANON 1972. Deep Sea drilling project in
the Arabian Sea. Geotimes 17(7): 22-24.
BAKER, H.B. 1912. The origin of contin-
ental forms. Mich. Acad. Sci. Ann-Rep.:
116-141.
BEUF, S.; B. BIJU-DUVAL; A. MAUVIER; P.
LEGRAND 1968. Nouvelles observations
sur le "Cambro-Ordovicien" du Bled el
Mass (Sahara Central) Serv. Géol. Algérie
Bull. 38: 39-52.
BUNCE, E.T.; M.G. LANGSETH; R.L. CHASE;
M. EWING 1967. Structure of the Western
Somali Basin. J. Geophys. Res. 72:2547-
2555.

DARRACOTT, B.W. 1974. The structure of
Speke Gulf, Tanzania, and its relation to
the East African Rift System. Tectonophys. 23: 155-175.
DINGLE, R.V.; R.A. SCRUTTON 1974. Continental breakup and the development of
Post-Palaeozoic sedimentary basins around
Southern Africa. Bull. Geol. Soc. Am. 85:
1467-1474.
---; S.W. GOODLAND; A.K. MARTIN 1978.
Bathymetry and stratigraphy of the northern Natal Valley (S.W. Indian Ocean) : A
preliminary account. Marine Geol. 28:
89-106.
DIXEY, F. 1960. The Geology and Geo-
morphology of Madagascar, and a comparison
with eastern Africa. Quart. J. Geol. Soc.
116: 255-268.
EMBLETON, B.J.B.; M.W. MCELHINNY 1975.
The palaeoposition of Madagascar: Paleo-
magnetic evidence from the Isalo Group.
Earth Planet. Sci. Lett. 27: 329-341.
FISHER, R.L.; J.G. SCLATER; D.P. MCKENZIE
1971. Evolution of the Central Indian
Ridge, Western Indian Ocean. Bull. Geol.
Soc. Am. 82: 553-562.
FLORES, G. 1970. Suggested origin of
the Mozambique channel. Trans. Geol. Soc.
S. Africa. 73: 1-16.
FLOWER, M.F.J.; D.F. STRONG 1969. The
significance of sandstone inclusions in
lavas of the Comores Archipelago. Earth
Planet. Sci. Lett. 7: 47-50.
FÖRSTER, R. 1975. The geological history
of the sedimentary basin of southern Moz-
ambique and some aspects of the origin of
the Mozambique Channel. Palaeogeog.,
Palaeoclim., Palaeoecol. 17: 267-287.
FRANCIS, T.J.G.; D. DAVIES; M.N. HILL
1966. Crustal structure between Kenya
and the Seychelles. Phil. Trans. R. Soc.
A 259: 240-261.
GREEN, A.G. 1972. Seafloor spreading in
the Mozambique Channel. Nature, Phys. Sci.
236: 19-21.
KEAST, A. 1973. Contemporary Biotas and
the Separation Sequence of the Southern
Continents. In D.H. Tarling; S.K. Runcorn
(Eds.), Implications of Continental Drift
to the Earth Sciences. Acad. Press, London
1. pp. 309-343.
KENT, P.E. 1972. Mesozoic history of the
East Coast of Africa. Nature 238: 147-148.
---; J.T. O'B. PERRY 1973. The develop-
ment of the Indian Ocean margin in Tan-
zania. In Sedimentary Basins of the African Coast, G. Blant (Ed.), pp. 113-131.
MCELHINNY, M.W.; B.J.J. EMBLETON; L. DALY;
J.P. POZZI 1976. Palaeomagnetic evidence
for location of Madagascar in Gondwanaland.
Geology 4 (8): 455-457.
SCHUCHERT, C. 1928. The hypothesis of
continental displacement. Am. Ass. Petrol.

Geol.: 104-144.

SCRUTTON, R.A. 1973. Structure and evolution of the sea floor south of South Africa. Earth Planet. Sci. Lett. 19: 250-256.

--- 1978. Davie Fracture Zone and the movement of Madagascar. Earth Planet. Sci. Lett. 39: 84-88.

---; R.V. DINGLE 1976. Observations on the processes of sedimentary basin formation at the margins of Southern Africa. Tectonophys. 36: 143-156.

---; A.DU PLESSIS 1973. Possible marginal fracture ridge south of South Africa. Nature 242: 180-182.

SIMPSON, E.S.W.; H.D. NEEDHAM 1972. The floor of the Southeast Atlantic Ocean : A Review. Eos 52(2): 168.

SMITH, A.G.; A.HALLAM 1970. The fit of the southern continents. Nature 225: 138-144.

SOWERBUTTS, W.T.C. 1969. Crustal structure of the East African Plateau and rift valleys from gravity measurement. Nature 223: 143-146.

TALWANI, M. 1962. Gravity measurements on H.M.S. Acheron in South Atlantic and Indian Oceans. Geol. Soc. Am. Bull. 73: 1171-1182.

TARLING, D.H. 1978. The Geological-Geophysical Framework of Ice Ages. In J. Griffin (Ed.), Climatic Change. Cambridge Univ. Press. pp. 3-24.

--- 1979. Palaeomagnetic Reconstructions and the Variscan Orogeny. Proc. Ussher Soc. 4: 233-260.

---; in press. The geologic evolution of South America during the last 200 million years. In R.L. Ciochon; A.B. Chiarelli (Eds.), Evolutionary Biology of the New World Monkeys and Continental Drift. Plenum, N.Y.

---; P.E. KENT 1976. The Madagascar controversy still lives. Nature 261: 304-5.

TUCHOLKE, B.E.; R.E. HOUTZ; J. DIEBOLD; D. M. BARRETT 1978. Continental crust beneath the Agulhas Plateau in the southwestern Indian Ocean. Abs. with Prog. Geol. Soc. Am. 10(7) 506-507.

WEGENER, A. 1914. Die Entsehung der Kontinente und Ozeane. Braunschweig, p.94.

Seafloor constraints on the reconstruction of Gondwana (Abstract)

J. J. VEEVERS, C. McA. POWELL & B. D. JOHNSON
Macquarie University, North Ryde, Australia

Indian Ocean and South Atlantic M-series magnetic anomalies are used to test a new fit
of Gondwana in which Madagascar adjoins NW India, and southern South America fits
without deformation into the Weddell Sea. The first spreading stage, 150 m.y. to 125
m.y., was between East and West Gondwana, during which Madagascar rotated half-way to
its present position. The second stage, 125 m.y. to 105 m.y., began with the separation
of South America and Africa, and Antarctica-Australia from India-Madagascar, and ended
with Madagascar reaching its final position with respect to Africa. In the third stage,
105 m.y. to just before 90 m.y., spreading continued except between India-Madagascar
and Africa. A major reorganisation of spreading systems at the end of the third stage
saw India separate from Madagascar. A slight mismatch between the modelled transform
faults and the observed fracture zone on the eastern side of the Dronning Maud set
suggests that only small readjustments of the new reconstruction may be required.

Mesozoic ammonite faunas of Antarctica and the break-up of Gondwana

M.R.A.THOMSON
British Antarctic Survey, Cambridge, UK

While Greater Antarctica remained continental throughout Mesozoic time, the Antarctic Peninsula was partly covered by sea and now contains fossiliferous marine sequences of the era. Present data suggest that the break-up of Gondwana was initiated in the Middle to Late Jurassic with the first significant lateral movements in the Early Cretaceous. During this period, the Antarctic Peninsula area occupied an important paleogeographical position, near the junction of major new seaways that were opening between South America and Africa, along the eastern coast of Africa, and between India and the Antarctic-Australasian block. The biogeographical affinities of ammonite faunas in the Antarctic Peninsula are briefly reviewed in the light of these changing paleogeographical conditions.

Good ammonite assemblages of Late Jurassic to Late Cretaceous age are represented. Although the opening of the sea between Africa and India/Antarctica is not unambiguously reflected by the ammonite record, it may have been a migrational route from at least Kimmeridgian times. The Cretaceous faunas contain a significant proportion of pandemic species, in addition to geographically restricted forms which seen to represent "polar" or high-latitude faunas.

INTRODUCTION

In Gondwana the Antarctic Peninsula occupied a key, if somewhat enigmatic, biogeographical position on the Pacific margin of the supercontinent and at the junction of the seaways which were to open up between South America, Africa, Madagascar, India and Greater Antarctica. Its local paleogeography during the Mesozoic has been discussed elsewhere (Thomson, in press); suffice to say that throughout most of the Jurassic and Early Cretaceous it had the form of an elongate volcanic arc. Fewer data are available for the Late Cretaceous, but recent mapping by the British Antarctic Survey suggests that the arc remained active during Late Cretaceous and perhaps much of Cenozoic time.

Understanding the position of the Antarctic Peninsula within Gondwana has proved to be one of the most difficult problems in the study of the supercontinent. If it is assumed that the peninsula always held the same geographical relationship to Greater Antarctica that it does at the present day, then the preferred positions and orientations for the latter within Gondwana usually result in problematical overlaps of the peninsula with South America, the Falkland Plateau or South Africa. Under such constraints, only a reconstruction which results in the Antarctic Peninsula being placed alongside the coast of southern Chile solves this problem. Alternatively, the possibility that lesser Antarctica may consist of a number of separate continental blocks (Antarctic Peninsula, Thurston Island area, Ellsworth Mountains and Marie Byrd Land), which previously held different relative positions within Gondwana, has received considerable attention (e.g. Barker & Griffiths 1977; DeWit 1977). Expanding-earth reconstructions (e.g. Owen 1976) also require relative displacement between Greater and Lesser Antarctica.

Southern Hemisphere climatic history suggests that Drake Passage remained closed until about 23 m.y. ago (Barker & Burrell, in press). The present study indicates that at least a shallow-water connection existed between the South Pacific and South Atlantic from Late Mesozoic time. This may have been around the southern end of the Antarctic Peninsula or through the Drake Passage area, or both.

Middle Jurassic to Late Cretaceous ammonite faunas in the Antarctic Peninsula region (Table 1, Fig. 1) are the only ones known from the whole of Antarctica. The faunal distribution data are presented on maps of Smith and Briden (1977), notwithstanding their obvious shortcomings in relation to the Antarctic Peninsula. Paleogeographical data are from Harrington (1962) and papers in Moullade and Nairn (1978).

MIDDLE JURASSIC

Middle Jurassic faunas have been identified from one small area in the southern

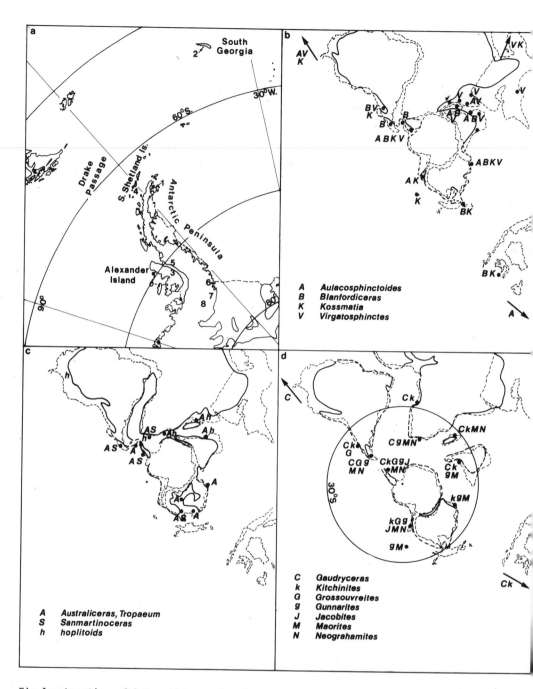

Fig.1a--Location of Antarctic ammonite faunas discussed in text; locality numbers correspond to Table 1.   b-d. Paleogeographical sketch maps for Gondwana during the Tithonian, Aptian and Campanian stages, respectively, showing the distributions of selected key ammonite genera.   The maps are based on those of Smith and Briden (1977).   Fine pecked lines denote existing continental shorelines; heavier lines in-dicate probable paleo-coastlines.

Table 1--Principal ammonite faunas of the Scotia arc and
Antarctic Peninsula area.

	LOCATION	AGE	REFERENCE
1	James Ross I. area	Late Cretaceous	Kilian & Reboul 1909; Spath 1953; Howarth 1958, 1966
2	South Georgia (Annenkov I.)	Early Cretaceous	Thomson et al., in press
3	Southeastern Alexander I.	(?)Kimmeridgian --Albian	Howarth 1958; Thomson 1974, 1979
4	S. Shetland Is.	Oxfordian-- Valangian	Tavera 1970; Covace- vich 1976; Smellie et al.; author, new data
5	Carse Point, Palmer Land	Tithonian	Thomson 1975
6	Lassiter Coast	Kimmeridgian --Tithonian	R. W. Imlay, unpubl.
7	Orville Coast	Late Jurassic	Author, new data
8	Behrendt Mountains	Middle - Late Jurassic	Quilty 1970; author, new data

Behrendt Mountains of southernmost Antarctic Peninsula region. A Bajocian fauna
of stephanoceratids and a possible sphaero-
ceratid is as yet poorly known, whereas the
Callovian has been identified on the basis
of some fragments of macrocephalitids.
According to Quilty (1970) these faunas are
consistent with a circum-Pacific migration
route with a connection to Europe via
Mexico.

LATE JURASSIC

Generalized paleogeography. Gondwana
was still more or less entire (Fig. 1b),
although rifting between Africa and India-
Antarctica probably had begun. Marine
incursions were restricted to its margins
except for a shallow sea which extended
southwards over the horn of Africa onto
northern and western Madagascar and East
Africa. When a southerly extension of
this sea reached the Antarctic Peninsula
region is uncertain; as early as the Toarcian (Hallam 1973) seems somewhat speculative, and the paleontological evidence for
the presence of Late Jurassic marine rocks
in southern Africa is equivocal (McLachlan
et al. 1976). The Antarctic Peninsula was
more or less isolated as an elongate land
mass (Thomson, in press). Evidence from
the Orville Coast (7 in Fig. 1a) indicates
that the Latady Fm sea passed distally into
a relatively deep-water facies, and the
presence of a sea between South Africa,
the Antarctic Peninsula and Dronning Maud
Land is a strong possibility, although its
size and shape are uncertain.
Biogeography. Marine faunas of Tithonian age are well represented in the Antarctic Peninsula region, although those of
Oxfordian and Kimmeridgian age are as yet
poorly known (cf. Howarth 1958; Stevens
1967; Quilty 1970). The recent discovery

of an ammonite resembling the Oxfordian
genus *Epimayites* on Low Island, South Shetland Islands (author, new data) is significant because it fills a gap in the known
circum-Gondwana distribution of the group
(Stipanicic et al. 1975). In the Latady
Fm of SE Palmer Land and Orville Coast,
ammonites are uncommon and mostly poorly
preserved. Nevertheless, new collections
from the Orville Coast contain rare specimens of the Kimmeridgian-Early Tithonian
Aspidoceras, and a variety of perisphinctids,
including possible examples of *Katroliceras*
and *Subplanites,* with Early Tithonian or
older affinities.
    The Tithonian faunas of Alexander Island
are characterized by *Virgatosphinctes* and
Aulacosphinctoides (Thomson 1979). Also
present are *Pterolytoceras,* haploceratids
and *Uhligites.* A single specimen of
Substeueroceras has also been obtained,
whereas species of *Kossmatia* are known
from western Palmer Land (Thomson 1972) and
the Latady Fm only. The youngest Tithonian
ammonite fauna is probably represented by
an assemblage of berriasellids with
Blandfordiceras. At the species level,
the Tithonian faunas show both South
American and Himalayan affinities, with
species close to *Virgatosphinctes andesensis,*
V. mexicana and *V. lenaensis* indicating the
former, and *Pterolytoceras exoticum,*
Uhligites aff. *kraafti, V. denseplicatus* and
V. frequens and *Aulacosphinctoides* aff.
sparsicosta and *A. smithwoodwardi* (?) the
latter. *Kossmatia* is a somewhat variable
genus with species recorded from the Himalayas, Indonesia, Antarctica, New Guinea,
Australia, and possibly New Zealand, as well
as many Northern Hemisphere localities;
fragments from the Orville Coast appear to
belong to the Himalayan *K. tenuistriata.*
If the andean *Pseudoblanfordia* is a synonym
of *Blanfordiceras* then this group shows a
remarkable distribution concentrated in Himalaya, Pakistan and Madagascar on the one
hand, and the antarctandean region on the
other (Fig. 1b).

    The observed faunal relationships of
Antarctic Late Jurassic ammonite assemblages
may be explained either by a circum - Gondwana
migration route, and/or by migration through
a trans-Gondwana seaway along the eastern
coast of Africa initiated at some time during
this period. The possible occurrence in
Antarctica of such genera as *Katroliceras*
and *Subplanites,* typical of the East Africa-
Cutch region, could argue for the presence of
the East African seaway in Kimmeridgian
times. A similar interpretation could be
placed on the distribution of such genera
as *Virgatosphinctes, Aulacosphinctoides*
and *Blanfordiceras.* Nevertheless, the

271

occurrence of *Subplanites* and *Blanfordiceras* in Indonesia and of *Aulacosphinctoides* in New Zealand, indicates that some at least may have migrated via the long route. Furthermore, it certainly appears that this was the case for such benthonic bivalves as the inoceramids and buchiids, which are absent from the east Africa-Cutch region but are widely distributed around the Tethyan and Pacific margins of Gondwana. At present the opening of the east African seaway is not clearly reflected by the ammonite record.

EARLY CRETACEOUS

Generalised paleogeography. The breakup of Gondwana was well under way. The South Atlantic had begun to open as an ocean, although it was still blocked to the north in the Brazil-Nigeria region and there were probably shallow-water barriers at the Rio Grande-Walvis ridge and the Falkland plateau. A long seaway was present along the eastern coast of Africa, and Madagascar was probably moving relatively eastwards. Marine beds in the Upper Gondwanas of the Coromandel Coast, southern India (Spath 1933) indicate a narrow marine incursion between India and Antarctica. Much of central Australia became flooded by an epicontinental sea from the north, but marine incursions onto other Gondwana areas were largely confined to the coastal margins.

Biogeography. The most diverse Early Cretaceous faunas occur in Alexander Island, and less diverse faunas are present in South Georgia and in the South Shetland Islands (Table 1). The oldest contain the widely distributed *Spiticeras* s.l., together with *Bochianites* and the distinctive Himalayan-Indonesian *Haplophylloceras*. Although *Himalayites* may range into the Berriasian in Alexander Island, its association with *Blanfordiceras* and *Spiticeras* in the South Shetland Islands probably marks a very Late Tithonian or Jurassic-Cretaceous boundary fauna. The Valangian is represented in the same area by *Neocomites* and *Bochianites*, but the most varied faunas are those designated as Aptian-Albian in Alexander Island. The precise stratigraphical age of these is in doubt. Although they contain a variety of supposedly characteristic Aptian aconeceratids (particularly *Sanmartinoceras*), large heteromorphs of *Tropaeum* and *Australiceras* types, and the later *Eotetragonites*, associated small heteromorphs resembling *Acrioceras* and European crioceratitids suggest a late Neocomian age, and *Sanmartinoceras* is now known in the Barremian of South Africa (Kennedy & Klinger 1979). These faunas are succeeded by problemati-

cal silesitids, which resemble Barremian and Lower Aptian European species, and finally by *Hamites*.

The spiticeratids and *Bochianites* of the earliest Cretaceous faunas maintain the close relationships with Himalayan-Indonesian forms like their late Tithonian forerunners; suggested South American affinities of the South Shetland Islands faunas (Tavera 1970) were based on poor material and were perhaps overstated. Although Covacevich (1976) argued that the affinities of the Valanginian faunas were closest to the east African region, free interchange of faunas with Pacific North America, Mexico and Europe is also implied by the presence of *Neocomites* and *Bochianites*. The *Tropaeum-Australiceras-Sanmartinoceras* fauna has counterparts in Australia (despite the great differences in prevailing oceanographical conditions), South Africa and Patagonia (Fig. 1c). Similar faunas are present in the Northern Hemisphere where they are also associated with hoplitoids (Day 1969). Although the latter occur along the margins of the east African seaway they are unknown farther south than South Africa where cheloniceratids are present in the Aptian (Kennedy & Klinger 1975). The small heteromorphs so typical of the Early Cretaceous achieved a wide distribution, which is a little surprising in view of their generally supposed benthonic mode of life. Possibly they were widely tolerant and able to take advantage of current drift during a planktonic larval stage (cf. Kennedy & Cobban 1976). Interchange with European faunas probably occurred via the Pacific coast of South America and a Mexican seaway, or via the east African seaway.

LATE CRETACEOUS

Generalized paleogeography. By Late Cretaceous times the fragments of Gondwana were becoming well separated by oceanic areas formed as a result of sea-floor spreading (Fig. 1d). The North and South Atlantic oceans had been connected since mid-Cretaceous times. However, Sclater *et al.* (1977) have suggested that a deep-water connection from the Antarctic Ocean northwards would still have been impeded by the Falkland plateau and the Rio Grande-Walvis ridge. Antarctica and Australasia were still attached, although a shallow arm of the sea appears to have extended eastwards along the greater part of the southern coast of Australia (cf. Ludbrook, in Moullade & Nairn 1978). New Zealand was beginning to break away at this time (Hayes & Ringis 1973). Marine incursions onto the Gondwana areas were confined largely to the continental margins, except for a shallow epicontinental sea covering much of the Sahara region (Furon 1963).

Biogeography. In Antarctica, Late Cretaceous faunas are known only from the NE coast of Graham Land (James Ross Island area). At present pre-Campanian faunas have been identified on the basis of inoceramid bivalves alone (Crame, in press) but rich ammonite faunas are present in the Campanian (Table 1). Cenomanian to Campanian inoceramids show strong European affinities, both at group and species levels. Strong relationships with Madagascan and Indian species suggest migration through the eastern Mediterranean and along the east African seaway or proto-Indian Ocean (Crame, in press).

The Campanian ammonite faunas are characterized by the kossmaticeratid genera, *Maorites, Jacobites, Neograhamites, Grossouvreites* and *Gunnarites*. All five are represented in New Zealand (Henderson 1970), but only three species are common to the two regions. Judging by faunal lists in Katz (1963) and Hünicken *et al.* (1975), and descriptions by Lahsen and Charrier (1972), correspondence with Patagonia is good, even at the specific level, with eight species in common. However, *Jacobites* has yet to be recorded from South America. Kossmaticeratid faunas are present also in South Africa (Kennedy & Klinger 1975 and references therein), Madagascar (Collignon 1955), southern India (Stoliczka 1863-6) and western Australia (Brunnschweiler 1966), but correspondence with the Antarctic faunas is poorer. Although other kossmaticeratid genera occur in Japan, and two Maestrichtian forms (*Pseudokossmaticeras* and *Brahmaites*) in Europe, the group is largely confined to "Gondwana". The five Antarctic genera have an even more restricted distribution (Fig. 1b), being found only at localities apparently within about 60° paleolatitude of the south pole. The reason cannot be that there were geographical barriers to their migration because they are unknown in the Campanian faunas of Angola, although other elements of the Antarctic fauna are present there (see below). Possibly the distribution of the kossmaticeratids restricted to Antarctica was climatically controlled.

By contrast, many other constituents of the faunas achieved wide distributions. The pandemic distribution of the tetragonitid *Pseudophyllites* has been commented on elsewhere (Kennedy & Cobban 1976), a distribution which is all the more remarkable when it is remembered that the species *P. indra* (Forbes) has been reported from Alaska, British Columbia, southern France, Zululand, Madagascar, southern India, Japan,

western Australia and Brazil, as well as Antarctica. The pachydiscids are represented by the widely-distributed *Pachydiscus, Eupachydiscus* and *Patagiosites*. *Gaudryceras* and *Kitchinites* have similar distributions to the kossmaticeratids in the fauna and are also found in Japan and Angola, and the former in California and Alaska. As yet the heteromorph components of the Antarctic faunas are not well known.

SUMMARY

Interrelated fossil marine faunas around the Tethyan and Pacific margins of Gondwana indicate free migration. As break-up proceeded, cross-cutting seaways opened up new areas for colonization and trans-Gondwana migration. Evidence from the key area of the Antarctic Peninsula suggests that, whereas ammonites probably took advantage of such new seaways from at least Late Jurassic times, some benthonic bivalves kept to the old marginal route. Although a deep-water connection between the South Atlantic and Pacific oceans was not formed until the Late Tertiary, Mesozoic faunas migrated between the two, probably via relatively shallow connections south of the peninsular or between the peninsula and South America. The Antarctic fossil faunas included both pandemic species and species with latitudinal restrictions which may have been climatically controlled. In approximately Aptian times there was a widespread Southern Hemisphere assemblage which largely mirrored that of the European area and western North America, and in Campanian times a kossmaticeratid fauna that was confined to within about 60° of the pole.

ACKNOWLEDGMENTS

The author expresses his apologies to all those workers on related fauna whose papers have been regrettably excluded from mention because of shortage of space. Discussions with my colleague, Dr J.A. Crame, are gratefully acknowledged.

REFERENCES

BARKER, P.F.; J. BURREL. In press. The influence upon Southern Ocean circulation, sedimentation and climate of the opening of Drake Passage. In C. Craddock (Ed.), Antarctic geoscience. Madison: University of Wisconsin Press.

BARKER, P. F.; D.H. GRIFFITHS 1977. Towards a more certain reconstruction of Gondwanaland. Phil. Trans. R. Soc., B. 279: 143-59.

BRUNNSCHWEILER, W. A. 1966. Upper Cretaceous ammonites from the Carnarvon Basin of Western Australia. I: The heteromorph Lytoceratina. BMR Bull., Geol. Geophys. 58: 58 pp.

COLLIGNON, M. 1955. Ammonites néocretacées du Menabe. III: Les Kossmaticeratidae. Ann. Géol. Serv. Mines, Madagascar 22: 54 pp.

COVACEVICH, V. C. 1976. Fauna Valanginiana de Peninsula Byers, Isla Livingston, Antártica. Rev. Geol. Chile 3: 25-46.

CRAME, J. A. in press. Upper Cretaceous inoceramids (Bivalvia) from the James Ross Island group and their stratigraphical significance. Brit. Antarc. Surv. Bull.

DAY, R. W. 1969. The Lower Cretaceous of the Great Artesian Basin. In K.S.W. Campbell (Ed.), Stratigraphy and paleontology. Essays in honour of Dorothy Hill. Canberra: Australian National University Press. Pp 140-73.

DE WIT, M. J. 1977. The evolution of the Scotia Arc as a key to the reconstruction of southwestern Gondwanaland. Tectonophysics 37: 53-81.

FURON, R. 1963. Geology of Africa. Edinburgh & London: Oliver & Boyd Ltd.

HALLAM, A. 1973. Distributional patterns in contemporary terrestrial and marine animals. Spec. Pap. Palaeontol. 12: 93-105.

HARRINGTON, H. J. 1962. Paleogeographic development of South America. AAPG Bull. 46: 1773-814.

HAYES, D. W.; J. RINGIS 1973. Seafloor spreading in the Tasman Sea. Nature (London) 243: 454-8.

HENDERSON, R. A. 1970. Ammonoidea from the Mata Series (Santonian-Maastrichtian) of New Zealand. Spec. Pap. Palaeontol. 6: 82 pp.

HOWARTH, M. K. 1958. Upper Jurassic and Cretaceous ammonite faunas of Alexander Island and Graham Land. Falkland Is. Dep. Surv. Sci. Rep. 21: 16 pp.

--- 1966. Ammonites from the Upper Cretaceous of the James Ross Island group. Brit. Antarc. Surv. Bull. 10: 55-69.

HÜNICKEN, M. A.; R. CHARRIER; A. LAHSEN 1975. Baculites (Lytoceratina) de la Provincia de Magallanes, Chile. Actas 1er Cong. Argentina, Paleontol. Bioestratig. 2: 115-40.

KATZ, H. R. 1967. Revision of Cretaceous stratigraphy in Patagonian Cordillera of Ultima Esperanza, Magallanes Province, Chile. AAPG Bull. 47: 506-24.

KENNEDY, W. J.; W. A. COBBAN 1976. Aspects of ammonite biology, biogeography and biostratigraphy. Spec. Pap. Palaeontol. 17: 94 pp.

---; H. C. KLINGER 1975. Cretaceous faunas from Zululand and Natal, South Africa. Introduction, Stratigraphy. Bull. Brit. Mus.

(Nat. Hist.) Geol. 25: 263-315.

---;--- 1979. Cretaceous faunas from Zululand and Natal, South Africa. The ammonite superfamily Haplocerataceae Zittel, 1884. Ann. S. African Mus. 77: 85-121.

KILIAN, W.; P. REBOUL 1909. Les céphalopodes néocrétacées des îsles Seymour et Snow Hill. Wiss. Ergeb. Schwed. Südpolarexped. 3(3): 75 pp.

LAHSEN, A.; R. CHARRIER 1972. Late Cretaceous ammonites from Seno Skyring-Strait of Magellan area, Magallanes Province, Chile. J. Paleontol. 46: 520-32.

McLACHLAN, I.R.; P. W. BRENER; I. K. McMILL 1976. The stratigraphy and micropalaeontology of the Cretaceous Brenton Formation and the PB-A/1 Well, near Knysna, Cape Province. Trans. Geol. Soc. S. Africa 79 341-70.

MOULLADE, M.; A. E. M. NAIRN (Eds.) 1978. The Phanerozoic geology of the World: II. The Mesozoic, A. Amsterdam: Elsevier.

OWEN, H. G. 1976. Continental displacement and expansion of the Earth during the Mesozoic and Cenozoic. Phil. Trans. R. Soc. 281: 223-91.

QUILTY, P. G. 1970. Jurassic ammonites Ellsworth Land, Antarctica. J. Paleontol 44: 110-16.

SCLATER, J. G.; S. HELLINGER; C. TAPSCOTT 1977. The paleobathymetry of the Atlantic Ocean from the Jurassic to the present. Geol. 85: 509-52.

SMELLIE, J. L.; R. E. S. DAVIES; M. R. A. THOMSON. In press. Geology of a Mesozoic intra-arc sequence on Byers Peninsula, Livingston Island, South Shetland Islands. Brit. Antarc. Surv. Bull.

SMITH, A. G.; J. C. BRIDEN 1977. Mesozoic and Cenozoic paleocontinental maps. Cambridge: Cambridge University Press. 63 pp.

SPATH, L. F. 1933. Revision of the Jurassic cephalopod fauna of Kachh (Cutch). Mem. Geol. Surv. India, Palaeontol. indica N.S. 9 (2): 659-945.

---; 1953. The Upper Cretaceous cephalopod fauna of Graham Land. Falkland Is. Dep. Surv. Sci. Rep. 3: 60 pp.

STEVENS, G. R. 1967. Upper Jurassic fossils from Ellsworth Land, West Antarctica, and notes on Upper Jurassic biogeography of the South Pacific region. N.Z. J. Geol. Geophys. 10: 345-93.

STIPANICIC, P. H.; G. E. G. WESTERMANN; A. C. RICCARDI 1975. The indopacific ammonite Mayites in the Oxfordian of the southern Andes. Ameghiniana 12: 281-305.

STOLICZKA, F. 1863-66. Ammonitidae, with revision of the Nautilidae etc. In H.F. Blanford; F. Stoliczka. The fossil Cephalopoda of the Cretaceous rocks of

southern India.   Mem. Geol. Surv. India,
Palaeontol. indica, 1:   41-216.
TAVERA, J. J. 1970.   Fauna Titoniana-
Neocomiana de la Isla Livingston, Islas
Shetland del Sur, Antártica.   Ser. Cient.
Inst. Antárct. Chileno 1:   175-86.
THOMSON, M. R. A. 1974.   Ammonite faunas of
the Lower Cretaceous of southeastern Alexan-
der Island.   Brit. Antarct. Surv. Sci.
Rept. 80:   44 pp.
--- 1975.   Upper Jurassic Mollusca from
Carse Point, Palmer Land.   Brit. Antarct.
Surv. Bull. 41 & 42:   31-42.
--- 1979.   Upper Jurassic and Lower Cre-
taceous ammonite faunas of the Ablation
Point area, Alexander Island.   Brit. Antarct.
Surv. Sci. Rep. 97:   37 pp.
---.   In press.   Mesozoic paleogeography
of western Antarctica.   In C. Craddock (Ed.),
Antarctic geoscience.   Madison: University
of Wisconsin Press.
---; P. W. G. TANNER; D. C. REX.   In press.
Fossil and radiometric evidence for ages of
depositon and metamorphism of sedimentary
sequences on South Georgia.   In C. Craddock
(Ed.), Antarctic geoscience. Madison:
University of Wisconsin Press.

The distribution of tetrapods and the break-up of Gondwana

EDWIN H. COLBERT
Museum of Northern Arizona, Flagstaff, USA

Our concept of Gondwana relies in part upon the distributions of fossil terrestrial vertebrates. The distribution of late Triassic tetrapods indicates that although the rifting of Gondwana may have been initiated then, there were nonetheless open avenues for the movements of land-living vertebrates across the breadth of the ancient supercontinent. In late Jurassic time a connection between Africa and South America probably still existed, and this seems to have persisted into the Cretaceous, even though there was an opening of the South Atlantic Ocean. Antarctica and Australia, connected with each other and probably with South America, may have provided a passage for the migration of early marsupials until well into Eocene time. Peninsular India, so frequently depicted as drifting through much of the Mesozoic, may have been connected to Africa until very late Cretaceous time, after which it may have drifted at a rapid rate toward its collision with Asia. The distributions of various Mesozoic reptiles, particularly dinosaurs, and of early Cenozoic mammals uphold these conclusions.

INTRODUCTION

Our modern concept of Gondwana is based upon many lines of evidence, of which the distributions of fossil tetrapods, especially the remains of terrestrial, fresh-water and estuarine vertebrates deserve particular attention. Crucial though the concept of Plate Tectonics may be to our understanding of the earth, and important though the physical evidence may be that has given such strong support to the theory, due attention must be given to all aspects of earth science, and a fossil accompanied by documented stratigraphic and geographic information cannot be ignored.

If fossils are of importance in determining land connections, they are equally important in determining the time when such land connections were severed. Therefore, it is here maintained that the distributions of fossils, particularly those of land-living vertebrates have much to tell us not only about the constitution of Gondwana, but also about its break-up (Cracraft 1974; Cox 1974; Hallam 1966, 1973).

PERMO-TRIASSIC TRANSITION

It is perhaps safe to say that there is now general agreement as to the concept of an intact Gondwana at the end of Paleozoic and the beginning of Mesozoic time. So far as fossils are concerned the discoveries of late Paleozoic terrestrial tetrapods supports this.

The Karroo Permo-Triassic beds in South Africa form a classic stratigraphic-paleontological sequence in which there are successive tetrapod faunas of unexcelled

variety and abundance (Kitching 1977) (Fig. 1). These faunas are dominated in an overwhelming fashion by the remains of therapsid reptiles, the so-called mammal-like reptiles, some of which, especially the carnivorous theriodonts, had approached very closely the threshold between mammal and reptile. Other therapsids, such as the dicynodonts, were highly specialized along their own lines, having beaked jaws and skulls characteristically distinguished by a pair of upper tusks. The lesser elements in the Karroo faunas include procolophonids - "lizard-like" cotylosaurian reptiles of generally primitive aspect, eosuchians (predecessors of the true lizards), and labyrinthodont amphibians, the solid-skulled amphibians so typical of late Paleozoic and Triassic time.

Beyond South Africa the Permian vertebrate record in Gondwana is sparse. A procolophonid and a couple of eosuchians are known from Madagascar. The recent discovery in South America of *Endothiodon*, a dicynodont very characteristic of the Upper Permian of South Africa, is especially significant, and future discoveries in South America might well strengthen the paleontological evidence for the close physical relationship between that continent and Africa during Permian time.

There are, however, abundant Lower Triassic tetrapods throughout Gondwana. One of the characteristic fossils in the Karroo sequence is *Lystrosaurus* (Fig.2), a highly specialized dicynodont with a flat-roofed skull and a strongly down-turned snout. Associated with it is *Thrinaxodon*, a small, advanced carnivorous therapsid,

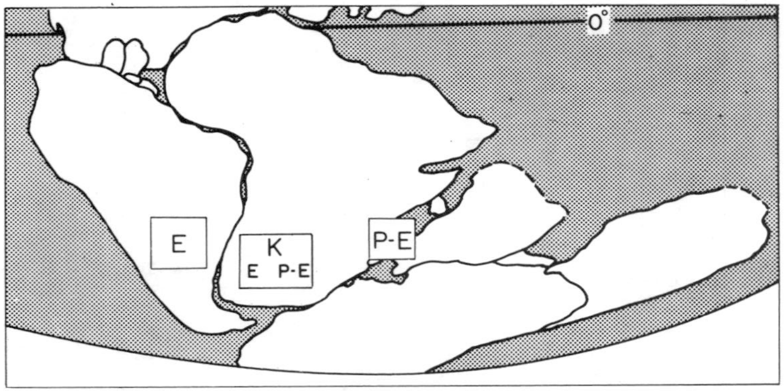

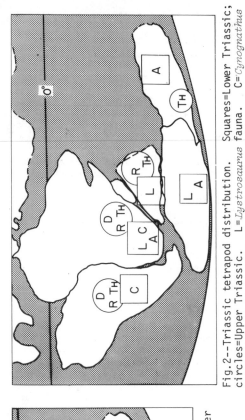

Fig. 1--End of the Permian; tetrapod distribution. K = upper Permian Karroo sequence with amphibians and reptiles. E=Endothiodon. P-E=procolophonids and eosuchians.

Fig. 2--Triassic tetrapod distribution. Squares=Lower Triassic; circles=Upper Triassic. L=*Lystrosaurus* fauna. C=*Cynognathus* fauna. Th=theco-donts. A=labyrinthodont amphibians. D=theropod dinosaurs. R=rhynchosaurs.

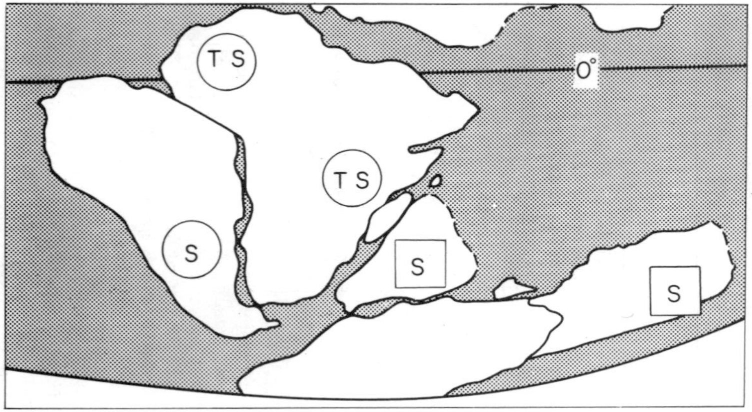

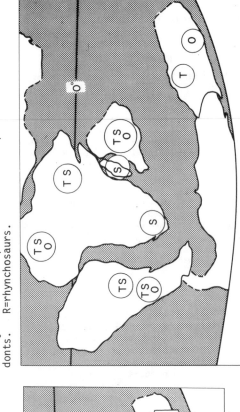

Fig. 3--Jurassic dinosaur distribution. Squares=Lower Jurassic; circles=Upper Jurassic. T=theropods. S=sauropods.

Fig. 4--Late Cretaceous dinosaur distribution. T=theropods. S=sauropods. O=ornithischians.

278

Procolophon, a persistent cotylosaurian reptile, and *Prolacerta*, a "lizard-like" eosuchian. In addition there is a small labyrinthodont amphibian, *Lydekkerina* (Kitching 1968).

In recent years these same genera of reptiles (for some even the same species) have been found in Antarctica (Colbert 1974; Colbert and Kitching 1975, 1977). A small lydekkerinid is also present on the south polar continent (Colbert and Cosgriff 1974). *Lystrosaurus* occurs in the Panchet beds of peninsular India, which reinforces evidence of the Indian peninsula's having been wedged between Africa and Australia in early Gondwana (Tripathi and Satsangi 1963). *Lystrosaurus* and its associated reptiles have not been found in Australia, but labyrinthodont amphibians closely related to those of the Lower Triassic part of the Karroo Series are in the Blina shales along its NW coast (Cosgriff 1965, 1969). Likewise, although *Lystrosaurus* and other members of its fauna are not presently known from South America (although indications of a *Lystrosaurus* Zone recently have been discovered), there are in the Puesto Viejo Fm of Argentina *Cynognathus*, an advanced carnivorous therapsid, and *Kannemeyeria*, a large dicynodont, both typical of the Lower Triassic *Cynognathus* Zone of the Karroo, immediately above the *Lystrosaurus* Zone (Bonaparte 1967). These and other Argentinian therapsids closely related to those of South Africa furnish paleontological evidence, in accordance with geophysical data, for the connection at the beginning of Mesozoic time of the two continents now separated by the South Atlantic Ocean.

END OF THE TRIASSIC

The passage from Lower to Upper Triassic sediments in the Karroo Series is marked by a dramatic decline of the once-dominant therapsid reptiles and the equally dramatic appearance of those reptiles that were to rule the continents during the remainder of Mesozoic time-particularly the dinosaurs and crocodilians, the archosaurian reptiles. Evidently there was an exchange of dinosaurs between Gondwana and Laurasia at this stage; *Syntarsus* from the late Triassic Forest Sandstone of Rhodesia is remarkably close to *Coelophysis* from the Chinle Formation of southwestern North America, both genera being small, carnivorous coelurosaurian dinosaurs (Raath 1969); *Plateosauravus* of the Stormberg beds of South Africa is a southern correlative of *Plateosaurus* of central Germany and of *Lufengosaurus* of western China, these being rather large, heavy prosauropod dinosaurs (von Huene 1956); and *Fabrosaurus* also from Stormberg beds is matched by a small, ancestral ornithischian dinosaur from Arizona (Colbert, in press).

Dinosaurs in Gondwana were certainly spread across much of the ancient supercontinent at the close of the Triassic Period. In South America, there are primitive saurischian dinosaurs of late Triassic age, notably *Plateosaurus*, *Riojasaurus* a melanorosaurid, *Pisanosaurus*, an early ornithischian dinosaur, and several other genera, whose exact relationships are still to be determined (Sill 1969).

In the Upper Triassic Maleri Formation of India are the remains of rhynchosaurs, characterized by beaked skulls, and with jaws having on the sides batteries of small, button-like teeth, perhaps for crushing tubers (Chatterjee 1974). These rather large, heavily-built reptiles, related to the modern rhynchocephalians, are found also in the Middle and Upper Triassic beds of Africa and South America, and the close relationships of the several involved genera point clearly to lines of tetrapod distributional communication within Gondwana at the close of Triassic time.

Other tetrapod remains reinforce this picture. Certain thecodont reptiles, these belonging to the characteristic Triassic order from which dinosaurs and other archosaurs were derived, are found in Africa, South America, India and Australia (Kitching 1977; Bonaparte 1966; von Huene 1940; Camp and Banks 1978). Therefore, we conclude that Gondwana was still one land mass at the close of Triassic history. The physical evidence would seem to show that although Gondwana was intact as the Triassic Period drew to a close, Pangaean rifting had been initiated by the beginning of a separation between North America and Africa-South America (Windley 1977).

JURASSIC

Lower Jurassic sediments are notably lacking in the remains of terrestrial tetrapods. There are, however, some significant fossils from this stage of earth history (Fig.3), of which particular mention should be made of the sauropod dinosaur, *Barapasaurus*, from the Kota Formation of peninsular India (Jain *et al.* (1975). It not only demonstrates quite clearly that the sauropods had become giants at the beginning of Jurassic history but also shows that at this date peninsular India must have been in close contact with another continent, perhaps with some other Gondwana area.

That Gondwana was still essentially intact in Middle Jurassic time also is indicated by the presence of the sauropod *Rhoetosaurus* in Queensland, Australia. This, together with various dinosaur trackways, also found in Queensland, implies a free passage of these gigantic tetrapods into Australia perhaps from Africa by way of Antarctica (R.E. Molnar, pers. comm. 1979).

In Upper Jurassic sediments tetrapods are numerous, dominated by the fossils of gigantic dinosaurs associated with other land-living vertebrates such as crocodilians, lizards and early mammals. One of the most notable of late Jurassic assemblages is in the Tendaguru beds in eastern Africa, containing dinosaurs closely related to and in some cases identical with those of the North American Morrison Formation. The extraordinarily large sauropod, *Brachiosaurus*, is found not only in Africa but also in Utah--an indication of wide movements of this tetrapod across Pangaea during the closing years of Jurassic history (Colbert 1965). The very recent discovery of sauropod dinosaurs in Upper Jurassic sediments of Argentina lends additional weight to the view that Gondwana was still largely intact at the end of Jurassic history, even though the opening of the South Atlantic ocean perhaps had begun (Bonaparte 1979). In this connection we note geophysical evidence indicating that the spreading ridge axis of the South Atlantic crust formed by 127 m.y. ago in earliest Cretaceous time (Windley 1977). Thus there is no significant discrepancy between the geophysical evidence and that of tetrapod distribution.

CRETACEOUS

Cretaceous rocks of continental origin are notable for the richness of their included fossil tetrapods throughout the world, yet this has often been given less than full attention. Some Cretaceous maps show South America as an island continent, separated from Africa by a rather broad South Atlantic Ocean, and India also as an island (as in the Triassic and Jurassic Periods) far removed from Africa and Asia (Dietz and Holden 1970). The Antarctic-Australian block likewise is commonly indicated as a separate land mass. Yet we know that large dinosaurs, represented by bones, skeletons, and trackways, lived in South America, peninsular India, Australia and New Zealand during the Cretaceous period (R.E. Molnar, pers. comm. 1979). They show close relationships to dinosaurs in other regions, and unless one believes in an unprecedented expression of parallel evolution, it seems obvious that there were overland lanes of communication within Gondwana during the later phases of Mesozoic history.

For example the sauropod dinosaurs were widely distributed (Fig.4) throughout Gondwana during Cretaceous time, with such genera as *Titanosaurus*, *Laplatasaurus* and *Antarctosaurus* in the Neuquén beds of Argentina and the Lameta Formation of central India (von Huene 1929; von Huene and Matley 1933). *Algoasaurus*, a titanosaurid, is present in the Uitenhage beds of southern Africa (Broom 1904), and *Laplatasaurus* in Madagascar as well as Argentina and India (Romer 1966). It can be argued that the apparent wide distribution of these genera may be in part illusory, owing to unduly broad taxonomic identifications. Yet even if generic identities are not involved in all of the above-cited examples, the fossils are nonetheless so closely related as to be significant. Large sauropod dinosaurs obviously were able to move back and forth through Gondwana until well into Cretaceous time.

The same was true for other dinosaurs as well, particularly the gigantic carnivorous megalosaurs, and some of the armoured dinosaurs (Colbert and Merrilees 1967). My argument centres upon the dinosaurs, since they necessarily required land connections order to become so widely distributed as they were, but other reptiles, notably the turtles and crocodilians (Romer 1966; Buffetant 1977) were also prevalent.

Thus a picture emerges of a largely intact continent until at least late Cretaceous time, even though rifting had developed to such an extent that the North and South Atlantic oceans were widely open in their respective southern regions (Larson and Ladd 1973). seems possible that during early Cretaceous time the eastern bulge of Natal, Brazil, was still in contact with the western reentrant Africa (probably as shown by Windley 1977). If the rift between these continental blocks took place along the lineaments of the Benue Trough, as seems likely, the date of rifting must have been late Cretaceous.

In the face of the fossil evidence one can hardly reconcile a prolonged Mesozoic drift of peninsular India toward Asia, as depicted by Dietz and Holden in 1970 and by others. The Lameta Formation of Central India, in which are found the dinosaurs mentioned above as well as carnivorous and armoured forms, is late Cretaceous, indicating a late Cretaceous separation of the Indian peninsula from its long-established position in Gondwana and a rapid drifting toward the Asiatic mainland, much as has been outlined on the basis of geophysical evidence by Molnar and Tapponnier (1977).

Antarctica and Australia evidently still were joined to each other in late Cretaceous time, and it is very probable that by this date Antarctica had some contact with South

America by way of the Antarctic Peninsula (Elliot 1972). These are the final pieces of the picture puzzle of a late Cretaceous Gondwana with most of the elements still in contact.

BREAK-UP

When did a significant break-up take place? Very probably this series of events occurred during the transition from Mesozoic to Cenozoic time. We know from the fossil evidence that South America was an island continent during much of Cenozoic history, yet at the very beginning of the Cenozoic, and again at the end of this era it did have connections with other parts of the world. The presence of marsupials and of primitive condylarths, ancient hoofed mammals, in South and North America indicates the passage of these animals back and forth during some part of the Paleocene Epoch. Soon, however, this link was broken--as shown by the independent evolution of mammals in South America-- not to be reestablished until late in the Pliocene epoch. There was probably a connection between South America and other parts of Gondwana in late Cretaceous time, probably extending into the beginning of the Cenozoic era and making Antarctica an intermediate link between South America and Australia. Some students of early mammals think that it was by way of such a ligation that the marsupials entered Australia from a New World centre of origin (Cox 1973). If this is correct it may be assumed that the link was broken after the marsupials invaded Australia but before the placental mammals could make the crossing.

After that Australia and Antarctica remained connected, probably into Eocene times, following which Australia, with its cargo of marsupials, drifted to its present position.

The history of early mammals is a blank in peninsular India, leading to the conclusion that during the first part of the Cenozoic Era this mass probably *was* an island drifting to the northeast, following its late Cretaceous separation from Gondwana. Some upper Eocene mammals have been found in Jammu and Kashmir, and mammals appear in peninsular India in sediments of mid-Tertiary age. This accords with Molnar and Tapponier (1977), who state that the subcontinent became connected with the Asiatic mainland about 40 m.y. ago--that is during late Eocene-early Oligocene time.

In summary, the evidence of land-living tetrapods, paralleling for the most part modern geophysical evidence, indicates that the break-up of Gondwana, although initiated during the Triassic-Jurassic transition, did not reach its culmination until the close of Cretaceous time. As Windley has remarked: "It takes a long time to fragment a supercontinent. The break-up of Pangaea began in the early Jurassic, 180 m.y. ago, and the last continent was not separated until the early Tertiary (45 m.y. ago)." (Windley 1977, p. 213.)

The distributions of fossil tetrapods show that South America probably broke away from Africa during the later stages of Cretaceous history but retained a connection with the Antarctica-Australian segment until the end of the Cretaceous Period or possibly into the advent of Cenozoic time. Australia and Antarctica remained as a single land mass until the Eocene Epoch. It was probably during late Cretaceous years that New Zealand separated from Australia. Perhaps the most notable discrepancy between the geophysical and vertebrate paleontological evidence concerns the timing of the separation and drift of peninsular India from Gondwana and toward the Asiatic mainland. It seems that the resolution of this discordance may be obtained by postulating a late Cretaceous rifting of India from its earlier connection and a very rapid movement to the north. Africa, in effect the centre of Gondwana, remained as the great relatively stationary remnant of the former supercontinent, diminished slightly by the separation of Madagascar. This schedule for the break-up of Gondwana accords with the known distributions of land-living tetrapods; it can be viewed as a realistic scenario for the fragmentation of the once vast southern hemisphere supercontinent.

REFERENCES

BONAPARTE, J.F. 1966. Chronological survey of the tetrapod-bearing Triassic of Argentina. Breviora 251: 1-13.
---1967. New vertebrate evidence for a southern Transatlantic connexion during the Lower or Middle Triassic. Palaeontology 10: pt. 4, pp. 554-563.
---1979. Dinosaurs: A Jurassic Assemblage from Patagonia. Science 205: 1377-1379.
BROOM, Robert 1904. On the occurrence of an opisthocoelian dinosaur (Algoasaurus bauri) in the Cretaceous beds of South Africa. Geol. Mag. (5) 1: 445-447.
BUFFETANT, Eric 1979. The evolution of the Crocodilians. Sci. Am. 241: 130-144.
CAMP, C.L.; Maxwell R. BANKS 1978. A proterosuchian reptile from the Early Triassic of Tasmania. Alcheringa 2; 143-158.
CHATTERJEE, S. 1974. A rhynchosaur from the Upper Triassic Maleri Formation of

India. Philos. Trans. Soc. London, Ser B 267, 884: 209-261.

COLBERT, Edwin H. 1965. The Age of Reptiles. Weidenfeld and Nicolson: London.

---1974. Lystrosaurus from Antarctica. Am. Mus. Novit. 2535: 1-44.

---(in press) A primitive ornithischian dinosaur from the Kayenta Formation of Arizona. Mus. N. Arizona.

---; John W. Cosgriff 1974. Labyrinthodont amphibians from Antarctica. Am. Mus. Novit. 2552: 1-30.

---; James W. Kitching 1975. The Triassic reptile Procolophon in Antarctica. Am. Mus. Novit. 2566: 1-23.

---;--- 1977. Triassic cynodont reptiles in Antarctica. Am. Mus. Novit. 2611: 1-30.

---; Duncan MERRILEES 1967. Cretaceous dinosaur footprints from Western Australia. J.R. Soc. West. Aust. 50: 21-25.

COSGRIFF, John W. 1965. A new genus of Temospondyli from the Triassic of Western Australia. J.R. Soc. West. Aust. 48: 65-90.

--- 1969. Blinasaurus, a brachyopid genus from Western Australia and New South Wales. J.R. Soc. West. Aust. 52 (3): 65-88.

COX, C. Barry 1973. Systematics and plate tectonics in the spread of marsupials. In Organisms and Continents through Time, Palaeont. Assoc., Special Papers in Palaeontology 12, pp. 113-119.

--- 1974. Vertebrate palaeodistributional patterns and continental drift. J. Biogeog. 1: 75-94.

CRACRAFT, Joel 1974. Continental Drift and Vertebrate Distribution. Ann. Rev. Ecol. Syst. 5: 215-261.

DIETZ, Robert S.; John C. HOLDEN 1970. The breakup of Pangaea. Sci. Am. 223 (4): 30-41.

ELLIOT, David H. 1972. Aspects of Antarctic geology and drift reconstructions. In Antarctic Geology and Geophysics. Raymond J. Adie (Editor). Int. Union Geol. Sci Ser. B(1): 849-857.

HALLAM, Anthony 1967. The bearing of certain palaeozoogeographical data on continental drift. Palaeogeog., Palaeoclimatol., Palaeoecol. 3: 201-241.

---(Ed.) 1973. Atlas of Palaeobiogeography. Elsevier: Amsterdam, New York. 531 pp.

JAIN, S.L.; T.S. KUTTY; T. ROY-CHOWDHURY; & S. CHATTERJEE 1975. The sauropod dinosaur from the Lower Jurassic Kota Formation of India. Proc. R. Soc. London, Ser. A 188: 221-228.

KITCHING, J.W. 1968. On the Lystrosaurus zone and its fauna with special reference to some immature Lystrosauridae. Palaeon-

tol. Afr. 11: 61-76.

---1977. The distribution of the Karroo vertebrate fauna. Bernard Price Inst. Palaeontol. Res., Mem. 1.

LARSON, Roger L.; John W. LADD 1973. Evidence for the opening of the South Atlantic in the Early Cretaceous. Nature (London) 246 (5430): 209-212.

MOLNAR, Peter; Paul TAPPONNIER 1977. The collision between India and Eurasia. Sci. Am. 236 (4): 30-41.

RAATH, M.A. 1969. A new coelurosaurian dinosaur from the Forest Sandstone of Rhodesia. Arnoldia 4 (28): 1-25.

ROMER, Alfred Sherwood 1966. Vertebrate Paleontology. Univ. of Chicago Press: Chicago. viii + 468 pp.

SILL, William D. 1969. The tetrapod-bearing continental Triassic sediments of South America. Am. J. Sci. 267: 805-821.

TRIPATHI, C.; P.P. SATSANGI 1963. Lystrosaurus fauna of the Panchet Series of the Raniganj coalfield. Palaeontol. Indica, N.S. 37: 1-65.

von HUENE, Friedrich 1929. Los Saurisquios y Ornitisquios del Cretaceo Argentino. Ann. Mus. La Plata. Ser. 2a (3): 1-196.

---1940. Die Saurier der Karroo-, Gondwana- und verwandten Ablagerungen in faunistischer, biologischer und phylogenetischer Hinsicht. Neues Jahrb. Mineral., Geol., Palaeontol., Abt. B 83: 246-347.

---; C.A. MATLEY 1933. The Cretaceous Saurischia and Ornithischia of the central provinces of India. Palaeontol. Indica 21: 1-74.

WINDLEY, Brian F. 1977. The Evolving Continents. John Wiley: New York. xviii+ 385 pp.

Jurassic Ferrar Supergroup tholeiites from the Transantarctic Mountains, Antarctica, and their relationship to the initial fragmentation of Gondwana

PHILIP R. KYLE, DAVID H. ELLIOT & JOHN F. SUTTER
Ohio State University, Columbus, USA

The Ferrar Supergroup consists of basalt lava flows (Kirkpatrick Basalt Gp), dolerite sills and dikes (Ferrar Dolerite Gp) and layered intrusive rocks of the Dufek intrusion (Forrestal Gabbro Gp). They are widespread along 3000 km of the Transantarctic Mountains, covering 1.0×10^5 km^2, with a volume of 5.0×10^5 km^3.

New incremental ^{40}Ar/^{39}Ar age determinations are reported here for six intrusive samples and a basalt flow. The best age estimate for the Ferrar Supergroup is 179 ± 7 m.y., although a younger event at 165 ± 2 m.y. is also likely.

The Ferrar Supergroup is characterised by extremely high ^{87}Sr/^{86}Sr ratios, which may originate from the subcontinental lithospheric mantle or by crustal contamination. If the mantle model can be confirmed, it is possible that Ferrar magmatism resulted from a rise in the mantle geotherms due to an increase of heat-producing K, U, and Th, elements which are unusually abundant in the Ferrar Supergroup.

Mantle processes leading to igneous activity acted independently of any plate boundary. Rifting that accompanied Ferrar magmatism was in part controlled by pre-existing zones of weakness. Evidently the Ferrar rift was a major structural feature adjacent to which the microplates of West Antarctica were rearranged.

INTRODUCTION

The first indication of the impending break-up of Gondwana was voluminous eruption of tholeiitic flood basalts and emplacement of dolerite sills, dikes and large layered intrusions. The cause of the fragmentation is still unknown but was likely a consequence of perturbations, thermal and chemical, within the mantle. Geochemical analyses of the basalts and associated intrusions may provide information on the subcontinental lithospheric mantle. The age, distribution and volume of tholeiites give indirect information about mantle events and the extent and rate of crustal extension and rifting.

DISTRIBUTION AND GEOCHEMISTRY OF THE FERRAR SUPERGROUP (Fig. 1)

Jurassic tholeiites in Antarctica have been geochemically subdivided into two major provinces (Faure *et al.* 1979): The Ferrar Supergroup, from North Victoria Land to the Pensacola Mountains, has high ^{87}Sr/^{86}Sr ratios, whereas basalt flows and dolerite intrusions from Dronning Maud Land have lower ^{87}Sr/^{86}Sr typical of other continental tholeiitic provinces (Brooks & Hart 1978; Faure *et al.* 1979).

Kirkpatrick Basalt Gp. Basalt lava flows occur mainly in three places. In the Central Transantarctic Mountains (CTM) the flows may have a stratigraphic thickness of over 600 m, though the thickest individual section is 525 m (Elliot 1970, 1972). In South Victoria Land (SVL) Kyle (in press a) found the thickest sequence to be a series of 20 flows, 380 m, at Brimstone Peak. Gair (1967) reported a 1380 m thick sequence in North Victoria Land (NVL).

Initial ^{87}Sr/^{86}Sr ratios of basalts from the CTM range from 0.7092 to 0.7142 (mean, 0.7113 (± 0.0012); Kyle in press b). Four analyzed basalts from NVL have similar initial ^{87}Sr/^{86}Sr ratios.

Ferrar Dolerite Gp. Dolerite (diabase) sills and dikes are the most widespread Ferrar Supergroup rocks, occurring from NVL to the Pensacola Mountains. The best exposures of the sills are found in SVL (Gunn & Warren 1962; Gunn 1962, 1963, 1966; Hamilton 1965). Sills in the area are typically 100-300 m thick but locally they can be up to 1000 m thick and show layering such as at the Warren Range (Grapes & Reid 1971).

Butcher Ridge near the head of the Darwin Glacier showed the first silicic rocks, other than granophyres, within the Ferrar Supergroup. Tentatively the rocks are assigned to the Ferrar Dolerite Gp on the basis of their inferred intrusive nature, but there is some doubt about the exact mode of emplacement. Rocks were classified on the basis of silica content: basaltic andesite (58-62% SiO_2), andesite (62-66% SiO_2), dacite (66-69% SiO_2) rhyodacite (69-72%). This also reflects natural divisions in the chemical trends. The rocks are believed to have a tholeiitic affinity.

Silicic rocks at Butcher Ridge are exposed over at least 4 km; the stratigraphic sequences shows the oldest unit is a massive andesitic to dacitic pitchstone intruded by a sheeted andesite dike complex. This is overlain by layered rhyodacites which pass gradationally into weakly layered fine grained basaltic andesite. The total stratigraphic thickness probably exceeds 300 m. The layered rhyodacite shows considerable deformation with large folds which may have formed while the rocks were still plastic. Folding may have resulted from deformation associated with intrusion of later doleritic bodies.

Initial $^{87}Sr/^{86}Sr$ ratios for Ferrar dolerites have a similar range to the basalts (Kyle, in press b). For sills in SVL the mean initial $^{87}Sr/^{86}Sr$ of 37 samples is 0.7117 (± 0.0011).

Forrestal Gabbro Gp. The Dufek intrusion, comprising the Forrestal Gabbro Gp has been subdivided into four formations and a number of members. It is a differentiated stratiform complex composed, in the exposed parts, mainly of cumulate gabbros with less anorthosite and pyroxenite cumulates (Ford 1976).

Exposure is confined to a 1.8 km thick lower section in the Dufek Massif and a 1.7 km upper section in the Forrestal Range; an intrusive contact to the body is exposed in the southern Forrestal Range (Ford 1976). Recent aeromagnetic surveys indicate the Dufek intrusion has a minimum area of 50,000 km^2, similar to the Bushveld complex (Behrendt *et al.* 1979). If Ford's estimated total thickness of 8 to 9 km is confirmed, then the body has a minimum volume of 400,000 km^3.

The initial $^{87}Sr/^{86}Sr$ ratios of Dufek samples are uniform in the layered mafic rocks at 0.709, but in the granophyric upper section they systematically increase with height and have a maximum value of 0.714 near the roof (C. Hedge, pers. comm.).

Discussion. Ferrar Supergroup rocks are widespread for 3000 km along the Transantarctic Mountains. Conservative estimates suggest an area 1.0 x 10^5 km^2 was affected by Ferrar magmatism, during which 5.0 x 10^5 km^3 of intrusive and extrusive rocks were emplaced.

Geochemically the most distinctive feature is the unusually high $^{87}Sr/^{86}Sr$ ratios (0.7086 - 0.7153). Basaltic rocks typically have ratios of 0.702 to 0.706 (Faure & Powell 1972). There is complete overlap in the $^{87}Sr/^{86}Sr$ ratios between the basalts, dolerites and gabbros, which supports

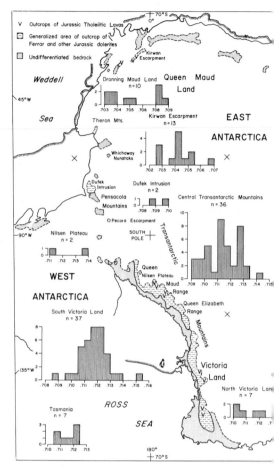

Fig.1--Distribution of Jurassic tholeiites in Antarctica and frequency histograms of $^{87}Sr/^{86}Sr$ ratios of the Ferrar Supergroup and tholeiites from Dronning Maud Land and Tasmania (after Kyle, in press b).

a comagmatic origin for all the Ferrar Supergroup.

RADIOMETRIC AGE DETERMINATIONS

Old Analyses (Table 1). Conventional K/Ar age determinations on the Kirkpatrick Basalt Gp and Ferrar Dolerite Gp show a range from 114 to 223 m.y. However, much of the scatter is probably due to excess, or to loss of, argon and/or potassium following and during cooling (Fleck *et al.* 1977; Elliot *et al.*, in press). (All age determinations have been recalculated from the published values, using the new constants of Steiner & Jager 1977.)

Incremental $^{40}Ar/^{39}Ar$ age determinations allow samples, in which the argon system is disturbed, to be recognised. Using $^{40}Ar/^{39}Ar$ dates, Elliot *et al.* (in press)

284

Table 1--Range of radiometric ages for the Ferrar Supergroup.

Location	Rb/Sr m.y.	K/Ar m.y.	$^{40}Ar/^{39}Ar$ m.y.	Best Estimate m.y.	References
Dufek intrusion	-	98-174	148-190	172±4	1
Pecora Escarpment Ferrar Dolerite	-	178-223		179±5	1
Central Trans-antarctic Mts					
Kirkpatrick Basalt	173±6	134-193	132-179	179±5	2,3,4
Ferrar Dolerite	-	114-205	166	?	3,6
South Victoria Land					
Kirkpatrick Basalt	-	153-190	176-186	181±5	5,6
Ferrar Dolerite	151±18	154-174	163-191	165±2,175	6,7
North Victoria Land Ferrar Dolerite	-	139-199	-	?	3
Tasmania	-	156-183	-	174±8	8

References: 1--Ford & Kistler, in press; 2--Faure *et al.*, in press; 3--Elliot *et al.*, in press; 4--Fleck *et al.* 1977; 5--Hall *et al.*, in press; 6--This paper (Table 2); 7--Compston *et al.* 1968; 8--Schmidt & McDougall 1976.

suggest that in the Queen Alexandra Mountains the Kirkpatrick Basalt Gp is 179 ± 4 m.y. Basalt samples from Carapace Nunatak in SVL have $^{40}Ar/^{39}Ar$ plateau ages of 184 m.y. (Hall *et al.*, in press).

Ford and Kistler (in press) estimate the age of the Forrestal Gabbro Gp as 172 ± 4 m.y. from conventional K/Ar age determinations on plagioclase separates.

New Data (Table 2). Conventional K/Ar dates for the Ferrar Dolerite Gp are widely scattered, indicating possible disturbance of the argon system. To get a better indication, incremental $^{40}Ar/^{39}Ar$ age determinations were made on four dolerite samples from CTM and SVL and two acidic intrusive rocks from the Butcher Ridge. A basalt flow from Gorgon Peak was also analysed.

Samples with undisturbed release spectra define two groups, around 165 and 175 m.y. old. Two dolerite samples from the region between the Mawson and David Glaciers have a mean $^{40}Ar/^{39}Ar$ age of 165 ± 2.4 m.y. The two silicic rocks from Butcher Ridge have a mean inferred age of 175.0 ± 2.7 m.y. Basalt from Ambalada Peak has an inferred age of 175.8 ± 3.0 m.y. which is analytically indistinguishable from other SVL basalts at Carapace Nunatak (Hall *et al.*, in press). A full discussion of the data, including the mean inferred age of the Butcher Ridge samples, will be presented elsewhere.

Discussion. It is apparent that the Kirkpatrick Basalts, Ferrar Dolerites and Forrestal Gabbros are coeval and were erupted over a short time, possibly less than 15 m.y. The best age estimate for the various lithological units of the Ferrar Supergroup is about 179 ± 7 m.y. There is also some evidence of limited Ferrar dolerite intrusion about 165 ± 2 m.y. ago. Incremental $^{40}Ar/^{39}Ar$ age determinations are apparently the only reliable method for estimating ages of the tholeiites.

STRATIGRAPHIC NOMENCLATURE

A feature which has emerged from the review above is the comagmatic and coeval nature of Ferrar Dolerite, Kirkpatrick Basalt and the Dufek intrusion. It is therefore proposed to elevate the Kirkpatrick Basalt and Ferrar Dolerite to Group status and to collectively name them along with the Forrestal Gabbro Gp, the Ferrar Supergroup. This proposal gives the Ferrar Dolerite and Kirkpatrick Basalt equal status.

TECTONIC SETTING OF THE FERRAR SUPERGROUP AND INITIATION OF FRAGMENTATION OF GONDWANA

It is becoming increasingly apparent that most of the Ferrar Gp was emplaced at about 179 ± 7 m.y. This falls within the time of the Karroo magmatism and predates the Parana, Rajmahal and Deccan traps, all of which are associated with Gondwana fragmentation. Ferrar and Karroo rocks are about 40 m.y. older than the oldest known seafloor magnetic anomaly associated with spreading between Gondwana fragments (anomaly M22, 140 m.y., in the Mozambique Basin; Norton & Sclater 1979).

The Ferrar and Dronning Maud Land tho-

Table 2--^{40}Ar/^{39}Ar age determinations for Ferrar Supergroup rocks from the Transantarctic Mountains.

Sample	Location	Plateau Data			Total Fusion
		Steps	% ^{39}Ar$_K$	Age (m.y.)	Age (m.y.)
Dolerite (HB1)	Halfmoon Bluff (175°10'W, 85°14'S)	5	100.0	165.9 ± 6.3	-
Intrusive Pitchstone (79028)	Butcher Ridge (155°50'E, 79°12'S)	No Plateau			153
Basaltic Andesite (79016)	Butcher Ridge (155°51'E; 79°11'S)	3	57.5	174.4 ± 3.4	175.2
Dolerite (77060)	Wright Valley (161°15'E; 77°16'S)	3	59.9	190.8 ± 6.6	181.2
Dolerite (78034)	The Mitten (160°35'E; 75°00'S)	5	100.0	167.3 ± 2.5	-
Dolerite (78220)	Gorgon Peak (158°20'E; 75°57'S)	4	94.1	162.8 ± 2.3	159.7
Basalt (78217)	Ambalada Peak (158°31'E; 75°57'S)	5	100.0	175.8 ± 3.0	-

leiites form a linear belt that, on the scale of resolution of available data, was emplaced synchronously over nearly 4000 km. These tholeiites link up geographically on Gondwana reconstructions with the distribution of the Karroo tholeiites, which themselves are widespread in South Africa and locally are concentrated along the Lebombo Monocline and northwards to various centres such as Nuanetsi, where alkaline rocks are important. The Karroo tholeiites show a significant spread in radiometric ages that may be reduced, as in the case of the Ferrar, on acquisition of ^{40}Ar/^{39}Ar data.

Reactivation of pre-existing zones of weakness has been convincingly argued for the Lebombo Monocline and belts of Mesozoic magmatism to the north. There is no clearly defined belt of weakness for the Karroo tholeiites, sensu stricto, nor is there for the Dronning Maud Land tholeiites. However, the Ferrar appears to be located over the Beardmore and Ross Orogenic belts, respectively, of late Precambrian and early Paleozoic age. These belts were close to but on the continental side of the plate boundary during the late Paleozoic-Early Mesozoic Gondwanian orogeny which immediately preceded, if not actually overlapped, the Ferrar magmatism. The locus of Mesozoic tholeiitic igneous activity was adjacent to the Gondwana plate boundary in Antarctica but far removed in Africa.

The rifts that accompanied magmatism were, in some cases, the sites of subsequent continental separation; however neither the Lebombo Monocline nor the rift zone of the Ferrar developed into an oceanic spreading center. Nevertheless the Ferrar rift was a major structural feature adjacent to which the microplates of West Antarctica were subsequently rearranged.

Undoubtedly thermal perturbation built up before eruption of lavas or emplacement of sills and dikes. During that time the Pacific margin of Gondwana was an active plate boundary. Because the Ferrar Supergroup is close to that boundary, and Dronning Maud Land and the Karroo are not, it has been suggested that the geochemical anomalies of the Ferrar are somehow related to subduction processes. If so, it requires contemporaneous enrichment in highly radiogenic strontium, or enrichment in rubidium at least in early Paleozoic time, or both. Alternatively, the geochemical anomalies may reflect mantle processes (Kyle, in press b) or granitic contamination (Faure et al. 1974).

ACKNOWLEDGMENTS

This study was supported by National Science Foundation grant DPP7721590. Discussions with Art Ford on the Dufek intrusion and the inclusion of the Forrestal Gabbro Group within the Ferrar Supergroup were extremely useful. Unpublished $^{87}Sr/^{86}Sr$ data for the Dufek intrusion was kindly provided by Carl Hedge.

REFERENCES

BEHRENDT, J.C.; D. DREWRY; E. JANKOWSKI; A.W. ENGLAND 1979. Aeromagnetic and radar ice sounding survey of the Dufek intrusion, Antarctica (Abstract). Geol. Soc. Am., Abst. Progr. 11; 386.

BROOKS, C.; S.R. HART 1978. Rb-Sr mantle isochrons and variations in the chemistry of Gondwanaland's lithosphere. Nature 271: 220-223.

COMPSTON, W.; I. MCDOUGALL; K.S. HEIER 1968. Geochemical comparison of the Mesozoic basaltic rocks of Antarctica, South Africa, South America and Tasmania. Geochm. Cosmochim. Acta 32: 129-149.

ELLIOT, D.H. 1970. Jurassic tholeiitic basalts of the central Transantarctic Mountains, Antarctica. In Eastern Washington State College, Cheney, Washington, March 1969. Proc. 2nd Columbia R. Basalt Symp., E.H. Gilmour, D. Stradling (Eds.).

--- 1972. Major oxide chemistry of the Kirkpatrick Basalts, central Transantarctic Mountains. In R.J. Adie (Ed.), Antarctic Geology and Geophysics, Oslo, Universitetsforlaget, 413-418.

--- 1975. Tectonics of Antarctica: A review. Am. J. Sci. 275 A: 45-106.

---; R.J. FLECK; J.F. SUTTER in press. Radiometric dating of Ferrar Group Rocks, Central Transantarctic Mountains. Am. Geophys. Un., Antar. Res. Ser.

FAURE, G.; J.R. BOWMAN; D.H. ELLIOT 1979. The isotopic composition of strontium of the Kirwan volcanics, Queen Maud Land, Antarctica. Chem. Geol. 26, 77-90.

---; J.L. POWELL 1972. Strontium isotope geology. Springer-Verlag Berlin 188 p.

---; J.R. BOWMAN; D.H. ELLIOT; L.M. JONES 1974. Strontium isotope composition and petrogenesis of the Kirkpatrick Basalt, Queen Alexandra Range, Antarctica. Contrib. Mineral. Petrol. 48: 153-169.

---; K.K. PACE; D.H. ELLIOT, in press. Systematic variations of $^{87}Sr/^{86}Sr$ ratios and major element concentrations in the Kirkpatrick Basalt of Mt. Falla, Queen Alexandra Range, Transantarctic Mountains. In C. Craddock (Ed.), Antarctic Geosciences, University of Wisconsin Press.

FLECK, R.J.; J.F. SUTTER; D.H. ELLIOT 1977. Interpretation of discordant $^{40}Ar/^{39}Ar$ age-spectra of Mesozoic tholeiites from Antarctica. Geochim. Cosmochim. Acta 41: 15-32.

FORD, A.B. 1976. Stratigraphy of the Dufek

intrusion, Antarctica. U.S. Geol. Surv. Bull. 1405-D: 1-36.

---; R.W. KISTLER in press. K/Ar age, composition and origin of Mesozoic mafic rocks related to Ferrar Group, Pensacola Mountains, Antarctica. N.Z.J. Geol. Geophys.

GAIR, H.S. 1967. The geology from the upper Rennick Glacier to the coast, northern Victoria Land, Antarctica. N.Z. J. Geol. Geophys. 10: 309-44.

GRAPES, R.H.; D.L. REID. 1971. Rhythmic layering and multiple intrusion in the Ferrar Dolerite of South Victoria Lane, Antarctica. N.Z. J. Geol. Geophys. 14: 600-604.

GUNN, B.M. 1962. Differentiation in Ferrar Dolerites, Antarctica. N.Z. J. Geol. Geophys. 5: 820-863.

--- 1963. Layered intrusions in the Ferrar Dolerites, Antarctica. Mineral. Soc. Am., Spec. Pap. 1: 124-133.

--- 1966. Modal and element variation in Antarctic tholeiites. Geochim. Cosmochim. Acta 30: 881-920.

---; G. WARREN 1962. Geology of Victoria Land between the Mawson and Mulock Glaciers, Antarctica. N.Z. Geol. Surv. Bull. 71: 157 p.

HALL, B.A., J.F. SUTTER, H.W. BORNS in press. The inception and duration of Mesozoic volcanism in the Allan Hills-Carapace Nunatak Area, Victoria Lane, Antarctica. In C. Craddock (Ed.), Antarctic Geosciences, University of Wisconsin Press.

HAMILTON, W. 1965. Diabase sheets of the Taylor Glacier region, Victoria Lane, Antarctica. U.S. Geol. Surv. Prof. Pap., 456-B: 1-71.

KYLE, P.R., in press a. Investigations of Ferrar Group rocks from south Victoria Land. Antar. J. U.S.

--- In press b. Development of heterogeneities in the subcontinental mantle: Evidence from the Ferrar Group, Antarctica. Contrib. Mineral. Petrol.

NORTON, I., J.G. SCLATER 1979. A model for the evolution of the Indian Ocean and the breakup of Gondwanaland. J. Geophys. Res. 84: 6803-6830.

SCHMIDT, P.W.; I. MCDOUGALL 1977. Paleomagnetic and potassium-argon dating studies of the Tasmanian dolerites. J. Geol. Soc. Austr. 25: 321-328.

STEIGER, R.H.; E. JAGER 1977. Subcommission on geochronology: Convention on the use of decay constants in geo- and cosmochronology. Earth Planet. Sci. Lett. 36: 359-362.

Institute of Polar Studies Contribution #394

Evolution of the Tasman Rift: Apatite fission track dating evidence from the southeastern Australian continental margin

M. E. MORLEY, A. J. W. GLEADOW & J. F. LOVERING
University of Melbourne, Parkville, Australia

Fission track ages of apatites from Paleozoic granitic rocks within about 150 km of the continental margin in southeastern Australia have been profoundly affected by processes related to continental rifting and the opening of the Tasman Sea. The ages decrease from about 360 m.y. at a distance of 100 km inland to about 80 m.y. along the coast, the youngest ages correlating with the initiation of sea-floor spreading in the Tasman Sea. Apatite ages >200 m.y. are representative of samples unaffected by rifting processes.

Two complementary mechanisms are though to have caused partial to complete resetting of the apatite ages at the time of continental breakup. The first is a strong heating of the continental margin due to increased heatflow at the time of rifting. Magmatic activity at this time provides independent evidence of this thermal event. The second mechanism of uplift and erosion on the flanks of the developing rift has also been significant in determining the apatite ages.

INTRODUCTION

Fission-track dating gives an estimate of the last time a particular mineral cooled through its closure or track retention temperature. Track retention temperatures vary widely in the U-bearing accessory minerals best suited to fission-track dating. The most recent estimates for typical geological cooling rates are 250 ±50°C for sphene, 200±50°C for zircon, and 100±20°C for apatite (Gleadow & Brooks 1979). The track retention temperature for apatite is the best known and is based on close agreement between annealing studies (Naeser & Faul 1969) and age measurements on deep borehole samples (Naeser 1979). Because of the relatively low stability of fission tracks in apatite, it is uniquely sensitive to low-temperature thermal events and the effects of uplift and erosion.

Various models proposed for continental breakup (e.g. Falvey 1974) involve increased heat flow in the region of the rift and some type of domal uplift centred over the site of the rift before breakup. Apatites in basement rocks adjacent to the rift and undergoing either of these processes would be heated or rapidly brought to the surface. Either should produce a decrease in apatite ages towards rifted continental margins. Studies of King Is in Bass Strait (Gleadow & Lovering 1978a) and in Greenland (Gleadow 1978; Gleadow & Brooks 1979) indicate a relationship between apatite ages in marginal basement rocks and continental rifting. This led to the present detailed work which clearly demonstrates a strong association between apatite fission track ages measured in granites from Paleozoic basement rocks in SE Australia to thermal and tectonic events associated with the initiation of the Tasman rift.

The major granite outcrops in SE Australia are outlined in Fig. 1. They are all within the Lachlan Mobile Zone, in which several cycles of marine sedimentation were disrupted by three phases of deformation, each accompanied by intrusion of granitic bodies and acid volcanism with the centre of igneous activity moving progressively east with time. Radiometric ages for these granites range from 421±8 m.y. for the Corryong Batholith (Rb/Sr whole rock; Brooks & Leggo 1972) to 379-397 m.y. for the Bega and Moruya Batholiths (Chappell & White 1976; Rb/Sr whole rock, biotite and K/Ar biotite, hornblende ages recalculated with the new constants of Steiger & Jäger 1977).

Apatites from suitable granite samples were separated using conventional electromagnetic and heavy liquid techniques and dated as described by Gleadow & Brooks (1979). Apatite fission track ages are shown in Fig.1 and were calculated using the following constants:

$$^{238}\lambda_f = 6.9 \times 10^{-17} \text{ yr}^{-1}, \quad \lambda_D - 1.551 \times 10^{-10} \text{ yr}^{-1}, \quad I = 7.253 \times 10^{-3},$$

$$^{235}\sigma = 5.802 \times 10^{-22} \text{ cm}^2.$$ One standard deviation errors are 2-4%, using the method of Naeser *et al.* 1978. Full analytical results may be found elsewhere (Morley, Gleadow, & Lovering, in prep.).

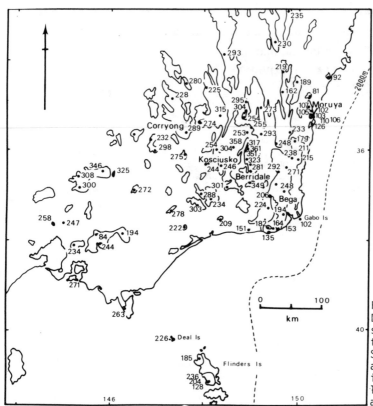

Fig. 1--SE Australia--
Distribution of granites
sampled and measured apatite fission track ages.
Statistical uncertainty
associated with these ages
is 2-4%. The major batholiths and 2000 m isobath
are also shown.

APATITE AGES AND CONTINENTAL RIFTING

The measured apatite ages range from 81-361 m.y. and are between 56 and 384 m.y. younger than known Rb/Sr and K/Ar ages so that clearly none is directly related to emplacement and crystallization of the granites. The ages increase regularly and steeply inland from the coast (Fig.2), and their contours tend to parallel the topographic contours in the uplifted highland region. The trough defined by the 250 m.y. contour in southern Victoria is broadly associated with the Mesozoic-Tertiary Gippsland Basin.

To plot the variation of apatite age with distance from the rifted continental margin, the 2000m isobath was used as an arbitrary reference line (Fig. 3). Despite some scatter, the points plotted define two distinct trends:

(1) Beyond about 150 km from the 2000 m isobath, the graph defines a rough plateau. The scatter in this area is mainly due to large differences in topographic elevation of the sampling localities, the oldest ages in general corresponding to the highest altitudes. The variation is not well defined but suggests that apatite ages may decrease by about 70 m.y./km in the upper few kilometres of the crust. An increase with elevation is well known from uplifted areas in other parts of the world (Wagner *et al*. 1977; Naeser 1979), but further discussion of this is beyond the scope of this paper. However, apatite ages greater than about 200 m.y. are found only in granitic rocks remote from the continental margin in SE Australia and appear to represent those samples which were unaffected by breakup events. The 200 m.y. contour in Fig. 2 can therefore be taken to define the approximate limit of disturbance by the Tasman rift system.

(2) Up to about 150 km from the reference line, the ages increase rapidly with distance at a rate of about 2.2 m.y./km. Samples closest to the rifted margin are consistently youngest (∿80-100 m.y.), and nearly correspond to the onset of seafloor spreading in the Tasman Sea (82 m.y.; Hayes & Ringis 1973). This implies that some of the apatite ages were completely reset at the time of rifting.

The rapid decrease in apatite ages towards

290

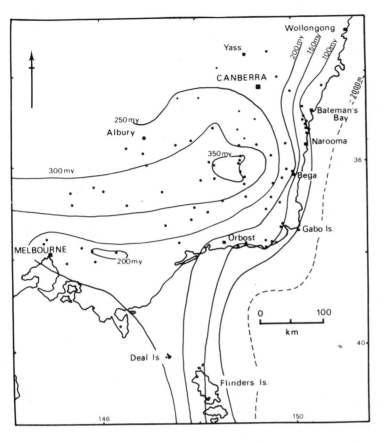

Fig. 2--SE Australia--
apatite age contours,
sample localities, and
2000 m isobath. Conti-
nuity (rather than clo-
sure) of the 300 m.y.
contour is based on
Gleadow & Lovering
(1979b).

the continental margin in SE Australia is
thought to reflect two different but related
processes: the rate at which the granites
were being uncovered by uplift and erosion,
and the extent to which they were affected
by increased heat flow at the continental
margin during rifting. As these two factors are complementary in determining the
apatite ages, estimates of the maximum
significance of each will be made independently assuming the other to be constant.

At one extreme, the apatite age pattern
might be explained by differential erosion
along the uplifted margin bordering the
Tasman rift. A constant thermal gradient
of 25°C/km was assumed, this being the
present average value in SE New South Wales
(Lilley *et al.* 1978). According to this
model an age of about 80 m.y. means that
the dated sample was at a depth of 3.4 km at
that time. As the rock is now at the surface the uplift model requires that 3.4 km
of overburden has been removed by erosion
in 80 m.y. In contrast Wellman (1979) has
estimated that total uplift over the last 90
m.y. in the Milton-Moruya area, where the
youngest ages of 80-100 m.y. are found, was
only about 0.2 km. Wellman's data however

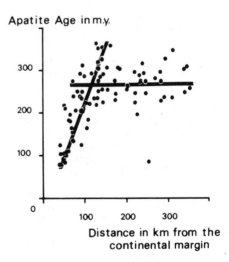

Fig.3--Variation of apatite fission track
ages with distance from the 2000 m isobath.

291

do not provide severe constraints on the amount of uplift and erosion before about 45 m.y.

Any domal uplift which occurred along the Tasman rift is presumed to be related to heating associated with convective upwelling in the upper mantle (Falvey 1974). The above model of uplift and erosion and its basic assumption of uniform heat flow therefore does not seem sufficient to explain the observed pattern of apatite ages.

Increased heat flow during incipient rifting is evidenced by the extrusion of mid-ocean ridge basalts into the opening Tasman Sea floor and Mesozoic alkaline igneous activity about the SE New South Wales coast (McDougall & Wellman 1976; Wass & Kleeman 1980). The Mt Dromedary complex near Narooma, for example, has a K/Ar biotite age of 97.9 m.y. (McDougall & Roksandic 1974; recalculated per Steiger & Jäger 1977) and is only marginally younger than the ∿120 m.y. recorded by apatites in nearby granits. On the other side of the rift, McDougall & van der Lingen (1973) have recorded 96 m.y. (new constants) for a rhyolite extruded onto the Lord Howe Rise.

A maximum estimate of the increased geothermal gradient at the time of incipient rifting (taken to be 80 m.y.) assumes no doming at that time, i.e. that the total uplift data of Wellman (1979) are correct for present-day coastal areas. For samples yielding ages of 80-100 m.y., $100°C$ (the apatite closure temperature) at 0.2 km (Wellman's total uplift estimate) implies a transient local thermal gradient of $425°C/km$ (ambient $T = 15°C$).

Geothermal gradients calculated in this way for samples near the SE New South Wales coast are extreme estimates and unlikely to be correct, since marginal uplift and increased heat flow are both believed to have influenced the measured apatite ages. Thus, Wellman's model, based on the evolution pattern of stream gradients during the last 45 m.y. and extrapolated backwards in time, does not include uplift along the continental margin before rifting. Pre-rift doming may have begun as early as the late Triassic-Jurassic (Gunn 1975) when the initial uplift of the eastern highlands is thought to have occurred (Wellman 1974). No evidence is presently available concerning the amount of vertical uplift and erosion which may have occurred over the developing Tasman rift before breakup, and such uplift must have been a significant contributing factor to the observed reduction in apatite ages. Until more is known about the erosion which occurred as a result of uplift marginal to the rift, the relative effects of increased heat flow and uplift in resetting the apatite ages cannot be properly gauged.

The apatite fission track ages presented here, however, provide clear evidence of a thermal and tectonic event in SE Australia associated with the initiation of the Tasman Rift. Coincidence of the youngest apatite ages with the oldest magnetic lineations known from the Tasman Sea floor suggests a thermal maximum at the time of initiation of sea floor spreading at about 82 m.y.

ACKNOWLEDGMENTS

This work was supported by grants from the Australian Research Grants Committee and the Australian Institute of Nuclear Science and Engineering, and by a Commonwealth Postgraduate Research Award for M.E.M.

REFERENCES

BROOKS, C.; M. D. LEGGO 1972. The Local Chronology and Regional Implications of a Rb-Sr Investigation of Granitic Rocks from the Corryong District, Southeastern Australia. J. Geol. Soc. Aust. 19: 1-19.

CHAPPELL, B. W.; A. J. R. WHITE 1976. Plutonic Rocks of the Lachlan Mobile Zone. Excursion Guide 13C, 25th Int. Geol. Cong., Sydney.

FALVEY, D. A. 1974. The Development of Continental Margins in Plate Tectonic Theory. Aust. Petrol. Exp. Assoc. J.: 95-106.

GLEADOW, A. J. W. 1978. Fission Track Evidence for the Evolution of Rifted Continental Margins. In R.E. Zartman (Ed.), Short Papers 4th Int. Conf. Geochronology, Cosmochronology, Isotope Geology. U.S. Geol. Surv. Open File Rep. 78-101: 146-148.

---; C. K. BROOKS 1979. Fission Track Dating, Thermal Histories and Tectonics of Igneous Intrusions in East Greenland. Contrib. Mineral. Petrol. 71: 45-60.

---; J. F. LOVERING 1978a. Geochronology of King Island, Bass Strait, Australia: relationship to continental rifting. Earth Planet. Sci. Lett. 37: 429-437.

---; --- 1978b. Thermal History of Granitic Rocks from Western Victoria: a fission track dating study. J. Geol. Soc. Aust. 25: 323-340.

GUNN, P. J. 1975. Mesozoic-Cainozoic Tectonics and Igneous Activity: Southeastern Australia. J. Geol. Soc. Aust. 22: 215-221.

HAYES, D. E.; J. RINGIS 1973. Seafloor Spreading in the Tasman Sea. Nature 243: 454-458.

LILLEY, F. E. M.; M. N. SLOANE; J. H. SASS

1978.   A Compilation of Australian Heat
Flow Measurements.   J. Geol. Soc. Aust.
24: 439-446

McDOUGALL, I.; Z. ROKSANDIC 1974.   Total
Fission ^{40}Ar/^{39}Ar ages Using HIFAR Reactor.
J. Geol. Soc. Aust. 21: 81-90.

---; P. WELLMAN 1976.   Potassium-argon ages
for some Australian Mesozoic Igneous Rocks.
J. Geol. Soc. Aust. 23: 1-9.

NAESER, C. W. 1979.   Fission-Track Dating
and Geologic Annealing of Fission Tracks.
In E. Jäger and J. C. Hunziker (Eds.)
Lectures in Isotope Geology.   Heidelberg:
Springer-Verlag. pp.154-169.

---; H. FAUL 1969. Fission Track Annealing in Apatite and Sphene. J. Geophys. Res.
74: 705-710.

---; N. M. JOHNSON; V. E. McGEE 1978.   A
Practical Method of Estimating Standard
Error of Age in the Fission Track Dating
Method.   In R.E. Zartman (Ed.), Short
Papers 4th Int. Conf. Geochronology, Cos-
mochronology, Isotope Geology.   U.S. Geol.
Surv. Open File Rep. 78-101: 303-305.

STEIGER, R. H.; E. JÄGER 1977.   Subcommission on Geochronology: Convention on
the Use of Decay Constants in Geo- and
Cosmochronology.   Earth Planet. Sci. Lett.
36: 359-362.

WAGNER, G.A.; G. M. REIMER; E. JÄGER 1977.
Cooling Ages Derived by Apatite Fission-
Track, Mica Rb-Sr and K-Ar Dating: the
uplift and cooling history of the central
Alps.   Mem. Ist. Geol. Mineral. Univ.
Padova 30: 27 pp.

WASS, S. Y.; J. D. KLEEMAN 1980.   Evidence
for late Cretaceous Crustal Extension in
South-eastern New South Wales.   Earth
Planet. Sci. Lett. (in press).

WELLMAN, P. 1974.   Potassium-Argon Ages
on the Cainozoic Volcanic Rocks of Eastern
Victoria, Australia.   J. Geol. Soc. Aust.
21: 359-376.

--- 1979.   On the Cainozoic Uplift of the
Southeast Australian Highland.   J. Geol.
Soc. Aust. 26: 1-10.

Early Cretaceous volcanism and the early breakup history of southeastern Australia: Evidence from fission track dating of volcaniclastic sediments

A. J. W. GLEADOW & I. R. DUDDY
University of Melbourne, Parkville, Australia

Non-marine sandstones of the 3 km thick early Cretaceous Otway Group were deposited during the initial rift-valley stage in the breakup of Australia and Antarctica. The rocks contain a high proportion of volcanogenic material often to the virtual exclusion of non-volcanic detritus.

Fission track dating of detrital minerals from these rocks gives early Cretaceous ages between 126 and 103 m.y. Concordance of ages in apatite, sphene and zircon, which have very different track stabilities, and the survival of older zircon ages shows that none of the ages have been reset. The fission track ages thus represent the original ages of the detrital components and show that a major volcanic episode occurred in southeastern Australia during the early Cretaceous. The volcanism probably occurred along the developing rift system that led to opening of the Southern Ocean.

INTRODUCTION

The Otway Basin in SE Australia formed during the early stages of continental rifting between Australia and Antarctica and marks the site of later separation of these two continents during the Eocene. The initial depositional phase of the Otway Basin is a 3 km thick sequence of early Cretaceous non-marine sediments known as the Otway Gp. These deposits consist of virtually undisturbed lenticular channel sandstones interbedded with a spectrum of finer-grained sediments including minor coal seams and soil horizons, and with locally abundant plant fossils. Fig.1 shows the extent of the Otway Basin, the areas of outcrop of the Otway Gp sediments and sample localities.

Much of the detritus is volcanogenic and has previously been attributed to either the erosion of pre-existing Paleozoic volcanics (Edwards & Baker 1943) or contemporaneous volcanic activity (Dellenbach 1966; Sutherland & Corbett 1974). Evidence for the former is based on the widespread occurrence of Devonian volcanic rocks of dominantly acid composition in central and eastern Victoria. Direct evidence of early Cretaceous volcanic activity in Victoria is unknown.

Petrographic studies (Duddy, in prep.) show that the Otway Gp has abundant rounded grains of altered glass and glassy volcanic rock fragments with fresh microphenocrysts of dominantly andesine feldspar, amphibole, pyroxene and biotite. In mineralogy and texture the rock fragments are quite unlike the Devonian volcanic rocks in Victoria. The majority are of alkaline intermediate composition, which contrasts with the acidic Devonian rocks. The volcanogenic material is extremely abundant and in many samples occurs to the virtual exclusion of non-volcanic detritus (e.g. R17685). These features argue against derivation of the Otway Gp from the Devonian volcanics and strongly suggests that volcanism was effectively synchronous with deposition.

A puzzling aspect of this interpretation is the complete absence of both lavas and pyroclastics from the entire early Cretaceous sequence in the Otway Basin and surrounding areas. Reported occurrences of igneous rocks and tuffs in the Otway Gp are not convincing; these rocks are almost certainly volcanogenic sediments also. The closest volcanic rocks of known early Cretaceous age are a small trachyandesite lava flow and associated minor intrusives ∿500 km away at Cape Portland in NE Tasmania (Fig.1). Sutherland and Corbett (1974) quote minimum ages of 95 ± 1 (K-Ar) to 99 ± 22 m.y. (Rb-Sr) for these rocks. The early Cretaceous is taken here to be the interval 96 - 143 m.y. (Armstrong 1978).

Alkali basalts and trachytes of Jurassic age are known in the Merino High area of western Victoria, and these probably correlate with basic volcanic rocks near the base of the Otway Basin sequence in several wells to the south. K-Ar whole rock ages of 161 ± 5 and 167 ± 6 m.y. respectively have been reported for a basic lava near the base of Casterton No.1 well and a trachyte from Coleraine .(Douglas *et al.* 1976). Monchiquite dykes with similar

Table 1--Combined fission track ages[1]

Sample No.	Mineral	N	R	ρ_s x10^6 (cm^{-2})	ρ_i x10^6 (cm^{-2})	n x10^{15} (n·cm^{-2})	U (ppm)	Age[2] (m.y.)
Otway Group sandstones:								
R17685	Sphene	10	0.970	4.49 (3856)3	10.4 (4472)	4.52 (2623)	92	118 ± 2
	Apatite	10	0.988	0.609 (1397)	2.44 (2803)	8.07 (2723)	12	121 ± 3
R19124	Sphene	9	0.951	6.24 (2817)	18.6 (4189)	5.91 (2475)	126	120 ± 3
	Apatite	10	0.958	0.630 (593)	3.14 (1478)	9.70 (2313)	13	118 ± 3
R17686	Sphene	11	0.979	7.18 (3868)	15.4 (4162)	4.26 (2623)	145	120 ± 2
	Apatite	9	0.969	0.926 (511)	3.49 (963)	8.00 (2723)	17	128 ± 3
Cape Portland trachyandesite:								
R19125	Apatite	10	0.933	0.192 (276)	1.07 (772)	9.03 (2313)	5	98 ± 4

1. N = number of grains counted, R = correlation coefficient, ρ_s = fossil track density, ρ_i = induced track density, n = neutron dose, U = uranium content.

2. $^{238}\lambda_f = 6.9 \times 10^{-17}$ yr^{-1}, $\lambda_d = 1.551 \times 10^{-10}$ yr^{-1}, $^{235}\sigma = 5.802 \times 10^{-22}$, $I = 7.253 \times 10^{-3}$.

3. brackets show total number of tracks counted.

Table 2--Fission track ages of single sphene and apatite grains

Sample	Sphene:*			Apatite:*		
	Grain No.	Fossil/Induced	Age (m.y.)	Grain No.	Fossil/Induced	Age (m.y.)
R17685	1	422 / 538	107 ± 7	1	50 / 111	110 ± 19
	2	518 / 584	121 ± 7	2	74 / 121	149 ± 22
	3	310 / 401	106 ± 8	3	127 / 246	126 ± 14
	4	523 / 566	126 ± 8	4	240 / 469	125 ± 9
	5	491 / 531	126 ± 8	5	111 / 228	119 ± 16
	6	480 / 526	124 ± 8	6	86 / 173	121 ± 16
	7	192 / 219	120 ± 12	7	234 / 515	111 ± 9
	8	326 / 376	118 ± 9	8	199 / 381	127 ± 11
	9	188 / 207	124 ± 12	9	115 / 252	111 ± 13
	10	406 / 524	106 ± 7	10	161 / 307	128 ± 12
R19124	1	371 / 524	126 ± 9	1	44 / 152	85 ± 14
	2	215 / 311	123 ± 11	2	29 / 70	121 ± 27
	3	262 / 372	126 ± 10	3	97 / 228	125 ± 15
	4	380 / 608	112 ± 7	4	69 / 182	111 ± 16
	5	239 / 368	116 ± 10	5	30 / 48	182 ± 42
	6	381 / 559	122 ± 8	6	86 / 162	155 ± 21
	7	407 / 545	133 ± 9	7	26 / 73	104 ± 24
	8	236 / 401	105 ± 9	8	141 / 380	109 ± 11
	9	326 / 501	116 ± 8	9	39 / 124	92 ± 17
				10	32 / 59	158 ± 35
R17686	1	403 / 425	122 ± 8	1	110 / 188	141 ± 17
	2	179 / 185	125 ± 13	2	39 / 72	131 ± 26
	3	232 / 240	124 ± 12	3	90 / 143	152 ± 20
	4	481 / 551	112 ± 7	4	41 / 76	130 ± 25
	5	413 / 429	124 ± 9	5	60 / 120	121 ± 19
	6	470 / 532	114 ± 7	6	38 / 83	111 ± 22
	7	364 / 418	112 ± 8	7	69 / 142	117 ± 17
	8	271 / 288	121 ± 10	8	30 / 77	94 ± 19
	9	441 / 414	137 ± 9	9	34 / 62	132 ± 28
	10	428 / 476	116 ± 8			
	11	186 / 204	117 ± 12			

* neutron doses as in Table 1

Table 3--Fission track ages of individual zircon grains

R17685	n = 1.23 x 10^{15} n·cm^{-2}		R17686	n = 1.21 x 10^{15} n·cm^{-2}	
Grain No.	Fossil/Induced	Age (m.y.)	Grain No.	Fossil/Induced	Age (m.y.)
1	298 / 103	108 ± 12	1	307 / 44*	495 ± 80
2	902 / 129	257 ± 24	2	266 / 30	319 ± 61
3	248 / 89	104 ± 13	3	191 / 73*	190 ± 26
4	311 / 105*	218 ± 25	4	162 / 50	118 ± 19
5	190 / 57	124 ± 19	5	272 / 85	117 ± 15
6	699 / 263*	196 ± 14	6	288 / 55	190 ± 28
7	252 / 93	101 ± 13	7	276 / 24	411 ± 87
8	281 / 156*	133 ± 13	8	159 / 28*	406 ± 83
9	436 / 314*	103 ± 8	9	323 / 33	351 ± 64
			10	816 /	327 ± 7

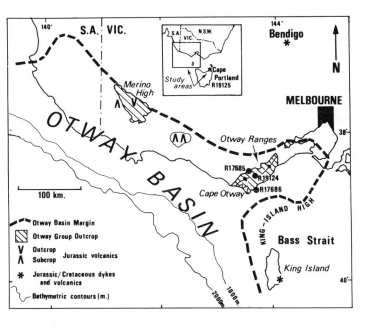

Fig.1--Location map

Jurassic K-Ar whole-rock ages of 153 ± 4 and 163 ± 4 m.y. occur in the Bendigo area of central Victoria (McDougall & Wellman 1976). On King Island in Bass Strait, McDougall and Leggo (1965) obtained a slightly younger K-Ar age of 143 ± 3 m.y., near the Jurassic-Cretaceous boundary, for biotite from a lamprophyre dyke. These ages have all been recalculated using the constants of Steiger and Jäger (1977). Although related, none of these Jurassic volcanic rocks have compositions which make them suitable source rocks for the bulk of the Otway Gp volcanogenic sandstones.

FISSION TRACK DATING

Fission track ages are extremely resistant to the effects of weathering and can give useful information on provenance of sedimentary rocks (Gleadow & Lovering 1974; McGoldrick & Gleadow 1977). A great advantage of the fission track method for this purpose is that ages can be measured on very small (<100μm) mineral grains, enabling discrimination of different age components amongst them. Here we report fission track ages of detrital minerals from the Otway Gp sandstones with the aim of proving the existence of contemporaneous volcanism in the early Cretaceous.

Like other mineral dating methods fission track dating records the time a mineral cooled through a characteristic temperature interval. Three uranium-bearing minerals suitable for fission track dating occur in the Otway Gp. The temperatures below which

tracks are stable in these minerals over geological time are discussed elsewhere (Gleadow & Lovering 1978; Gleadow & Brooks 1979) and the values accepted here are 250 ± 50°C for sphene, 200 ± 50°C for zircon, and 100 ± 20°C for apatite.

These minerals were separated from the 90 - 250 μm size fraction by conventional heavy liquid and magnetic techniques and dated using methods described by Gleadow and Brooks (1979) and Gleadow and Lovering (1978). Sphenes were mounted in epoxy and etched in 50N NaOH. The external detector method, where fossil tracks are counted in the mineral grains to be dated and induced tracks in an adjacent muscovite detector, was used. This method enables ages to be calculated for single grains, and about ten of the most suitable were chosen for counting in each mount.

To compare sphene and apatite ages a combined age was calculated from the individual grain data as if all the crystals came from a single source (Table 1). Errors were determined by the method of Naeser et al. (1978), which takes into account the high correlation between fossil and induced counts. Table 1 also includes an apatite age for the trachyandesite lava from Cape Portland. Tables 2 and 3 show ages for the individual grains. Errors for these ages were calculated from the numbers of tracks counted and are probably overestimated as no measure of the correlation coefficient for fossil and induced counts is possible for a single grain. All errors

297

are quoted at the level of one standard deviation.

In general the sphenes produced the most precise ages because of their relatively large grain-size and their high and uniform uranium concentrations. Apatites had fewer tracks, resulting in a greater scatter of ages, and many grains had interfering dislocations which reduced the number of suitable grains for counting. Relatively few zircons gave satisfactory results as they were less abundant, and many grains were at least partly metamict making them unsuitable for dating.

DISCUSSION

Tables 1-3 show that most of the fission track ages obtained for sphenes, apatites and, to a lesser extent, zircons, fall within the early Cretaceous (96 - 143 m.y.). Fig. 2 shows the distribution of all the single grain ages as well as that of only the more precise results, those with errors <12 m.y. (about 10%). No sphene or apatite ages are older than Jurassic and none of the more precise ages lies outside the early Cretaceous. This is also true for just over half of the zircon ages, the remainder being much older, 218 - 495 m.y. Sphene and apatite ages have a very similar distribution, but the apatite ages are more scattered.

The close coincidence of ages for sphene and apatite (Table 1), which have effective geological track retention temperatures of about 250° and 100°C respectively, argues strongly that these ages have remained un-

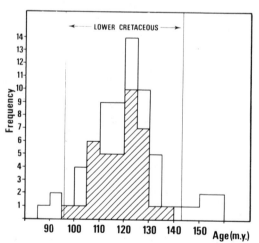

Fig.2. Distribution of fission track ages. Larger histogram shows results (Tables 2, 3) for all grains of sphene, zircon and apatite; hatched area shows only results where calculated error was ⩽ 12 m.y.

changed during sedimentation, diagenesis and any subsequent event. This is furthe. supported by the close agreement of the younger zircon ages and the preservation o: Paleozoic ages in many of the zircons. The early Cretaceous ages are therefore interpreted to represent the time of volca: ic activity which produced the abundant vol canogenic detritus in the Otway Gp.

The simplest possible explanation of the results is that they represent a single and very short major volcanic episode. It would be dated closely by the combined results for sphene and apatite (Table 1) and the major peak in Fig.2, which define an age of about 120 m.y. According to this explanation the dispersion of single grain ages shown in Fig. 2 would be due entirely to experimental uncertainty. However, the more precise single grain ages suggest a more complex situation with a bimodal distribution of ages representing pulses of volcanism at about 106 and 123 m. y. within a longer episode, perhaps 20 m.y. These pulses may correlate with peaks in th abundance of apatite and sphene in a number of wells in the western Otway Basin (Fig.3)

Apatites from the trachyandesite lava at Cape Portland give an early Cretaceous age of 98 ± 4 m.y. (Table 1), in close agreement with K-Ar and Rb-Sr ages (Sutherland & Corbett 1974), and younger than nearly all the grains dated from the Otway Gp sandstones.

The older zircon grains (Table 3) appear to belong to two separate groups. One has ages ranging from 190 - 257 m.y. with a mea and standard deviation of 210 ± 29 m.y. and is found in both the zircon samples. Thes ages could represent either early Jurassic volcanism or cooling ages from basement roc The second group is found only in sample R17686 and has Paleozoic ages ranging from 319-495 m.y. with a mean of 387 ± 65 m.y. This sample occurred near the margin of the basin and has a relatively high content of non-volcanic detritus presumably derived from the adjacent basement high culminating in King Island. Zircon fission track dating often gives very similar results to K-A dating of biotites, and McDougall and Leggo (1965) reported biotite ages for granitic rocks from King Island ranging from 352 - 467 m.y. with a mean of 398 ± 38 m.y. These are consistent with the King Island High's being the source of the old zircons.

In accordance with stratigraphic position (Fig.3), there is a progressive decrease in the youngest age observed in each of the three samples. Only in the stratigraphically youngest sample, R17685, is there a definite bimodal distribution of the mineral ages. The proposed younger pulse of volcanism at about 106 m.y. must

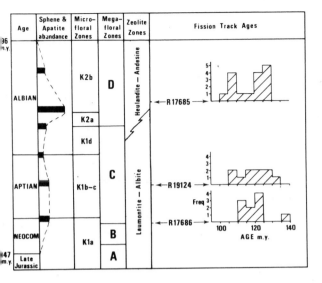

Fig. 3 Distribution of fission track ages compared with stratigraphic subdivision. Thickness of zones is approx. proportional to stratigraphic thickness. Grain ages used had error of ≤ 12 m.y. Sphene and apatite data are from Kalangadoo No. 1 well (see Rochow 1971).

therefore be represented between samples R17686 and R19124. Abundant fresh volcanic material in R17685 suggests that this sample is close to the horizon representing the time of the younger pulse which would therefore be middle Albian in age.

The relative thickness of microfloral (Evans 1971) and megafloral (Douglas 1969) zones shown in Fig. 3 indicates that most of the Otway Gp lies above R17686 (late Neocomian). Less than 500 m of sediment are known below this in megafloral zones A and B. It is possible that this thickening of the sedimentary sequence from late Neocomian time may coincide with the onset of major early Cretaceous volcanism. Fig.2 and Table 2 show a very rapid increase in the number of fission track ages at about 126 m.y. suggesting a correlation of this age with the late Neocomian.

SOURCE OF THE VOLCANICS

The volume of volcanic detritus in the Otway Gp is probably >50 000 km^3 (Duddy, in prep.). The early Cretaceous was thus one of the most significant volcanic episodes in the history of SE Australia, and it is remarkable that no volcanic source rocks are known. Two source areas seem possible, either a volcanic arc to the east or along the developing rift system between Australia and Antarctica.

Derivation from a volcanic arc to the east would explain the abrupt cessation of supply of volcanic detritus at the end of the early Cretaceous. Opening of the Tasman Sea at that time would have isolated such an arc source from Australia and pre-

vented any more volcanic material from reaching the Otway Basin. However, an easterly arc source would require very long transportation of the sediment--by way of the Gippsland Basin and over the Mornington Peninsula tectonic high. It is difficult to reconcile such a route of more than 1000 km with the lack of contamination, very large volume and freshness of the volcanic material preserved in the sediments. Further the slightly alkaline composition of the volcanogenic detritus is more consistent with a continental rather than a volcanic arc source.

The most likely source of the detritus in the Otway Gp is therefore thought to be volcanism along the continental rift system which was developing in early Cretaceous time. This period of volcanism, previously signalled by scattered igneous activity in the Jurassic, was a major feature of the early breakup history of Australia and Antarctica.

ACKNOWLEDGMENTS

This work was supported in part by a grant from the Australian Institute for Nuclear Science and Engineering and a Commonwealth Postgraduate Research Award to one of us (IRD). We thank Dr O.P. Singleton for encouragement and helpful discussions on this project.

REFERENCES

ARMSTRONG, R. L. 1978. Pre-Cenozoic Phanerozoic time scale - computer file of critical dates and consequences of new and in-progress decay constant revisions. In

G.V. Cohee, et al. (Eds.), AAPG Stud. 6: 73-91.

DELLENBACH, J. 1965. A petrological study of sediments from Beach Petroleum N.L. Geltwood Beach Well No.1, Otway Basin, South Australia. BMR Record 1964/77 (Unpublished).

DOUGLAS, J. G. 1969. The Mesozoic floras of Victoria. Geol. Surv. Victoria Mem. 28.

---; C. ABELE; S. BENEDEK; M. E. DETTMAN; P. R. KENLEY; C. R. LAWRENCE 1976. Mesozoic. In J.G. Douglas; J. A. Ferguson (Eds.), Geol. Soc. Aust. Spec. Pub. 5: 143-176.

EDWARDS, A. B.; G. BAKER 1943. Jurassic arkose in southern Victoria. Proc. R. Soc. Victoria 55: 195-228.

EVANS, P. R. 1971. Palynology In M. A. Reynolds (Ed.). A review of the Otway Basin. BMR Rep 134.

GLEADOW, A. J. W.; C. K. BROOKS 1979. Fission track dating, thermal histories and tectonics of igneous intrusions in East Greenland. Contrib. Mineral. Petrol. 71: 45-60.

---; J. F. LOVERING 1974. The effect of weathering on fission track dating. Earth Planet. Sci. Lett. 22: 163-168.

---; --- 1978. Thermal history of granitic rocks from western Victoria: a fission track dating study. J. Geol. Soc. Aust. 25: 323-340.

MCDOUGALL, I.; P. J. LEGGO 1965. Isotopic age determinations on granitic rocks from Tasmania. J. Geol. Soc. Aust. 12: 295-332.

---; P. WELLMAN 1976. Potassium-Argon ages for some Australian Mesozoic igneous rocks. J. Geol. Soc. Aust. 23: 1-9.

MCGOLDRICK, P. J.; A. J. W. GLEADOW 1977. Fission-track dating of Lower Palaeozoic sandstones at Tatong, north-central Victoria. J. Geol. Soc. Aust. 24: 461-464.

NAESER, C. W.; N. M. JOHNSON; V. E. MCGEE 1978. A practical method of estimating standard error of age in the fission-track dating method. U.S. Geol. Surv. Open File Report 78-701: 303-304.

ROCHOW, K. A. 1971. Subdivision of the Otway Group based on a sedimentary study and electric log interpretations. In H. Wopfner and J. G. Douglas (Eds.), Geol. Surv. S. Aust. Victoria, Spec. Bull.: 155-176.

STEIGER, R. H.; E. JÄGER 1977. Subcommission on Geochronology: Convention on the use of decay constants in geo- and cosmochronology. Earth Planet. Sci. Lett. 36: 359-362.

SUTHERLAND, F. L.; E. B. CORBETT 1974. The extent of Upper Mesozoic igneous activity in relation to lamprophyric intrusions in Tasmania. Pap. Proc. R. Soc. Tasmania 107: 175-190.

India, SE Asia and Australia in Gondwana fit in the light of evolution

of the Himalaya-Indoburma-Indonesian mobile belts (Abstract)

S. K. ACHARYYA
Geological Survey of India, Dimapur, Nagaland

The continental intracratonic Gondwana sediments (Permian-Early Cretaceous) of the Indian peninsular shield have no physical connection and continuity with the Late Paleozoic island-arc rock associations in the Lesser, the Tibetan and the Trans-Himalayan belts. Much of this wide transverse scatter is an effect of Himalayan nappe movement. Paleozoic geosynclinal sedimentation in the Himalayas with varied litho-facies and Paleozoic-Early Mesozoic magmatic-metamorphic-thermal events there are broadly comparable and coeval with those of south-central Tibet, Burma-Malaya and eastern Australia. Broad coincidence in time and space between Late Paleozoic reworking of granitoid and metamorphic rocks and calc-alkaline volcanism along this belt possibly suggests partial melting of thickened and shortened continental crust close to an active plate margin. Late Paleozoic volcanics petrographically similar to those of the Burma-Malaya organic belt are also widespread in Indochina.

Permian and Late Mesozoic to Tertiary paralic shelf sediments, basin tectonics with normal faulting etc., and Early-Cretaceous-Tertiary marine faunas along the Atlantic type southeast coast of India are similar to those of western Australia but not eastern Antarctica. Early Permian marine faunas from Gondwana and its margins like central and western India, Salt Range, central Afghanistan, Himalaya, south Tibet and Thailand are similar to those of western and eastern Australia. This indicates open marine connection and close paleo-latitudinal span for these localities.

The Mesozoic-Cenozoic island-arc zone is represented along south-central Tibet-Indoburma-Indonesia. The Burma-Malayan and south-central Tibetan Paleozoic-Mesozoic orogenic rocks, occurring in the hinterland of the Himalayan-Indonesian mobile belts, were reactivated by Late Cretaceous-Tertiary alpine orogenesis along their border regions. Orogenesis and upheaval soon triggered off pre-Tertiary hinterland nappes directed towards the convexity of these mobile arcs. The Mesozoic-Cenozoic miogeosyncline was essentially covered in the Himalayan sector. Fossiliferous Late Mesozoic-Early Neogene shelf-miogeosynclinal sediments and alpine flysch-molasse sediments are juxtaposed sporadically as scales and wedges within the lower nappes of Paleozoic and older rocks of the Lesser Himalayan windows located up to 80 km north of the frontal schuppen zone. The Paleozoic-Mesozoic Tibetan Himalayan shelf sediments with intercalations of *Glossopteris* and *Ptillophyllum* floras may represent exposed parts of the miogeanticline. Klippen of hinterland metamorphics also occur in close tectonic associations with the ophiolites along the gravity negative Naga-Yoma belt of the Indoburmese range.

In view of these geological and geodynamic constraints a modified Gondwana reconstruction for the Permo-Triassic period is proposed between India, SE Asia and Australia. An active plate-margin condition possibly existed along the periphery of Gondwana. Pre-Permian fragmentations of Gondwana possibly created several micro-continents and a proto-Indian Ocean between Africa, eastern Antarctica, and western Australia. Late Paleozoic island-arc zones of Himalaya-Tibet, Burma-Malaya, and eastern Australia located between Gondwanic India, Australia and Cathaysia/Laurasia land masses might have provided passageways for floral and land faunal mixtures and encroachments recorded from Himalaya, Tibet, Thailand, Laos, New Guinea, and also Iraq.

Fragmentation of Gondwana and its bearing on the evolution
of the Indian coast line (Abstract)

C.S. RAJA RAO & N.D. MITRA
Geological Survey of India, Calcutta

The separation of India from the African continent began before the Cretaceous; its movement from the combined Antarctica-Australian continent in a direction perpendicular to the east coast of India is defined by a complex set of longitudinal rifts with intervening ridges which run parallel with the NE-SW trending structural grain of the basement rocks. This phase of rifting is represented by the Cauvery, Palar, Krishna-Godavari and Athgarh basins along the continental margin and extensive basalts in the Rajmahal-Bengal basin. Rifting began in early Cretaceous times, accompanied by marginal subsidence and marine transgression from a proto-Indian ocean. Along the west coast of India, an open marine shelf was established in the Kutch area as early as Jurassic. This followed a marine transgression that might have extended to the East African coast with the inception of the proto-Indian ocean. Sedimentation along the west coast of India was interrupted by the eruption of basalts between 65 and 35 m.y. ago. With the beginning of intracontinental rifting the eruption of plateau-type basalts occurred in Neocomian-Aptian times in Mozambique and Madagascar. Later major centres of eruption of plateau basalt began close to the west coast of India. In contrast to the east coast of India, there is no evidence of marginal faulting along the west coast, and a likely reason for this is that the palaeoshelf lay farther to the west.

Crustal structures and tectonic significance of Antarctic rift zones

(from geophysical evidence)

V. N. MASOLOV, R. G. KURININ & G. E. GRIKUROV

SEVMORGEO, Leningrad, USSR

Recent geological and geophysical investigations have revealed many rift zones in different structural provinces, especially in shelves and adjacent areas underlain by transitional-type crust. In Antarctica, many of the rift grabens have been preserved by covering ice. Some have been sufficiently well studied by geophysical methods to be recognised at three different crustal levels--i.e., the bedrock surface, the surface of solid crust, and the surface of the Moho discontinuity. The rift zones are commonest in the lows of bedrock relief related to extensive intra- and pericontinental structural depressions. This suggests that the Antarctic shelf, which is about 40% of the total area of the Antarctic landmass, was largely formed by rifting, which has been the main process causing destruction of Antarctic continental shelf. A similar mechanism is believed to have been responsible for the fragmentation of all Gondwana and consequent formation of the Southern Ocean.

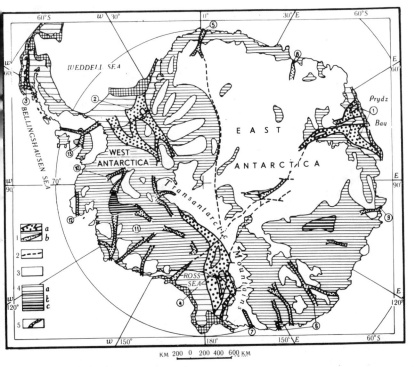

KM 200 0 200 400 600 KM

Fig. 1--Distribution of rift zones in Antarctica. 1: Major rift zones (a) from geological, geomorphological, geophysical data; (b) inferred. 2: Other faults. 3: Areas with bedrock rising a.s.l. 4: Areas with bedrock lying b.s.l. (a) relatively well-studied parts of open shelf; (b) ice-covered shelf inland; (c) deepest parts of subglacial basins. 5: Continental slope. Figures in circles = rift zones named after the following: 1 - Lambert Glacier; 2 - Filchner Ice Shelf; 3 - Bransfield Strait; 4 - Ross Sea; 5 - Jutulsträumen; 6 - Victoria & Wilkes Lands; 7 - Rennick Glacier; 8 - Shirase Glacier & Lutzow Holm; 9 - Denman Glacier; 10 - Rutford & Minnesota Glaciers; 11 - Marie Byrd Land; 12 - Abbott Ice Shelf; 13 - George VI Sound & Palmer Land.

INTRODUCTION

Recent geological and geophysical studies carried out in the world's oceans have revealed an important role of rift-forming processes in the formation of Atlantic-type continental margins and related deep ocean basins. In the last ten years, British, Soviet, and United States geophysical surveys have disclosed many rift structures under the Antarctic ice. Airborne radio-echo-sounding has been a particularly valuable technique. Paradoxically, the ice cover has proved to be advantageous because it has prevented the deposition of clastic sediments that obscure graben morphology in rift areas elsewhere in the world.

Soviet geophysical studies, including seismic, gravity, and magnetic surveys aimed at deep crustal structures, have been mainly in the Lambert Glacier and Weddell Sea shelf areas. This paper is concerned, however, with the distribution and significance of all the rift zones (Fig. 1) in Antarctica, using the Lambert Glacier rift as a model for their interpretation.

THE LAMBERT GLACIER RIFT ZONE

Using geophysical methods, the principal tectonic structure in the Lambert Glacier area can be recognised in the bedrock topography, in the solid crust surface relief, and in the Moho discontinuity (Fedorov *et al.* in press).

The bedrock surface relief is a graben-like depression occupied by the Lambert Glacier and the Amery Ice Shelf. The depression is over 700 km long and about 100 km wide in the north, narrowing to <20 km in the south (Fig. 2a). The bottom is locally 1500 m b.s.l. The Lambert Glacier is flanked on the west by the Prince Charles Mountains with elevations up to 1500-2000 m a.s.l. but decreases towards the Lars Christensen Coast and continues offshore to about 250 m b.s.l. In the south, the eastern side of the Lambert Glacier is flanked by the Mawson Escarpment with bedrock rising to about 1500 m, while in the NE, along the Ingrid Christensen Coast, it and the Amery Ice Shelf are bounded by low (<200 m) nunataks between 70° and 71° S. The hypsometric amplitude of bedrock topography is about 1500-2000 m west of the Lambert Glacier graben and 3000-3500 m within the graben zone. To the east bedrock relief is lower and smoother, except for the Grove Mountains, represented by scattered nunataks reaching 2000 m to the SE.

The Lambert graben is better defined in the solid crust surface relief (Figs. 2b, 3), which consists of a number of narrow subgrabens within the major graben. (Surface of the solid (or consolidated) crust

is deduced from magnetic evidence (Kurin & Grikurov, in press) as the surface of magnetic basement, or the surface of the upper magnetic layer of the crust. This layer in the Lambert Glacier area can in most cases be reliably identified as consisting of the Archean crystalline basement locally overlain by supracrustal folded sequences of Proterozoic to early Paleozoic age.)

The "keyboard" of blocks retains general sublongitudinal orientation and is about 100 km wide in the north, 130 km in the central part of the region, and 80 km in the extreme south. The most prominent tectonic feature coincides with the Mawson Escarpment, which emerges as a median horst splitting the major graben in two branches. The eastern branch lies between the Mawson Escarpment and the Grove Mountains, where indications of its presence are practically absent in either bedrock or ice surface topography; in spite of that, in the solid crust surface relief this eastern branch appears to continue the major linear "step" of the Lambert graben. Large displacement of solid crust surface up to 10 km may be reflected in much smaller displacements of the bedrock surface.

The Prydz Bay basin NE of the Lambert graben, and with the solid crust surface up to 12 km deep, is about 100 km wide and extends NE towards the ocean for approximately 200 km. To the west of the Lambert graben is a symmetrical counterpart to the Prydz Bay basin complicated by numerous inliers. The deepest (up to 5 km) parts of this depression are localised within a relatively narrow (20-30 km) northwesterly trough passing into the Sodruzhestvo Sea shelf.

The relief of the Moho discontinuity also reveals distinct features related to the presence of a rift (Figs. 2c, 3). Crust thickness is greatest at the flanks of the Lambert graben, where it exceeds 40 km under high mountain massifs but ranges between 35 km in the areas where bedrock elevations are closest to sea level. Under the deepest parts of the Lambert graben the crust is only 22-25 km thick, the Moho bulging upward along its whole length while descending to the south. Seismic boundary velocities at the Moho discontinuity do not exceed 7.8-7.9 km/sec. In the south the mantle bulge bifurcates into two separate crests under the western and eastern branches. The Mawson Escarpment is underlain by 40 km of crust, the adjacent grabens by 30 km of crust. The Conrad interface is usually about 10 km above the Moho and has a similar shape. At the Ingrid Christensen Coast the lower part of the "granitic" layer appears to have an abnormally high density (2.85-2.95 g/cm^3).

304

Fig. 2--Structure of Lambert Glacier rift zone at different crustal levels. A = bedrock relief; B = solid crust relief; C = Moho & Conrad discontinuity relief. 1: Bedrock surface countours (m) (a) sea level; (b) a.s.l.; (c) b.s.l.; (d) isobaths. 2: Hypsometric position of solid crust surface (km) (a) a.s.l.; (b) sea level - -3; (c) -3 - -5; (d) -5 - -10; (e) -10 - -15 and deeper. 3: Surfaces of deep-seated interfaces (a), (b) contours of Moho and Conrad distontinuity surfaces, respectively; (c) depths to deep-seated interfaces att. to Conrad (C) and Moho (M) (Adams et al. 1971). All values in km. 4: Lines of schematic cross-sections of Fig. 3. 5: Coastline and/or ice shelf front.

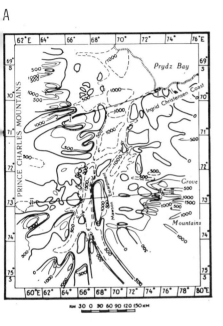

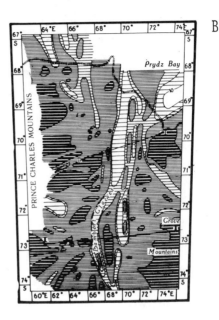

THE FILCHNER ICE SHELF RIFT ZONE

The crustal structure below the Weddell Sea shelf and adjacent mountains is not well known. In the Filchner Ice Shelf area, the bedrock surface contains a meridional depression approximately 400 km long and 100 km wide, with bottom depths from 1200-1400 m. At 80°S the depression is joined by an E-W low, locally as deep as 1000 m b.s.l., occupied by the Slessor Glacier. East of the meridional depression mean bedrock elevations are about 700 m a.s.l., with a few peaks as high as 1100-1700 m. To the west the bedrock lies close to sea level under Berkner Island. The hypsometric gradient of bedrock topography reaches 3000 m to the east of the Filchner Depression and no more than 1500 m to the west. The depression can be traced northwards on the ocean floor as a bathymetric deep about 350 km long, striking NNE and widening towards the continental slope.

The structure of the Filchner graben and its northern continuation appears more complex in the solid crust surface--i.e., the lower Precambrian basement surface and supracrustal sequences. This surface is reliably identified only in the eastern part of

305

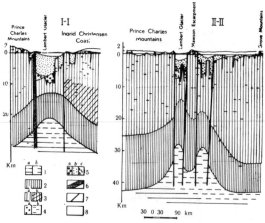

Fig. 3--Schematic representation of crustal structure across Lambert Glacier rift zone. (Section lines in Fig. 2c). 1: Mantle (a) normal, (b) abnormally low density. 2:Lower ("basaltic") layer of crust. 3: Upper ("granitic") layer of crust (a) normal, (b) abnormally high density. 4: Calculated positions of magnetic sources (a) upper, (b) lower boundaries. 5: Sequences filling rift grabens & other depressions of solid crust surface: (a) Permian coal measures; (b) predom. Cenozoic molasse, glacial & marine-glacial deposits; (c) low-density volcanic rocks. 6: Zone of deep fault intruded by high-density (ultramafic?) rocks. 7:Other faults. 8: Ice.

the region where the lower Precambrian has been mapped and recognised as the most likely source of magnetic anomalies.

The main "stem" of the Filchner rift zone is clearly defined in the magnetic basement topography as a graben 400 km long, more than 100 km wide, and more than 10 km deep. The bifurcation at 80°S is well defined. The western branch passes the southern side of Berkner Island and opens to the Ronne Ice Shelf, reaching 15 km. The eastern branch extends south for about 240 km along 38°W, its eastern boundary formed by the Shackleton Range, Theron Mountains, and Luitpold Coast, with total gradients in the solid crust relief of 10 km in the south and 15 km in the north. The Slessor and Recovery glaciers cut the eastern "shoulder" of the graben and mark supplementary offshoots with depths of the solid crust surface at 10 and 5 km respectively. The western "shoulder" is elevated relative to its bottom only for 2-3 km and lies approximately within the -10 km isoline contour.

Northward the Filchner graben passes into a NE-trending depression in which the depths of magnetic basement exceed 15 km.

The SE part of the Weddell Sea shelf is thus characterised by subsidence of the solid crust surface most likely associated with down-faulting of the craton. The amplitude of the down-faulting probably increases to the W and NW.

The position of the Moho discontinuity determined from the gravimetric information and the correlation between regional Bouguer anomalies and the Moho relief established for the Lambert Glacier area (Kurinin & kurov, in press). In the hypothetical crustal model (Fig. 3), the observed Bouguer anomaly was plotted as the mean value of discrete gravity readings obtained within the band 60 km wide (30 km N and S of 78°S), then the model was compiled as the best of observed and computed graphs. According to that model, the Filchner graben may be underlain by a crustal opening in which sediments about 15 km thick and 2.5 g/cm³ in density overlie uplifted mantle, possibly covered by a thin "basaltic" layer.

RIFT ZONES OF THE BRANSFIELD STRAIT, THE ROSS SEA SHELF, AND OTHERS

The Bransfield Strait rift zone is a sea bottom graben 500-600 km long, 40-50 km wide and 1500-2000 km deep (Davey 1972; Ashcroft 1972). The steep NW side is almost linear between Deception and Clarence islands in the South Shetlands. The SE side along the Antarctic Peninsula is less pronounced. The bottom is nearly flat, with a few elevations 500-600 m high. The graben is marked by sharp positive Bouguer and isostatic anomalies reaching 150 and 80 mgal respectively. Seismic evidence indicates a 5-km deep solid crust surface associated with a mantle rise responsible for crust attenuation to 14-15 km. The mantle density within this rise is unusually low, with a boundary velocity of 7.7 km/sec at the Moho discontinuity. The velocity values increase rapidly to 8.0-8.5 km/sec under both "shoulders" of the graben. Crustal models (Ashcroft 1972; Davey 1972) indicate a "basaltic" layer approximately 10 km thick overlain by a volcanic and a sedimentary layer, each about 2 km thick, in the rift.

Most of the rift zones in the Ross Sea are on its western side, adjacent to the Transantarctic Mountains. The most extensive zone can be traced for almost 1200 km under the Ross Ice Shelf and continues for 500 km farther in the open part of the Ross Sea. At the southern end of the Transantarctic Mountains, between the Leverett Glacier and the Nimrod Glacier, it strikes between NW and W, its width averaging 80-100 km, with the bedrock surface locally as low as 800-900 m b.s.l. (Robertson et al. in press).

Between the Nimrod and Byrd glaciers, the rift zone widens to almost 300 km and is divided into two branches by a bedrock horst; at the same place the strike changes to between NNE and N, and the greatest depth to bedrock increases to 1300 m. The change may be due to transection of the rift zone by a system of transcontinental faults which extend from the Lambert Glacier via the sub-ice Gamburtsev Mountains to the Nimrod-Byrd glaciers area, where they cross the Transantarctic Mountains and extend into the Ross Sea shelf. Near Ross Island the rift zone's western branch splits into a series of smaller blocks with relative hypsometric amplitude reaching 800-1000 m (Hayes & Davey 1975).

Geophysical information for the Ross Ice Shelf area (Bentley *et al.* in press; Robertson *et al.* in press) provides no direct evidence on the solid crust surface topography and the Moho relief. The graben adjacent to the southern and central Transantarctic Mountains has a relative gravity high reaching 20 mgal. This fact suggests a rise of the masses responsible for isostatic compensation, possibly a sub-graben mantle swell. In the open part of the Ross Sea shelf the eastern branch of the rift zone is underlain by a graben in the solid crust surface that is down-faulted at least 2-3 km relative to the adjacent blocks (Hayes & Davey 1975). A crustal model implies shallow high-density (subcrustal?) masses below the graben. The eastern branch has no topographic expression on the sea floor.

Geophysical data suggest that rift structures may also be present in other parts of the Ross Sea shelf, especially near the Siple Coast and Roosevelt Island. These rifts and the major complex rift zone described above are probably integral parts of a rift system of much greater extent.

Other rift zones (Fig. 1) have been distinsuished from geomorphology and in a few cases supported by scarce geological and/or geophysical evidence.

In East Antarctica such zones are limited in number. The deep graben of Jutulstraumen is probably a rift structure that divides the mountains of Queen Maud Land approximately along the Greenwich meridian. A se-

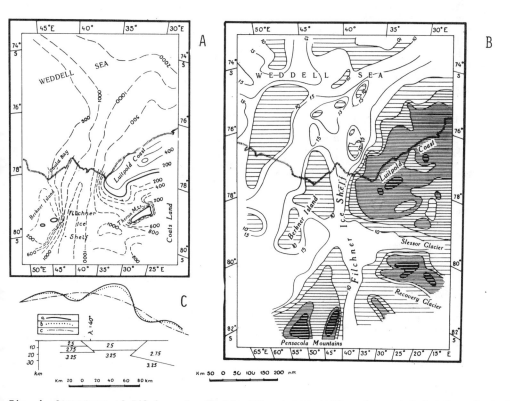

Fig. 4--Structure of Filchner Ice Shelf rift zone at different crustal levels. A: bedrock relief; B: solid crust relief; C: crust model compiled for section along 78°S from gravity & magnetic data (a) obs. Bouguer anomaly; (b) computed Bouguer anomaly; (c) regional anomaly. Designations for A and B as in Fig. 2.

ries of sub-ice grabens in the bedrock surface between 130° and 150°E have steep sub-ice topographic gradients with a relief of 1500-2000 m (Steed & Drewry, in press) and can be regarded as rifts. The Rennick Glacier graben is equally well displayed in the topography. Other rift zones believed to be present in East Antarctica (such as the Shirase Glacier-Lutzow Holm Bay zone or the Denman Glacier zone) are less prominent morphologically, but they may be better expressed at deeper crustal levels.

Inferred rift zones are much more numerous in West Antarctica. Those NW of the Ellsworth Mountains, in Marie Byrd Land along 140°W and between the Whitmore and the Horlich Mountains are not only apparent in the bedrock topography but are also reflected in the regional isostatic gravity anomalies by a moderate high (over +20 mgal) in the first area, a sharp low (above -50 mgal) in the second, and an abrupt gradient (from -20 to +15 mgal) in the third. (Bentley & Robertson, in press). Extensive rift grabens in central and eastern Marie Byrd Land are deduced from a close association of deep depressions in sub-ice relief with adjacent alkaline basalt terrane and major stratovolcanic structures (Bentley & Chang 1971). The rift zones bounded by the Sentinel Range (Rutford Glacier), Thurston Island (Abbott Ice Shelf), and Alexander Island (George VI Sound) are recognised from geomorphology, at places confirmed by local alkali-basaltic volcanism.

DISCUSSION

Most of the rift zones are in West Antarctica (Fig. 1), which emerges as a vast rift belt comprising a few large rift systems. The main system borders the Transantarctic Mountains and can be traced almost continuously between the continental slopes of the Ross and Weddell seas. In central West Antarctica this system splits into two branches encompassing the Ellsworth-Whitmore Mountains block. A few rift zones of variable orientations extending from Marie Byrd Land appear to be related to the same main system. A separate rift system aligned with the island archipelagoes can probably be recognised in the Bellingshausen Sea area.

In East Antarctica the main transcontinental system of faults and rifts extends from the Lambert Glacier on the Indian Ocean Coast to the Nimrod-Byrd glaciers area in the Transantarctic Mountains. Many rift zones also occur in the ice-covered ridge of Wilkes Land (130°-150°E) that separates the two largest subglacial bedrock depressions in this part of the continent.

A correlation between bedrock hypsometric levels and the distribution of rift zones is apparent. In the sector between 0° and 73° areas with bedrock surface below sea level are practically absent, and rift zones are very few. This block is bounded by the Lambert-Nimrod glaciers fault zone to the east of which the bedrock surface descends below sea level and the number of rift zones noticeably increases. To the west there is probably a similar down-stepping of the bedrock surface along the fault zone extending from the Jutulstraumen towards the South Pole; a prominent drop certainly occurs in the Filchner rift zone.

The largest morphostructural features in Antarctica are related to the transcontinental Ross-Weddell rift system. A spectacular escarpment along the Transantarctic Mountains, marking the boundary of East and West Antarctica, represents the high "shoulder" of the rift. This system is responsible for the greatest bedrock surface relief (locally up to 5000-7000 m) and the deepest bedrock subsidence (locally more than 2000 m b.s.l.) known in Antarctica. The most extensive areal bedrock depressions, including the Ross Sea-Weddell Sea and Byrd subglacial basins, are apparently associated with the same rift system.

The rift zones are clearly concentrated in the areas of intra- and pericontinental depressions. Those in high mountains (e.g. the Lambert, Jutulstraumen, Rennick Glacier rift zones) incorporate only the least developed (youngest) "upper reaches" of rift. The more-mature "down-stream" parts always stretch out into the continental shelf. One may speculate that the high areas now subject to rifting will eventually undergo fragmentation and subsidence. All present day bedrock depressions with numerous rift zones are not necessarily reversed former highlands. The fragmentation of up-arched crust is probably only one variety of the rifting process. The general cause is more likely to be related to the existence of certain epochs of global extension of the lithosphere.

The evidence suggests that rifting has formed most of the Antarctic shelf which constitutes, with its vast ice-covered regions, at least 40% of Antarctica. This conclusion is best documented for the Ross Sea, Weddell Sea, and Prydz Bay shelves, which are all modern sedimentary basins at the junctions of intracontinental rift zones systems) with the pericontinental (circum-Antarctic) fault--i.e., the continental slope escarpment. Their funnel-like shape broadening oceanward, suggests that the basins occupy the sites of most active destruction of continental crust by rift-forming processes.

CONCLUSIONS

(1) Antarctica is characterised by more numerous rift zones than elsewhere in the Circum-Pacific mobile belt. The unique intensity of rift-forming processes in the southernmost part of the Pacific sets the West Antarctic rift belt apart as a distinct feature in the Earth's crust.

(2) The relation of East Antarctic rift systems to global crustal structures is more problematic. The Nimrod-Lambert glaciers rift system may continue into the Indian Ocean towards the Kerguelen Plateau, where continental crust is associated with Cenozoic alkaline igneous activity. Concentration of rift zones in Antarctica between 130° and 150°E may be related to a system of longitudinal fracture zones transverse to the Australian-Antarctic Ridge between 120° and 130°E.

(3) Antarctica is the only continent whose shelf forms as much as 40% of its total area. This feature is apparently related to unusually intense rift-forming processes in the southern polar region of the Earth.

(4) The vast Antarctic shelf (about 6.5 x 10^6 km^2) is rather homogeneous in origin. It probably all resulted from destruction of a once much larger and monolithic continental mass believed to have been an integral part of Gondwana. In the authors' opinion, rifting apparently caused the degradation of the Antarctic continent and its gradual assimilation into the ocean. A consequent implication is that destructive tectogenesis on a global scale responsible for the Gondwana break-up and the formation of the Southern Ocean has also resulted largely from rift-forming processes. The leading role of these processes in the disassembling of large continental masses seems indisputable, whether plate tectonics, oceanisation, or general expansion of the Earth is the primary mechanism.

REFERENCES

ADAMS, R.D. 1971. Reflections from discontinuities beneath Antarctica. Bull. Seismol. Soc. Am. 61(5): 1441-1451.

ASHCROFT, W.A. 1972. Crustal structure of the South Shetland Islands and Bransfield Strait. Sci. Rep. Br. Antar. Surv. 66. 43 pp.

BENTLEY, C. R.; F. K. CHANG 1971. Geophysical exploration in Marie Byrd Land, Antarctica. In A. P. Crary (Ed.), Ant. Res. Ser. Washington; A. Geophys. Un. 16: 1-38.

---; J. D. ROBERTSON. In press. Isostatic gravity anomalies in West Antarctica. In C. Craddock (Ed.), Antarctic geosiences. Madison: Univ. of Wisconsin Press.

---; ---; L. L. GREISCHAR. In press. Isostatic gravity anomalies of the Ross Ice Shelf, Antarctica. In C. Craddock (Ed.), Antarctic geosciences. Madison: Univ. of Wisconsin Press.

DAVEY, F. J. 1972. Marine gravity measurements in Bransfield Strait and adjacent area. In R. J. Adie (Ed.), Antarctic geology and geophysics. Oslo: Universitetsforlaget. Pp. 39-45.

FEDOROV, L.V.; G. E. GRIKUROV: R. G. KURININ; V. N. MASOLOV. In press. Crustal structure of the Lambert Glacier area from geophysical data. In C. Craddock (Ed.), Antarctic geosciences. Madison: Univ. of Wisconsin Press.

HAYES, D. E.; F. J. DAVEY 1975 A geophysical study of the Ross Sea. In, Init. Rep. DSDP XXVIII: 887-907.

KURININ, R. G.; G. E. GRIKUROV. In press. Crustal structure of part of Antarctica. In C. Craddock (Ed.), Antarctic geosciences. Madison: Univ. of Wisconsin Press.

ROBERTSON, J. D.; C. R. BENTLEY; J. W. CLOUGH; L. L. GREISCHAR. In press. Sea bottom topography and crustal structure below the Ross Ice Shelf, Antarctica. In C. Craddock (Ed.), Antarctic geosciences. Madison: Univ. of Wisconsin Press.

STEED, R.H.N.; D. J. DREWRY. In press. Radio echo-sounding investigations of Wilkes Land, Antarctica. In C. Craddock (Ed.), Antarctic geosciences. Madison: Univ. of Wisconsin Press.

The Late Mesozoic fragmentation of the New Zealand segment of Gondwana

M. G. LAIRD

N. Z. Geological Survey, Christchurch, New Zealand

Following compression, uplift, and erosion of the rocks of the Rangitata Orogen (Carboniferous-Lower Cretaceous) a change to extensional tectonics occurred in mid-Cretaceous times throughout the New Zealand region.

It is best represented by a widespread basal or intra-Albian unconformity which separates indurated structurally complex sparsely-fossiliferous strata of the Rangitata Orogen from younger less indurated structurally simpler, and relatively fossiliferous rocks. A rift system developed on the West Coast of the South Island, while a tectonically-controlled basin developed on the east coast of both islands. This early extensional phase was accompanied by volcanism ranging from rhyolitic to basaltic.

In the Late Cretaceous much of the New Zealand area was peneplained before widespread Campanian and Maastrichtian transgression. This transgression and pene-contemporaneous extrusion of alkali basalts over much of the New Zealand area and renewed onset of rifting on the West Coast were related to the opening of the Tasman Sea and the beginning of separation from Antarctica.

INTRODUCTION

The cessation of sedimentation (Carboniferous - Early Cretaceous) on the site of the Rangitata Orogen was accompanied by widespread compressional deformation, uplift, erosion, and, on the western foreland, calc-alkaline plutonism. This brought to a close the main phase of the Rangitata Orogeny (Bradshaw *et al.* this volume).

Immediately-following tectonic events affected the deposition and distribution of mid and Late Cretaceous rocks, which crop out over large areas of New Zealand (Fig. 1), and have been less easy to categorise. At many places mid-Cretaceous rocks overlie an angular unconformity separating highly deformed and indurated sparsely fossiliferous older strata from markedly less deformed and indurated more fossiliferous strata of late Aptian to Albian age (Stevens & Speden 1978). Even where an unconformity is not obvious, the change in deformation and induration is evident and appears to represent a major change in tectonic regime. In latest Cretaceous times, a major transgression associated with an underlying unconformity or lithofacies boundary was also widespread throughout New Zealand. The evidence for these two events is detailed below and in Fig. 2 , and tectonic interpretations are made in the light of the penecontemporaneous events affecting the SW Pacific sector of Gondwana.

MID-CRETACEOUS EVENTS AND SEQUENCES

West Coast, South Island. West of the Alpine Fault in the South Island, coarse continental deposits of Albian age unconformably overlie Early Paleozoic sedimentary rocks or granites of mid-Cretaceous or older age. In the northern part the continental sediments have been deposited in N to NNE-oriented grabens or fault-angle depressions with rapidly-rising interfluves (writer's observations). Locally restricted acid volcanics, between 105-109 m.y., are associated with the basal deposits but may be separated from them by an unconformity (Adams & Nathan 1979).

Otago and Southland. From Albian or Cenomanian times, coarse continental sediments were deposited east of the Alpine Fault in the southern half of the South Island in fault-angle depressions (Mutch & Wilson 1952; Bishop & Laird 1976) and lie with marked angular unconformity on folded, metamorphosed, and peneplained rocks of the Rangitata Orogen. In western Southland, limited exposures of non-marine and marine sandstone of late Albian age (on pollen - Dr I. Raine, pers. comm.) rest unconformably on older rocks.

Canterbury. Late Albian to Cenomanian subaerial calc-alkaline rhyolite, andesite, and dacite with K-Ar ages between 98 and 92 m.y. (Oliver *et al.* 1979) rest on peneplained Jurassic and Triassic rocks of the Rangitata Orogen.

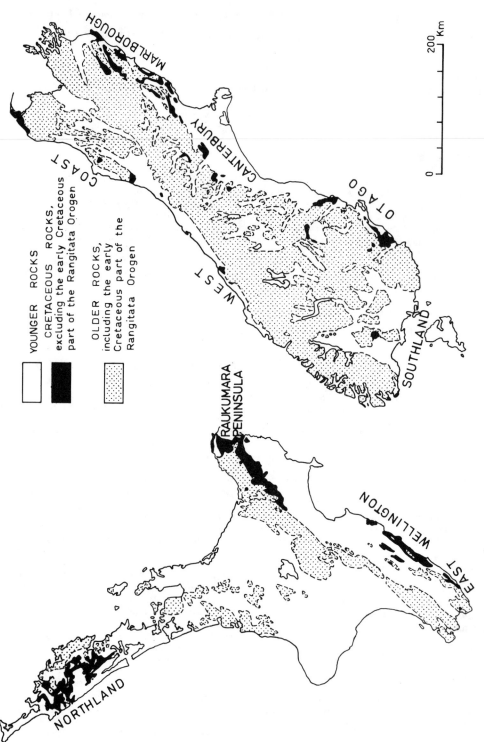

Fig. 1--Maps of North and South Islands, showing distribution of Cretaceous rocks.

Marlborough and East Wellington. Marine Albian and Cenomanian sediments were deposited in an apparently continuous east coast basin which occupied much of Marlborough, east Wellington, and the Raukumara Peninsula. In the first two areas, rocks of late Albian age rest with marked unconformity or clear induration/lithofacies disjunction on highly-deformed strata of similar age (Stevens & Speden 1978; Moore & Speden 1979), and it is evident that the unconformity, although recording a period of strong deformation, uplift, and erosion, represents only a very short time interval. The oldest sediments overlying the unconformity in both Marlborough and east Wellington are mainly olistostromes passing upwards into flysch-like units of alternating sandstone and mudstone, although marginal marine deltaic deposits occur in southern Marlborough (writer's observations). In Marlborough, several eruptive centres, aligned in a NE-trending belt, can be identified. Two large layered gabbro bodies occur, and chemically and petrologically similar basalt flows are interbedded with Cenomanian sediments. The eruptions may have occurred along NE-trending rifts (Nicol 1977). In east Wellington basalt sills, tuffs, and volcanogenic sediments and some acid tuffs of Albian age occur (Moore & Speden 1979).

Raukumara Peninsula. A marked angular unconformity is underlain by Late Jurassic strata. Pockets of thin, late Aptian shallow marine sediments overlie it locally and are unconformably overlain by Albian conglomerate and breccias of an inshore facies derived from an actively rising and eroding landmass (Speden 1975; Moore 1978). Early Albian to perhaps Coniacian sedimentation was probably controlled, at least in part, by N-NE-striking faults (Moore 1978). Basic volcanic rocks occur within middle to late Albian sediments at several localities (Speden 1976; Stevens & Speden 1978).

Northland. In northern Northland indurated, thin-bedded sandstone and interbedded basic and acidic lavas of late Aptian to Albian age (Hay 1975) are overlain unconformably by keratophyre breccia and indurated flysch-like sandstones and mudstone inter-fingering with Albian basic volcanics. Basaltic lava becomes more common upwards and forms large masses associated with Albian to Turonian sediments. The Albian and younger rocks change progressively upwards from flysch-like to shelf sediments thought to have filled a subsidiary trough which formed during eversion of the Rangitata Orogen (Hay 1975). Elsewhere in Northland, only small and fault-bounded

outcrops of late Albian to Cenomanian sediments are known.

Chatham Islands. On Pitt Island in the Chathams, 800 km east of the South Island, weakly indurated mudstone, fine sandstone, and minor conglomerate of freshwater and shallow marine facies of Albian and Cenomanian age (Grindley *et al.* 1977) are overlain conformably by Late Cretaceous tuffs and basic lavas. The base is not seen but is inferred to rest unconformably on sediments of probable Jurassic age (Austin *et al.* 1973).

LATE CRETACEOUS EVENTS AND SEQUENCES

Although in some areas, such as on the Chathams, and along parts of the east coast of North and South Islands, sedimentation continued without major break from Albian to late Campanian times, elsewhere uppermost Cretaceous strata (where present) rest either on peneplained rocks of the Rangitata Orogen and older sequences, or unconformably on mid-Cretaceous rocks (Fig. 2).

Otago and Southland. In east Otago and eastern Southland Maastrichtian coal measures overlie older rocks with marked unconformity. In both areas they were formed in fault-angle depressions or grabens.

West Coast. Late Cretaceous rocks occur in three main areas west of the Alpine Fault (Fig. 3). In the south coal measures and marine sandstone of late Campanian and Maastrichtian age disconformably overlie mid-Cretaceous coarse clastic deposits (Nathan 1977; Dr I. Raine, pers. comm.). The sedimentary sequence is conformably overlain by late Maastrichtian to early Paleocene alkaline basalt. Individual vents have not been located, but a linear NE-trending belt of high amplitude magnetic anomalies associated with the volcanics can be traced for 80 km along or close to the Cape Foulwind Fault Zone, at least parts of which were active in earliest Tertiary times (McNaughton & Gibson 1970). This fault zone extends NE offshore and is parallel to the coast for approximately 560 km (Norris 1978; Fig. 3).

In the central portion of the West Coast sediments of Late Cretaceous age, which overlie Albian strata unconformably, are preserved only along or near the NNE- to NE-trending Paparoa Tectonic Zone, which extends N for 200 km from the Alpine Fault near Hokitika to the coast 45 km NE of Westport (Fig. 3). It has been periodically active since at least Late Cretaceous times and probably earlier (Laird 1968). A fault-controlled trough, probably about 10 km wide, within the zone accumulated

313

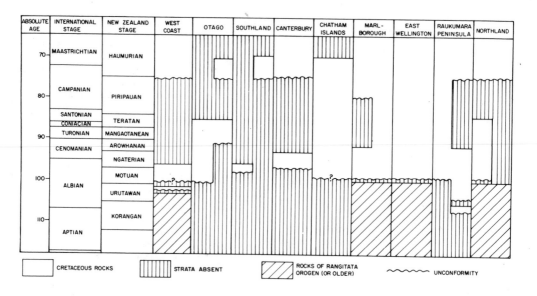

ABSOLUTE AGE	INTERNATIONAL STAGE	NEW ZEALAND STAGE	WEST COAST	OTAGO	SOUTHLAND	CANTERBURY	CHATHAM ISLANDS	MARL-BOROUGH	EAST WELLINGTON	RAUKUMARA PENINSULA	NORTHLAND
70	MAASTRICHTIAN	HAUMURIAN									
80	CAMPANIAN	PIRIPAUAN									
	SANTONIAN										
	CONIACIAN	TERATAN									
90	TURONIAN	MANGAOTANEAN									
	CENOMANIAN	AROWHANAN									
		NGATERIAN									
100	ALBIAN	MOTUAN									
		URUTAWAN									
110		KORANGAN									
	APTIAN										

☐ CRETACEOUS ROCKS ▥ STRATA ABSENT ▨ ROCKS OF RANGITATA OROGEN (OR OLDER) 〰 UNCONFORMITY

Fig. 2--Correlation of Cretaceous units in New Zealand. Time-scale based on Odin (1978); international and New Zealand stage correlations based on Stevens & Speden (1978).

late Campanian to early Paleocene non-marine sediments from the adjacent uplifted blocks, and was again active in the Eocene. Contemporaneous basalt flows and volcanic breccias were erupted at several centres along the southern portion. A probably earlier period of igneous activity is represented by swarms of lamprophyre, basalt, and trachyte dikes, dated in the north at 78 - 84 m.y. (Adams & Nathan 1978). They are in a belt aligned NNE, truncated by the Alpine Fault to the SW. Two are associated with magnetic anomalies and are inferred to relate to major bodies of lamprophyre magma emplaced at shallow depth (Hunt & Nathan 1976), slightly east of the Paparoa Tectonic Zone (Fig. 3). Apparently similar lamprophyre dike swarms lie east of the Alpine Fault in west Otago (Wellman & Cooper 1971), but they are not associated with known Cretaceous rocks, and their correlation is uncertain (Hunt & Nathan 1976).

In the NW, non-marine sediments accumulated in NE- to NNE- oriented grabens or fault-angle depressions during Maastrichtian and early Paleocene times (Titheridge 1977; Pilaar & Wakefield 1978). Onshore, thick clastic deposits ranging from breccia to sandstone and thin coal seams which unconformably overlie Lower Paleozoic rocks west of the Wakamarama Fault (Fig. 3) were deposited in response to movements on this fault during Late Cretaceous times (Titheridge 1977). Sediments of similar age

and lithology have been recognised in offshore commercial wells up to 160 km north of the South Island, where they appear to have been deposited in NE- or NNE-trending grabens and fault-angle depressions (Pilaar & Wakefield 1978; Fig. 3).

Canterbury and Marlborough. Sea transgressed over the South Island during Late Cretaceous times. In Marlborough, the transgressive deposits are marine glauconitic sandstones of Campanian and early Maastrichtian age overlying older Cretaceous while in Canterbury, thin marine Maastrichtian sediments, often with a basal non-marine phase, unconformably overlie highly-deformed rocks of the Rangitata Orogen.

East Wellington and Raukumara Peninsula. Sediments of Senonian age are well-developed and stratigraphically continuous with Early Tertiary sediments in many areas. In the western Raukumara Peninsula, a major regional event involving deformation and erosion occurred before deposition of late Campanian to Maastrichtian sediments, and strata of this age unconformably overlie older rocks (Stevens & Speden 1978). To the E, Campanian sediments conformably overlie rocks of Santonian age. Sedimentary evidence suggests that during late Santonian to Maastrichtian times a double-sided basin formed in response to block faulting with actively-eroding landmasses both to the W and the E or SE (Stevens & Speden 1978). In east Wellington, deposition was in many

314

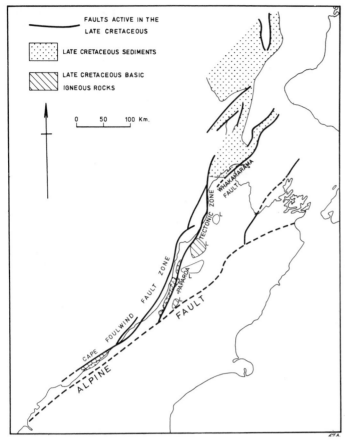

Fig. 3--Fault systems inferred to have been active during the Late Cretaceous in the western portion of New Zealand. Distribution of Late Cretaceous sediments and igenous rocks (west of the Alpine Fault) also shown. Data from outcrop, drillhole, and geophysical study. Offshore distribution of faults and Late Cretaceous sediments north of the South Island after Pilaar & Wakefield (1978). L = basic dike swarms; hatched areas = lamprophyre bodies inferred at depth. Data from Hunt & Nathan (1976).

instances continuous or semi-continuous from mid-Cretaceous times (Moore & Speden 1979). Dolerite sills and basalt flows occur in Campanian to Maastrichtian rocks (Moore, in press).

Northland. Maastrichtian beds comprise almost the whole of the outcropping Cretaceous and unconformably overlie older rocks in many areas (Stevens & Speden 1978). Isolated inliers of older beds represent all of the pre-Maastrichtian Late Cretaceous stages. The stratigraphy is complicated because much of the Northland Cretaceous is incorporated in either the Late Oligocene Oherahi Chaos Breccia or the Early Oligocene Northland Allochton (Ballance & Spörli 1979).

Chatham and Campbell Islands. On Pitt Island (Chathams) a thin sequence of Albian to late Cenomanian sediments is overlain conformably by basaltic tuffs which range in age up to Campanian. These are succeeded conformably by alkaline basaltic volcanics radiometrically dated at 75-80 m.y. (Grindley *et al.* 1977). The Maastrichtian has not been recognised, and Paleocene sedi-

ments rest unconformably on the Campanian rocks. The only other New Zealand offshore island where Cretaceous rocks are exposed is Campbell Island, approximately 700 km south of the South Island. Here thin Maastrichtian to Paleocene largely fine-grained sediments unconformably overlie low-grade metamorphic rocks and are conformably overlain by younger Tertiary sediments (Beggs 1978).

TECTONIC SYNTHESIS

The major regional tectonic event following subduction and compression along the eastern margin of the SW Pacific segment of Gondwana in the late Paleozoic and early Mesozoic was the separation of the New Zealand micro-continent from Australo-Antarctica at about 80 m.y. B.P. (Hayes & Ringis 1973; Christoffel & Falconer 1973). Patterns of arching, rifting, and igneous activity accompanying extensional tectonics preceding and associated with continental separation have been recorded in S and SE Australia during Early and Late Cretaceous times.

It is here suggested that the Cretaceous record outlined earlier also reflects the initiation of extensional tectonics and local patterns of rifting in the New Zealand region. In parts of New Zealand, deposition appears to have been continuous between mid and Late Cretaceous times (e.g., parts of the east coast of both islands, and the Chatham Islands), but over much of New Zealand where Cretaceous rocks are preserved, a major transgression occurred in late Campanian or Maastrichtian times, and latest Cretaceous non-marine or shallow marine sediments rest unconformably on older rocks. The Upper Cretaceous rocks were deposited at or soon after the time of onset of sea-floor spreading in the region as a result of partial foundering of the New Zealand block. The Chatham Island alkaline basic volcanics radiometrically dated at between 75-80 m.y. are related to the onset of separation from West Antarctica (Grindley *et al*. 1977). Alkaline basic igneous rocks of Maastrichtian age on the east coast of the North and South Islands may also be related to continental separation.

On the West Coast of the South Island the Upper Cretaceous sediments and volcanics are clearly structurally controlled. In the north, Upper Cretaceous coal measures lie within or close to two NNE- oriented graben complexes (Fig. 3) which were actively subsiding until at least early Paleocene times. Maastrichtian basic volcanism was associated with one of them (the Paparoa Tectonic Zone). The belt of early Campanian basic dikes lying to the east of the Paparoa Tectonic Zone may represent the results of extension along a further NNE-trending tectonic lineament, particularly as Maastrichtian non-marine beds have recently been reported from a drillhole (Aratika No. 2) within the belt (pers. comm., T. Haskell, Petrocorp Exploration). In the south of the West Coast, Upper Cretaceous non-marine and shallow marine sediments and uppermost Cretaceous - lowermost Paleocene basaltic volcanics lie on or close to the Cape Foulwind Fault Zone. These belts, along which igneous acitivity was concentrated, and which acted as depocentres for thick accumulations of sediment, are here considered to be elements of a major rift system which was active from Campanian to at least Early Paleocene times and which formed in response to the same tectonic events which resulted in the opening of the Tasman Basin. The intrusion of lamprophyre dike swarms within or close to the graben complexes contemporaneously with the beginning of sea-floor spreading on the site of the Tasman Basin at about 80 m.y. is also regarded as supporting evidence.

This zone of rifting--the West Coast Rift System--has a tectonic and sedimentary history closely similar to that of the rift systems developed along the southern and eastern boundaries of the Australian continent. However, continental separation was not completed, and the rift system is inferred to be a "failed arm", along which incipient continental separation occurred in Late Cretaceous times. The arm was probably an offshoot from a triple junction located SW of the South Island, the other two arms of which continued spreading to form the Tasman Basin. A similar tectonic history was postulated by Burke and Dewey (1973) for the Gippsland Basin in SE Australia. The probable relationship of the West Coast Rift System to the southern and eastern Australian Late Cretaceous rifts is shown in Fig. 4.

The other major Cretaceous tectonic/stratigraphic element in the New Zealand region post-dating the second phase of the Rangitata Orogeny in the Early Cretaceous is the widespread intra or basal Albian unconformity and the frequently associated overlying "orogenic" deposits. Although this unconformity is locally diachronous, it appears to represent both a change in style of deposition and of tectonics from the preceding compressional regime associated with the Rangitata Orogeny. Taking into account tectonic events occurring elsewhere in the SW Pacific sector of Gondwana (e.g. southern Australia), it is inferred to represent the evidence in the stratigraphic column of the change from a regime of compression to one of extension.

The most extensive and best documented areas of Albian and Cenomanian rocks occur on the West Coast and in the east coast basin where thick coarse continental and gravity emplaced marine sediments respectively rest on the unconformity. On the West Coast outcrop and known subsurface occurrences of mid-Cretaceous sediments are largely co-extensive with latest Cretaceous deposits, and infilled NNE- or N-trending grabens (see earlier). Although there is no direct evidence indicating under what tectonic regime the grabens and detritus-supplying horsts formed, on analogy with southern Australian events it is tempting to assume that arching and rifting occurred at much the same time in the New Zealand region. The unconformity-bounded Albian deposits west of the Alpine Fault are considered to have formed in response to initial arching and keystone faulting along the West Coast Rift System, and represent a pre-breakup phase of deposition (equivalent to the Otway Group of the Otway Basin).

On the east coast of New Zealand from

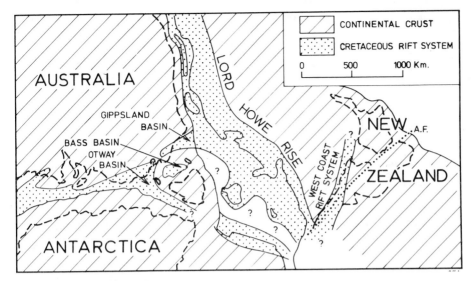

Fig. 4--Cretaceous rift systems in the Southwest Pacific segment of Gondwana. Lord Howe Rise rift system after Mutter & Jongsma (1978). Restoration of New Zealand before Late Cenozoic movement on the Alpine Fault after Bradshaw et al. (this vol.). A.F. = Alpine Fault.

Marlborough to east Wellington, where Albian sediments of commonly mass flow origin rest unconformably on highly deformed rocks also of Albian age, fault control of sedimentation along N-S or NE trends is evident. The association of these sediments with Albian and Cenomanian alkaline volcanics and intrusives following a NE structural trend in Marlborough, suggests the possibility that a further rift system developed along the east coast at this time. Reactivation of this structure in latest Cretaceous times resulted in a second period of extrusion of alkaline volcanics. The localised and spatially restricted unconformity-bounded late Aptian strata of the Raukumara Peninsula indicate that extensional movements may have begun here as early as the late Aptian.

Elsewhere in New Zealand structural control of sedimentation appears to have been more localised and no distinct pattern is evident in Northland or in the Chatham Islands. In Otago, Albian and Cenomanian deposits formed in fault-angle depressions against active faults oriented NW or NE, probably also reflecting local effects of extension. Non-marine deposits of Albian age in western Southland probably formed within the West Coast Rift System (prior to dislocation by the Alpine Fault - see reconstruction of New Zealand, Fig. 4).

ACKNOWLEDGMENTS

The ideas expressed above have benefitted from discussion with numerous colleagues and the manuscript from critical review by Mr S. Nathan and Dr R.P. Suggate. Unpublished data on Aratika No. 2 well, Westland, were kindly made available by Petrocorp.

REFERENCES

ADAMS, C. J.; S. NATHAN 1979. Cretaceous chronology of the Buller Valley, South Island, New Zealand. N.Z. J. Geol. Geophys. 21(4): 455-62.

AUSTIN, P.M.; R. C. SPRIGG; J. C. BRAITHWAITE 1973. Structure and petroleum potential of eastern Chatham Rise, New Zealand. A.A.P.G. Bull. 57(3): 477-497.

BALLANCE, P. F.; K. B. SPÖRLI 1979. Northland Allochthon. J. R. Soc. N.Z. 9(2): 259-275.

BEGGS, J. MAC. 1978. Geology of the metamorphic basement and Late Cretaceous to Oligocene sedimentary sequences of Campbell Island, Southwest Pacific Ocean. J. R. Soc. N.Z. 8(2): 161-177.

BISHOP, D. G.; M. G. LAIRD 1976. Stratigraphy and depositional environment of the Kyeburn Formation (Cretaceous), a wedge of coarse terrestrial sediments in central Otago. N.Z. 6(1): 55-71.

BRADSHAW, J. D.; C. J. ADAMS; P. B. ANDREWS (this volume). Carboniferous to Cretaceous on the Pacific margin of Gondwana: The Rangitata phase of New Zealand.

BURKE, K.; J. G. DEWEY 1973. Plume-generated triple junctions: key indications in applying plate tectonics to old rocks. J. Geol. 81: 406-433.

CHRISTOFFEL, D. A.; R. K. H. FALCONER 1973. Changes in the direction of sea floor spreading in the south-west Pacific. In R. Fraser (Ed.), Oceanography of south Pacific. 241-247. N.Z. Commission for UNESCO, Wellington.

GRINDLEY, G. W.; C. J. D. ADAMS; J. T. LUMB; W. A. WATTERS 1977. Paleomagnetism, K-Ar dating and tectonic interpretation of upper Cretaceous and Cenozoic volcanic rocks of the Chatham Islands, New Zealand. N.Z. J. Geol. Geophys. 20(3): 425-67.

HAY, R. F. 1975. Sheet N7. Doubtless Bay. Geological map of New Zealand 1:63,360. Government Printer: Wellington. 1st ed.

HAYES, D. E.; J. RINGIS 1973. Seafloor spreading in the Tasman Sea. Nature 243: 454-458.

HUNT, T.; S. NATHAN 1976. Inangahua magnetic anomaly, New Zealand. N.Z. J. Geol. Geophys. 19(4): 395-406.

LAIRD, M. G. 1968. The Paparoa Tectonic Zone. N.Z. J. Geol. Geophys. 11(2): 435-454.

McNAUGHTON, D.A.; F. A. GIBSON 1970. Reef play developing in New Zealand. Oil Gas J. 68(45): 89-95.

MOORE, P. R. 1978. Geology of western Koranga Valley, Raukumara Peninsula. N.Z. J. Geol. Geophys. 21(1): 1-20.

--- (in press). Late Cretaceous-Tertiary stratigraphy, structure and tectonic history of the area between Whareama and Ngahape, eastern Wairarapa, New Zealand. N.Z. J. Geol. Geophys.

---; I. G. SPEDEN 1979. Stratigraphy, structure, and inferred environments of deposition of the Early Cretaceous sequence, eastern Wairarapa, New Zealand. N.Z. J. Geol. Geophys. 22(4): 417-34.

MUTCH, A. R.; D. D. WILSON 1952. Reversal of movement of the Titri Fault. N.Z. J. Sci. Tech. 33(5): 398-403.

MUTTER, J. C.; D. JONGSMA 1978. The pattern of the pre-Tasman Sea rift system and the geometry of Breakup. Bull. Aust. Soc. Explor. Geophys. 9(3): 70-75.

NATHAN, S. 1977. Cretaceous and Lower Tertiary stratigraphy of the coastal strip between Buttress Point and Ship Creek, South Westland, New Zealand. N.Z. J. Geol. Geophys. 20(4): 615-54.

NICOL, E. R. 1977. Igneous Petrology of the Clarence and Awatere Valleys, Marlborough. Ph.D. thesis, Victoria University of Wellington.

NORRIS, R. M. 1978. Late Cenozoic geology of the West Coast shelf between Karamea and the Waiho River, South Island, New Zealand. N.Z. O.I. Memoir 81.

ODIN, G. S. 1978. Results of dating Cretaceous-Paleogene sediments, Europe.

Contributions to the geologic time scale. Am. Assoc. Petrol. Geol. Stud. Geol. 6. 127-141.

OLIVER, P. J.; T. C. MUMME; G. W. GRINDLEY; P. VELLA 1979. Paleomagnetism of the Upper Cretaceous Mount Somers volcanics, Canterbury, New Zealand. N.Z. J. Geol. Geophys. 22(2): 199-212.

PILAAR, W. F. H.; L. L. WAKEFIELD 1978. Structural and stratigraphic evolution of the Taranaki Basin, offshore North Island, New Zealand. J. Aust. Petrol. Explor. Assoc. 18(1): 93-101.

SPEDEN, I. G. 1975. Cretaceous stratigraphy of the Raukumara Peninsula. Parts I and II. N.Z. Geol. Surv. Bull. 91. 70p.

--- 1976. Geology of Mt Taitai, Tapuaeroa valley, Raukumara Peninsula. N.Z. J. Geol. Geophys. 19(1): 71-119.

STEVENS, G. R.; I. G. SPEDEN 1978. New Zealand. In M. Moullade, A. E. M. Nairn 1978 (Eds.). The Mesozoic. A. The Phanerozoic geology of the World II. Elsevier, Amsterdam.

TITHERIDGE, D. G. 1977. Stratigraphy and sedimentology of the upper Pakawau and lower Westhaven Groups (Upper Cretaceous-Oligocene), northwest Nelson. M.Sc. thesis, University of Canterbury.

WELLMAN, P.; A. F. COOPER 1971. Potassium-argon age of some New Zealand lamprophyre dikes near the alpine fault. N.Z. J. Geol. Geophys. 14(2): 341-50.

Lower Mesozoic position of southern New Zealand determined from paleomagnetism of the Glenham Porphyry, Murihiku Terrane, Eastern Southland

G. W. GRINDLEY, P. J. OLIVER & J. C. SUKROO
N. Z. Geological Survey, Lower Hutt, New Zealand

The Glenham Porphyry comprises an andesitic suite of terrestrial flows and small sub-volcanic intrusions of Late Triassic to early Jurassic age within Murihiku Supergroup sediments of Eastern Southland. The rocks have a stable, reversed-polarity thermoremanent magnatisation (TRM) component overprinted by a strong secondary component of normal polarity and low coercivity, that is removed in a demagnetising A.C. field of 30 mT. The mean TRM, corrected for tectonic tilt, is D = $343°$, I = $+78°$ (α_{95} = 5.6 K 29) giving a paleomagnetic north pole position at $24°$S, $162°$E (dp= $9.2°$, dm = $10.2°$), and a paleolatitude of $66°$. When rotated back to Antarctica, the reconstructed paleopole position, at $43°$S, $218°$E is close to the Jurassic pole positions of other Gondwana continents.

The paleomagnetic data places limits on subsequent finite rotation and translation of the Murihiku Terrane in Eastern Southland. Since the Murihiku Terrane is considered a primary fore-arc basin on the Pacific side of an andesitic volcanic arc, active since the Permian, on the margin of Gondwana, the amount of independent rotation and drift relative to the arc and its crystalline basement must be minimal.

GEOLOGICAL SETTING AND AGE

The Glenham Porphyry outcrops over a small area 35 km south of Gore and 40 km east of Invercargill (Fig. 1). Outcrops of Glenham Porphyry have been identified from semi-continuous outcrop, with consistent flow banding attitudes; from spheroidally weathered remnants of the Porphyry, still enclosed in a decomposed matrix, and from columnar-jointed flows and pyroclastics exposed in road cuttings.

The Glenham Porphyry has been previously mapped as intrusive into Murihiku Supergroup sediments of the Southland Syncline (Hatherton 1966; Watters *et al.* 1968), and has been associated with the Park Intrusives east of the Takitimu Mountains (Wood 1966). However, its petrography suggests terrestrial lava flows and not intrusives. Also, K/Ar dates on the volcanics are mainly older than the enclosing sediments (Fig. 2).

Many K/Ar dates on plagioclase separates cluster around 190 ± 5 m.y. with only one sample older at 230 ± 5 m.y. (J. Gabites, pers. comm.). An Early Jurassic age is tentatively accepted for the Glenham Porphyry, but the older, and apparently anomalous, date could be interpreted as the true eruptive age of the flows with the younger cluster of K/Ar dates recording a low-grade metamorphic resetting.

The Mesozoic sediments are shallow marine and non-marine beds with unconformities indicating periods of emergence above sea level.

Accepting the 190 m.y. date for the Porphyry, volcanism would coincide with a regional unconformity between sediments of uppermost Triassic (Otapirian Stage), and early Jurassic (Ururoan Stage) age. The 230 m.y. date would also coincide with a regional unconformity between middle Triassic (Ladinian) and Late Permian (Tatarian) beds. The thickness of Mesozoic sediments overlying the volcanics is probably less than 2000 m and the metamorphic grade is low (zeolite facies).

The contacts between Glenham Porphyry and the surrounding Murihiku Supergroup sediments are not exposed, but because the Porphyry is older, it is assumed that the contacts are either faults or unconformities. Evidence from mapping suggests steep faults along the northeast contact in the south and along the western contact in the north (Fig. 2). Flow banding and pyroclastic layers dip to the east or northeast, and the outcrops are mainly on the western edges of upthrown fault blocks, gently tilted to the east. The volcanics probably range in thickness between 50 and 200 m. Enclosing sediments are folded into a series of sub-parallel open anticlines and synclines striking NW and plunging gently to the SE (Fig. 2). Both Triassic and Jurassic rocks strike parallel to the regional fold axes of the Southland Syncline and typically dip less than $30°$ (Wood 1966; Watters *et al.* 1968). The folding is post-middle Jurassic and pre-early Cenozoic.

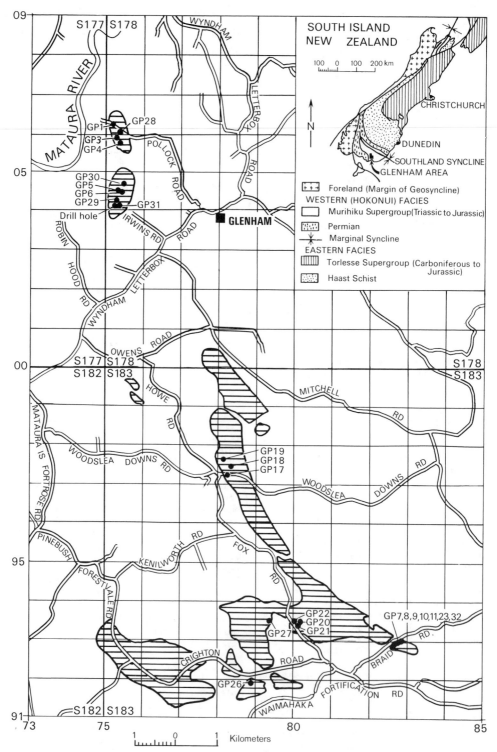

Fig.1--Map of paleomagnetic sample sites for the Glenham Porphyry. Grid reference lines from NZMS 1 Sheet S178 Wyndham and NZMS 1 Sheet S183 Tokanui.

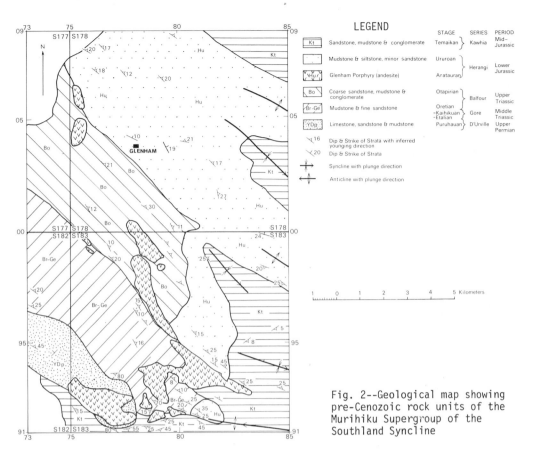

LEGEND

		STAGE	SERIES	PERIOD
Kt	Sandstone, mudstone & conglomerate	Temaikan	Kawhia	Mid–Jurassic
	Mudstone & siltstone, minor sandstone	Ururoan	Herangi	Lower Jurassic
Hu	Glenham Porphyry (andesite)	Aratauran		
Bo	Coarse sandstone, mudstone & conglomerate	Otapirian	Balfour	Upper Triassic
Br-Ge	Mudstone & fine sandstone	Oretian –Kaihikuan –Etalian	Gore	Middle Triassic
YDp	Limestone, sandstone & mudstone	Puruhauan	D'Urville	Upper Permian

Dip & Strike of Strata with inferred younging direction

Dip & Strike of Strata

Syncline with plunge direction

Anticline with plunge direction

1 0 1 2 3 4 5 Kilometers

Fig. 2--Geological map showing pre-Cenozoic rock units of the Murihiku Supergroup of the Southland Syncline

The Glenham Porphyry is a calc-alkaline andesite containing plagioclase and pyroxene phenocrysts with minor magnetite in a glassy matrix. Plagioclase phenocrysts (5 - 8 mm) are tabular, euhedral andesine (An_{40} - An_{55}). Pyroxene phenocrysts (1 - 2.5 mm) are euhedral to subhedral hypersthene and augite. Euhedral grains of magnetite (or titano-magnetite) up to 0.05 mm form up to 1% of the rock and are often associated with, or enclosed by, the phenocrysts. The groundmass is typical of flow rocks and consists of fresh undevitrified brown glass with scattered microlites of plagioclase, pyroxene and magnetite. The magnetite is less than 10 μm in diameter and forms up to 2% of the rock.

PALEOMAGNETIC MEASUREMENTS

Determination of Magnetisation Directions. Eight randomly chosen specimens were used for a pilot study to determine magnetic stability. These showed two types of behaviour on demagnetisation. The first type (GP5B.04 and GP8B.13) showed no change in direction on demagnetisation (Fig. 3b) and a very stable

relative intensity plot (Fig. 3c). The Zijderveld (1967) projections (Fig. 3d, e) also show only a single primary thermoremanent magnetisation (TRM) of reversed polarity.

The second type generally started with a steep inclination of normal polarity and ended (after demagnetisation) with a steep inclination reversed polarity (Fig. 3a, b). These samples contain a magnetically soft isothermal remanent magnetisation (IRM) component that was readily removed by alternating fields of 10 mT (Fig. 3c), and a smaller more stable secondary component that was not completely removed until a 35 mT field was applied (Fig. 3f - i). The magnetic properties are consistent with the presence of abundant single domain magnetite of low TiO_2 content, in accord with their geochemistry.

Paleomagnetic Pole Position. The tectonic corrections applied to the paleomagnetic directions consist of a regional dip derived from overlying Jurassic sediments of $25°NE$ and striking $300°$ for the southern main sites (Figs. 1, 2) and $20°NE$, strike $305°$, for the northern sites near Irwins

321

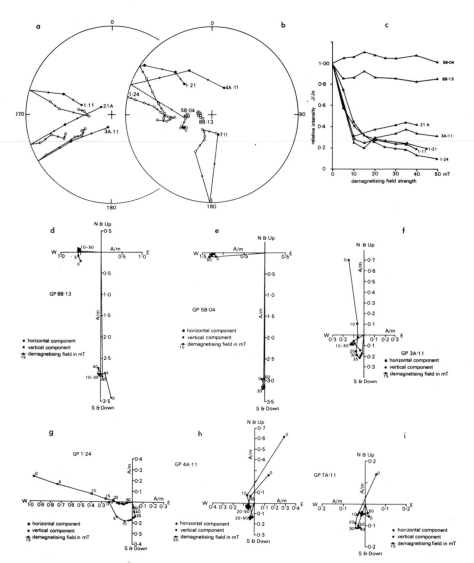

Fig.3-Alternating Field demagnetisation diagrams: (a) and (b). Equal area stereographic projections of magnetic directions during AF demagnetisation from 0-50 mT. Open circles represent reversed polarity, closed circles represent normal polarity. (c) Relative intensity curves for fields of 0 - 50 mT. (d) to (i) Zijderveld - type diagrams showing horizontal and vertical projections of the magnetic vectors on progressive step-wise demagnetisation from 0 to 50 mT.

Road (Figs. 1, 2). A correction was also made for the 5° SE plunge of the Southland Syncline, as reflected in the plunge of local synclines and anticlines adjacent to the Glenham Porphyry. Fig. 4 shows the directions on equal area stereographic plots. Table 1 lists NRM, TRM and TRM corrected for tectonic tilt, together with a mean paleomagnetic direction and the resultant paleomagnetic pole position at 24°S, 162°E

(polar errors: dp = 9°, dm = 10°). The reversed mean paleomagnetic inclination of +78° gives a paleolatitude of 66°.

POLAR WANDER PATHS

A reconstruction is provided (Fig. 5a), on the same projection (polar stereographic) as Figure 19 of Weissel and Hayes (1977), but with the Tasman Sea and Southern Ocean close back to Anomaly 33 (approx. 80 m.y. B.P.).

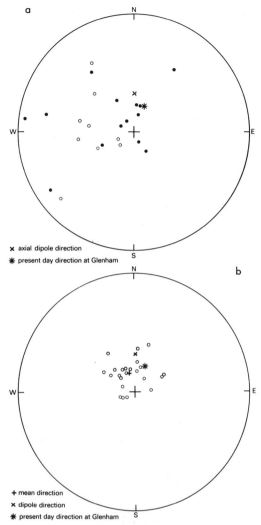

x axial dipole direction

✳ present day direction at Glenham

+ mean direction

x dipole direction

✳ present day direction at Glenham

Fig. 4--Equal area stereographic plots of magnetic directions. (a) NRM directions; (b) cleaned and tectonically corrected directions.

New Zealand is clearly separated into two parts along the line of the Alpine Fault which is considered to have originated as a plate boundary from an oceanic transform, formed during early spreading. The Tasman Sea has commenced to open, forming a small intracontinental rift basin connected via complex transforms and small rift zones to a subduction zone along the Pacific margin east of the Lord Howe Rise. Subduction has largely ceased in the New Zealand - West Antarctica region, except west of the Antarctic Penin. The Cenozoic APWP for Antarctica as determined from Australian paleomagnetic data by

McElhinny *et al.* (1974) has been tested by paleomagnetic data from Antarctica itself (Weissel *et al.* 1977, Fig. 11). New Zealand paleomagnetic data can also be used to test the validity of the reconstruction. The Cretaceous section of the APWP (Fig. 5a) is in fact best determined from New Zealand data from both east and west of the Alpine Fault (Grindley *et al.* 1977; Oliver *et al.* 1979; Grindley & Oliver 1979).

The Cretaceous-Cenozoic APWP (Fig. 5a) can be extended back into the early Mesozoic by comparing the paleomagnetic pole position for the Glenham Porphyry with the Australian and Antarctic early Mesozoic paleopoles. McElhinny (1973) estimates a mid-Jurassic pole for Antarctica at 55°S, 215°E using data from the Ferrar Dolerite (160-170 m.y.), in the Transantarctic Mountains. When Australia is rotated back to Antarctica (Weissel *et al.* 1977), the Australian late Triassic - middle Jurassic paleopole (Schmidt 1976; Table 5) at 47°S, 176°E, rotates to a position at 58°S, 206°E (α_{95} = 8.8).

This pole position lies close to the smooth curve connecting the Cretaceous section of the Australian - N.Z. - Antarctica APWP with the Glenham Porphyry pole at 43°S 218°E. The Antarctic Mesozoic pole at 55°S 215°E (α_{95} = 8.7°) occupies an intermediate position between the Australian and New Zealand poles (Fig. 5a).

The circles of confidence of the New Zealand and Antarctic poles overlap and are not statistically distinct. The Australian and New Zealand circles of confidence do not overlap, either due to known inconsistencies in the Australian data or to movement of New Zealand away from Gondwana (Fig. 5b).

GONDWANA REASSEMBLY

Early Mesozoic Poles. Mean paleomagnetic poles for the various continents and microcontinents were shown on the Gondwana reassembly of Smith and Hallam (1970) by Schmidt (1976, Fig. 6, poles given in Table 5). The New Zealand paleopole determined from the Glenham Porphyry, when rotated to Africa, lies at 73°S, 94°E, and is thus farthest from the Gondwana continental margin (Fig. 5b). Its circle of confidence (α_{95} = 9°) intersects those of the other paleopoles except for Australia and South America. It lies closest to the African pole, statistically the best determined (α_{95} = 3.9°), and essentially valid from early Triassic to middle Jurassic (Fig. 2 and Table III in Vandenberg 1979). One of the African pole positions, for the Nuanetsi Igneous Complex of Rhodesia, dated at 190 - 195 m.y., lies at 61°S, 89°E (Brock 1968), and records an interval of mainly reversed polarity. This Nuanetsi Reversed Zone, within the Graham Normal Inter-

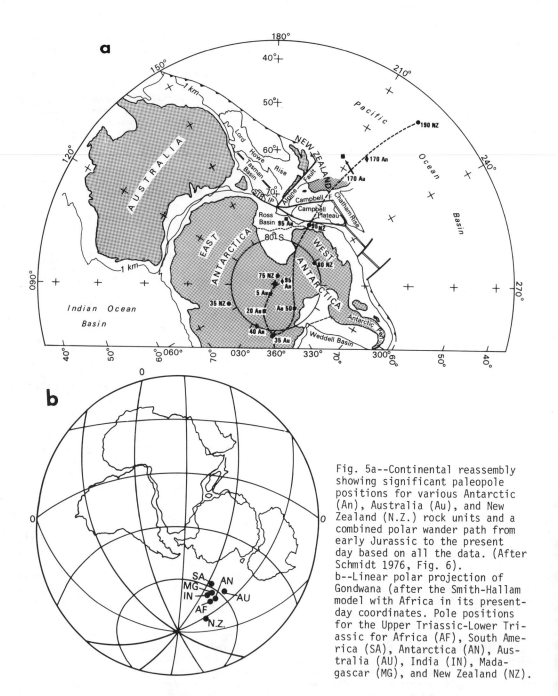

Fig. 5a--Continental reassembly
showing significant paleopole
positions for various Antarctic
(An), Australia (Au), and New
Zealand (N.Z.) rock units and a
combined polar wander path from
early Jurassic to the present
day based on all the data. (After
Schmidt 1976, Fig. 6).
b--Linear polar projection of
Gondwana (after the Smith-Hallam
model with Africa in its present-
day coordinates. Pole positions
for the Upper Triassic-Lower Triassic for Africa (AF), South America (SA), Antarctica (AN), Australia (AU), India (IN), Madagascar (MG), and New Zealand (NZ).

val (McElhinny & Burek 1971), may correspond
to the Glenham Porphyry reversed zone.

Tectonic Rotations. The tight clustering
of the Early Mesozoic paleopoles on the Gondwana reassembly must be considered excellent
evidence for the general validity of the
reconstruction. However, the Smith-Hallam
reconstruction does not take into account

independent rotations of Mesozoic-Cenozoic
oroclinal mountain belts such as the Antarctic Peninsula and New Zealand. Evidence
has been presented (Dalziel *et al.* 1973;
Kellogg & Reynolds 1978) to show that the
Antarctic Peninsula orocline formed before
the late Cretaceous. Similar evidence shows
that the New Zealand orocline is probably a

324

Table 1--Paleomagnetic directions and pole position, Glenham Porphyry.

Site	N	N.R.M.			Cleaned, uncorrected		Cleaned, corrected for tilt			
		D	I	J	D	I	D	I	K	α_{95}
GP 1	7	281	-25	0.68	273	+61	316	+70	83	6.7
GP 3	5	150	-83	0.64	223	+58	211	+79	36	13.0
GP 4	5	014	-72	0.67	237	+58	249	+79	88	8.2
GP 5	6	246	+62	1.96	265	+58	301	+70	290	3.9
GP 6	6	263	+50	0.89	264	+63	312	+75	106	6.5
GP 7	3	150	-75	0.29	195	+71	086	+78	500	5.5
GP 8	9	281	+51	2.19	261	+77	012	+73	301	3.0
GP 9	5	291	-80	0.60	260	+70	348	+75	29	14.5
GP 10	3	008	-71	0.65	309	+73	004	+60	118	11.4
GP 11	8	287	-6	0.27	277	+75	004	+70	222	3.7
GP 17	5	249	-65	0.61	254	+67	335	+76	29	14.6
GP 18	5	236	-15	0.40	228	+55	241	+80	42	11.9
GP 19	6	323	-39	0.78	264	+52	301	+65	129	5.9
GP 20	7	328	+33	0.16	335	+76	017	+56	86	6.6
GP 21	4	330	-66	0.29	225	+59	236	+84	23	19.8
GP 22	5	034	-39	0.20	162	+80	064	+68	220	5.2
GP 23	5	329	-81	0.44	287	+54	324	+57	29	14.5
GP 26	3	221	+77	1.50	162	+81	062	+68	392	6.2
GP 27	4	155	-4	0.40	252	+61	311	+76	132	8.0
GP 28	6	313	+52	0.90	271	+76	006	+76	148	5.5
GP 29	9	279	+59	4.23	277	+68	338	+73	117	4.8
GP 30	6	243	+78	1.35	243	+80	036	+79	343	3.6
GP 31	7	227	+18	3.45	247	+64	292	+81	268	3.7
GP 32	12	016	-78	0.24	263	+65	330	+72	326	2.4
MEAN	24				254	+70	343	+78	29	5.6

POLE POSITION: Lat. 24°S Long. 162°E dp = 9.2° dm = 10.2°

N = number of specimens; D = declination (degrees); I = inclination (degrees); J = magnetic intensity (A/m); K = Fisher's precision parameter; α_{95} = radius cone of 95% confidence

composite feature that developed during both the Early Cretaceous and Late Cenozoic (Oliver *et al.* 1979; Grindley & Oliver 1979).

The Murihiku Terrane of eastern Southland lies on the east-striking southern limb of the New Zealand Orocline (Fig. 2). The paleomagnetic results indicate relative stability of this terrane in the Gondwana reassembly. The permissible rotation can be gauged from the uncertainty in declination, calculated from the formula $\delta_{95} = \sin^{-1}$ (sin α_{95}/cos I) where I is the paleomagnetic inclination. The angle $2\delta_{95}$ is subtended at the sampling site by the tangents to the circle of confidence (radius α_{95}) of the paleopole. For the Glenham Porphyry δ_{95} is 28° and the total uncertainty in declination ($2\delta_{95}$) is 56°. The Murihiku Terrane could in theory have been rotated up to 28° from either direction to its present mean strike of 300°T. It could also have not been rotated at all.

Large-scale tectonic rotations of the Murihiku Terrane have been postulated by Austin (1975) and Spörli (1978), who suggested that the NE-striking segment of the New Zealand orocline is the original direction. This hypothesis is not supported by the evidence from the Glenham Porphyry. Nor does it support large-scale microplate drift towards Gondwana, combined with unspecified rotation of the Murihiku Terrane postulated by Howell (1979). In fact, the New Zealand paleopole lies farthest from the continental margin, suggesting that the Murihiku Terrane, if it moved at all, moved away from Gondwana rather than towards it.

ACKNOWLEDGMENTS

Thanks are due to Prof. D. Christoffel and Prof. P. Vella for the use of equipment, to colleagues at the Geological Survey for draughting and typing, and to Dr. T. Hatherton, Dr. I. G. Speden, and Prof. P. Vella for editorial comments.

REFERENCES

AUSTIN, P.M. 1975. Paleogeographic and Paleotectonic Models for the New Zealand

Geosyncline in Eastern Gondwanaland. Geol. Soc. Am. Bull. 86: 1230-1234.

BROCK, A. 1968. Paleomagnetism of the Nuanetsi Igneous Province and its bearing upon the sequence of Karroo Igneous Activity in Southern Africa. J. Geophys. Res. 73(4): 1389-1397.

DALZIEL, I.W.D.; R. LOWRIE; R. KLIGFIELD; N. D. OPDYKE 1973. Paleomagnetic data from the southernmost Andes and the Antarctandes. In: Tarling, D.H.; & S.K. Runcorn 1973. Implications of Continental Drift to the Earth Sciences. New York, N.Y. Academic Press I: 87-101.

GRINDLEY, G. W.; C. J. D. ADAMS; J. T. LUMB; W. A. WATTERS 1977. Paleomagnetism, K-Ar Dating and Tectonic Interpretation of Cretaceous and Cenozoic Volcanic Rocks from Chatham Islands, New Zealand. N.Z. J. Geol. Geophys. 20(3): 425-467.

---; P. J. OLIVER 1979. Paleomagnetism of Upper Cretaceous dikes, Buller Gorge, North Westland in relation to the bending of the New Zealand Orocline. In: The Origin of the Southern Alps (R.I. Walcott, M. Cresswell, Eds). Roy. Soc. N.Z. Bull. 18: 131-147.

---; F. J. DAVEY 1980. The reconstruction of New Zealand, Australia and Antarctica (a review). In: Antarctic Geoscience (C. Craddock, Ed.). Univ. Wisconsin Press. Madison, Wis.

HATHERTON, T. 1966. A geophysical study of the Southland Syncline. Bull. N.Z. D.S.I.R. 168: 49 pp.

HOWELL, D. G. 1979. Mesozoic microplates of New Zealand. Geol. Soc. N.Z. Newsl. 47: 24-25.

KELLOGG, K.S.; R. L. REYNOLDS 1978. Paleomagnetic results from the Lassiter Coast, Antarctica and a test for oroclinal bending of the Antarctic Peninsula. J. Geophys. Res. 83: 2293-2300.

McELHINNY, M.W. 1973. Paleomagnetism and Plate Tectonics. Cambridge University Press, 358 pp.

---; P. J. BUREK 1971. Mesozoic paleomagnetic stratigraphy. Nature 232: 98-102.

---; B.J.J. EMBLETON; P. WELLMAN 1974. A synthesis of Australian paleomagnetic results. Geophys. J. R. Astron. Soc. 36(1); 141-151.

OLIVER, P. J.; T. C. MUMMÉ; G. W. GRINDLEY; P. VELLA 1979. Paleomagnetism of the Upper Cretaceous Mt Somers Volcanics, Canterbury, New Zealand. N.Z. J. Geol. Geophys. 22(2): 199-212.

SCHMIDT, P. W. 1976. The non-uniqueness of the Australian Mesozoic paleomagnetic pole position. Geophys. J. R. Astron. Soc.47: 285-300.

SMITH, A. G.; A. HALLAM 1970. The fit of the Southern Continents. Nature 225: 139-144.

SPÖRLI, K. B. 1978. Mesozoic tectonics, North Island, New Zealand. Geol. Soc. Am. Bull. 89: 415-425.

VANDENBERG, J. 1979. Paleomagnetic data from the western Mediterranean a review. Geol. en Mijn. 58(2): 161-174.

WATTERS, W. A.; I. G. SPEDEN; B. L. WOOD 1968. Sheet 26 - Stewart Island. In, Geological Map of New Zealand 1:250 000. D.S.I.R., Wellington, N.Z.

WEISSEL, J. K.; D. E. HAYES 1977. Evolution of the Tasman Sea reappraised. Earth Planet. Sci. Lett. 36: 77-84.

---;---; E. M. HERRON 1977. Plate tectonics synthesis: the displacements between Australia, New Zealand and Antarctica since the Late Cretaceous. Marine Geol. 25: 231-277.

WOOD, B. L. 1966. Sheet 24 - Invercargill. Geological Map of New Zealand 1: 250 000. D.S.I.R., Wellington, N.Z.

ZIJDERVELD, J. D. A. 1967. AC Demagnetisation of Rocks: Analysis of Results. In, Methods in Paleomagnetism, D. W. Collinson et al. (Eds.). Elsevier: Amsterdam.

Notes on a 100 m.y. reconstruction for eastern Gondwana (Abstract)

H. W. WELLMAN
Victoria University of Wellington, Australia

For the reconstruction of New Zealand itself, the single most useful feature is the eastern (Dun Mountain) member of the Stokes Magnetic Anomaly (dotted line in figure). It has been traced for 2200 km from North Cape to the Antipodes Islands, is continuous except for a 500 km dextral displacement at the Alpine Fault and a 400 km dextral displacement at a parallel fault off the east coast of the South Island, and has a true length of 1300 km.

For New Zealand, Australia, and Antarctica the simplest reconstruction is made by taking a suitable map (say, one on an equidistant polar projection), cutting out all the seafloor younger than 80 m.y. (seafloor Anomaly 36 to the south of New Zealand, seafloor Anomaly 34 elsewhere), and then reassembling the remainder.

First build up a greater Australia. Trace the -3 km contour around the aouth and east coasts of Australia, not forgetting the South Tasmanian Ridge. Then attach Australia to that part of New Zealand northwest of the Apline Fault by cutting out the Tasman Sea between the pair of 80 m.y.-old seafloor anomalies. Also cut out the narrow seafloor strip of the Norfolk Island Trough. Now fit the greater Australia on to the -3 km contour around Greater Australia, setting the South Tasmanian Ridge over the Iseline Bank of Antarctica.

Next, using the seafloor transforms, fit the continental part of New Zealand south of the Alpine Fault Zone to Lesser Antarctica. The southeast corner of Campbell Plateau fits in against the eastern side of Iselin Bank, but because the 80 m.y. magnetic anomaly (mapped as Anomaly 36) is 400 km south of the south margin of the Chatham Rise, a corresponding gap must be left in the reconstruction.

There is not an appoximate fit that gives the general position of continents and continental fragments for 80 m.y., but the resulting tectonic shape of New Zealand is improbable and the total length available for the Stokes Magnetic Anomaly about 400 km too short. But in making the reconstruction, some 7000 km of seafloor was cut out, and the discrepancies are thus relatively small.

A hypothetical Transantarctic Fault with a sinistral displacement of 500 km was then postulated to provide room for the Stokes Magnetic Anomaly.

In order to make the shape and size of the various pieces correct as they would be when fitted together on a sphere, successive projections were then calculated from successive reconstructiona. From an improved 80 m.y. reconstruction, a "two-plate" attempt was made to get the complete closure shown in the figure. The gap between Chatham Rise and Lesser Antarctica, the gap between Bollons Seamount and the Campbell Plateau, and the gap represented by the Bounty Basin were closed. With the closure the granite of Bounty Island was moved to the south side of the Stokes Magnetic Anomaly, and the anomaly itself was jointed across the fault to the east of the South Island. Almost all the small faults on the reconstruction relate to the inferred two-plate 80 - 100 m.y. movement.

Irrespective of details, the main feature shown by the reconstruction is growth from 300 - 100 m.y. of the 200 km wide belt of Pacific-facing greywacke and schist that extends for at least 4000 km from New Caledonia to Chatham Rise.

Most of the many references relevant to this note have been given by H. M. McCracken under 'Plate Tectonics' 1980, for the "Geological and Mineral Atlas of Australia," Bureau of Mineral Resources, Canberra.

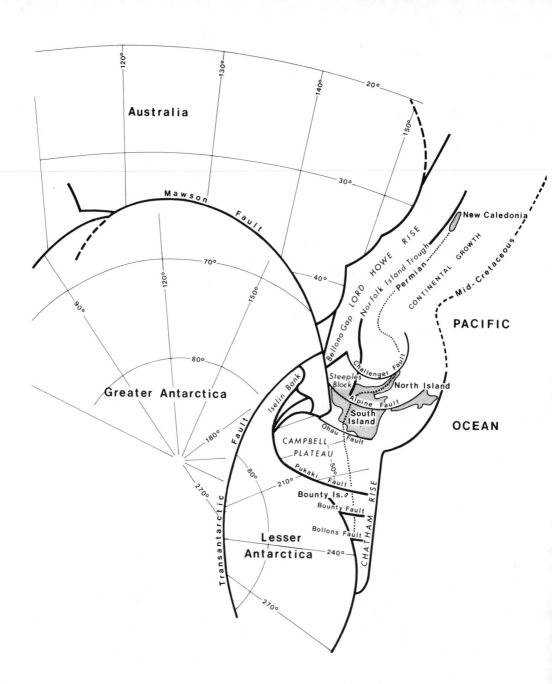

The importance of continental fragmentation history to petroleum accumulation

MURRAY HOWARD JOHNSTONE
Geological Survey of Western Australia, Perth

As the search for oil moved offshore in the 1950's and 1960's, extensive seismic surveys and the drilling of deep oil test wells revealed details of the structure and stratigraphy of the continental shelves for the first time. It was found that the shelves of the rifted fragments of Gondwana had remarkably similar histories, each recording a sequence of events which led to the eventual drifting apart of the continental masses. First, a gentle, linear downwarp develops, accompanied in some places by faulting, tilting, and erosion of older sedimentary rocks as a reaction to the new stress regime. Later, a rift valley forms within the downwarp. Finally, the continents drift apart and the newly formed continental shelf adjusts isostatically to its position marginal to the continent.

Accumulations of hydrocarbons are associated with the unconformities generated by each of these tectonic events. Whether the accumulations are of oil or gas is related to the nature of the plant material from which the hydrocarbons were derived. If the debris comes from land plants (in sediments deposited within the rift valley sequence) it will normally produce gas. Oil is more likely to be generated from the marine plants associated with major transgressions (at the beginning of the initial downwarp and drift phases). The volume of organic debris depends on the volume of plant material which in turn depends on paleoclimate (paleolatitude) and distance from the open sea (paleogeography). The most prolific fields found on rifted continental margins are those in aborted rifts such as the North Sea and Bass Strait.

INTRODUCTION

Much of the work on Gondwana has concentrated on faunal and lithologic correlations between coastal basins which are thought to be part of pre-drift megastructures. Many of the early reconstructions of "Gondwana" were based on the fits of either coastlines or the 200-m bathymetric contours of landmasses. The great majority of the early workers on Gondwana problems only had access to material from outcrop and bores within onshore basins, so it is natural that projections of data were not made too far offshore. In the past decade, a wealth of sophisticated geophysical surveys has been carried out into ever deeper water surrounding the continents by both oil companies and governments. These are tied to deep bores drilled by the companies in the search for hydrocarbons, and the integration of this structural and stratigraphic data has led to a much clearer understanding of the processes which occur at continental margins preceding, during, and after continental breakup. Around the Australian continent, a series of seismic lines shot on behalf of the Commonwealth Bureau of Mineral Resources (Hogan and Jacobson 1975) has clearly indicated the presence of thick sequences of sedimentary rocks resting on continental basement out to bathymetric depths of up to 4000 m. These sequences are in turn overlain by post-breakup sediments which extend further offshore into abyssal depths where they can be seen to be deposited on oceanic crust. The extension of the marginal basins of the continents out to depths of 4000 m and the interpretation of the structure and stratigraphy of these basins are only now becoming known outside the oil exploration industry. Pioneer papers releasing many of the concepts derived from this early evaluation of oil company data were those by Beck and Lehner (1974), Falvey (1974), and Deighton *et al.* (1976).

Australia is a natural laboratory for the investigation of continental-margin phenomena. The oceanic crust off the northwest coast of the continent was formed in the Late Jurassic (Fig.1), from the western side of the Exmouth Plateau down to the SW corner of the continent in the Neocomian, off the eastern coast in the Late Cretaceous, and along the south coast in the Late Eocene and later times. A detailed search for oil has been carried out in the coastal basins of Australia for 25 years (and in the offshore portions of these basins for 15 years).

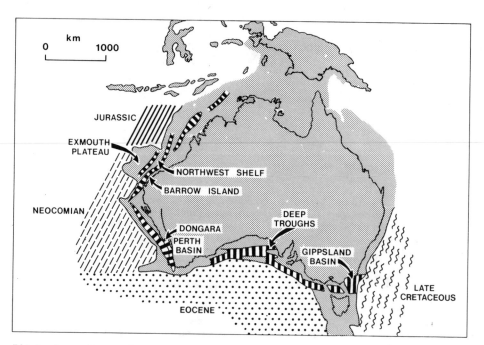

Fig.1--Ages of oceanic crust around Australia and adjacent pre-rift basins, and with main localities mentioned in the text.

The data from the Western Australian basins illustrate the relationship between the various phases of continental separation and the generation of hydrocarbons in basins along coasts of separation.

THE DEVELOPMENT OF RIFTED MARGINS

One of the best-documented theories of the mode of formation of rifted margins is that proposed by Falvey (1974) and elaborated on by Deighton *et al.* (1976), exemplified by the basins along the southern margin of Australia. The basin originated as a graben which underwent further fracturing until the continent broke and new oceanic crust formed between the two parts. The process took 50 to 60 m.y. from initial graben formation to breakup. The Red Sea and the Great Rift Valley of Africa are examples of the early stages. An excellent account of the development of the southern marginal basins of Australia is given by Boeuf and Doust (1975).

The marginal basins along the W and NW coasts of Australia have a longer history with an early phase of passive downwarp which lasted for 60 m.y. Many wells and closely-spaced seismic lines provide a reliable basis for detailed structural and stratigraphic analysis.

The first phase of the cycle of separation before continental rupture was the development of a linear downwarp. At many places, the downwarp was associated with little or no faulting even though it was filled with an enormous volume of shallow-marine and deltaic sediments, implying considerable uplift and erosion of areas marginal to the downwarp. The coarse and immature nature of some of the sediments indicated rapid erosion from high areas close to the downwarp.

Whether the west coast basins are unique in their early downwarp phase or whether this phase is not yet recognised in other basins is uncertain. However, it provides additional source and reservoir beds for hydrocarbons in the West Coast basins.

The crustal mechanisms responsible for the beginning of gentle downwarping are not clear. The sediments immediately underlying the downwarped sequence provide evidence of a heat pulse and minor tectonism (faulting and block tilting followed by uplift and minor erosion) preceding the downwarp. Such large linear downwarps and inferred flanking upwarps require mass material transfer in the base of the crust. Falvey (1974) postulated a heat flow from the mantle which metamorphosed the deeper crust into high density granulite which would behave isostatically like the mantle. With the evidence we already have of a heat pulse in the pre-downwarp sediments,

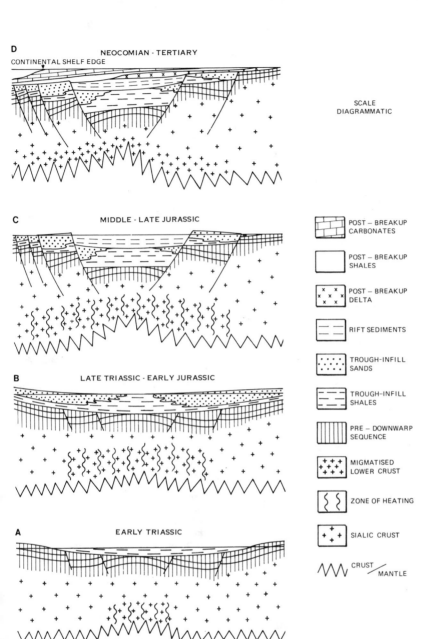

D NEOCOMIAN - TERTIARY

CONTINENTAL SHELF EDGE

SCALE
DIAGRAMMATIC

C MIDDLE - LATE JURASSIC

	POST – BREAKUP CARBONATES
	POST – BREAKUP SHALES
	POST – BREAKUP DELTA
	RIFT SEDIMENTS
	TROUGH-INFILL SANDS
	TROUGH-INFILL SHALES
	PRE – DOWNWARP SEQUENCE
	MIGMATISED LOWER CRUST
	ZONE OF HEATING
	SIALIC CRUST
	CRUST / MANTLE

B LATE TRIASSIC - EARLY JURASSIC

A EARLY TRIASSIC

Fig.2--Development of a "Western Australian Type" rifted margin: A. Trough-onset event: Initial heating deep in the crust along a linear zone causes uplift and truncation of any pre-trough sediments (and sometimes mild heating during a "heat pulse") before gentle down-warping, flooding by sea, and marine shale deposition. B. Trough-infill phase: Broadening of the heated zone of widening and deepening of the downwarp. Readjustment of deep crustal and subscrustal material causes upwarp of basin margins and deposition of large delta-sys-tems which cover the marine shales. C. Rift-infill phase: Crust finally ruptures. Deeper grabens fill with further marine and paralic sediments; horsts are eroded or stay at base level to form potential reservoirs. This sequence often contains high-quality organic ma-terial. D. Post breakup phase: Deltas again encroach over the basin and now-dormant faults. A marine shale deposited at time of breakup covers entire basin with a substantial caprock. As continental margin subsides, a prograding wedge of sediment builds out from the conti-nent depressing much of the organic-rich section into generative temperatures.

331

such a process would explain the development of the linear downwarp. Continued heating over a long period would result in the progressive conversion of lower crust to dense granulite and continued deepening of the trough. Only when the crust became sufficiently thin and brittle (60 to 90 m.y. after the initiation of the trough in the Western Australian basins) would fracturing occur on a large scale and a classic rift develop. The long "pre-rift" trough development distinguishes the "Western Australian model" proposed herein from the often-quoted "Red Sea model" (Dewey & Bird 1970).

STAGES IN THE DEVELOPMENT OF THE RIFTED MARGINS OF WESTERN AUSTRALIA

Trough onset event. The "pre-breakup sequence" of sediments and crystalline rocks was warped, faulted, and eroded in response to isostatic readjustments, and the sediments were subjected to a heat pulse which raised the level of metamorphism of their contained organic matter (Fig. 2A). If the downwarp depressed the trough below sea level, the trough-onset unconformity would be onlapped by a sequence of shallow marine shales.

As the heated zone broadened deep in the crust, the trough itself would have deepened, but faulting remained minimal (Fig. 2B). The large volumes of sediments eroded from adjacent uplifted areas deposited as deltaic complexes which interfingered with and covered the marine shales of the early transgression (Fig. 2B). Coarse sands several hundred kilometres from the margins of the trough indicate no interruption of river flow into the centre of the downwarp, and it is unlikely that there were any major faults in the trough.

Rift onset event. Continued metamorphism of the deep crust would finally result in the weakening of the crust sufficiently to allow major brittle fracture to occur, producing a series of horsts and grabens (Fig. 2C). Rapid deepening of the main graben permitted a new influx of the sea, but some of the horsts remained permanently above sea level. The "rift onset unconformity" truncates successively older rocks towards the tilted and eroded basin margins but is weakly developed or absent in the centres of the deep grabens. Because the infill sediments of the rift phase climbed up on and progressively engulfed the horsts, the age of the sediments immediately above the unconformity varies widely.

With waning fault activity at the end of the "rift" phase, a deltaic sequence often reasserted itself. Once more, bodies of deltaic sand attempted to cross the rift but were terminated by the incoming post-breakup shale (Fig. 2D).

Breakup event. In the Carnarvon Basin of Western Australia (Fig. 2D), separation did not occur within the central graben, but some distance to the NW (left) of the section. The late-stage sediments of the central graben and the over-riding delta were onlapped by the marine shales of the post-drift sequence. As the thinned marginal crust achieved isostatic balance by steady subsidence, it was onlapped first by a sequence of shallow-marine shales and sands, and, next, by a thick seaward-prograding wedge of carbonates.

At the time of breakup, the high southern latitudes of the Australian basins prevented the development of evaporite deposits in the basins formed by the early marine incursions. Other parts of Gondwana were in warm temperate to subtropical paleolatitudes during the rift and breakup phases of separation. Under favourable conditions of restricted access of marine waters, extensive deposits of salt formed, and where they were mobilized into salt diapirs, they provided most important loci for hydrocarbon traps.

FACTORS AFFECTING HYDROCARBON GENERATION AND ACCUMULATION

Four conditions are now known to be essential for the development of commercial hydrocarbon deposits (Tissot & Welte 1978):

Burial of sapropelic (marine algal) or humic (land plant) material in sediments.

Deposition of this material in reducing bottom conditions.

Conversion of sapropel by heating at 50°-200° and of humic material at 100°-200° to usable hydrocarbons.

Association of source beds reservoirs and blanket sealing shale, with a suitable structural history.

Basins with no marine beds, and only humic material, may contain gas fields. Basins with marine beds and sapropelic material may contain oil fields.

HYDROCARBON ACCUMULATIONS ON RIFTED CONTINENTAL MARGINS

Although many portions of rifted continental margins contain signs of hydrocarbons, commercially viable deposits are not common. Whereas much of the exploration of these marginal basins has been disappointing, some "giant fields" have been found--Gippsland Shelf in SE Australia, the gas fields of Australia's Northwest Shelf, and the oil and gas fields of the North Sea.

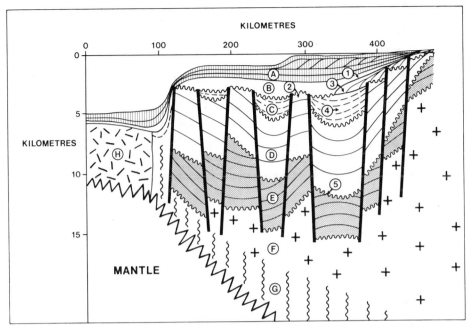

Fig.3--The "Western Australian Model" continental margin (modified after Falvey) showing the relationship of the various hydrocarbon occurrences to the depositional and tectonic sequences. Tectono-sedimentary sequences: A. Post-breakup carbonates. B. Post-breakup marine shales. C. Rift-infill sequence. D. Trough-infill sequence. E. Pre trough-onset sequence. F. Sialic crust. G. Migmatised basal crustal material. H. Oceanic crust. Hydrocarbon occurrences: 1. Shoals within Unit B (Barrow I.). 2. Horsts draped with Unit B (North Rankin, Marlin). 3. Post-breakup deltaic sands (Barrow I., Frigg). 4. Hydrocarbons forming in the generative zone (deep in Barrow I., structures deep in the Perth Basin). 5. Transgressive sands on the trough-onset unconformity (Dongara).

These three areas share features with other marginal basins on pull-apart coasts--abundant organisms due to suitable climates, suitable depositional environment for the preservation of organic matter, and suitable potential reservoirs in the correct structural situation for trapping migrating hydrocarbons--but they have one particular feature in common: they are all in aborted rifts. Each is protected from the direct influence of the mid-oceanic ridge by a large mass of sialic crust--Tasmania provided protection for the Gippsland Shelf, the Exmouth Plateau protected the Northwest Shelf, and the British Isles protected the North Sea. Whether the structures in such aborted rifts are not damaged during the pull-apart phase and hence retain their hydrocarbons, or whether there is more chance of the development of silled basins and hence more chance for the preservation of large amounts or organic matter is not clear. All that can be shown is that aborted rifts have a much better chance of containing

significant hydrocarbon accumulations than do "open" rifted margins.

Although based on Western Australian examples, Figure 3 summarizes the commonest mode of occurrence of commercial hydrocarbons in rifted-margin basins. Almost all the known commercial accumulations are closely associated with the unconformities. In the North Sea, the Ekofisk Field occurs in a porous carbonate bank at the base of the Tertiary carbonates, but the shallowest producing field in Australia is at Barrow Island where production is from shoaling sands just above the base of the post-breakup marine transgression (1 in Fig.3). The most prolific production comes from within horst blocks formed during the rifting phase which are sealed by the overlying marine shales at the base of the post-breakup marine incursion (2 in Fig.3). Here, it is the compactional draping of the shales over the horst which creates the trap. Examples are the Ninian Field in the North Sea and, in Australia, the Marlin Field in Gippsland Basin and the North Rankin Gas

333

Field on the Northwest Shelf.   The deltaic sequences associated with the breakup
unconformity provide a combination of
reservoir beds and source which give rise
to accumulations (3 on Fig. 3).   Turbidite
deposits in deeper waters in front of the
deltas often host prolific fields like
Frigg in the North Sea.   Within the rift-
valley source sediments, the zone of generation of hydrocarbons (the "thermal
window") may contain further hydrocarbon
accumulations but these are often complicated by high pressures caused by release
of water from clays during the primary migration phase of the hydrocarbons (4 on
Fig. 3).   Finally, we have in Western
Australia two occurrences of hydrocarbons
controlled by the earliest phase of the
cycle--the trough-onset unconformity (5 on
Fig. 3).   These are the Dongara Gasfield
in the northern Perth Basin where gas and
minor oil is produced from the basal sand
of the Triassic transgression which overlies eroded Permian sediments, and in the
extreme north of the state where there was
a spectacular gas blowout from an uppermost Permian sandstone beneath the Triassic
marine shales.   In both of these occurrences, the bulk of the gas is most likely
derived from the underlying tilted and
eroded Permian sediments, with a minor
contribution from the sapropel-rich basal
portion of the marine shale which forms the
caprock of the accumulation.

ACKNOWLEDGMENTS

The author owes much to numerous discussions with many geologists from companies
who were involved in exploring this area.
My thanks are especially due to the
management and staff of West Australian
Petroleum for 25 years of stimulating
association.   My more immediate thanks go
to Dr P.E. Playford and Dr H.T. Moors of
the Geological Survey of Western Australia
for constructive criticism of this paper and
to the drafting and stenographic sections of
the Mines Department for their able assistance in its preparation.
   This paper is published with the permission of the Director, Geological Survey of
Western Australia.

REFERENCES

BECK, R. H.; LEHNER, P. 1974.   Oceans,
new frontier in exploration.   AAPG Bull.
58/3:   376-395.
BOEUF, M. G.; H. DOUST 1975.   Structure
and development of the southern margin of
Australia.   J. Aust. Pet. Explor. Assoc.
15/1:   33-43.

DEIGHTON, I.; D. A. FALVEY; D. J. TAYLOR
1976.   Depositional environments and geotectonic framework: southern Australian
continental margin.   J. Aust. Pet. Explor.
Assoc. 16/1:   25-36.
DEWEY, J. F.; J. M. BIRD 1970.   Mountain
belts and the new global tectonics.   J.
Geophys. Res. 75/14:   2625-2647.
FALVEY, D.A. 1974.   The developments of
continental margins in Plate Tectonic
theory.   J. Aust. Pet. Explor. Assoc. 14/1
96-106.
HOGEN, A. P.; E. P. JACOBSON 1975. Geophysical results from the northwest margin.   Aust. BMR Rec. 1975/101 (unpubl.)
KLEMME, H. D.   1971.   The giants and super-
giants, Part 2.   Oil Gas J. March 8,
1971:   103-110.
TISSOT, B. P.; D. H. WELTE   1978.   Petroleum Formation and Occurrence.   Berlin:
Springer-Verlag.   538 pp.

Conference registrants (attending and non-attending)

ARGENTINA

C. Azcuy
Dept Cs Geologicas
Ciudad Universitaria
Pabellon 2, Nunez 1428
Buenos Aires

C. Cingolani
Calle 59, nº 716
1900 La Plata

Z.B. de Gasparini
Fac. de Ciencias Nac. y
Museo de la Plata
Paseo del Bosque
La Plata

R.F.N. Page
Acassuso 1969
1636 Olivos
Buenos Aires

AUSTRALIA

M. Apthorpe
Woodside Petroleum
  Development Pty., Ltd.
Box D188
Perth 6001, W.A.

S. W. Carey
24 Richardson Ave.
Dynnyrne 7005, Tas.

D. Christophel
Dept. Botany
Univ. Adelaide
Adelaide, S.A.

M. J. Clarke
Geol. Surv. of Tasmania
Box 124B
Hobart, Tas.

J. M. Dickins
Bureau Mineral Resources
Box 378
Canberra, A.C.T.

B.J.J. Embleton
CSIRO Div. Mineral Physics
Box 136
North Ryde 2133, N.S.W.

N. Farmer
Geol. Surv. of Tasmania
Box 142B
Hobart, Tas.

G. L. Frazer
4 Bradley Drive
Carlingford 2118, N.S.W.

A.J.W. Gleadow
Univ. of Melbourne
Parkville 3052, Vic.

H. J. Harrington
Geol. Dept.
Univ. of New England
Armidale 2351, N.S.W.

R. A. Henderson
43 Wentworth Ave.
Mundingburra
Townsville, Qld.

R. S. Hill
Botany Dept.
Jas. Cook Univ.
Townsville, Qld.

J. B. Jago
School Applied Geology
S. A. Inst. Technology
North Terrace
Adelaide, S. A.

N. J. de Jersey
Geol. Surv. Queensland
2 Edward Street
Brisbane, Qld.

M. H. Johnstone
Geol. Surv. W. Australia
66 Adelaide Terrace
Perth 6000, W.A.

C. J. Klootwyk
Res. School Earth Sci.
Inst. Adv. Studies
ANU
Box 4
Canberra 2600, A.C.T.

J. F. Lovering
Dept. Geology
Univ. Melbourne
Parkville, Vic.

N. H. Ludbrook
110 Watson Ave
Toorak Gardens 5065, S.A.

M. W. McElhinny
Res. School Earth Sci.
Inst. Adv. Studies
ANU
Box 4
Canberra 2600, A.C.T.

B. McElvey
Dept. Geology
Univ. New England
Armidale 2351, N.S.W.

E. W. Milligan
Box 19
Woden 2606, A.C.T.

R. E. Molnar
Queensland Museum
Gregory Terrace
Fortitude Vy 4006, Qld.

A. Partridge
Esso Australia Ltd.
Box 4047
Sydney 2001, N.S.W.

V. L. Passmore
Box 378
Canberra 2601, A.C.T.

N. S. Pledge
S. Australian Museum
Adelaide 5000, S.A.

J. F. Rigby
Geol. Surv. Queensland
2 Edward Street
Brisbane 4000, Qld.

F. Robbins
81 Mackenzie St.
Bendigo, Vic.

B. Runnegar
Dept. Geology
Univ. New England
Armidale 2351, N.S.W.

V. Scheibnerova
Geol. Mining Museum
36-64 George St.
Sydney 2000, N.S.W.

P. W. Schmidt
CSIRO Div. Mining Physics
Box 136
North Ryde 2113, N.S.W.

G. A. Thomas
Dept. Geology
Univ. Melbourne
Parkville 3052, Vic.

D. J. Titheridge
Geol. Dept.
Univ. Wollongong
Box 1144
Wollongong 2500, N.S.W.

E. M. Truswell
Bureau of Mineral Resources
Box 378
Canberra 2601, A.C.T.

J. J. Veevers
Dept. Earth Sci.
Macquarie Univ.
North Ryde 2113, N.S.W.

A. A. Warren
Dept. Zoology
La Trobe Univ.
Bundoora 3083, Vic.

R. E. Wass
Geol. Dept.
Univ. Sydney
Sydney 2006, N.S.W.

J. B. Waterhouse
Dept. Geology
Univ. Queensland
St Lucia, Brisbane, Qld.

H. E. Wilkinson
Dept. Minerals & Energy
Box 522
Bendigo 3550, Vic.

G. E. Williams
BHP Melbourne Res. Lab.
Box 264
Clayton 3168, Vic.

M. Woollands
BP House
1 Albert Road
Melbourne 3004, Vic.

G. Young
Bur. Mineral Resources
Box 378
Canberra 2601, A.C.T.

BELGIUM

P. Pierart
22 Place du Parc
7000 Mons

BRAZIL

A.C.R. Rocha-Campos
Inst. Geociencias
Univ. Sao Paulo
C.P. 20899
Sao Paulo

CANADA

E. Farrar
Dept. Geology
Queens University
Kingston, Ont.

CHINA

Cheng Zhengwu
Chinese Acad. Geol. Sci.
c/o Minister of Geology
Beijing

Wu Shunbao
Wuhan College of Geology
Hubei

Yang Zunyi
Beijing Graduate School
Wuhan College of Geology
Beijing

DENMARK

K. R. Pedersen
Geol. Institut
Univ. Aarhus
Ole Wormsalle
800 Aarhus

FRANCE

L. Beltan
Inst. Paleontologie
8 Rue de Buftou
Paris

M. Waterlot
Univ. de Lille
Sci. de la Terre
B.P. 36
59650 Villeneuve d'Ascq

WEST GERMANY

R. Förster
Bayer. Straatssammlung
  für Paleontol. & Hist.
  Geol.
R-Wagnerstr. 10/1
8000 München

J. Mehl
Weiherhofstr. 14
D7800 Freiburg

H. Miller
Gievenbeckerweg 64
4400 Münster

U. Rosenfeld
Westf. Wilhelms-Univ.
Geol.-Paleontol. Inst.
Gievenbeckerweg 81
4400 Münster

H. Wopfner
Geol. Inst. Univ. Köln
Zulpicherstr. 49
5 Köln

INDIA

S. Acharyya
Geol. Surv. India
4 Chowringhee Lane
Calcutta 16

M. Bannerjee
Dept. Botany
Univ. Calcutta
35 Ballygunge Circ. Rd.
Calcutta

T. N. Basu
Central Mine Planning Inst.
Gondwana Place
Ranchi 834008 Bihar

N. Bhattacharya
Sedimentology Lab.
9 Kaulagarh Rd.
Dehra Dun 248195 U.P.

S. M. Casshyap
Geology Dept.
Aligarh Muslim Univ.
Aligarh 202001

N. D. Mitra
Coal Wing
Geol. Surv. India
29 Jawaharlal Nehru Rd.
Calcutta 700016

A. K. Mukherjee
Central Fuel Ins.
Dhanbad 828108 Bihar

C. S. Raja Rao
Coal Wing
Geol. Surv. India
29 Jawaharlal Nehru Rd.
Calcutta 700016

C. Ramachandran
Geol. Surv. India
15 Park Street
Calcutta 700016

S. C. Shah
Geol. Surv. India
29 Jawaharlal Nehru Rd.
Calcutta 700016

B. Shrivastava
Central Mine Planning Inst.
BPA Building
P.O. Dhansar
Dhanbad 828106

R. P. Verma
Central Mine Planning Inst.
BPA Building
P.O. Dhansar
Dhanbad 828106

INDONESIA

J. Aspden
Jr. Ir. H. Juanda 458
Bandung

MALAGASY

H. J. Rakotoarivelo
B.P. 906
Univ. Tananarive

THE NETHERLANDS

H. Wensink
Geological Institute
Oude Gracht 320
Utrecht

NEW CALEDONIA

J.-P. Paris
BRGM - B.P. 56
Noumea

NEW ZEALAND

C. J. Adams
Inst. Nuclear Sciences
DSIR, Private Bag
Lower Hutt

J. Anderson
Geology Dept.
VUW, Private Bag
Wellington

P. Andrews
N.Z. Geol. Survey
DSIR, Box 30368
Lower Hutt

P. Ballance
Geology Dept.
Univ. Auckland
Auckland

P. J. Barrett
Geology Dept.
VUW, Private Bag
Wellington

C. Bliss
11 Bank Street
Whangarei

J. D. Bradshaw
Geology Dept.
Univ. Canterbury
Christchurch 1

M. A. Bradshaw
Centerbury Museum
Rolleston Avenue
Christchurch

R. Brathwaite
N.Z. Geol. Survey
DSIR, Box 30368
Lower Hutt

N. J. Brown
R.D. No. 8
Frankton

C. J. Burgess
Geology Dept.
VUW, Private Bag
Wellington

H. Campbell
N.Z. Geol. Survey
DSIR, Private Bag
Lower Hutt

J. D. Campbell
Dept. Geology
Univ. Otago
Box 56
Dunedin

M. P. Cave
Dept. Geology
Univ. Canterbury
Christchurch 1

G. A. Challis
N.Z. Geol. Survey
DSIR, Box 30368
Lower Hutt

C. Chonglakmani
Geology Dept.
Univ. Auckland
Auckland

D. Christoffel
Inst. Geophysics
VUW, Private Bag
Wellington

C. L. Clark
Water & Soil Bureau
Min. of Works
Wellington

R. H. Clark
Dept. Geology
VUW, Private Bag
Wellington

J. D. Collen
Dept. Geology
VUW, Private Bag
Wellington

A. R. Crawford
Geology Dept.
Canterbury Univ.
Christchurch 1

M. Crundwell
Dept. Geology
VUW, Private Bag
Wellington

F. J. Davey
Geophysics Division
DSIR, Box 1320
Wellington

M. Dearborn
Chamberlain St. R.D. 2
Upper Moutere
Nelson

A. J. Eggo
Dept. Geology
VUW, Private Bag
Wellington

R. H. Findlay
Antarctic Division
DSIR, Box 2110
Christchurch

C. A. Fleming
Dept. Geology
VUW, Private Bag
Wellington

P. Froggatt
Dept. Geology
VUW, Private Bag
Wellington

J. Gabites
Inst. Nuclear Sci.
DSIR, Private Bag
Lower Hutt

V. C. Grabham
640 Main Road
Stoke, Nelson

J. P. Graham
Box 246
Hastings

J. A. Grant-Mackie
Dept. Geology
Univ. Auckland
Auckland

R. Grapes
Dept. Geology
VUW, Private Bag
Wellington

M. R. Gregory
Dept. Geology
Univ. Auckland
Auckland

G. W. Grindley
N.Z. Geol. Survey
DSIR, Box 30368
Lower Hutt

F. Harmsen
Dept. Geology
VUW, Private Bag
Wellington

J. Harper
Dept. Mathematics
VUW, Private Bag
Wellington

F. Hasibuan
Dept. Geology
Univ. Auckland
Auckland

D. Haw
BP Oil Exploration
Box 892
Wellington

J. Hogan
Dept. Geology
VUW, Private Bag
Wellington

N. deB. Hornibrook
N.Z. Geol. Survey
DSIR, Box 30368
Lower Hutt

J. T. Hunter
122 Mallard Drive
Rotorua

M. R. Johnston
N.Z. Geol. Survey
DSIR, Private Bag
Nelson

P.J.J. Kamp
Dept. Earth Sciences
Univ. Waikato
Hamilton

J. R. Kayal
Inst. Geophysics
VUW, Private Bag
Wellington

M. G. Laird
N.Z. Geol. Survey
Univ. Canterbury
Christchurch 1

R. P. Lynch
Science Information Div.
DSIR, Box 9741
Wellington

D. MacFarlane
Dept. Geology
Univ. Otago
Box 56
Dunedin

J. M. McLennan
Dept. Geology
Univ. Canterbury
Christchurch 1

D. C. Mildenhall
N.Z. Geol. Survey
DSIR, Box 30368
Lower Hutt

T. Montague
Dept. Geology
Univ. Canterbury
Christchurch 1

P. Moore
N.Z. Geol. Survey
DSIR, Box 30368
Lower Hutt

P. Morris
Dept. Geology
VUW, Private Bag
Wellington

N. Newman
Dept. Geology
Univ. Canterbury
Christchurch 1

P. J. Oliver
N.Z. Geol. Survey
DSIR, Box 30368
Lower Hutt

A. Palmer
Dept. Geology
VUW, Private Bag
Wellington

F. Pirajno
Box 435
Nelson

J. I. Raine
N.Z. Geol. Survey
DSIR, Box 30368
Lower Hutt

B. Robinson
Physics Engr. Lab.
DSIR, Private Bag
Lower Hutt

P. H. Robinson
Dept. Geology
VUW, Private Bag
Wellington

G. Roder
N.Z. Geol. Survey
DSIR, Box 30368
Lower Hutt

J. E. Scanlon
Shell & BP Todd Oil
Private Bag
New Plymouth

P. C. Scott
104 Main North Road
Christchurch 5

I. Speden
N.Z. Geol. Survey
DSIR, Box 30368
Lower Hutt

K. B. Spörli
Dept. Geology
Univ. Auckland
Auckland

R. D. Stanley
12 Tatar Ave.
Matamata

G. R. Stevens
N.Z. Geol. Survey
DSIR, Box 30368
Lower Hutt

J. Thornton
29 Carlton St
Wellington 3

P. Vella
Dept. Geology
VUW, Private Bag
Wellington

D. Waghorn
Dept. Geology
VUW, Private Bag
Wellington

S. Weaver
Dept. Geology
Univ. Canterbury
Christchurch 1

H. Wellman
Dept. Geology
VUW, Private Bag
Wellington

J. Wiffen
138 Beach Road
Haumoana, Hawkes Bay

A. Wright
N.Z. Geol. Survey
DSIR, Box 61012
Otara, Auckland

SOUTH AFRICA

J. Anderson
Botanical Res. Inst.
Private Bag X101
Pretoria 0001

M. De Wit
Bern. Price Inst. Geophys.
Univ. Witwatersrand
1 Jan Smuts Ave
Johannesburg 2001

T. W. Gevers
Dept. Geology
Univ. Witwatersrand
Johannesburg

A. Keyser
Geol. Surv. S. Africa
Pretoria

L. King
7 Ribston Place
Westville 3630
Natal

J. G. McPherson
Dept. Geology
U.C.T.
C. P. 7799
Rondebosch

J.N.J. Visser
Dept. Geology
Univ. Orange Free State
Box 339
Bloemfontein 9300

V. von Brunn
Dept. Geology
Natal University
Pietermaritzburg

U.S.S.R.

A. Dralkin
Sevmorgeo
120 Moyka 190121
Leningrad

V. L. Ivanov
Sevmorgeo
120 Moyka 190121
Leningrad

E. Kamenev
Sevmorgeo
120 Moyka 190121
Leningrad

V. Masolov
Sevmorgeo
120 Moyka 190121
Leningrad

UNITED KINGDOM

D. H. Tarling
Dept. Geophysics
The University
Newcastle-upon-Tyne

M.R.A. Thomson
British Antarctic Surv.
N.E.R.C. Madingly Road
Cambridge CB30 ET

R. H. Wagner
Dept. Geology
Univ. Sheffield
Brookhill, Sheffield S3 7HF

U.S.A.

R. S. Babcock
Dept. Geology
W. Washington Univ.
Bellingham, Wn 98225

E. H. Colbert
Museum N. Arizona
Route 4, Box 720
Flagstaff, Ariz. 86001

J. W. Collinson
2706 Ione Street
Worthington, Ohio 43085

J. W. Cosgriff
Dept. Biology
Wayne State Univ.
Detroit, Mich. 48202

J. C. Crowell
Dept. Geol. Sci.
UCSB
Santa Barbara, Calif.
93106

I.W.D. Dalziel
Lamont-Doherty Geol. Obs.
Palisades, N.Y. 10964

D. Davis
Gulf Oil Company
Houston, Texas

D. Elliot
Inst. Polar Studies
Ohio State Univ.
Columbus, Ohio 43210

R. Fields
Dept. Geology
Univ. Montana
Missoula, Mont. 59801

R. Forsythe
Lamont-Doherty Geol. Obs.
Palisades, N.Y. 10964

G. S. Hamill
Gulf Oil Company
Houston, Texas

J. Herrlin
Dept. Geology
Univ. Montana
Missoula, Mont. 59801

K. S. Kellogg
U.S. Geol. Survey
Federal Center
Denver, Colo. 80225

P. Kyle
Inst. Polar Studies
Ohio State Univ.
Columbus, Ohio 43210

J. Pojeta
U.S. Geol. Survey
E-501 U.S. Nat. Museum
Washington, D.C. 20560

E. Stump
Dept. Geology
Arizona State Univ.
Tempe, Ariz. 95281

K. Swisher
Dept. Geology
Univ. Montana
Missoula, Mont. 59801

P. Tasch
Dept. Geology
Wichita State Univ.
Wichita, Kan. 62208

C. L. Vavra
Inst. Polar Studies
Ohio State Univ.
Columbus, Ohio 43210

URUGUAY

A. Mones
Dept. Paleontol.
Museo Nac. de Hist. Nat.
C.C. 399
Montevideo